Ocean Dumping
of Industrial Wastes

MARINE SCIENCE

Coordinating Editor: Ronald J. Gibbs, *University of Delaware*

Recent Volumes:

A Continuation Order Plan is available for this series. A continuation order will bring delivery of each new volume immediately upon publication. Volumes are billed only upon actual shipment. For further information please contact the publisher.

Ocean Dumping
of Industrial Wastes

Edited by

Bostwick H. Ketchum

Woods Hole Oceanographic Institution
Woods Hole, Massachusetts

Dana R. Kester

University of Rhode Island
Kingston, Rhode Island

and

P. Kilho Park

National Oceanic and Atmospheric Administration
Rockville, Maryland

PLENUM PRESS • NEW YORK AND LONDON

Library of Congress Cataloging in Publication Data

International Ocean Dumping Symposium, 1st, University of Rhode Island, 1978.
   Ocean dumping of industrial wastes.

   (Marine science; v. 12)
   "Proceedings of the First International Ocean Dumping Symposium, held October
10-13, 1978, at the University of Rhode Island, West Greenwich, Rhode Island."
   Includes bibliographical references and index.
   1. Waste disposal in the ocean—Congresses. 2. Factory and trade waste—Congresses.
3. Marine pollution—Congresses. I. Ketchum, Bostwick H., 1912-   II. Kester,
Dana R. III. Park, Paul Kilho, 1931-   IV. Title.
TD763.I58 1978   363.7'28   80-29598
ISBN-13:978-1-4684-3907-6   e-ISBN-13:978-1-4684-3905-2
DOI: 10.1007/978-1-4684-3905-2

Based on the proceedings of the First International Ocean Dumping Symposium,
held October 10–13, 1978, at the University of Rhode Island,
West Greenwich, Rhode Island

© 1981 Plenum Press, New York
Softcover reprint of the hardcover 1st edition 1981
A Division of Plenum Publishing Corporation
227 West 17th Street, New York, N.Y. 10011

PREFACE

In recent years there has been an increased realization that the casual disposal of wastes can lead to a deterioration in environmental quality with substantial impacts on society. The management of waste disposal practices must consider the various alternatives of discharging and decomposing wastes on land, in the atmosphere, and in the marine environment. Up until 1972 ocean dumping was used increasingly to dispose of sewage sludge, industrial wastes, and dredged material. In subsequent years regulations were developed to reduce and minimize ocean dumping. These regulations were prompted often by ignorance of the possible effects of waste disposal in the ocean rather than by knowledge that such ocean dumping was detrimental to the marine environment or to man. The relationship between waste disposal and the oceans can be viewed in either of two ways. One may want to assure that waste disposal procedures do not alter adversely the marine environment, or one may choose to utilize the ocean as a waste depository to reduce the burden placed on the continental ecosystem and on the atmosphere. From either perspective it is essential that there be an adequate base of technical information to assess the fate and effects of wastes introduced to the ocean.

A series of original technical papers has been compiled in this book to present some of the recent results of research on industrial waste disposal in the ocean. Most of these papers are based on work which was presented and discussed at the First International Ocean Dumping Symposium which was held in October 1978 at the University of Rhode Island, W. Alton Jones Conference Center. The manuscipts submitted for this publication were reviewed and in most cases they were revised prior to final editing and retyping for camera ready copy.

We would like to express our sincere and deep appreciation to Marilyn A. Maley for her work in the style editing and much of the retyping of the manuscripts. She was assisted by Janice Millar, James Fontaine, and Susan Gelsomino. The efforts and diligence of these four people are largely responsible for the uniformity we have attempted to achieve in this typewritten manuscript. This work was supported in part by NOAA grant 04-8-M01-192.

Bostwick H. Ketchum
Dana R. Kester
P. Kilho Park

November 1980

CONTENTS

I

INTRODUCTION

OCEAN DUMPING RESEARCH:

HISTORICAL AND INTERNATIONAL DEVELOPMENT

P. Kilho Park and Thomas P. O'Connor

National Oceanic and Atmospheric Administration
Rockville, Maryland 20852

The ocean is our birthplace.
From its primordial abyss
Life began aeons ago.
Effectively it has stabilized
Our climate, water-cycle, lands.
It has been renewing itself.

The ocean is our resource;
It is not our garbage depot.
It is our life-sustaining reservoir.
If man alters it useless once
We may then return back to
Abiotic, God-less world.

The ocean enables man to live on.
Let us keep the ocean alive,
Thereby the miracle of humanity
May go on on this planet.
Let us diligently study the ocean
In order to live intelligently.

-Momiji-

ABSTRACT

Plans for systematic and comprehensive research and monitoring
of the effects of ocean dumping were prepared in 1970, followed by
their execution in recent years. Forty-three nations have joined the
international convention on the "Prevention of Marine Pollution by
Dumping of Wastes and Other Matter" that became effective on August
30, 1975. At present, substances internationally prohibited from
being ocean dumped include organohalogens, mercury, cadmium and their
compounds, persistent plastics, high-level radioactive wastes, and
biological and chemical warfare agents.

INTRODUCTION

The final receptacle of many wastes is the ocean. Industrial
and municipal liquid wastes, after being treated, run into the ocean.
Agricultural pollutants, such as pesticides and fertilizers, are
often carried into the ocean by runoff. Oil spills are common occur-
rences in recent years. Air pollutants, such as lead, rain over the
ocean which covers 70 percent of the Earth's surface. Because the
ocean occupies the lowest topographic domain in the hydrological cycle of the Earth's surface, it is not possible for us to transfer
pollutants from the ocean to other domains. A single critical mis-
take by man could be devastating.

The ocean is an international resource. It is an important food
source of many countries. It stabilizes the climate, the oxygen-car-
bon dioxide balance in the atmosphere, and it provides water for the
Earth's hydrologic cycle.

The resource concept conflicts with the use of the ocean as a
receptacle for waste. Within the ocean's assimilative capacity, in
theory, these two uses may coexist. However, unassimilated substances, such as long-life radioactive wastes and synthetic toxic
organic substances, will accumulate in the ocean since nature is not
capable of altering them. Degradable organic and inorganic substances, given sufficient time and dilution, can become incorporated
into the natural oceanic background of material concentrations.

Strategies to control man's wastes are 1) to detoxify them by
such means as incineration, 2) to diffuse and disperse for assimila-
ble substances, and 3) to isolate and contain persistent toxic
wastes. Proper use of these strategies with contingency plans to
cope with accidents and to abandon dumpsites when necessary is an intelligent direction to take. This International Symposium strives to
establish an adequate scientific basis for such an approach to disposal problems.

HISTORICAL SKETCH

On December 30, 1675, Governor Edmund Andros, second English Governor of the Colony of New York, forbade any person to "cast any dung, dirt, refuse of ye city or anything to fill up ye harbor or among ye neighbors under penalty of forty schillings" (Forsythe, 1977). While the British colonial government followed by U.S. municipal and federal governments have had over 300 years to follow up on this prohibition, the New York harbor still is considered very polluted. Contaminant inputs into the harbor and the Bight have exceeded recycling and removal.

Also during the 17th century, the French explorers were very impressed by the beauty of Lake Huron of North America; they named it "The Sweet Sea." Today that designation appears ironic. The byproducts of modern technology and a large population have polluted the Great Lakes that include "The Sweet Sea." Eutrophication of Lake Erie is a notable example.

The above examples illustrate the persistence and magnitude of pollution. Recognizing the importance of waste disposal, U.S. President Richard M. Nixon directed the Council on Environmental Quality (CEQ) to study ocean dumping. In his April 15, 1970, message to the U.S. Congress, he asked CEQ to prepare research, legislative, and administrative recommedations (USCEQ, 1970). The CEQ report recommended a comprehensive national policy on ocean dumping of wastes to end unregulated ocean dumping; it recommended the prohibition of ocean disposal of any materials harmful to the marine environment. The

Table 1. Existing and Proposed International Ocean Dumping Conventions.

Name of Convention	Area of Concern	Date of Adoption	Date of Entry into Force
Oslo Convention	North Sea and Northeastern Atlantic Ocean	15 February 1972	7 April 1974
London Convention	World-wide	19 December 1972	30 August 1975
Helsinki Convention	Baltic Sea	22 March 1974	—
Barcelona Convention	Mediterranean Sea	16 February 1976	12 February 1978

Ocean Dumping Act, Public Law 92-532, was enacted, along with the
Clean Water Act, Public Law 92-500, in 1972. A historical summary
between The River and Harbor Act of 1899 and of other U.S. legisla-
tions up to 1972 is given by Pararas-Carayannis (1973).

The London Ocean Dumping Convention which became effective in
1975 is a milestone of international cooperation on ocean pollution
control. The convention is designed to be applicable to all over the
world; at present forty-three nations have joined the convention.
Other important regional conventions are Oslo Convention that deals
with the pollution of North Sea and Northeastern Atlantic, Helsinki
Convention for the Baltic Sea and Barcelona Convention for the Mediterranean Sea. Table 1 summarizes these four conventions.

To offer an overview with emphasis on the scientific issues associated with ocean dumping, both the U.K. and U.S. efforts will be
described chronologically followed by the details of the London Ocean
Dumping Convention.

U.S. CEQ Report (1970)

The U.S. CEQ study on ocean dumping concluded that there was a
critical need for a national policy on ocean dumping. Though this
study was concerned mainly with the U.S., the results are applicable
internationally. The following are digests from the CEQ report:

International Action. The CEQ report recognizes the international character of ocean dumping. Unilateral action by the United
States can deal with only a part of the problem. Effective international action will be necessary if damage to the marine environment
from ocean dumping is to be averted.

Regulation. 1. Ocean dumping of harmful materials should
cease.

2. When existing information on ocean dumping effects is inconclusive, yet the best indicators show adverse effects, such dumping
should be phased out.

3. Dumping criteria should consider present and future impacts,
irreversibility, volume and concentration of the wastes, and location
of disposal.

4. Biologically critical areas should be protected.

5. Undigested sewage sludge dumping should be stopped.

6. Digested or stablized sludge dumping should be phased out.

7. Polluted dredge spoil dumping should be phased out. Navigational benefits should be weighed carefully against damages before
dredging.

8. High-level radioactive waste ocean dumping should be prohibited.

9. Low-level liquid radioactive waste discharge to the ocean
should be controlled nationally and internationally.

10. Toxic industrial waste dumping should be stopped as soon as
possible.

   Research Needs   1. Pathways of waste materials in the marine
ecosystem along with the origin and ultimate fate of pollutants
should be studied.
   2. Representative marine ecosystems should be protected so
that man-induced changes may be evaluated.
   3. Basic physical and chemical processes in the ocean with emphasis on estuaries and coastal areas should be studied.
   4. Lethal and sublethal toxicity and long-term effects of toxic
material on marine life should be studied. Information should be obtained on toxicant persistence, chemo- and bio-degradation, radioactivity effects, and on assimilation capacity of the ocean.
   5. Public health risk information should be gathered. Pathogen
pathways should be studied. Effective methods of measuring public
health dangers are needed.
   6. Alternative methods to ocean dumping should be developed as
well as waste recycling. In addition to technical problem solving,
the social, institutional, and economic aspects of waste management
should be studied thoroughly.
   7. Effective national and international monitoring systems
should be developed. Early detection of pollution as well as effective data coordination should be developed and implemented.

   From the end of World War II, 1945, to 1970, ocean dumping
activity in the United States increased rapidly. In 1968, about 48
million tons of wastes were dumped at sea. Upon enactment of the
Ocean Dumping Act, PL 92-532, in 1972, this trend was reversed.
Quantities and types of wastes for 1968 are given below:

Waste Type	Amount (Million tons)	Percentage (%)
Dredge spoils	38.4	80
Industrial wastes	4.7	10
Sewage sludge	4.5	9
Others	0.6	1

   The waste type "Others" includes construction and demolition debris, solid wastes including paper, wood, plastics, and rubber, many
of which can float to the sea surface, explosive and chemical muni-
tions, and radioactive wastes.

U.S. Clean Water Act (1972)

   On October 18, 1972, the U.S. Congress (1972a) enacted a public
law, PL 92-500, entitled "Federal Water Pollution Control Act Amendment of 1972." It is commonly called the "Clean Water Act." The objective of this act is to restore and maintain the chemical, physical, and biological integrity of the nation's waters. One of the
goals is that a major research and demonstration effort be made to
develop technology necessary to eliminate the discharge of pollutants
into the navigable waters, waters of the contiguous zone, and the

ocean. Pipeline discharges of pollutants into navigable waters, which include its territorial sea, became unlawful by this law without a permit.

Excerpts pertinent to the control of marine pollution from the Act are as follows:

"RESEARCH, INVESTIGATION, TRAINING, AND INFORMATION

"Sec. 104.(n) (1) The [Environmental Protection Agency (EPA)] Administrator shall, in cooperation with the Secretary of the Army, the Secretary of Agriculture, the Water Resources Council, and with other appropriate Federal, State, interstate, or local public bodies and private organizations, institutions, and individuals, conduct and promote, and encourage contributions to, continuing comprehensive studies of the effects of pollution, including sedimentation, in the estuaries and estuarine zones of the United States on fish and wildlife, on sport and commercial fishing, on recreation, on water supply and water power, and on other beneficial purposes. Such studies shall also consider the effect of demographic trends, the exploitation of mineral resources and fossil fuels, land and industrial development, navigation, flood and erosion control, and other uses of estuaries and estuarine zones upon the pollution of the waters therein.

"(2) In conducting such studies, the Administrator shall assemble, coordinate, and organize all existing pertinent information on the Nation's estuaries and estuarine zones; carry out a program of investigations and surveys to supplement existing information in representative estuaries and estuarine zones; and identify the problems and areas where further research and study are required.

"IN-PLACE TOXIC POLLUTANTS

"Sec. 115. The [EPA] Administrator is directed to identify the location of in-place pollutants with emphasis on toxic pollutants in harbors and navigable waterways and is authorized, acting through the Secretary of the Army, to make contracts for the removal and appropriate disposal of such materials from critical port and harbor areas. There is authorized to be appropriated $15,000,000 to carry out the provisions of this section, which sum shall be available until expended.

"OCEAN DISCHARGE CRITERIA

"Sec. 403. (a) No permit under section 402 of this Act for a discharge into the territorial sea, the waters of the contiguous zone, or the oceans shall be issued, after promulgation of guidelines established under subsection (c) of this section, except in compliance with such guidelines. Prior to the promulgation of such guidelines, a permit may be issued under section 402 if the Administrator determines it to be in the public interest.

"(b) The requirements of subsection (d) of section 402 of this Act may not be waived in the case of permits for discharges into the territorial sea.

"(c) (1) The Administrator shall, within one hundred and eighty days after enactment of this Act (and from time to time thereafter), promulgate guidelines for determining the degradation of the waters of the territorial seas, the contiguous zone, and the oceans, which shall include:

"(A) the effect of disposal of pollutants on human health or welfare, including but not limited to plankton, fish, shellfish, wildlife, shorelines, and beaches;

"(B) the effect of disposal of pollutants on marine life including the transfer, concentration, and dispersal of pollutants or their byproducts through biological, physical, and chemical processes; changes in marine ecosystem diversity, productivity, and stability; and species and community population changes;

"(C) the effect of disposal, of pollutants on esthetic, recreation, and economic values;

"(D) the persistence and permanence of the effects of disposal of pollutants;

"(E) the effect of the disposal at varying rates of particular volumes and concentrations of pollutants;

"(F) other possible locations and methods of disposal or recycling of pollutants including land-based alternatives; and

"(G) the effect on alternate uses of the oceans, such as minerals exploitation and scientific study.

"(2) In any event where insufficient information exists on any proposed discharge to make a reasonable judgment on any of the guidelines established pursuant to this subsection no permit shall be issued under section 402 of this Act."

U.S. Ocean Dumping Act (1972)

On October 23, 1972, the U.S. Congress (1972b) enacted a public law, PL 92-532, entitled "Marine Protection, Research, and Sanctuaries Act of 1972." It is commonly called the "Ocean Dumping Act." The Congress declared that it is the policy of the United States to regulate the dumping of all types of materials into ocean waters which would adversely affect human health, welfare, or amenities, or the marine environment, ecological systems, or economic potentialities.

To implement the U.S. policy, the Act regulates the transportation of material from the United States for dumping into ocean waters, and the dumping of material, transported from outside the United States, if the dumping occurs in ocean waters over which the United States has jurisdiction or over which it may exercise control, under accepted principles of international law, in order to protect its territory or territorial sea.

The Act prohibits the dumping of high-level radioactive wastes and all biological, chemical, and radiological warfare agents into the ocean. The dumping of other wastes, except dredge spoils regulated by the U.S. Army Corps of Engineers, is to be strictly regulated by the U.S. Environmental Protection Agency.

The Title II of the Ocean Dumping Act is called "Comprehensive Research on Ocean Dumping." It reads as follows:

"TITLE II--COMPREHENSIVE RESEARCH ON OCEAN DUMPING

"Sec. 201. The Secretary of Commerce, in coordination with the Secretary of the Department in which the Coast Guard is operating and with the EPA Administrator shall, within six months of the enactment of this Act, initiate a comprehensive and continuing program of monitoring and research regarding the effects of the dumping of material into ocean waters or other coastal waters where the tide ebbs and flows or into the Great Lakes or their connecting waters and shall report from time to time, not less frequently than annually, his findings (including an evaluation of the short-term ecological effects and the social and economic factors involved) to the Congress.

"Sec. 202. (a) The Secretary of Commerce, in consultation with other appropriate Federal departments, agencies, and instrumentalities shall, within six months of the enactment of this Act, initiate a comprehensive and continuing program of research with respect to the possible long-range effects of pollution, overfishing, and man-induced changes of ocean ecosystems. In carrying out such research, the Secretary of Commerce shall take into account such factors as existing and proposed international policies affecting oceanic problems, economic considerations involved in both the protection and the use of the oceans, possible alternatives to existing programs, and ways in which the health of the oceans may best be preserved for the benefit of succeeding generations of mankind."

To implement the Section 201 mandate, National Oceanic and Atmospheric Administration (NOAA) in the Department of Commerce established the Ocean Dumping Program on October 1, 1976. It was elevated to Ocean Dumping and Monitoring Division in the National Ocean Survey in NOAA in January 1979.

U.K. Dumping at Sea Act (1974)

An Act to control dumping in the sea for the United Kingdom
(1974) was enacted on June 17, 1974, requiring ocean dumping license.
In determining whether to grant a license, the licensing authority,
in England, The Ministry of Agriculture, Fisheries and Food, considers conditions in such a license to protect the environment and
its resources from ocean dumping.

In the United Kingdom, ocean dumping of radioactive waste is
governed by "Radioactive Substances Act, 1960" which was enacted on
June 2, 1960. This Act regulates the keeping and use of radioactive
material, and to make provision as to the disposal and accumulation
of radioactive waste. The power to grant authorization in respect to
the disposal of radioactive waste in England is given to the Minister
of Housing and Local Government and the Minister of Agriculture,
Fisheries and Food, in Scotland, the same authority is given to the
Secretary of State, in Northern Ireland, to The Minister of Health
and Local Government and to the Minister of Commerce.

Internationally, since the United Kingdom is a signatory member
state of both the Oslo and London Conventions, any procedure which
has been developed for the effective application of these conventions
is an accepted procedure between U.K. and the government of any convention states concerned.

U.S. National Academy of Sciences Case Study (1975)

The United States National Academy of Sciences (1975) issued a
report entitled "Assessing Potential Ocean Pollutants." Case studies
conducted by the Academy found that both transuranic elements and
hexachlorobenzene represent potential hazards. No demonstrable hazard to open ocean ecosystems at present levels of release is given by
low molecular weight chlorinated hydrocarbons, acrylonitrile, iron
and copper wastes, open ocean litter, tetracycline and technetium.
Aromatic hydrocarbons can give some local adverse effects.

The report singled out the four following parameters that are
critically needed to evaluate a material's being toxic to the ocean.
They are:
1. Toxicant production and rate of its release;
2. Residence time;
3. Bioaccumulation;
4. Toxicity.

The major recommendations given include:
1. Releases of transuranic elements should be kept to an absolute minimum.
2. Waste should not be discharged into biologically active
marine areas.

The International Ocean Dumping Convention (1975)

    The International Convention on the "Prevention of Marine Pollu-
tion by Dumping of Wastes and Other Matter" became effective on August 30, 1975. It is commonly called as London Ocean Dumping Convention. Intergovernmental Maritime Consultative Organisation (IMCO)
(1975) serves as its secretariat. As of October 1979, the following
43 governments have ratified or acceded to the Ocean Dumping Convention. These states are listed below:

Afghanistan	Morocco
Argentina	Netherlands
Byelorussian SSR	New Zealand
Canada	Nigeria
Cape Verde	Norway
Chile	Panama
Cuba	Philippines
Denmark	Poland
Dominican Republic	Portugal
Federal Republic of Germany	South Africa
Finland	Spain
France	Sweden
German Democratic Republic	Switzerland
Guatemala	Tunisia
Haiti	Ukrainian SSR
Hungary	United Arab Emirates
Iceland	United Kingdom
Jordan	United States
Kenya	USSR
Libyan Arab Jamahiriya	Yugoslavia
Mexico	Zaire
Monaco	

    Consultative meetings are being held annually at IMCO Headquar-
ters in London, U.K., to implement and to improve the convention
text. Because of the far-reaching importance of the London Ocean
Dumping Convention for humanity, reading of its entire text is en-
couraged. The Annex I substances, ocean dumping of which is prohibited, and the Annex II substances, which need special care, are listed below:

"Annex I

1. Organohalogen compounds.
2. Mercury and mercury compounds.
3. Cadmium and cadmium compounds.
4. Persistent plastics and other persistent synthetic materials, for example, netting and ropes, which may float or may remain in
suspension in the sea in such a manner as to interfere materially
with fishing, navigation or other legitimate uses of the sea.

5. Crude oil, fuel oil, heavy diesel oil, and lubricating oils, hydraulic fluids, and any mixtures containing any of these, taken on board for the purpose of dumping.

6. High-level radio-active wastes or other high-level radioactive matter, defined on public health, biological or other grounds, by the competent international body in this field, at present the International Atomic Energy Agency, as unsuitable for dumping at sea.

7. Materials in whatever form (e.g., solids, liquids, semiliquids, gases or in a living state) produced for biological and chemical warfare.

8. The preceding paragraphs of this Annex do not apply to substances which are rapidly rendered harmless by physical, chemical or biological processes in the sea provided they do not:

(i) make edible marine organisms unpalatable, or

(ii) endanger human health or that of domestic animals.

The consultative procedure provided for under Article XIV should be followed by a Party if there is doubt about the harmlessness of the substance.

9. This Annex does not apply to wastes or other materials (e.g., sewage sludges and dredged spoils) containing the matters referred to in paragraphs 1-5 above as trace contaminants. Such wastes shall be subject to the provisions of Annexes II and III as appropriate."

"Annex II

The following substances and materials requiring special care are listed for the purpose of Article VI(1)(a).

A. Wastes containing significant amounts of the matters listed below:

> arsenic   )
> lead   )
> copper   )   and their compounds
> zinc   )
> organosilicon compounds
> cyanides
> fluorides
> pesticides and their by-products not covered in Annex I.

B. In the issue of permits for the dumping of large quantities of acids and alkalis, consideration shall be given to the possible presence in such wastes of the substances listed in paragraph A and to the following additional substances:

> beryllium)
> chromium   )   and their compounds
> nickel   )
> vanadium   )

C. Containers, scrap metal and other bulky wastes liable to sink to the sea bottom which may present a serious obstacle to fishing or navigation.

D. Radioactive wastes or other radioactive matter not included

in Annex I. In the issue of permits for the dumping of this matter,
the Contracting Parties should take full account of the recommendations of the competent international body in this field, at present
the International Atomic Energy Agency."

The Convention text still requires exact scientific definition
on such terms as "harmlessness", "trace contaminants", "significant
amount", "high- and low-level radioactive wastes", and any future
amendments on Annexes I and II substances.

Oceanographic Assessment of Ocean Disposal (1976)

An objective assessment on waste disposal options among air,
land, and water in order not to protect one sector of the environment
at the costs of other sectors was conducted by the U.S. National Research Council (1976). The ocean may be an attractive sink for some
residuals after a thorough comparative study. This report considers
ocean disposal in terms both of the need for ocean disposal, and of
the capacity of the marine ecosystem to receive wastes. Specifically, this report describes the current legislation and regulations,
followed by the source terms such as the amount and properties of the
wastes, and the processes that affect their distribution and fate in
the marine environment.

On the dumpsite selection, the report stresses that characterization of proposed dumpsites should include geological descriptions
of bottom morphology, the nature of the substrate, and the rates of
sedimentation or erosion, geotechnical characteristics and seismic-
ity, in addition to physical, chemical, and biological processes occurring at the dumpsite. '

An important conclusion from this report was that a dumpsite may
have to be abandoned because of mismanagement leading to overloading,
because of changes in legislation or in the waste itself, or because
of changes in the value of the disposal area for other purposes.
Thus, the selection of a dumpsite must consider the possibility of
future abandonment.

A slate of oceanographic research recommendations has been made
in the report. The specific research areas recommended are dispersion, bioassay, air-sea interactions, environmental effects at and
reclamation of existing sites, chemical processes associated with
synthetic organic compounds, petroleum hydrocarbons and sulfides,
geotechnical studies and geological hazards, and the effects of rare
events.

U.S. Environmental Protection Agency Ocean Dumping Regulations and
Criteria (1977)

On January 11, 1977, The U.S. Environmental Protection Agency

(EPA) issued "Ocean Dumping: Final Revision of Regulations and Criteria (1977)." The EPA rules and regulations describe in detail, considering the state of art of the oceanographic and technological knowledge, the operational procedures to be followed when an ocean dumping permit is sought.

Of special scientific interest is the specific criteria for dumpsite selection. The factors considered include:
1. Geographic location;
2. Location in relation to breeding, spawning, nursery, feeding, or passage areas of living resources in adult or juvenile;
3. Location in relation to amenity areas such as swimming beaches;
4. Types, quantities, packing, methods of release of wastes;
5. Feasibility of surveillance and monitoring
6. Diffusion, dispersion, mixing;
7. Previous dumping effects including cumulative effects;
8. Interference with shipping, fishing, recreation, mineral extraction, desalination, aquaculture, areas of special scientific importance, and other legitimate uses of the ocean;
9. Water quality and ecology of the site;
10. Potentiality for the development or recruitment of nuisance species at the site;
11. Cultural or historical site.

The factor (5) which considers the feasibility of surveillance and monitoring is an important consideration to refute the validity of the "out of site, out of mind" attitude.

U.S. Comptroller General's Report (1977)

The U.S. General Accounting Office (1977) prepared a report entitled "Problems and Progress in Regulating Ocean Dumping of Sewage Sludge and Industrial Wastes" to the U.S. Congress. This report concluded that:
1. Some wastes containing harmful substances that exceeded safety levels were dumped in the ocean;
2. The wastes were dumped too rapidly to be assimilated by the marine environment;
3. Some of the proposed alternatives to ocean dumping may be environmentally more harmful when the total environmental impact is considered.

The alternatives to ocean dumping pose difficult questions. It is difficult to ascertain exactly what the total environmental effect would be if wastes, formerly dumped in the ocean were to be transferred to other parts of the environment, such as air, groundwater, or land. Would other forms of disposal be less damaging to the environment than ocean dumping? Such comparative studies need sufficient data, scientific as well as socio-economical, for both land and ocean

disposal options. We should strive toward obtaining these data in
the near future.

This report cites the difficulty encountered in enforcing the
ocean dumping regulations. For instance, The U.S. EPA 1977 Criteria
and Regulations require that mercury and cadmium concentrations in
the solid phase of a waste must not exceed 0.75 mg/kg and 0.6 mg/kg
respectively or must be less than 50 percent greater than the average
content of these elements in natural sediments of similar lithologic
characteristics as those at the disposal site. Twenty-six municipal
permit holders in the New York-New Jersey area were once dumping sewage sludge containing either mercury or cadmium that sometimes exceeded by more than 100 times these established levels. Although it
may degrade the marine environment, EPA regulations allowed the dumping of mercury and cadmium in excess of safety levels if the materials were present in sewage sludge. It is, therefore, recommended
that municipal waste sources must be separated from any toxic waste
sources in order to maintain sewage sludge less toxic.

U.S. National Ocean Pollution Research and Development and Monitoring
Planning Act of 1978

On May 8, 1978, the above Act, PL 95-273, was enacted by the
U.S. Congress (1978). The purposes of the Act are as follows:
1. To establish a comprehensive 5-year plan for federal ocean
pollution research and development and monitoring programs.
2. To develop the necessary base of information to support, and
to provide for, the rational, efficient, and equitable utilization,
conservation, and development of ocean and coastal resources.
3. To designate NOAA as the lead federal agency for preparing
the comprehensive 5-year plan and to require NOAA to carry out a comprehensive program of ocean pollution, research and development, and
monitoring under the plan.

In the Act, NOAA may provide financial assistance in the form of
grants or contracts for research and development and monitoring projects or activities which are needed to meet priorities set forth in
the 5-year plan, if such priorities are not being adequately addressed by any federal department, agency, or instrumentality.

This Act will contribute definitely to obtain the needed scientific basis with which we can assess correctly the impact of oceanic
pollution.

FUTURE PERSPECTIVES

Persistent efforts of many nations and their concerned citizens
have resulted in the International London Ocean Dumping Convention
becoming effective in 1975. The ocean being man's common resource,

the approach we thus collectively have taken is commendable.

Each state must cope with its own unique problems in order to survive and prosper. For some states, land options for waste disposal are almost unavailable, thus ocean dumping may be the only available option. On the other hand, fisheries are very important for some other states, and the coastal waters around them must be protected from damage.

Nevertheless, the ocean is a common resource for man, and ocean dumping must not exceed the assimilative capacity of the ocean. In addition, it is an intelligent approach to establish the scientific basis now by studing present ocean dumping events.

What will be our future? Will future generations find a more contaminated ocean than the one we have enjoyed? Let us examine the trend in ocean dumping in the foreseeable future.

Dredge Spoils

Maintenance dredging must continue for shipping lanes and harbor facilities. The major issue here is how to dispose of heavily polluted dredged materials from such places as active commercial harbors. An effective method is the isolation and containment of the wastes, not returning them to the dynamic hydrosphere of the Earth. Some planners are considering the establishment of an offshore artificial island off New York for dredged material containment; the island also will be used for offshore oil transshipment and storage. At the mouth of the Chesapeake Bay, the U.S. Army Corps of Engineers practice the isolation and containment principle by depositing polluted dredged sediments within a man-made confinement called Crany Island.

Unpolluted dredged materials can be a resource. They are used for beach nourishment and for construction sand and gravel.

Industrial Wastes

The types of contaminants in industrial wastes dumped at sea vary greatly because of the diversity of industries and production processes involved. Highly toxic wastes are banned internationally from ocean dumping. For instance, substances listed in Annex I of the London Ocean Dumping Convention are prohibited.

Internationally, IMCO, in accord with the London Convention, has begun gathering information on ocean-dumped wastes. Most of the industrialized nations have been compiling their ocean-dumping data and they regularly report to IMCO. The data thus gathered will give a precise input with which oceanic budget calculations with respect to pollution may be made.

Industrial wastes by manufacturing process for the United States that were dumped in the ocean were compiled by Smith and Brown (1970) as shown below:

Types of Wastes	Estimated Tonnage	Percent
Waste Acids	2,720,500	58
Refinery Wastes	562,900	12
Pesticide Wastes	328,300	7
Paper Mill Wastes	140,700	3
Others	938,100	20

Let us consider that this dumping contained about 3×10^{10} moles of strong acid. Since the total oceanic buffer capacity is about 3×10^{18} equivalents, the U.S. contribution is one 100 millionth of the oceanic buffer capacity. If we further assume that the world's total contribution being ten times of the United States, the acid dumping alone, in theory, is in the order of 1/10,000,000 of the oceanic buffer capacity. Regionally, though, the change can be greater because the entire ocean cannot immediately interact with dumped material. Exemplified here is the need to better understand horizontal and vertical dispersive properties of the ocean.

Of special concern is the toxicity of synthetic organic compounds that are proven to be notoriously toxic but also bioaccumulated, and not biogradable. The Kepone contamination of the James River estuary and polychlorobiphenyl, PCB, in the Hudson River runoff are notable examples. They are banned from ocean dumping by the London Convention.

The ocean dumping of industrial wastes has been steadily decreasing in the U.S. water recently. Internationally, the same may not be true as many countries are becoming industrialized. In the future, it may become imperative to set international quotas for ocean dumping. Such decisions must be supported by solid scientific data. Therefore, the needed scientific information must be gathered now.

Sewage Sludge

In the United Kingdom sewage sludge ocean dumping is an important on-going practice. Notable dumpsites around England and Wales are Liverpool Bay, Bristol Channel, Plymouth, Exeter, Nab Tower, Thames, Roughs Tower, and Spurn Head.

In the United States, two active dumpsites are those of New York and Philadelphia. These municipalities have been asked to end their sludge dumping by the end of 1981. The Philadelphia sludge dumping history and future phase out schedule are shown in Fig. 1.

   In both U.K. and U.S.A. the sludge dumping impact is studied
intensively and extensively. Efforts are being made to understand
the short-, medium-, and long-term fate of the sludge and its persistent components, metals and non-degradable synthetic organic compounds, and its effects on the physical and chemical characteristics
of the receiving area, both water and sediments.

   It may be possible to cease the ocean dumping of sewage sludge
within the United States in the future. However, internationally,
the same may not be true, for the alternative to the ocean dumping is

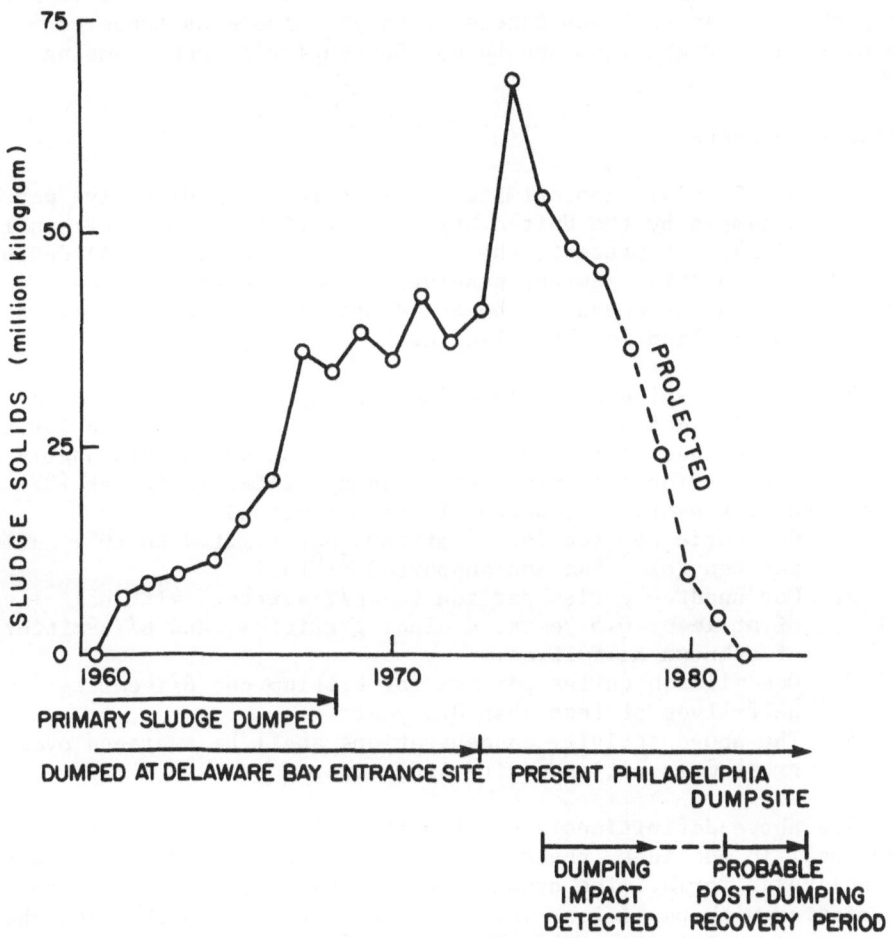

Fig. 1. Sewage sludge ocean dumping by Philadelphia. The city
         produces over 100 million kilograms of sludge solids a
         year.

severely limited in some states. In addition, ocean outfalls of domestic sewers are accepted practices including those in the United
States. A notable example is the Los Angeles regional outfall, but
there are many others on both the east and west coasts.

Since sewage sludge is dumped in shallow water, its benthic effects must be scrutinized thoroughly. Sludges may affect fish and
shellfish quality and be a hazard to human health because of pathogens and other disease causing agents.

Of special concern is the possibility of the separation of
municipal wastes from industrial wastes before sludge is prepared for
ocean dumping. This kind of source control minimizes the pollution
effect due to sludge dumping. The London Convention text states that
the prohibited Annex I substances in sewage sludge as trace contaminants are exempt from the London Convention's ocean dumping
prohibition.

Radioactive Wastes

During 1946-1970 approximately 10^5 curies of radioactive wastes
were ocean-dumped by the United States (Robert Dyer, EPA, personal
communication). At present, the Nuclear Energy Agency (NEA) dumps
low-level radioactive wastes, generated from several European
countries, into the ocean at the NEA dumpsite, 500 nautical miles
southwest of Ireland in the Atlantic.

High-level radioactive waste dumping is banned by the London
Convention. The definition of high-level radioactive wastes for the
purpose of the Annex I to the London Convention was provisionally defined by International Atomic Energy Agency, IAEA, in August 1978 as
follows (Robert Dyer, EPA, personal communication):
   1. One curie per ton for α-emitters but limited to 10^{-1} curie
       per ton for ^{226}Ra and supported ^{210}Po.
   2. One hundred curies per ton for β/γ-emitters with half-lives
       of at least 0.5 years, excluding tritium, and β/γ-emitters
       of unknown half-lives.
   3. One million curies per ton for tritium and β/γ-emitters with
       half-lives of less than 0.5 years.
   4. The above activity concentrations shall be averaged over a
       gross mass not exceeding 1000 tons.

The above definitions are scientifically vague, for they do not
give specific nuclides, their specific activities, and their ionic
and molecular forms. The dynamics of radioactivity in the ocean, its
association with particles, movement into and through the food chain,
and to some extent its physical mixing will depend on the specific
nuclides and chemical species involved. A better definition must
consider the bioaffinity of each nuclide in the ocean, such as ^{90}Sr
being incorporated into the tests of marine organisms.

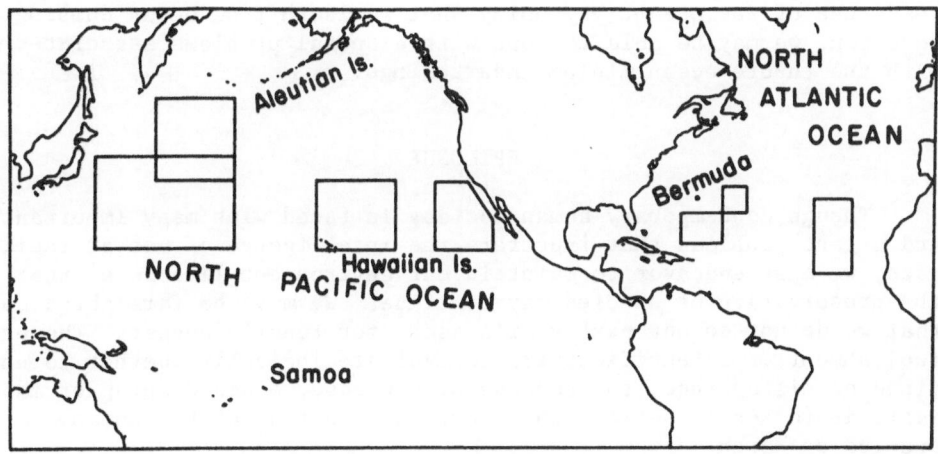

Fig. 2. Areas now under study by the radioactive waste seabed dis-
         posal program. Desirable characteristics for a repository
         beneath the sea floor include a water depth of 4000 meters
         or more, the presence of about 60 meters of red clay, and a
         continuous geologic record of a stable sediment environment.
         (Source: Charles Hollister, Woods Hole Oceanographic
         Institution).

Though banned from ocean dumping, highly radioactive, longlived
nuclear wastes have been piling up at storage sites around the world.
A group of researchers is considering carefully burial below the
ocean floor as a place to dispose of these nuclear wastes (Fig. 2).
The concept is that the marine sediments beneath the oceans would be
the geological barrier, not the ocean above them. Much scientific
and technological information must be established on the subseabed
option before man can accept it socially, institutionally, and economically. Politically, consideration of high-level nuclear waste
disposal in international waters is very uncertain. At the moment,
no provisions of the Law of the Sea documents would bar such activity, and deep-sea disposal is considered to be a serious option for
some states.

Ocean Mining Wastes

Manganese nodule recovery from the Pacific Ocean is considered
economically feasible by some states. Since the first generation
plants for nodule processing probably will be based on land, an important consideration now is the safe disposal of its wastes. Factory ships may emerge later, which may need to dump their wastes into
the ocean for economic reasons.

Other mining efforts may occur as man exploits the mineral

resources of the ocean. By using what we learn from ocean dumping research, we may be able to cope with disposal problems associated with the future ocean mining undertakings.

EPILOGUE

Though contemporary human society is faced with many important and urgent problems that frustrate the intelligence of man at rapid rate, we must endeavor to maintain our environment livable so that the preservation of species may continue. We must be farsighted so that we do not devour next year's seeds for today's hunger. The establishment of scientific basis to evaluate logically whether to continue or discontinue, to increase or decrease, ocean dumping of man's waste is to assist society to assure continuation of the human species on Earth.

REFERENCES

Forsythe, E. B. (1977) Ocean dumping authorization hearing state ment, March 9, 1977, Ninety-Fifth Congress, U. S. Government Printing Office, 94-4960, page 220.

Inter-Governmental Maritime Consultative Organization (1975) Convention on the prevention of marine pollution by dumping of wastes and other matter. 15 pp.

Kester, D. R., R. C. Hittinger, and P. Mukherji (1981) Transition and heavy metals associated with acid-iron waste disposal at Deep Water Dumpsite 106. In: Ocean Dumping of Industrial Wastes, B. H. Ketchum, D. R. Kester, and P. K. Park, editors, Plenum Press. This volume, pp. 215-232.

Paras-Carayannis, G. (1973) Ocean dumping in the New York Bight: An assessment of environmental studies, Tech. Memorandum No. 39, U. S. Coastal Engineering Research Center, 159 pp.

Smith, D. D., and R. P. Brown (1970) An appraisal of oceanic disposal of barge-delivered liquid and solid wastes from U. S. Coastal cities. Prepared by Dillingham Corporation for U. S. Department of Health, Education, and Welfare, Bureau of Solid Waste Management, under contract No. PH 86-68-203.

United Kingdom (1960) Radioactive Substances Act, 1960.

United Kingdom (1974) Dumping at Sea Act, 1974.

United States Congress (1972a) Federal water pollution control act amendments of 1972. Public law 92-500, 86 STAT. 1060.

United States Congress (1972a) Marine Protection, Research, and Sanctuaries Act of 1972. Public Law 92-532, 86 STAT. 1052.

United States Congress (1978) National Ocean Pollution Research and Development and Monitoring Planning Act of 1978. Public Law 95-273, 92 STAT. 228.

United States Council on Environmental Quality (1970) Ocean Dumping: A National Policy. U. S. Government Printing Office 0-404-547, 45 pp.

United States Environmental Protection Agency (1977) Ocean Dumping:
     Final Revision of Regulations and Criteria, U. S. Federal
     Register, Tuesday, January 11, 1977, Part VI, 42 (7), 2462-2490.
United States General Accounting Office (1977) Problems and Progress
     in Regulating Ocean Dumping of Sewage Sludge and Industrial
     Waste. Report to the Congress by the Comptroller General of the
     United States, CED-77-18, 61 pp.
United States National Academy of Sciences (1975) Assessing Potential
     Oceanic Pollutants, 438 pp.
United States National Research Council (1976) Disposal in the
     Marine Environment; an oceanographic assessment, 76 pp.

INDUSTRIAL OCEAN DUMPING IN EPA REGION II - Regulatory Aspects

Peter W. Anderson and Richard T. Dewling

Marine and Wetlands Protection Branch
Environmental Protection Agency (Region II)
26 Federal Plaza
New York, NY 10007

ABSTRACT

The Marine Protection, Research, and Sanctuaries Act of 1972 (PL 92-532) mandates that the Environmental Protection Agency "prevent or strictly regulate" the disposal, via vessel, of waste materials into the ocean.

The majority of ocean dumping, past and present, in the United States occurs at dump sites managed by EPA Region II. In 1977, for example, over 90 percent or 6.1 million metric tonnes of the U. S. ocean dumping occurred at sites managed by this Region.

EPA's implementation of this Act, beginning in April 1973, spurred industrial dumpers at sites managed by the Region to develop and construct land-based treatment facilities and/or implement other environmentally acceptable alternatives for the handling their wastes. Of the roughly 150 industrial waste generators in 1973, as of April 1979 only 13 remain. Ten are under EPA orders to cease ocean dumping by the end of 1981. The other three, Allied Chemical Corp., NL Industries, Inc., and E. I. duPont de Nemours & Co. (Grasselli Plant), have demonstrated that their wastes are not harmful to the marine environment and that no feasible alternative exists at present which is more environmentally acceptable than ocean dumping. Thus, the level of ocean dumping activities in Region II represents a significant reduction from that prior to the passage of the Act.

BACKGROUND

Region II of the U.S. Environmental Protection Agency (EPA) covers New York, New Jersey, Puerto Rico, the Virgin Islands, and

adjacent ocean waters. The ocean dumping of waste materials -- industrial wastes, sewage sludge, and dredged materials -- is a regional issue of intense public, scientific, and political interest and
concern. Much of this interest results, unquestionably, from the
fact that the majority of ocean disposal via vessel, past and present, in the United States occurs at dump sites managed by Region II.
For example, in 1977 about 87 percent by volume of all dumping of
waste materials, other than dredged materials, took place at dump
sites located in the New York Bight off the coasts of New York and
New Jersey. When industrial dumping activities off Puerto Rico are
included, this figure increases to 91 percent (Table 1). Dumping activities during earlier periods are reported for both Region II (U.S.
Environmental Protection Agency, 1978) and the nation (U. S. Environmental Protection Agency, 1978; U. S. Army Corps of Engineers, 1978).
The ocean dumping of industrial wastes in Region II takes place at
three dump sites; two in the New York Bight - the acid wastes and industrial wastes dump sites - and one off the north coast of Puerto
Rico (Fig. 1). EPA-approved ocean dump sites are published in the
Federal Register (U. S. Environmental Protection Agency, Jan. 11,
1977).

                            REGULATORY ASPECTS

     In the history of environmental protection, concern for the
ocean is relatively new. Before passage of the Marine Protection,
Research, and Sanctuaries Act (MPRSA) on October 23, 1972, there were
few direct legal controls, either inside or outside of the United
States, on dumping of wastes at sea. Earlier major federal legislation was concerned only with oil pollution resulting from exploration
or exploitation of marine resources.

     The MPRSA (PL 92-532) established for the first time a national
policy of strictly regulating ocean dumping by banning the dumping of
chemical, biological, or radiological warfare agents and high-level
radioactive wastes, and by authorizing a permit system for the

Table 1. Type and amounts of waste materials ocean dumped in 1977.

| Waste Types | Quantities (metric tonnes) | | |
	Region II	United States	% Reg II
Industrial	1,618,000*	1,673,000	97
Sewage sludge	4,119,000	4,658,000	88
Cellar dirt	344,000	344,000	100
Incineration	13,700	40,600	34
Other	0	90	0

*Includes 285,000 tonnes in Puerto Rico.

dumping of all other materials in ocean waters. Under the Act, EPA
is authorized to administer and enforce the entire ocean dumping program and to issue permits regulating the dumping of all materials except dredged materials; the latter materials are dumped under U. S.
Army Corps of Engineers (COE) permits issued consistent with EPA's
marine environmental impact criteria (U. S. Environmental Protection
Agency, Jan. 11, 1977).

Title I of this Act is designed to regulate all dumping and
transport for the purpose of dumping of waste materials within the
territorial seas and the contiguous zone of the United States. The
territorial seas are roughly defined as those waters lying within 5.5
kilometers of a baseline (roughly the coastline); the contiguous

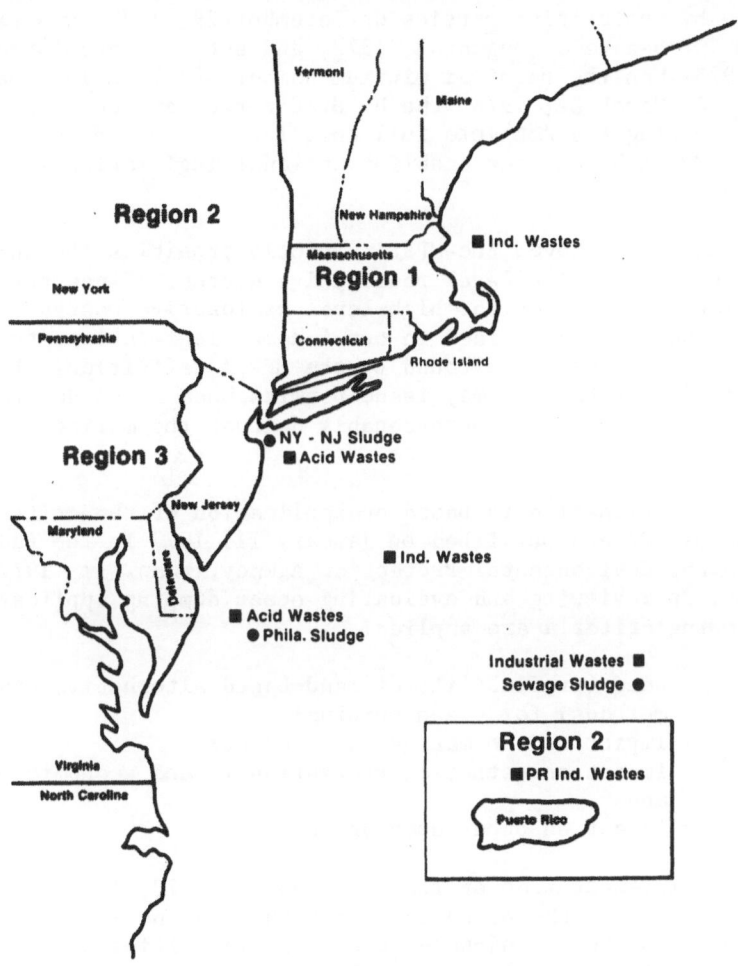

Fig. 1.   Ocean dumping sites regulated by the U.S. EPA and off the
          northeastern U.S.

zone, as those waters lying between 5.5 and 22 kilometers from the
baseline. In addition, under Title I EPA regulates all dumping any-
where in the world by any ship leaving from a U.S. port and by all
U.S. flag vessels. Thus, the MPRSA mandates EPA to administer and
enforce the regulation of ocean dumping, not only within the territorial seas and the contiguous zone, but also within international
waters.

Concurrent with the establishment of legislative controls by the
Congress, international negotiations were being conducted for the development of an international treaty to regulate the dumping of
wastes in the marine environment. These negotiations produced the
"International Convention on the Prevention of Marine Pollution by
Dumping of Wastes and Other Matter". This Convention was opened for
signature by contracting parties on December 29, 1972, was ratified
by the U. S. Senate on August 3, 1973, and entered into force in
August 1975 when the required minimum number of 15 nations had ratified it. On March 22, 1974, the U. S. Congress amended the MPRSA (PL
93-254) to bring the Act into full compliance with the Convention;
thus, the MPRSA became the enabling national legislation for the Convention.

As indicated above, the MPRSA strictly prohibits the dumping of
warfare agents and high-level radioactive wastes. Since the current
international definition for high-level radioactive wastes is less
stringent than that contained in our federal legislation, the United
States will continue to be bound by the MPRSA definition. For other
materials, EPA or the COE may issue permits when it is determined
that the dumping will not unreasonably degrade the marine environment.

This determination is based on application of the criteria established by EPA and published on January 11, 1977 in the Federal Register (U. S. Environmental Protection Agency, Jan. 11, 1977).
Basically, in reviewing and evaluating ocean dumping applications,
the following criteria are applied:

- need (availability of land-based alternative disposal
  methods) for ocean dumping;
- impact on the marine environment;
- impact on esthetic, recreational, and economic values;
  and
- impact on other uses of the ocean.

EPA's implementation of the MPRSA on April 23, 1973 spurred industrial dumpers in the NY-NJ metropolitan area and Puerto Rico to
develop and construct land-based treatment facilities and/or implement other environmentally acceptable alternatives for the handling
their wastes. Initially, dumpers were identified by EPA according to
the quantity and type of waste being handled. Site visits were then

made by EPA staff to determine each dumper's immediate need for continuing their practice of ocean dumping and the availability of environmentally acceptable alternatives. Based on these visits, 47 industries were immediately required to cease ocean dumping (U. S. Environmental Protection Agency, 1977). Where alternatives were not readily available, industries were issued permits that required all wastes, except acid wastes, to be dumped at the industrial waste sites, located either 160 kilometers due east of Cape May, New Jersey or 80 kilometers north of Arecibo, Puerto Rico. Such permits required the submission of detailed engineering studies outlining alternatives to ocean dumping, and:

- either the establishment of an EPA approved compliance schedule to implement an environmentally acceptable alternative disposal method on or before December 31, 1981; or
- demonstrate to EPA's satisfaction that no feasible land-based disposal method exists at present that is more environmentally acceptable than ocean dumping and that the waste proposed to be dumped is in full compliance with the environmental impact criteria (U. S. Environmental Protection Agency, Jan. 11, 1977), also on or before December 31, 1981.

This regulatory process has been fairly successful. Of the roughly 150 industrial ocean dumpers in 1973, only 13 remain as of April 1979. With the exception of Allied Chemical Corp., NL Industries, Inc., and E. I. duPont (Grasselli Plant), all remaining industrial dumpers in the New York Bight and Puerto Rico have enforceable compliance schedules within their permits which require the cessation of ocean dumping by the end of 1981.

NEW YORK BIGHT - ACID WASTES SITE

The acid wastes dump site was established by the COE (Pararas-

Table 2. EPA approved ocean dumpsites managed by Region II.

Primary Use	Coordinates (Lat., Long.)	Area (sq km)	Depth (m)
Acid Wastes (NY Bight)	40°16'N - 40°20'N 73°36'W - 73°40'W	41	24
Industrial (NY Bight)	38°40'N - 39°00'W 72°00'W - 72°30'W	1550	1800
Industrial (Puerto Rico)	19°10'N - 19°20'N 66°35'W - 66°50'W	363	6100

Carayannis, 1973) in April 1948 after coordination with several federal and State agencies. Since 1973, EPA (U. S. Environmental Protection Agency, Jan. 11, 1977) has described its location as found in
Table 2. Prior to 1973, a slightly different method was employed in
describing the location (Pararas-Carayannis, 1973).

    The site is currently used by two industries, NL Industries,
Inc. of Sayreville, New Jersey, and Allied Chemical Corp. of Elizabeth, New Jersey, to dispose of acid wastes. Prior to 1974, caustic
wastes generated at the E. I. duPont de Nemours & Co., Grasselli
Plant in Linden, New Jersey also were dumped at this site. These
wastes are now dumped at the industrial wastes site. Quantities
dumped by these three waste generators during the period 1973-78 are
tabulated in Table 3.

    By-product hydrochloric acid wastes are ocean dumped by Allied
Chemical Corp. These wastes are generated in the manufacture of Genetron R 12/11 (monofluorotrichloromethane and dichlorodifluorometh-
ane), Genetron R 22 (chlorodifluoromethane), and TFE (tetrafluoro-
ethylene) monomer acids. Manufacture of Genetron R 12/11 was shut
down in May 1977, as a result of decreasing sales volumes due to the
ozone controversy. Thus, these waste streams are not currently being
dumped. Chemical and biological characteristics of the barged wastes
are presented in Table 4. Authorized annual amounts under the current permit are 52,000 metric tonnes. As noted in Table 3, actual

Table 3. Quantities, in thousand metric tonnes (wet), of industrial
          wastes ocean dumped in EPA Region II since 1973.

Site/Year	1973	1974	1975	1976	1977	1978
Acid Wastes (NY Bight)						
–Allied Chemical	59	56	48	47	29	26
–NL Industries, Inc.	2300	1990	1840	1230	604	1230
–duPont (Grasselli)	142	78	—	—	—	—
	2501	2124	1888	1277	633	1256
Industrial Wastes (NY Bight)						
–duPont (Grasselli)	115	154	263	163	107	171
–duPont (Edge Moor)	—	—	—	—	379	371
–American Cyanamid	118	137	116	119	130	111
–Modern Trans. Co.	34	35	78	63	83	72
–Gen. Marine Trans.	—	—	—	5	2	—
–Chevron Oil Co.	25	26	22	—	—	—
–Hess Oil Co.	7	—	—	—	—	—
–Camden Sewage Sludge	—	—	—	—	48	54
	299	352	479	350	749	779
Industrial Wastes (Puerto Rico)						
–PCI International	35	210	229	327	285	327

quantities dumped between 1973 and 1978 are less than those permitted and ranged from 26-59 thousand metric tonnes.

By-product sulfuric acid-iron wastes generated in the manufacture of titanium dioxide are dumped by NL Industries, Inc. Authorized annual amounts currently are 1.36 million metric tonnes. Actual quantities dumped (Table 3) have ranged from 600-2300 thousand metric tonnes during the period 1973-78. The low quantity dumped during 1976-77 resulted from loss production during an extended strike which closed down the plant. The waste which consists of waste acid and an insoluble gangue solids slurry in a 12:1 ratio by volume, contains approximately 5 percent by weight sulfuric acid, about 35 grams per liter (g/l) of ferrous iron, and about 2 g/l of titanium. Chemical and biological characteristics of the barged waste are summarized in Table 4.

Allied's acid wastes are transported to the dump site aboard their specially constructed barge "AC-5". NL's acid-iron wastes are transported aboard two rubber-lined barges owned by Moran Towing Corp., the "Moran 102" and "Moran 108". The "Moran 102" is used intermittently as a backup or replacement. Waste acids are released

Table 4. Characteristics of wastes barged to the acid wastes dumpsite in the NY Bight. Based on sample collected in December 1978.

Parameter	Units	Allied Chem.	NL Ind.
Chemical			
−Arsenic	µg/l	1680	24
−Cadmium	"	< 10	< 25
−Chromium	"	360	12600
−Copper	"	50	320
−Lead	"	13	1900
−Mercury	"	.4	< 5
−Nickel	"	160	4750
−Zinc	"	60	23600
−TOC	mg/l	48	49
−Oil & Grease	"	11	.9
−Total Solids	"	104	−−
−Susp. Solids	"	< 10	9990
Physical			
−pH	SU	< 1.0	.5
−Spec. Grav.	g/ml	1.084	1.168
Bioassay (96 hr)			
−Menidia (TL50)	mg/l	425	1725
−Skeletonema (EC50)	"	102	420

below water level, while these barges are underway. Capacity, discharge rate, and trip frequency are summarized in Table 5.

NEW YORK BIGHT - INDUSTRIAL WASTES SITE

The industrial wastes dump site (also known as DWD 106) was established in 1965 (Interstate Electronics Corp., 1973) by the COE based on recommendations of the U. S. Fish and Wildlife Service and after coordination with several federal agencies. Since 1973, its location has been described by EPA (U. S. Environmental Protection Agency, Jan. 11, 1977) as in Table 2. Prior to 1973, a slightly different method was employed (Interstate Electronics Corp., 1973). The site was established following requests from several New Jersey industries for authorization to ocean dump toxic chemical wastes when State health authorities refused to permit inland disposal methods, such as landfilling or stream discharge, due to possible contamination of potable supplies.

Because of the high costs associated with dumping at this site, its use prior to 1974 was limited basically to toxic materials. Most industrial wastes, other than acid wastes, were dumped at the near-shore sewage sludge dump site (Fig. 1). In 1974, EPA required that all but three of these industrial wastes be dumped at the farther offshore site. Only the dumpers of biological sludges generated by three industries were permitted to remain at the nearshore sludge

Table 5. Summary of data on barges used by industrial ocean dumpers in EPA Region II.

Barge Name	Capacity (tonnes)	Discharge Rate (1) (tonnes/km)	Frequency of Trips
AC-5	1730	24.5	twice/month
Moran 108	3750	205	daily
Moran 102	2400	205	replacement
Edge Moor I	3790	76	twice/week
Sparkling Waters	3500	106 (2)	weekly
		61.5 (3)	weekly
Lisa, Maria, Forest	7200	265	6/year
Liquid Waste 1	2650	71.5	weekly
		20.5 (4)	weekly
Whitwater II	7200	71.5	replacement

(1) Maximum authorized discharge rate while barge is underway. Rate is based on bioassay toxicity and dispersion characteristics.
(2) Rate for duPont-Grasselli.
(3) Rate for American Cyanamid.
(4) Rate for Squibb.

site. In March 1977, EPA granted the request of E. I. duPont de Nemours and Co., Edge Moor Plant, that it be allowed to move its dumping activities from a site 65 kilometers due east of the Maryland-Delaware border to the industrial wastes site.

The site is currently (April 1979) used by the four industries listed below:

- American Cyanamid Co., Linden, New Jersey
- E. I. duPont de Nemours and Co., Grasselli, New Jersey
- E. I. duPont de Nemours and Co., Edge Moor, Delaware
- Merck and Co., Rahway, New Jersey

This is a significant reduction from the nearly 150 industrial waste generators which ocean dumped their wastes in the New York Bight prior to 1973 (U. S. Environmental Protection Agency, 1978; U. S. Environmental Protection Agency, 1977). It should be noted that duPont-Edge Moor is scheduled by EPA permit condition to cease ocean dumping by November 1, 1980, Merck and Co. by December 31, 1980, and American Cyanamid Co. by April 30, 1981. Quantities dumped at this site during the period 1973-78 are tabulated in Table 3.

Table 6.   Characteristics of wastes barged to the industrial wastes dumpsite in the NY Bight. Based on sample collected in December 1978.

Parameter	Units	Am. Cyan.	duPont (NJ)	duPont (DE)
Chemical				
-Arsenic	µg/1	879	10	375
-Cadmium	"	2.5	122	430
-Chromium	"	85	475	108000
-Copper	"	1750	224	1600
-Lead	"	6	70	26000
-Mercury	"	9.8	4.0	60
-Nickel	"	405	122	14400
-Zinc	"	55	382	178000
-TOC	mg/1	18800	3740	< 100
-Oil & Grease	"	380	12.4	< 1.0
-Total Solids	"	17300	122000	--
-Susp. Solids	"	142	969	37
Physical				
-pH	SU	5.9	12.2	2.2
-Spec. Grav.	g/ml	1.018	1.147	1.062
Bioassay (96 hr)				
-Menidia (TL50)	mg/1	66	1250	5000
-Skeletonema (EC50)	"	18	628	3670

Note that contributions by small industrial waste generators who have
since ceased dumping are summed and listed as Modern Transportation
Co., or General Marine Transport Corp., their waste transporter.

By-product hydrochloric acid-iron wastes generated in the production of titanium dioxide pigment are dumped by the duPont-Edge
Moor plant. Authorized 1979 amounts are 300,000 tonnes. The authorized amounts will decrease to 136,000 tonnes in 1980. Actual amounts
dumped in 1978 were 371,000 tonnes (Table 3). Characteristics of the
barged waste are tabulated in Table 6. The duPont-Edge Moor wastes
are transported to the dump site aboard the specially designed rubber-lined barge "Edge Moor I" (Table 5).

Wastewater ocean dumped by the duPont-Grasselli plant is generated in the production of anisole and n,o-dimethylhydroxylamine. It is
a 10-15 percent aqueous sodium sulfate solution generally containing
less than one percent of soluble, low molecular weight organic compounds, such as methanol, methylamines, and phenol. Authorized
amounts are 295,000 tonnes. Actual amounts dumped (Table 3) ranged
from 107 to 263 thousand tonnes during 1973-78. Note on Table 3 that
wastes from this plant were dumped at both the acid and industrial

Table 7. Characteristics of wastes barged to the industrial wastes
          dumpsite in Puerto Rico. Based on sample collected in
          December 1978.

Parameter	Units	PCI	Squibb
Chemical			
-Arsenic	µg/l	<2	<1
-Cadmium	"	130	80
-Chromium	"	500	<1
-Copper	"	800	60
-Lead	"	780	520
-Mercury	"	<.2	<.5
-Nickel	"	400	520
-Zinc	"	1400	180
-TOC	mg/l	42000	24700
-Oil & Grease	"	1200	121
-Total Solids	"	97100	57300
-Susp. Solids	"	52600	217
Physical			
-pH	SU	6.0	10
-Spec. Grav.	g/ml	1.031	--
Bioassay (96 hr)			
-Tripneustes (TL50)	mg/l	1270	740
-Skeletonema (EC50)	"	3600	320

wastes sites during 1973-74. Characteristics of the barged wastes
are tabulated in Table 6. The duPont-Grasselli wastes are generally
transported to the dump site aboard the barge "Sparkling Waters"
(Table 5), although several other small vessels are available as
backup when needed.

Merck's wastewaters result from an intermediate process in the
manufacture of thiabendazole. Authorized annual amounts are 36,500
tonnes. Actual amounts dumped in 1978 were 35,000 tonnes. Merck is
a customer of Modern Transportation Co. and is included as such in
the totals listed in Table 3. Characteristics of the barged wastes
are summarized in Table 6. The barges "Lisa", "Maria", and "Forest"
(Table 5) are used in transporting the Merck waste to the dump site,
although several other small barges can be used as replacements when
necessary.

The most toxic wastes presently ocean dumped are those generated
by American Cyanamid Co. The wastes are generated in the manufacture
of rubber chemicals, paper chemicals, water treating chemicals, non-
persistent organophosphate insecticides, mining chemicals, intermediates, and surfactants. Current authorized annual amounts are
123,000 tonnes. Actual amounts dumped (Table 3) during 1973-78 have
ranged from 111 to 137 thousand tonnes. Characteristics of the
barged wastes are tabulated in Table 6. The American Cyanamid wastes
are generally transported to the dump site aboard the "Sparkling
Waters" (Table 5), together with the duPont-Grasselli plant wastes.

All wastes dumped at the New York Bight industrial site are discharged below water level while the barge is underway. Capacity,
discharge rate, and trip frequency are summarized in Table 5.

PUERTO RICO - INDUSTRIAL WASTES SITE

The Puerto Rico industrial wastes ocean dump site was established by EPA in January 1972, after coordination with cognizant federal
and Commonwealth agencies. The site was established at the request
of Merck Sharp and Dohme Quimica de Puerto Rico, Inc. of Barceloneta
in order to permit the ocean disposal of wastes, rather than to discharge such wastes into the Rio Grande de Manati. The use of this
site was to be on a temporary basis and was planned to be discontinued upon completion of the Barceloneta Regional Wastewater Treatment
Plant (BRWTP). Changes in environmental regulations and delays in
construction of the plant resulted in the required use of this site
by several other Barceloneta area industries for the disposal of
"strong" waste streams, in lieu of "sink holes". The location of
this site is described in Table 2.

At present, there are seven industries which continue ocean
dumping at the Puerto Rico site. They are:

- Bristol Alpha Corp.
- Merck Sharp and Dohme Quimica de P. R., Inc.
- Pfizer Pharmaceuticals, Inc.
- Schering Corp.
- Cyanamid Agricultural de P. R.
- Upjohn Manufacturing Co.
- Squibb Manufacturing, Inc.

Three additional waste generators, Abbott Laboratories, P. R. Olefins and Oxochem Enterprises, ceased ocean dumping prior to 1979 (U. S. Environmental Protection Agency, 1978; U. S. Environmental Protection Agency, 1977). Bristol, Cyanamid, Merck, Pfizer, Schering, and Up-john, all located in the Barceloneta area, are scheduled to tie into the BRWTP and cease ocean dumping by September 30, 1981. Squibb is practicing ocean dumping temporarily, while replacing their existing treatment facility in Humacao. They are scheduled to cease dumping on April 30, 1980.

Quantities dumped at this site during the period 1973-78 are tabulated in Table 3. Note that all individual contributions are to-taled and listed as PCI International, Inc., their waste transporter. Authorized annual amounts for individual industries are listed below:

Bristol Alpha	11,000	metric	tonnes
Cyanamid	34,500	"	"
Merck	75,500	"	"
Pfizer	19,000	"	"
Schering	56,500	"	"
Squibb	7,600	"	"
Upjohn	182,000	"	"

Wastes generated result from the manufacture of pharmaceuticals, such as antibiotics, dewormers, and other drugs.

Two barges which discharge below water level while the barge is underway are employed in transporting the wastes to the dump site. The "Liquid Waste 1" is used most often. The larger barge, "Whit-water II", is used only occasionally. Capacity, authorized discharge rate, and frequency of trips are summarized in Table 5. Characteristics of mixed barged wastes are tabulated in Table 7.

SUMMARY

Since passage of the MPRSA in October 1972, EPA has brought all dumping of sewage sludge, industrial wastes, and dredged material in the ocean under strict regulatory control.

All dumpers of harmful wastes are on schedules which will as-sure that the dumping of such wastes, industrial and municipal, will

cease by the end of 1981. After that time, only those wastes which
comply with EPA's environmental impact criteria and which will not
cause unreasonable degradation in the marine environment will be permitted to be dumped in the ocean.

The greater majority of the dumping of industrial wastes in the
United States occurs at dump sites managed by EPA Region II and located in the New York Bight and off the north coast of Puerto Rico.
Of the roughly 150 industrial ocean dumpers in Region II in 1973,
only 13 remain and ten of these must stop ocean dumping before the
end of 1981.

REFERENCES

Interstate Electronics Corp. (1973) Ocean Waste Disposal in Selected
    Geographic Areas. IEC, Anaheim, CA, Rept. No. 4461C1557, 376
    pp.
Pararas-Carayannis, G. (1973) Ocean Dumping in the New York
    Bight: An Assessment of Environmental Studies. U. S. Army Corps
    of Engineers, Tech. Memo. No. 39, 159 pp.
U. S. Army Corps of Engineers (1978) 1977 Report to Congress on
    Administration of Ocean Dumping Activities, Public Law 92-532
    (MPRSA of 1972). Department of the Army, Washington, 47 pp.
U. S. Environmental Protection Agency (Jan. 11, 1977) Ocean Dumping,
    Final Revision of Regulations and Criteria. Federal Register,
    42 (7), 2485.
U. S. Environmental Protection Agency (1977) Ocean Dumping in the
    United States-1977, Fifth Annual Report of the Environmental
    Protection Agency on Administration of Title I, Marine
    Protection, Research, and Sanctuaries Act of 1972, as amended.
    U. S. EPA, Washington, 65 pp.
U. S. Environmental Protection Agency (1978) Final Environmental
    Impact Statement on the Ocean Dumping of Sewage Sludge in the
    New York Bight. U. S. EPA, Region II, New York, 245 pp.
U. S. Environmental Protection Agency (1978) Ocean Dumping in the
    United States, Sixth Annual Report of the Environmental
    Protection Agency on Administration of Title I, Marine
    Protection, Research, and Sanctuaries Act of 1972, as amended,
    January-December 1977. U. S. EPA, Washington, 53 pp.

OCEAN DUMPING - HISTORY, CONTROL AND

BIOLOGICAL IMPACT

Robert Johnston

Department of Agriculture and Fisheries for Scotland
Marine Laboratory
Aberdeen, Scotland AB9 8DB

ABSTRACT

The history, legislation and enforcement of measures to limit ocean dumping in many countries expresses the increasing global concern for guarding environmental quality. A wide range of persistent and non-persistent pollutants is regularly proposed for dumping which presents testing challenges to the managers who in turn must devise effective and practical solutions. Research on dumping involves not only laboratory toxicity testing but more realistic large-scale and field appraisals that assess behavioural and sub-lethal effects. Ocean dumping is an emotive subject which demands responsible case presentations on the part of the control authorities, dumping agencies, environmentalists and the media in order to communicate a true and understandable picture to the general public.

INTRODUCTION

Many centuries ago the wise and wealthy developed the water transport systems and primitive sewers in countries round the Mediterranean. In the course of time these influences reached northwards to the British Isles and much later Britain was to lead the rush into the industrial revolution that overwhelmed the finer senses for environmental quality. The tide of influence turned and in its wake the countries around the Mediterranean Sea lost their forests and their self sustaining agriculture in the uplands as the people adapted to the explosive growth of an economy and life-style based on industrial technology, intensive farming and tourist invasion. The Mediterranean which had once been a challenging expanse of sea to the ancient mariner was now "a small crowded lake, and a polluted one" (Henry, 1977).

The British Isles, mercifully surrounded by unconfined sea and
ocean, suffered equally from unleashed industrial growth and spoila-
tion of some of its countryside and rivers but escaped the worst effects on its shores and estuaries. For much of Europe the Oslo Convention to control offshore dumping and the Paris Convention to control pipeline discharges were 200 years too late. Too late to control what? Should one ignore the transient pollutants like sewage
and whisky wastes and think only about persistent obliterating solids
like debris from coal and shale, town refuse, slag from iron works,
major coastal reclamations and dredging?

It is generally agreed that non-persistent pollutants can be ignored unless their inputs are continuous and above the threshold that
the receiving waters can accept. Also, some persistent solids can be
gradually absorbed into the bottom sediments by biological activity
and, on land, remedial action can be undertaken to restore visual
amenity. In Scotland some measure of irreparable permanent damage
can be demonstrated in those estuaries subjected to massive pollution
with the consequent loss of the original benthos, seaweeds and associated fauna and in more than one instance herring spawning grounds
and salmonid fisheries. In Europe, truly oceanic dumping is a rare
event restricted to special cargoes; routine dumping is into waters
of less than 100 m.

REGULATION OF OCEAN DUMPING IN SCOTLAND

The Dumping at Sea Act, 1974 is the UK instrument implementing
the Oslo Convention. In essence the Fishery Departments of Scotland,
Northern Ireland and England and Wales have been given the duty of
enforcing this legislation using their existing administration, fishery inspectorate and marine scientists with minimal staff allocation.
The Act says "In determining whether to grant a licence the authority
shall have regard to the need to protect the marine environment and
the living resources which it supports from any adverse consequences
of the dumping" and goes on "The regulatory authority shall include
such conditions in a licence as appear to them to be necessary or ex-
pedient for the protection of that environment...."

Applicants for a Dumping at Sea Licence must pay a fee (of about
100 dollars) which just covers the administrative expenses for the
simple cases. In complex cases the actual costs to the Department
may be very much greater. Licences relate to all placements on land
submerged at mean high water spring tide and deeper whether for reclamation, sewage outfalls, pier construction or any operation con-
strued as the dumping of any matter in the sea except moorings, aids
to navigation or other deposit made by or with the written consent of
a harbour authority or lighthouse authority. It also does not now
apply to gas and oil pipelines or any materials proper to the normal
conduct of offshore petroleum operations (Petroleum and Submarine

Pipelines Act 1975).

The enforcement officers have full powers to search, take samples, supervise dumping operations including access to the ship's navigation instruments to verify the dumping position. If need be they can forbid or stop a dumping operation. The Act specifically includes all other responsibilities under other general or private legislation. There is an appeal procedure. If special chemical or toxicity testing is needed to enable the acceptability of a waste to be decided, such tests must be paid for by the applicant.

Life would be a little simpler if all dumping activities came under one legal framework. Discharges to the sea of oily waters, chemicals used in the oil industry including drilling muds, pipelines, pipeline shields and various other materials are governed by the Prevention of Oil Pollution Act, one of these unbelievably enthusiastic measures to prevent pollution that specifically and totally forbids any operational discharge of oil whatever by activities other than oil refining. No offshore oil activities are possible if zero oil emission has to be met so this legislation has had to be amended by piecemeal clauses tacked on to other legislation (Petroleum and Submarine Pipelines Act 1975) thus permitting controlled discharges to be made of treated oily ballast water and production waters from platforms. In legal terms this patchwork just about holds together but is clumsy. Fortunately in practice the authorities who have to implement the environmental aspects of these dumping and oil controls generally speaking involve the same people for both. It is mighty difficult to frame legal controls that are effective for current industrial activity and at the same time are flexible and can absorb major developments.

It would be too big a task to describe adequately all ocean dumping in European waters. There are excellent statements about inputs to the North Sea in a series of ICES Cooperative Research Papers Nos A 13 (1969), 39 (1974), 69 (1977), and there are useful summaries in the publications of the Interim Oslo and Paris Commissions (1974 on). The International Council for the Exploration of the Sea is a scientific body which considers together discharges from land and by dumping; this is a much wiser and more rational course than is pursued in these twin Commissions and in U. S. law which distinguishes between "ocean dumping" and ocean discharges. It is wrong to treat the two topics in isolation either in pollution management or in pollution research.

More is to be gained from looking at specific ocean dumping instances than by comparing how much pollution is taking place on either side of the Atlantic. I want to consider dumping as it affects marine life, especially commercial fish and shellfish species. A typical year's spectrum of applications provides a good example. In the UK there are usually about 600 applications of which 550 are

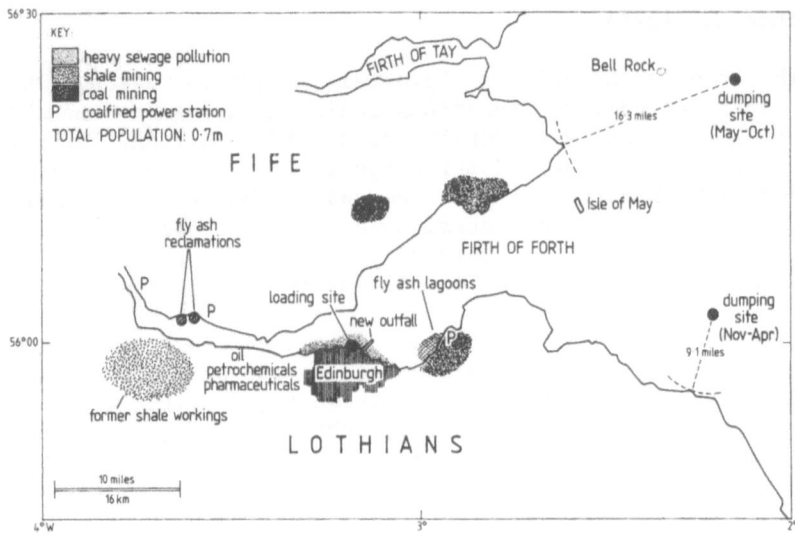

Fig. 1. Major pollution activities around the Firth of Forth.

pursued beyond the initial enquiry. Of these between 80 and 100 refer to Scottish waters (Table 1). There is a story in every line of
these applications; they are not all success stories.

EXAMPLES OF OCEAN DUMPING PROBLEMS

The Firth of Forth

     The most destructive dumping took place long before the Oslo
Convention and seriously affected the ecology of the Firth of Forth
(Fig. 1). Prior to 1880 this estuary was famous for its fisheries
for herring, sprats, sparling, haddock, whiting, cod and flatfish,
inland as far as it was tidal. There were also oyster beds that in
1978 money terms would have yielded $1 million a year. These and
progressively much of the usual seaweed species, the algal and rock
fauna have either gone entirely or changed over to worms and weeds.
The causes of these changes relate to pressures from population and
industry and include some element of over-fishing and bad fisheries
/shell fisheries management (Johnston and Davies, 1975).

     In the last 50 years it has been an uphill struggle to promote
river and estuary purification in Scotland and the rest of the United
Kingdom. Water quality is improving and pollution activities are now
better controlled but the estuary is burdened with a legacy of mil-
lions of tons of coal and shale debris, crude sewage residues, structural manipulations of the seabed by large scale dredging and

Table 1.   Licences under Dumping at Sea Act, 1974 (Scottish area).
Annual number issued, 80-100 (total).

---

1.   Maintenance and Capital Dredging

Cubic metres	$<10^3$	10^4	10^5	10^6	$>10^6$
Number	1	9	16	10	3

Total spoil for sea disposal 7×10^6

Land areas reclaimed (part by spoil, part by clean infill)
approx. 120 ha per year (300 acres)

---

2.   Other Important Activities

40 new or extended sewer outfalls
18 jetties, marinas, breakwaters and other structures
Sewage sludge from Glasgow area 1.1 Mt to Garroch Head
Sewage sludge from Edinburgh area 0.3 Mt to two sites used
    6-monthly
Wastes from explosives manufacture 500 t
Aluminium smelter slurry 75,000 M^3 containing 7,500 t
    solids (carbon, alumina, iron hydroxides and cyrolite)

---

3.   Odds and Ends

Allowed

Rock salt	Nitroglycerine
Nitric acid	*Coal debris and washings
Buckled oil pipeline	Milk residues
Seal carcases	Perimeter fencing
House refuse	Cement washings

Disallowed

Dock gates	Railway carriages
Whisky wastes	*Coal debris

*Increased pressure is being brought to bear to eliminate
unconfined disposal of coal debris and washings to the
foreshore.

---

reclamations together with unrelenting pressures from industry as it evolves and a constant or slowly increasing impact of river borne treated sewage.

Power generation is important for Scotland's industrial belt and the Forth has three major coal-burning electricity generating stations which begins to approach the acceptable limit for air pollution by SO_2. The ash is being used to create parkland areas and ultimately for housing or industrial development. This is progress but it reduces the area of intertidal sand and mud flats which are valuable for fish, duck, geese, waders and seabirds.

Some ornithologists even regret the recent removal of the sewage solids enriched with grain residues from large breweries which accumulated along the Edinburgh foreshore. In winter large flocks of migrant birds fed on these deposits but in summer the rotting organic matter created a much-talked-about smell and amenity nuisance. These solids are now dumped at sea; an operation accepted as a substantial improvement by the fisheries scientists and most fishermen.

Similar stories could be told about the Tyne, Tees, Humber, Thames, Mersey and Clyde; and sadly also the Baltic and Mediterranean Seas on a different scale.

Dredge Spoil Disposal

Dredging spoil dumping activity is a major form of ocean dumping in terms of tonnage but the environmental impact varies widely. In Loch Ryan in southwest Scotland this conflict is acute (Fig. 2). Dredging is required to promote a new, larger and more convenient vehicle and passenger ferry facility serving traffic between Scotland and Northern Ireland. This service is nationally owned. The 750,000 m^3 material being dredged is silt, mud and boulder clay. The resources at stake are Scotland's only remaining sizeable native oyster beds, a salmonid fish farm and flatfish grounds within the loch, and in possible dumping areas outside the loch are the main spawning ground for Clyde herring and a valuable white fish trawling area. Although scenically beautiful, the area and coast are not a major holiday attraction. The site to be dredged and the geometry of the loch necessitate long and therefore expensive sea passages for spoil barges if a dumping ground without any possible effect on these resources is regarded as the best option. It also means that the barges have to be capable of withstanding possibly severe open sea conditions. The only available but somewhat risky alternative would be to allow dumping in the stretch of water just inside the loch where the prevailing water movements are slack. Beyond the mouth of Loch Ryan the coastal current is strong and would carry silt and other fine particulates northward over the Ballantrae Banks herring spawning grounds. Inevitably some areas of oyster bed would be buried by the reclamation and additional areas along the enlarged

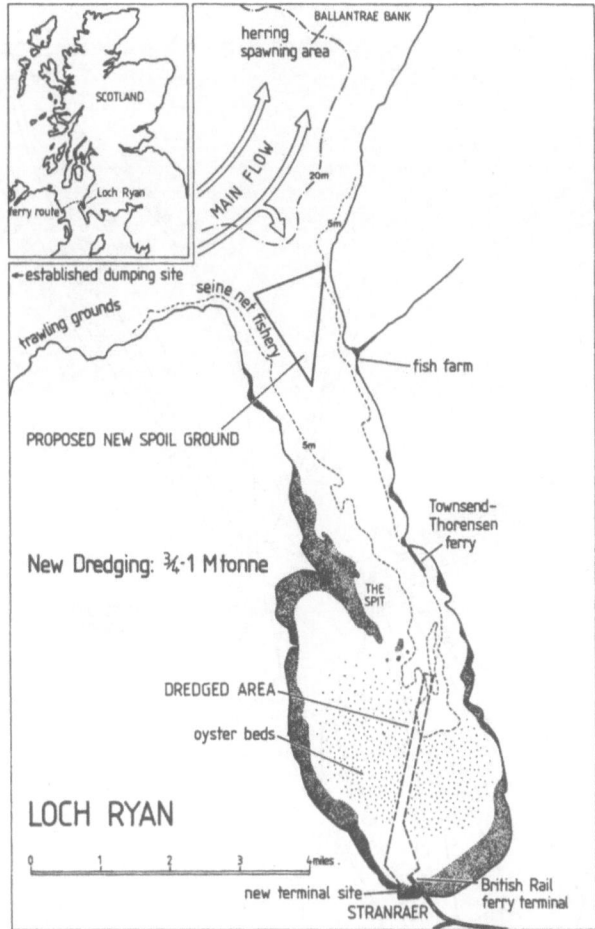

Fig. 2.   The dredging and dumping in relation to marine resources in
          Loch Ryan.

access channel and turning area would be seriously disturbed.   How
much more would be temporarily or permanently lost due to dispersed
silt and how much more by migration of silt from the dumping ground?
This question requires a field investigation.

Waste Disposal Related to Oil and Gas Operations

     North Sea oil has made a big impact on Scotland and has created
a lot of work in relation to dumping at sea and the effects of oily
water discharges.   Oil and gas platforms used in the North Sea are
gigantic creations in steel or concrete.   In the initial glow of an
anticipated North Sea oil bonanza the Government approved and

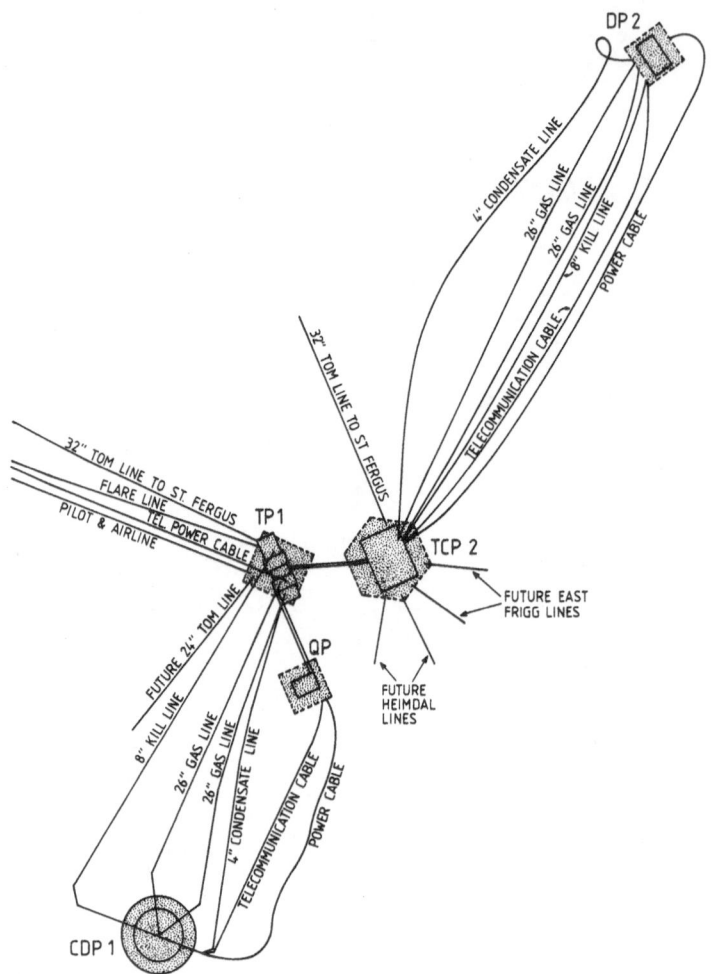

Fig. 3.   Interplatform network related to gas production (from a
          British Gas diagram).

partially financed more than twice as many platform construction
yards as were needed and many hundreds of acres of coast have been
unnecessarily spoiled and they resulted in correspondingly large
unnecessary spoil dumpings. These platforms are towed out to sea and
fixed firmly on the seabed. Layers of modules are added for accommo-
dation, work areas and oil and gas processing. Around them grows a
web of pipe works which renders several square kilometers of sea
inaccessible for fishing (Fig. 3).

Cryolite Sludge Disposal

One of our most troublesome wastes is, by all logical and scientific arguments, not a problem. I refer to the sludge from the
cryolite recovery plant at a large aluminium smelter. This sludge
consists of carbon, alumina, iron oxides and traces of cryolite. The
levels of soluble metals are acceptable and any cyanide has been neutralized. Toxicity tests showed that juvenile flatfish and shrimps
are unaffected even when there is so much carbon that the animals
were invisible. Even on prolonged exposure flatfish survive as well
as in clean sea water. Large cod wired with electrodes to monitor
respiration and heart beat show escape-type reaction to dilute sludge
plumes (about 100 to 200 ppm solids) but survived indefinitely. They
also display "coughing" and altered heart beat reactions to concentrations of this order. Water depth and movement in the dumping area
are adequate to secure dilutions to this amount except in the very
immediate plume. Irrespective of this evidence and of calculations
showing how even in "worst case" conditions the total solids over
many years could not create a significant build up on the bottom or
dispersed in the water column, there have been loud and persistent
claims of reduction of salmon catches. The dumping site was chosen
to avoid white fish grounds taking into account the neighbourhood of
moderately important salmonid fisheries by fixed nets from the shore
and by rod and line on several rivers. There are also nearby Neph-
rops fisheries. No evidence for effects on salmonids other than a
few years of catch returns has ever been produced but the opponents
to the dumping will not accept that any evidence of lack of toxicity
is valid and continue to claim that salmon movements have changed as
a result. This is a topic about which the pollution literature is
silent.

ASSESSMENT OF MARINE POLLUTION IMPACTS

This leads on quite naturally to two important topics (1) research on ocean dumping and (2) coping with justified and unjusti-
fied criticisms and publicity.

The research on ocean dumping in Scotland covers the whole spectrum from routine LC_{50} toxicity testing to very advanced multidisciplinary studies. In a highly cost conscious government fisheries
research institute we justify (or disguise) much of this challenging
work by relating it to regional, national and international commit-
ments; if possible all three simultaneously.

For prosaic toxicity testing we use shrimps and flatfish; always
two kinds of animals independently. These are important commercial
species relevant to Scottish waters.

For major pollution projects we also make use of herring eggs

and larvae which allows us to measure effects on (a) fertilisation,
(b) mortality or abnormal egg embryology (c) larval production (d)
time to hatching (e) survival during the yolk-sac stage. Here we
have the involvement of an important commercial species exposed at a
very sensitive stage in a bioassay lasting about three weeks.

We have used large plastic enclosures to study the fate of metallic pollutants (Hg, Cd, Pb, Cu) and also oil slicks and oily ballast water. This is one of the few meaningful methods of studying
marine ecosystems which include bacteria, phytoplankton, zooplankton
and fish larvae in relation to sublethal pollutant concentrations.
The fate of the pollutant as it changes its mode of occurrence and
chemical speciation can also be pursued. The sedimenting material
can be quantified and further studied by injection into bottom chambers for its effects on typical benthic communities. The time scale
used is in the range of 3 weeks to 3 months.

Inevitably controlled situations whether in test tube or large

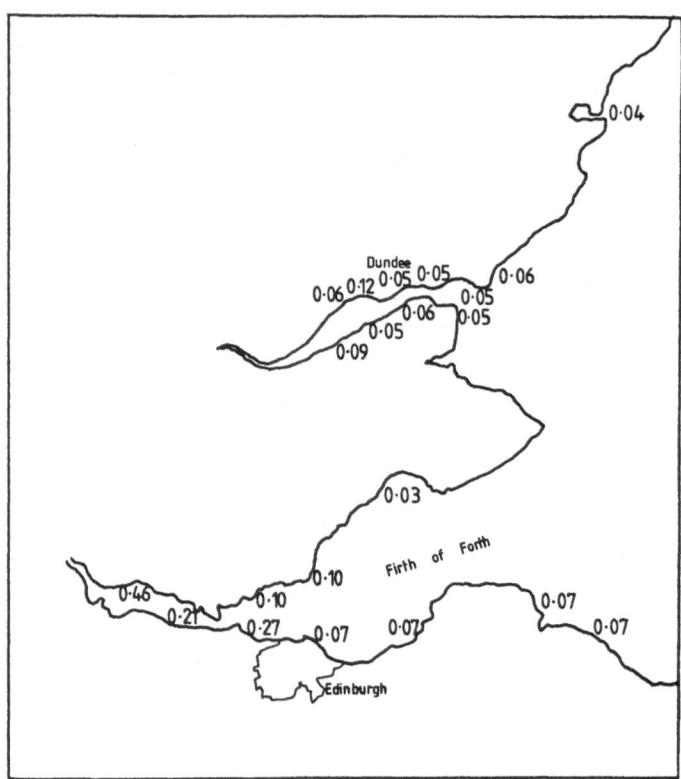

Fig. 4.   Mercury concentrations (μg g^{-1} wet weight) in mussels from
          part of the East Scottish coast.

plastic enclosure have major limitations in their ability to mimic the open sea and their valid time scales are related to the volume enclosed. It is not yet practically and economically feasible to study the life cycle of commercial species of fish and shellfish under controlled conditions of exposure to given pollutants.

A very useful and informative parallel attack makes use of field bioassays. We have used the "Mussel Watch" approach to establish relative levels of uptake of heavy metals, hydrocarbons and organo-chlorines into Mytilus edulis all around the Scottish coast (Davies and Pirie, 1978; Cowan, 1978). As back-up we have also exposed clean mussels in cages in the Firth of Forth to measure Hg and methyl Hg uptake (Davies and Pirie, 1977) which gives results that confirm and to some extent calibrate mercury in tissue in relation to mercury in the field. From such results one can estimate a mercury budget for the estuary. The degree of enhancement of Hg and Zn levels in Mytilus in a moderately industrialised area is well demonstrated in Figs. 4 and 5.

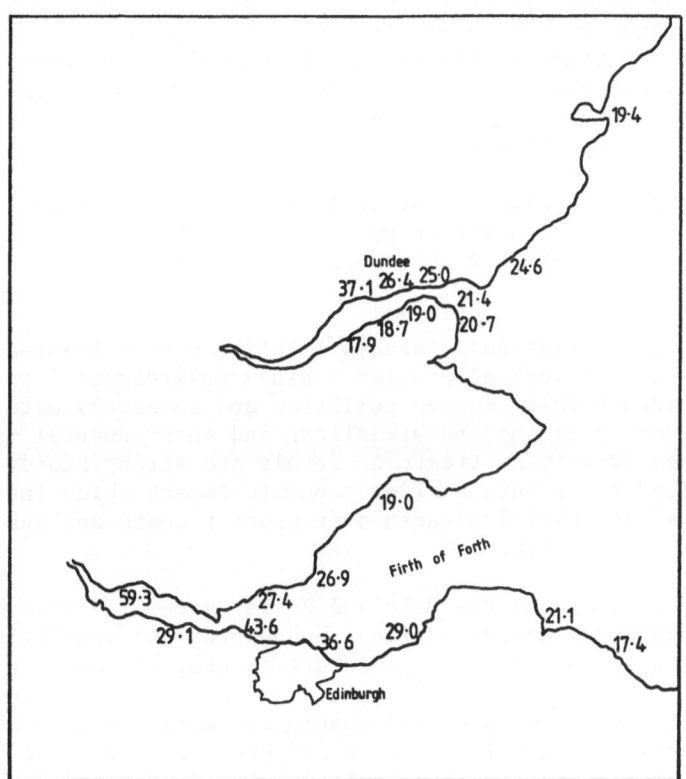

Fig. 5. Equivalent zinc concentration (μg g^{-1} wet weight) in mussels from the same positions.

For oily water discharges into coastal waters the focus of
attention is not on mortalities which are precluded by legislative
control, but on taints or off flavours arising from voluminous low
concentration discharges which mainly contaminate sedentary or near-
sedentary species such as clams and scallops. We assess the degree
of tainting by studying both native stocks and caged specimens set
out at various distances and by laboratory exposures to known dilutions. The study techniques include trained taste panels and detailed hydrocarbon analyses.

A lot of effort has also gone into studies on benthic animals in
relation to pollutants and on pilot scale tank studies on semi-
natural food web systems with and without pollutants (Saward et al.,
1975).

SUMMARY

In the USA ocean dumping is entirely a domestic problem but in
Europe the North Sea is everybody's fish pond and everybody's cess-
pool. Although the EEC countries are making enormous efforts to
rationalise the right to pollute and to establish the acceptable
degree of pollution, it is impossible to attain equal constraints for
all. In the past it has often been fishermen, especially the small
local operator, who lost out.

Pollution is a highly emotional subject and it is exceedingly
difficult for the scientist to put across to the general public a
balanced statement of fact unless he can persuade the communications
media to lend support.

Almost all pollution is caused locally, can be treated locally
and amounts to a parochial problem. Since environmental protection
demands efforts both to remove pollution and to survey water quality
there arise costs to the industrialists and environmental managers
that increase greatly as treatment levels are strengthened. Hence
these local problems have a wider economic impact which introduces
international and global concern over product costs and associated
environmental penalties.

EEC directives and the Oslo and Paris agreements have drawn
scientists together across national boundaries, by specifying the
quantities and composition of pollutant inputs, by measuring pollutants in all important fish and shellfish and by pooling thoughts on
productive lines of research and meaningful monitoring. When you
know the scale of pollution, how it affects our daily food and something about the associated biochemistry, physiology and social medicine, this establishes the true threat of marine pollution. It is
not an academic exercise; it is a substantial problem which affects
our society and its future.

REFERENCES

Cowan, A. A. (1978) Organochlorine compounds in mussels from Scottish
    coastal waters. ICES Marine Environmental Quality Committee
    C.M1978/ E:39 (mimeo.)

Davies, I. M. and J. M. Pirie (1977) The use of the mussel Mytilus
    edulis as a bioassay organism for mercury in sea water. ICES
    Fisheries Improvement Committee: C.M1977/E:36.

Davies, I. M. and J. M. Pirie (1978) Trace metals in mussels from the
    Scottish coast. ICES Marine Environmental Quality Committee
    C.M1978/E:33 (mimeo.)

Henry, P-M. (1977) The Mediterranean: A threatened microcosm. Ambio,
    (6), 300-307. (and other articles in this special number
    devoted to the pollution of the Mediterranean Sea).

Saward, D., A. Stirling, and G. Topping (1975) Experimental studies
    on the effects of copper on a marine food chain. Mar. Biol.,
    29, (4), 343-361.

THE EFFECTS OF OCEAN DUMPING ON THE

NEW YORK BIGHT ECOSYSTEM

Harold M. Stanford, Joel S. O'Connor and
R. Lawrence Swanson

Marine EcoSystems Analysis (MESA) Program
New York Bight Project
State University of New York
Stony Brook, NY 11794

ABSTRACT

The effects of ocean dumping on a marine ecosystem must be considered in the context of the character and quantity of all contaminants reaching the system. The highly variable nature of natural
factors within the ecosystem and of waste inputs to it make the direct determination of cause-effect relationships of anthropogenic materials quite difficult to discern. Indirect methods for such determination are possible, obviating the need for exhaustive study of
waste materials and the ecosystem. All of these factors are addressed in a case study of the New York Bight ecosystem.

INTRODUCTION

Passage of the Marine Protection, Research and Sanctuaries Act
of 1972 (Public Law 92-532, commonly referred to as the "Ocean
Dumping Act") reflected widespread concern for the effects of ocean
dumping on the marine environment. The Act calls for the initiation
of "... a comprehensive and continuing program of monitoring and
research regarding the effects of the ocean dumping of material into
ocean waters..."

The National Oceanic and Atmospheric Administration (NOAA) es-
tablished the Marine EcoSystems Analysis (MESA) Program in 1972. The
New York Bight Project was the first project initiated under the MESA
Program and undertook some of the responsibilities NOAA acquired

53

under the Act. Begun in mid-1973, this project was designed to determine the condition of the New York Bight ecosystem, and to identify the practical significance of the environmental problems facing
the Bight and its users. The consequences of ocean dumping in the
New York Bight were to be examined in the context of its overall
quality and other contaminant sources to the Bight. A secondary aim
of the MESA Project was to provide a sound rationale for assessing
ocean dumping in other areas.

Results of the MESA New York Bight Project to date are sufficient to allow 1) tentative assessment of the effects of ocean dumping on the New York Bight ecosystem, in the context of other contaminant inputs, and 2) an understanding of some of the anticipated or
projected ecosystem impacts of ocean dumping in other areas, together
with a general approach for assessing their degree.

GENERAL ASPECTS

Several kinds of material are dumped in the coastal waters of
the New York Bight: municipal wastes (sewage sludge); dredged material (from both freshwater and saline environments); industrial wastes;
cellar dirt (excavation and construction materials); and wrecks. The
last two are generally considered to be relatively inert in the marine environment and of little consequence to the ecosystem, outside
the small areas of these two dumpsites. Therefore, they will not be
considered in detail here.

Each waste has a character dependent upon that of its sources.
Dredged material is as diverse as the myriad of inputs to a harbor or
estuary. Sewage sludge composition reflects the makeup of the area
served by a particular treatment plant, and includes fractions of urban runoff and industrial waste. The quality and quantity of industrial wastes, disposed via sewage systems or dumped directly into the
ocean, are often dependent upon the cyclical and periodic nature of
manufacturing runs. Effects of these materials on the marine ecosystem into which they are dumped are functions of:
    1. The physical and chemical composition, and amounts of the
wastes.
    2. The chemical forms of the contaminants in the dumped wastes
available to the biota.
    3. The general oceanography and meteorology affecting the fate
of the contaminants during dumping and after their introduction into
the marine environment.
    4. The morphology of the bottom in the vicinity of the ocean
dumping site (affects the initial deposition, accumulation, and any
later redistribution of material).
    5. The additional effects of other anthropogenic materials
reaching the same areas as the dumped wastes.

In addition to ocean dumping, significant amounts of other waste products reach the Bight directly and/or through the estuaries, from runoff, raw sewage and effluent outfalls, industrial disposal, atmospheric fallout and groundwater seepage. As the New York Bight oceanographic regime and its driving forces are complex, it is extremely difficult to separate observationally the effects of individual contaminant sources. The ecosystem effects of ocean dumping in the New York Bight are only part of the overall effects of various contaminant sources.

THE NEW YORK BIGHT

Geographically, the New York Bight (Fig. 1) can be considered to be that portion of the continental shelf south of Long Island and east of New Jersey. With an average width of some 165 km (90 n mi),

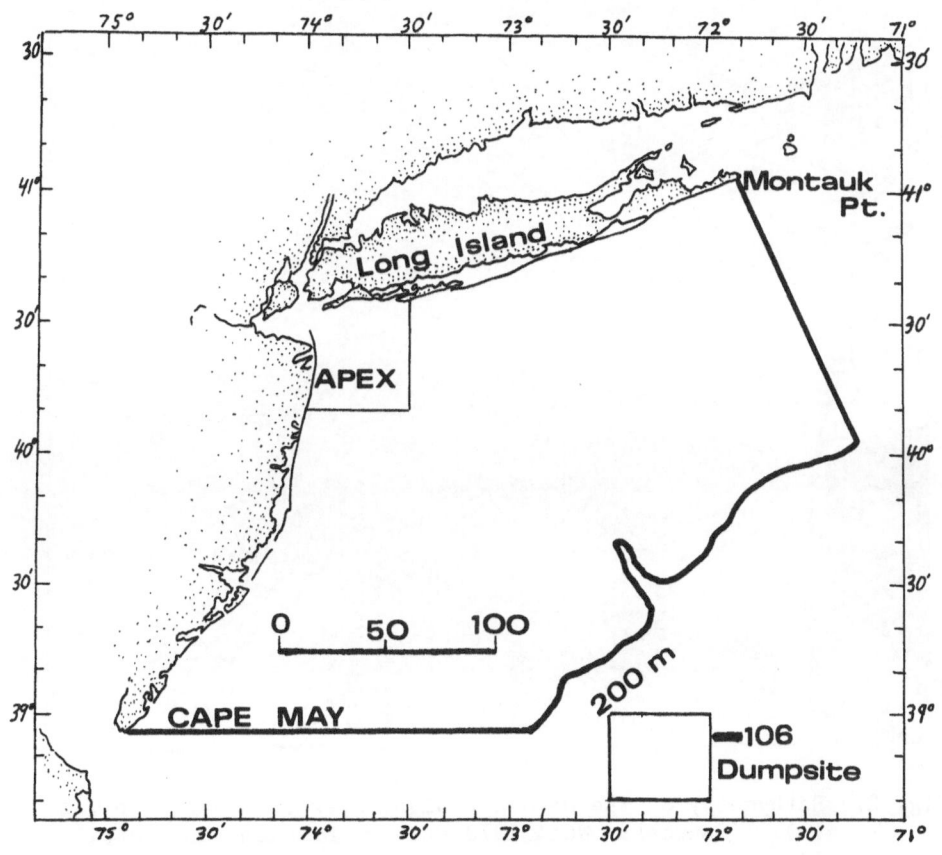

Fig. 1. New York Bight

the Bight has an areal extent of approximately 39,000 km² (15,000 n mi²).

    Most contaminant inputs, including those from ocean dumping, are to the New York Bight Apex (Fig. 2). The Deep-Water Dumpsite, located approximately 196 km (106 n mi) from Ambrose Tower (Fig. 2) is being addressed at length in other papers in this volume, and is not covered here.

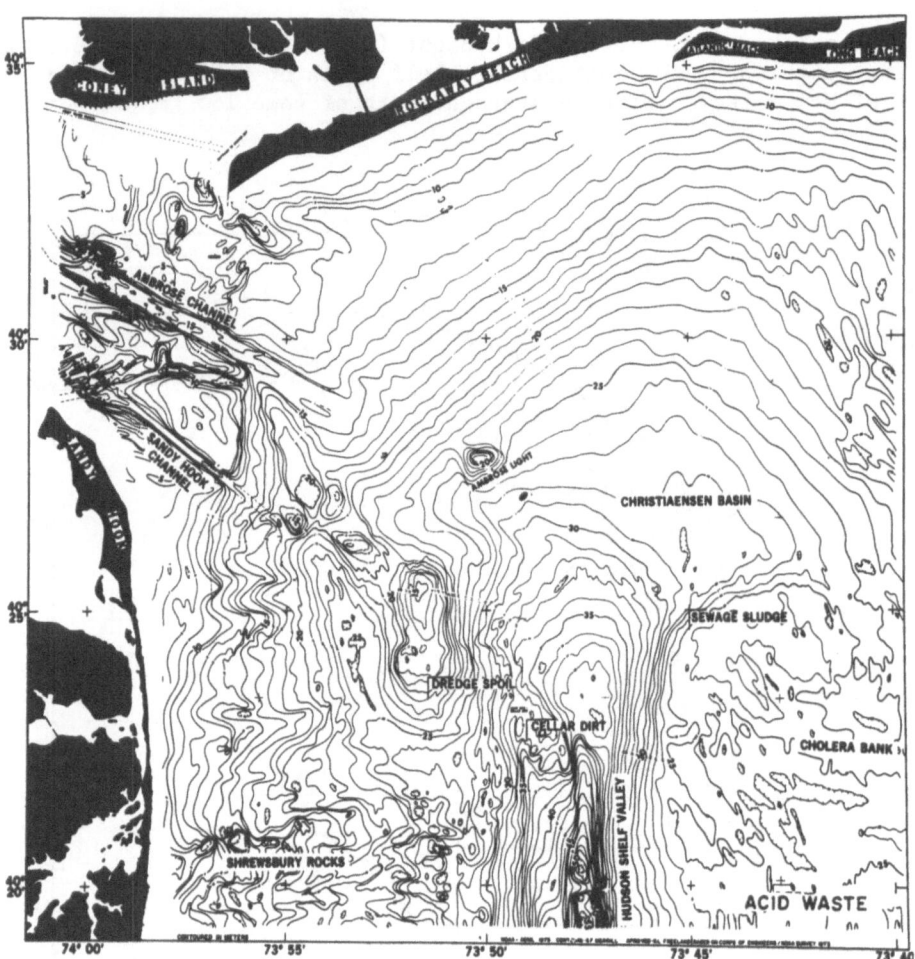

Fig. 2.   Bathymetry of the New York Bight Apex (contour interval 1m). (Sources: NOAA 1973 survey; Freeland and Merrill, 1977).

Extensive sampling of surficial sediments has allowed preparation of a fairly-detailed map of grain size distribution in the New
York Bight Apex (Fig. 3). Muds occur in low areas, namely in the
axis of the Hudson Shelf Valley (and its extension toward the Sandy
Hook-Rockaway Point transect) and in the Christiaensen Basin. Small
mud patches also occur in shallower water, near the Long Island
shore. The patches are up to tens of meters in diameter, tend to be
elongated in the predominant flow direction, and are generally only a
few centimeters or so thick. The underlying sediment is often characterized by stacked sequences of such layers, separated by layers of
coarser materials. The New Jersey platform and Cholera Bank are
floored with sand. Coarse sand and sandy gravel occur in the shoal
areas of northern Cholera Bank and Shrewsbury Rock (Swift _et al._,
1974).

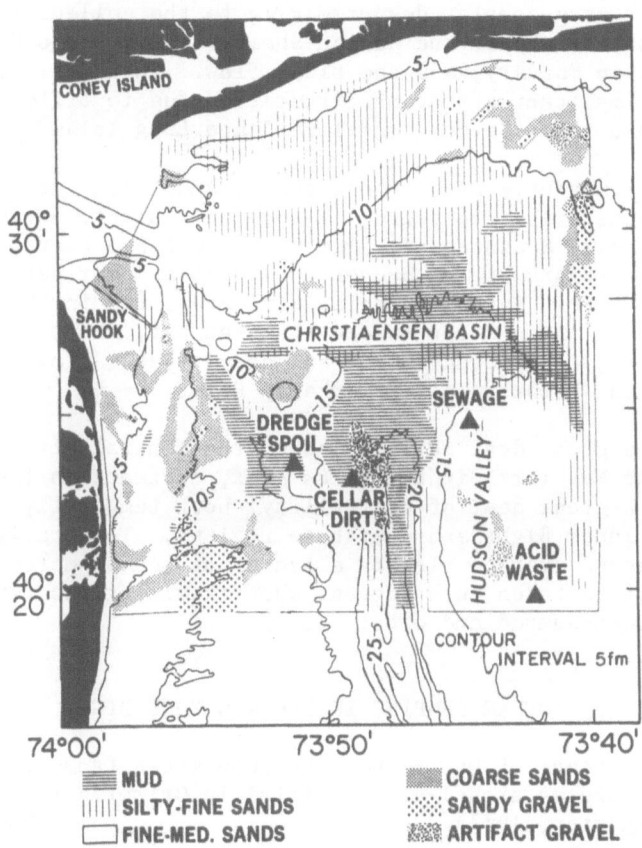

Fig. 3. Distribution of surficial sediment in the New York Bight
Apex.

Muds accumulate in relatively quiet deep areas such as the Hudson Shelf Valley and the Christiaensen Basin. Nearshore mud patches
accumulate in comparatively intense energy environments because of
the high concentrations of suspended sediment available for deposition during any relatively calm period. The period of survival of
any deposited mud is dependent upon the time to the next period of
intense wave and current energy sufficient to overcome the forces
tending to stabilize the mud deposit.

Average currents in the outer Bight flow southwesterly at speeds
of about 4-5 cm/sec (2 n mi/dy) at the surface, decreasing to one-
half or less of that speed near the bottom. This general pattern is
subject to alteration, even to total reversal, for extended periods
(as long as 2-3 months), especially during summer, and off New Jersey, a relatively shallow shelf area (Hansen, 1977).

Circulation in the inner Bight (approximated by the defined
Apex) is far more complex due primarily to the collective influences
of the coastline shape, the Hudson Shelf Valley, the outflow from the
Hudson-Raritan Estuary, and the tidal flow. In addition, there is a
subtle but important effect on circulation due to differences in
water depth off New Jersey (shallower) and Long Island (deeper).

Current measurements suggest that a clockwise eddy exists, at
least in the statistical sense, in the inner Bight, during periods of
sustained flow. Measurements also show a net shoreward flow in at
least some portions of the Hudson Shelf Valley over periods of several weeks or more. Further complexities are shown in the generally
estuarine-type circulation toward the transect, where surface water
flows seaward, along Sandy Hook, and deeper water flows toward the
Hudson-Raritan Estuary, along Rockaway Point.

While a great deal is known about the general circulation features of the New York Bight and its Apex, currents in the New York
Bight have a great deal of variability, both temporally and spatially, in the inner Bight and nearshore regions. Temporally variable
flow can exceed average flow by a factor of ten or more. This variability must be taken into account when examining the fate and effects of ocean-dumped and other wastes.

OCEAN DUMPING IN THE NEW YORK BIGHT

The locations of New York Bight dump sites (some defined as
points and some as areas) are specified in Criteria for Ocean Dumping
(Federal Register, 1977).

The locations for some of the ocean dumping in the New York
Bight have changed during the past few years, as have compositions
and amounts of the several generic materials dumped, and the

Table 1. Establishment of original ocean dumpsites in the New York
          Bight.

Waste Material	Year
Dredged Material	1888
Wrecks	1889
Cellar Dirt	1908
Sewage Sludge	1924
Acid Waste	1948
Toxic Chemicals	1965

From Achrem (1973)

periodicity of their dumping. As examples, until 1974 some waste
chemicals now dumped at DWD-106 were dumped at the then designated
Apex acid waste site (EPA, 1976). The volume of waste acid dumped
has decreased substantially since 1970 (EPA, 1975), and upgrading of
sewage treatment plants in the New York-New Jersey metropolitan area
will result in changes in the composition of dumped sewage sludge, as
well as an increase in the volume (EPA, 1976).

     The River and Harbor Act of 1888, as amended 12 July 1952, des-
ignated the Army District Engineer, New York as Supervisor of the
Harbor, and provided that he designate ocean waste disposal areas and
administer a permit program for transporting the waste materials
(Achrem, 1973). Table 1 shows the year in which the original dump
sites in the New York Bight for the several wastes were established.
Since implementation of the Marine Protection, Research and Sanctuaries Act of 1972 (as amended), on 23 April 1973, the Environmental
Protection Agency has had the responsibility of issuing permits for
ocean dumping, and for regulating procedures for applying for, issuing and denying permits. The same act continues to require permits
and regulations for ocean dumping of dredged material to be issued by
the Army Corps of Engineers, with review authority by the Environmental Protection Agency.

     Pararas-Carayannis (1973) provided an early attempt to identify
and quantify the general types of wastes dumped in the New York Bight
covering sewage sludge, dredged material, waste acid, cellar dirt,
toxic chemicals, and wrecks. The data have been updated and extended
by Achrem (1973), and EPA (1973, 1975, 1976). Table 2 provides a
compilation of such data for the years 1973 to 1977.

CHARACTER OF OCEAN-DUMPED WASTES

     The earliest published data on the composition of any of the
wastes dumped in the Bight were analyses of acid wastes (Redfield and

Table 2.   Annual (calendar year) ocean dumping inputs to the New York
            Bight apex.

	Sewage Sludge (wet tons)	Acid Wastes (wet tons)	Cellar Dirt (yd^3)	Dredged Material (yd^3)
1973	4.578×10^6	2.762×10^6	455×10^3	8.3×10^6
1974	4.203×10^6	2.338×10^6	360×10^3	10.82×10^6
1975	4.270×10^6	2.083×10^6	185×10^3	6.4×10^6
1976	4.377×10^6	1.412×10^6	147×10^3	9.31×10^6
1977	4.487×10^6	0.698×10^6	40×10^3	5.29×10^6

From P. Anderson (EPA Region II, personal communication, 15 February
1978)

Walford, 1951). Gross (1970) reported chemical analyses of various
wastes, primarily sewage sludge, from the New York-New Jersey metropolitan region. These early analyses suggested large differences in
the composition of the same kinds of wastes from several sources.
For example, the data in Gross (1970) show significant differences in
major and minor elements in sewage sludges from five sewage treatment plants in New York City. Subsequently, Mueller et al. (1976)
compiled an extensive set of chemical composition data for wastes
dumped in the New York Bight during 1973 which confirmed this compositional variation. For example, compositional data from 28 individual sources of dredged material [from an Army Corps of Engineers
(1971) unpublished survey, cited in Mueller et al. (1976)], in terms
of total load and percentages are presented in Table 3.

The relative importance of the mass loadings of different dumped
materials is shown in Table 4, and the average annual metal concentrations for several metals in some New York City treatment plant
sludges for 1973 and 1974 are shown in Table 5.

The general trends indicate that, in terms of mass loads,
dredged material is probably a larger source for most of the parameters examined than are the other dumped wastes. However, this assessment does not address the important problem of the availability
of any contaminant in the wastes to the biota.

The degree of importance that can be attributed to the estimated
magnitudes of mass loads of chemicals/parameters dumped into the New
York Bight must include consideration of the other sources of materials. Mueller et al. (1976) used various methods for approximating

Table 3.   Composition of dredged material from New York-New Jersey
           metropolitan area.

Project number	Average concentrations[a]						
	TVS %	COD %	TKN %	O&G %	Hg ppm	Pb ppm	Zn ppm
1	8.48	10.1	0.26	2.07	1.05	238	245
2	3.23	4.99	0.15	0.79	0.36	133	184
4	9.48	8.88	0.31	0.74	2.00	243	434
5	5.45	5.70	0.10	0.98	2.80	233	176
6	9.68	14.9	0.24	0.72	<1	95	255
7	9.83	9.70	0.43	1.48	0.65	74	149
8	9.33	13.0	0.30	2.03	0.09	412	675
9	2.55	4.07	0.90	0.035	1.1	57	63
10	3.24	2.80	0.16	0.83	0.21	122	145
11	7.54	7.93	0.27	1.27	0.31	162	263
12	13.9	16.4	0.36	2.90	0.08	579	835
13	10.5	13.5	0.45	1.64	0.28	295	420
14	6.99	7.91	0.26	1.88	0.17	151	206
15	13.3	14.9	0.48	2.80	0.69	338	391
16	8.85	5.95	0.15	2.39	0.22	393	296
17	12.7	17.3	0.39	0.65	<1	365	505
18	26.0	20.1	0.20	0.99	<1	1,450	10,400
19	5.70	4.26	0.19	0.32	<1	85	141
20	12.4	13.9	0.44	2.36	0.38	100	186
23	2.82	2.73	0.085	0.69	0	35	29
24	0.41	0.33	0.03	0	0	10	1
25	0.20	0.02	0.01	0.21	0.17	0	5
26	0.33	0.13	0.03	0.03	0	10	1
27	4.41	4.12	0.16	0.30	0.36	30	47
28	0.93	0.15	0.017	0.35	0	11	14

a.   Data from an Army Corps of Engineers (1971) unpublished survey.

From Mueller et al. (1976)

the mass loads of elements/chemical parameters to the Bight from
other sources, including atmospheric fallout, industrial and municipal wastewater, river runoff, urban runoff and groundwater seepage.
Table 6 compares the relative importance of these sources to the
Bight. The data are broad approximations; therefore, several important matters should be considered:
     1. Only a single year, 1973, is considered, and, in some cases,
data for only a part of that year are extrapolated to the entire
year.

Table 4. Mass loads of wastes dumped in the New York Bight Apex.

Parameter	Load, metric tons/day	Percentage				
		Dredge spoils	Sewage sludge	Acid waste	Chemical waste	Rubble
V, 10^6yd^3/yr	21.7[c]	53	26	15	3	3
SS	15,000	86	3	0.7	0.05	11
ALK	45		71		29	
BOD_5	430	49	46		5	
COD	3,200	65	34	0.1	1	
TOC	660	82	17		1	
MBAS[d]						
O & G	330	92	7		0.9	
NH_3-N	50	74[a]	20	0.04	6	
Org-N	35	74[a]	20		6	
TKN	85	74	20		6	
NO_2+NO_3-N	0.086		53		47	
Total N	85	74	20		6	
Total P[d]						
Total P	69	92	7	0.3	0.8	
Cd	2.0	98	2	0.1	0.06	
Cr	2.5	93	3	4	0.1	
Cu	7.1	89	10	0.7	0.07	
Fe[b]	180					
Hg	0.026	50	50	0.04	0.7	
Pb	5.6	85	13	3	0.03	
Zn	9.3	78	19	2	0.7	

Fecal coliform = 10^8[e] organisms/day

Total coliform = 3 x 10^8[e] organisms/day

a. Using sewage sludge TKN ratio   b. From average Fe/Cu ratio for raw and digested sludge of 25, Mueller (1972).   c. 18.6 cfs.   d. Assumed negligible.   e. From Mueller et al. (1976)

Table 5.   Average annual metal concentrations in sewage sludge for
           selected New York City treatment plants.

1973	Hg (µg/l)	Cd(µg/l)	Pb (mg/l)
HUNTS POINT	530	250	30
JAMAICA	880	640	51
PORT RICHMOND	220	190	23
TALLMAN ISLAND	450	560	24
26TH WARD	720	4,900	99
NEWTOWN CREEK	470	4,400	160
1974			
HUNTS POINT	50	340	26
JAMAICA	220	430	38
PORT RICHMOND	260	90	16
TALLMAN ISLAND	630	240	84
26TH WARD	480	1,200	100
NEWTOWN CREEK	150	1,500	280

From EPA (1975).

2.   Table 6 expresses only total loads and neglects their uneven
distribution to the Bight. The relative importance of any and all
sources is a function of the geographical location/area of concern of
their input, and their redistribution within the ecosystem.
3.   All the wastes input to the estuary do not reach the Bight.
Lack of knowledge of the Hudson-Raritan Estuary and other estuaries
precludes estimating that portion, perhaps a very significant portion, which does not leave the estuary.
4.   Dredged material is, in effect, considered twice -- once as
dredged material, per se, and once as source material, part of which
becomes deposits in the estuary which are dredged.
Any and all of these factors, as well as others, have the potential
for influencing considerably the assessment of the relative importance of the several major sources of wastes to the New York Bight.

Ketchum et al. (1951) have estimated the flushing time of Apex

Table 6. Total waste mass loads to the New York Bight (from Mueller et al., 1976). The river runoff is for that portion of the river-estuary system that is gauged for flow.

Parameter	Direct Bight		Percentage Contribution			Runoff	
	Barge	Atmospheric	Coastal Zone				
			Wastewater		River	Urban	Groundwater
			Municipal	Industrial			
FLOW	0.02	59	5	0.4	33	2	0.4
SS	63	5	4	0.2	16	12	Nil
ALK	1	Nil	35	0.3	59	5	0.03
BOD5	21	9	48	2	11	9	0.01
COD	32	10	35	1	13	9	0.01
TOC	25	12	29	1	18	15	0.02
MBAS			86		5	9	0.05
O&G	38		22	0.7	16	23	
NH3-N	24	4	55	3	10	4	0.04
ORG-N	19	9	45	2	21	5	0.02
TKN	21	6	51	2	15	5	0.02
NO2+NO3-N	0.07	33	6	0.3	60	0.6	0.7
TOTAL-N	16	13	40	2	25	4	0.2
ORTHO-P		1	72		18	9	Nil
TOTAL-P	50	0.7	35	1	9	4	Nil
Cd	82	2	5	0.6	5	5	0.001
Cr	50	1	22	0.8	10	16	Nil
Cu	51	3	11	9	10	16	0.006
Fe	79	3	5	0.5	6	6	0.01
Hg	9		71	2	13	5	
Pb	44	9	19	3	6	19	0.004
Zn	29	18	8	2	21	22	0.009
F.Coli-winter	<0.01	Nil	87	0.2	0.01	13	Nil
summer	<0.01	Nil	85	0.2	0.01	15	Nil
T.Coli-winter	<0.01	Nil	91	0.1	0.05	9	Nil
summer	<0.01	Nil	84	0.2	0.1	16	Nil

waters, from salt balance studies, to be about one week. In considering copper and zinc in the Apex, Segar and Cantillo (1976) show
residence times of as little as ten days. For the Raritan Bay portion of the Hudson-Raritan Estuary, Parker et al. (1976) indicate a
residence time of water of as long as 4.0 tidal cycles. Such residence times vary seasonally. It is relative to these time frames
that waste inputs to the Bight should be examined. In this context,
annual averages can often be misleading.

Dredging in the Hudson-Raritan Estuary does not occur continuously. When it does occur, large volumes of material are dumped over
short periods of time. In some instances the material is relatively
clean; in others it is highly contaminated (Table 3). Atmospheric
input is greatest during rains, as is runoff. Appropriate lag times
for river (days) and urban (hours) runoff and for groundwater seepage
must also be taken into account in examining input rates over a short
period of time. The waste inputs related to municipal waste treatment are also not constant. Storage facilities and transportation
schedules for sewage sludge result in most activity being concentrated during the work week, and little during the weekend.

Industrial wastewater input rates are tied directly to production run schedules which often occur only a few times a year for
particular wastes. In the New York-New Jersey metropolitan area,
significant amounts of industrial wastewaters are discharged directly
into sewer systems (Klein et al., 1974). Industrial and residential
inputs to the sewer systems are quite variable. Another factor to be
considered is that many of the treatment plants in the New York-New
Jersey metropolitan area serve combined storm and sanitary sewers.
Rain changes drastically the composition of material reaching the
treatment plants (Mytelka et al., 1973) and, thereby, the character
of the sewage sludge dumped in the Bight.

The use of specific tracers or indicators of the presence of
particular wastes in the sediments of the New York Bight has been examined. Concentrations of metals and organics, and ratios of particular metals and particular organics have been examined with little
success. Hatcher and McGillivary (1979) showed the potential utility
of coprostanol as an indicator of sewage-related contamination. However, any coprostanol found may be derived from sewage, raw discharges to the Hudson-Raritan Estuary, sewage sludge, sewage in
dredged material, or sewage discharged from ship sanitary systems,
etc. What has been learned is that the "mixing-bowl" nature of the
New York Bight transport system tends to mix the various waste materials entering the Bight so that it is difficult to distinguish the
sources of contaminants found at any given location in the Bight.

The form of the contaminant, both the physical form and the
chemical form (particularly whether or not it is in the dissolved or
particulate state), is an important determinant of its route and rate

of transport, and its fate. Both chemical and physical form may
change within the ecosystem. For example, a contaminant in the dissolved state at its source need not remain dissolved as it enters estuarine or Bight waters. The reverse is also true for the particulate state. For example, Hatcher et al. (1978) have found at some 25
minutes after dumping of sewage sludge that, after correcting for dilution, dissolved concentrations of iron and manganese decrease (suggesting their involvement in flocculation/precipitation), copper increases (suggesting its dissolution from particulate phases) and cadmium remains about the same.

EFFECTS OF OCEAN DUMPING ON THE NEW YORK BIGHT ECOSYSTEM

Effects on the Environment

    Gross (1970) examined the inputs of natural materials and wastes
to the sediments of the New York Bight, showing that the wastes comprise most of the total. He did not, however, indicate how much of
either kind of material remains on the bottom. From hydrographic
surveys of the Apex conducted in 1936 and 1973, Freeland and Merrill
(1976) find that net deposition during the intervening years occurred
only at dredged material dump sites. Otherwise, net erosion or no
net change were evident for other Apex areas. This is suggestive
that contaminant materials do not accumulate continuously in the
Apex. However, net bathymetric change data do not consider the degree of replacement or exchange of earlier-deposited materials with
more recently-deposited contaminated materials or the degradation of
materials. The accuracy and precision for the two surveys, while
meeting national charting standards, were such that significant erosion or deposition may not have been detected.

    Short-term effects (of the order of hours or so) on the quality
of the water column by ocean dumping have been noted by a number of
studies. Early studies (Ketchum et al., 1951; Redfield and Walford,
1951; Ketchum et al., 1958) focused on the acid wastes dumped in the
Bight. Later studies, including those reported by Sandy Hook Laboratory (1972), Segar and Cantillo (1976), Segar and Berberian (1976),
and Garside and Malone (1978) addressed the Apex as a whole. These
and other studies have found that direct impacts of ocean dumping on
the chemistry of the Bight Apex water column are localized and of a
short-term nature. These findings have been confirmed by site-spe-
cific studies related to sewage sludge and to dredged material. Callaway et al. (1976), using transmissometer measurements, found that
suspended material from sewage sludge dumping reached background
values in a few hours. Similar results have been found by Proni et
al. (1978) using acoustic tracking techniques, and by Hatcher et al.
(1978) on the basis of chemical analyses of discrete samples. These
investigators also found that some materials remain concentrated at

the pycnocline for somewhat longer times. Lee and Jones (1977) reported similar findings for dredged material dumped in the New York
Bight.

Local influences of sewage sludge dumping are often seen in concentration patterns of nutrients (silicate, nitrate, nitrite, phosphate). During times of the year when the Apex water column is
stratified, bottom waters often show higher concentrations of nutrients near the site. This same tendency is seen under non-stratified
conditions, but the concentration levels are generally reduced (MESA,
1975). Segar and Cantillo (1976) also report that the concentration
of dissolved zinc in the immediate vicinity of the sewage sludge dump
site is often anomalously high.

Drake (1976) and Nelson (1977) have shown that suspended particulate material from all sources is effectively mixed together and
distributed in the Bight, particulary in the Apex. Microscopic examination of surficial muds, reflective of that portion of suspended
material temporarily residing on the bottom, showed sewage-derived
materials to comprise less than about 3% of the total (Drake, 1974).
The influence of individual sources of these sewage-derived materials
could not be ascertained.

Ocean dumping has been suggested as a cause in the depletion of
dissolved oxygen in the inner New York Bight and the nearshore areas
along New Jersey and Long Island, particularly the severe depletion
and anoxia which occurred off New Jersey in summer 1976. However,
extensive analysis of the oxygen depletion problem shows that ocean
dumping is not the primary cause of the low dissolved oxygen concentrations (Segar and Berberian, 1976; Garside and Malone, 1978; Swanson and Sindermann, 1979). Ocean dumping does, however, add to the
severity of the problem, as do the inputs of other oxygen demanding
wastes which reach the New York Bight.

Visual observations of dredged material sewage sludge and acid
waste dumping reveal slicks, and accumulation of some floatable substances during actual dumping operations. For dumped acid wastes,
the slicks have remained identifiable for long periods of time (Charnell and Maul, 1973). There appear to have been no effects noted on
the zooplankton from the acid wastes (Weibe et al., 1973).

During the summer of 1976, large amounts of floatable wastes,
generally of a sewage-related origin, inundated the Long Island ocean
beaches. Investigations showed that ocean dumping was only a minor
contributor to the floatables found on the beaches. The major source
of floatables to the waters of the New York Bight -- which were then
stranded on the beaches because of persistent, consistent southerly
winds -- was determined to be the outflow from the Hudson-Raritan Estuary (NOAA, 1977; Swanson et al., 1978).

Table 7. Priority chemical contaminants of the New York Bight, June
          1977.

Category A. Major Perceived Threats That Require Continued Study
             Cadmium compounds
             Chlorinated pesticides*
             Mercury compounds
             Polynuclear aromatic hydrocarbons (PNAHs)***
             Polychlorinated biphenyls (PCBs)
             Plutonium

Category B. Potentially Significant Threats For Which Data Must Be
Collected and Interpreted
             Benzidenes
             Chlorobenzenes
             Chlorophenols
             Diphenylhydrazine
             Halogenated diphenyl esters
             Isophorone
             Low molecular weight halogenated hydrocarbons
                (LMHHs)**
             Petroleum hydrocarbons (PHCs, other than PNAHs)
             Tellurium

Category C. Substances Not Requiring Priority Attention At Present
On The Basis of Existing Information
             Arsenic compounds
             Chromium compounds
             Lead compounds
             Nitrobenzenes
             Nitrophenols
             Phenols
             Phthalates
             Selenium compounds
             Silver compounds

*Aldrin/Dieldrin, Chlordane (technical mixture and metabolites),
 DDT and metabolites, Endosulfan and metabolites, Endrin and
 metabolites, Heptachlor and metabolites, and hexachlorocyclohexane
 (all isomers), Toxaphene.

**Carbon tetrachloride, chloroform, chlorinated ethanes (includes
 1,2-dichhloroethane, 1,1,1-trichloroethane, and hexachloro-
 ethane), dichloroethylenes (1,1- and 1,2-dichloroethylene),
 halomethanes (other than specified), tetrachloroethylene,
 tricholoroethylene, and vinyl chloride.

***Aromatic compounds with unsaturated ring structures: Benzene,
 alkyl substituted benzenes, and polynuclear hydrocarbons with
 multiple alkyl substitutions.

Table 8.   PCBs and DDTs in sewage sludge from New York City sewage
treatment plants (ppb dry weight).

	Aroclor 1242	Aroclor 1248	Aroclor 1254	Aroclor 1260	Total PCB
Wards Island	60	190	680	300	1200
Hunts Point	700	1300	700	1300	4000
Passaic Valley	2600	2200	770	650	6200
Tallmans Island	650	430	980	240	2300
MEAN	1000	1030	780	620	3400
STD. DEV.	1100	910	140	840	2200

	o,p' DDE	p,p' DDE	o,p' DDD	p,p' DDD	o,p' DDT	p,p' DDT	Total DDT
Wards Island	*	68	110	180	20	*	380
Hunts Point	*	*	22	120	*	*	140
Passaic Valley	110	*	46	*	*	120	280
Tallmans Island	420	1800	60	*	*	*	2300
MEAN	130	470	60	80	5	30	780
STD. DEV.	200	890	37	90	–	–	1020

From West et al. (1977)

The long-term effects of wastes reaching the New York Bight
environment can best be seen in the sediments. High metal and
organics concentrations tend to coincide with the depths of low
areas, primarily the Christiaensen Basin, and the Hudson Shelf
Valley, near the Basin (Gross et al., 1971; Sandy Hook Laboratory,
1972; Hatcher and Keister, 1976). However, some higher concentrations are observed in the immediate vicinities of the dredged material and sewage sludge dump sites. The concentration patterns reflect:
accumulation of the heavier portions of dumped dredged material and
sewage sludge in the general vicinities of the locations where each
waste is dumped; and mixing of suspended particulate materials, from
natural and waste-related sources and settling out of some portions
of this mixture in topographic lows, particularly during quiescent
conditions. This latter mechanism is responsible for the formation
of near-shore mud patches off Long Island (Freeland et al., 1978),
which have, erroneously, been attributed to "creeping" of bottom
materials from the dump sites and Christiaensen Basin toward land.

The uptake of metals and artificial organic compounds by food
organisms is a source of concern to scientists and the public. A
recent study by a panel of experts sought to identify those chemical
contaminants which are, or are likely to be, the most problem in the
Bight. Three criteria were considered in this study.
     1. Weighed most heavily: existing or projected future hazard to
human health (by any mechanism: ingestion, bodily contact, genetic
modification of water-borne pathogens to resist antibiotics, etc.)
     2. Weighed heavily: existing or projected future hazard to ma-
rine species harvested by man (commercially or for recreation), and
     3. Weighed less heavily: existing or projected future hazard to
any other marine biota.
Results are summarized in Table 7.

     Prior to initiation of Panel efforts, few studies had addressed
the chemical contaminants listed under Categories A and B in wastes
which reach the Bight, or in biota, sediments or waters of the Bight,
with the exception of the metals cadmium and mercury, polychlorinated
biphenyls (PCBs) and some chlorinated pesticides. Sandy Hook

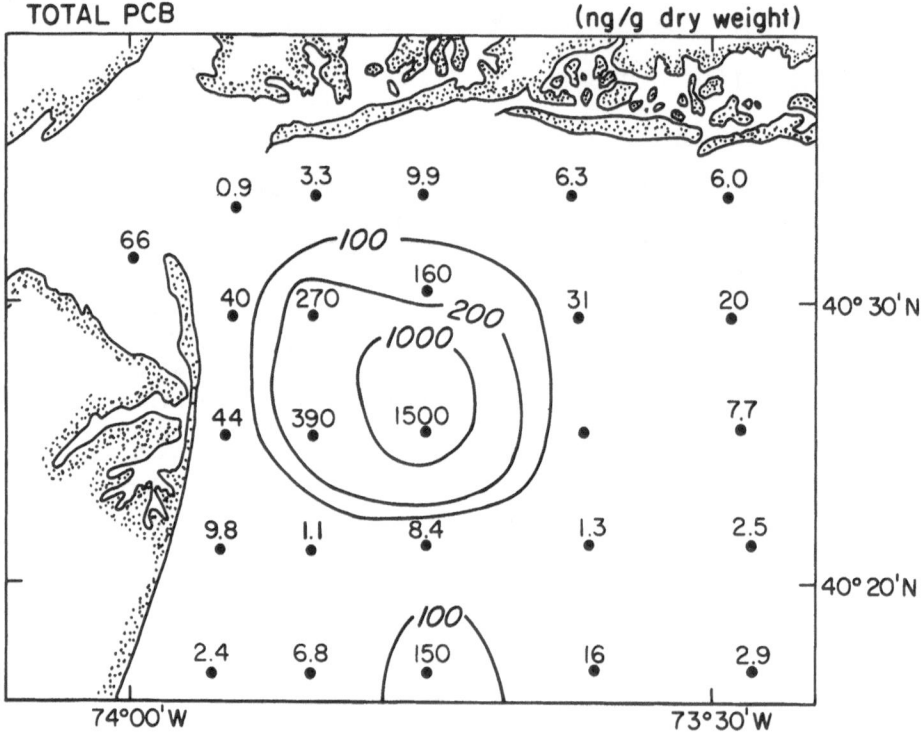

Fig. 4.   Total PCB distribution in sediments of the New York Bight
          Apex.

Table 9. Concentrations of aromatic hydrocarbons (in units of ng/ml for surface microlayer and sewage sludge in units of ng/g dry wt. for all other samples.)

	Surface Microlayer			Sewage Sludge Composite, 3 Stations	Dredged Material			Bottom Sediments		
	A	B	C		Newtown Creek	Gowanus Canal	Pierhead Channel	Lower Bay	Christ. Basin	Outer Bight
Naphthalene	<.01	.03	.03	100	120000	100	200	100	100	
2-methylnaphthalene	<.01	.06	<.01	200	2300	500	200	20	100	<.1
1-methylnaphthalene		<.01	<.01	50	1800	200	100	5	40	<.1
Biphenyl		<.01	<.01	100	700	300	200	40	50	20
Dibenzothiophene				50	2700	1000	100	40	50	
Phenanthrene	<.01	<.01	<.01	200	14600	1000	300	300	300	
Anthracene				50	9600	500	200	100	50	1
1-methylphenanthrene		<.01	<.01	50	1500	1000	100	50	40	
Fluoranthene		<.01	<.01	100	10200	2000	50	500	400	1
Pyrene		<.01	<.01	200	7200	3000	500	500	400	
Benz(a)anthracene	<.01			50	5600	3000	500	500	500	<.3
Chrysene				30	3000	2000	400	400	300	<.2
Benzo(e)pyrene			<.5	20	1000	1000	200	200	200	
Benzo(a)pyrene			<.5	10	1300	500	100	200	200	<.2
Perylene			<.5	<.5	400	300	50	100	50	
Latitude (°N)	39°41'	40°30'	40°12'	(Wards Island, Hunts Point,	40°44'	40°40'	40°40'	40°28'	40°25'	39°00'
Longitude (°W)	74°00'	73°56'	73°50'	Newtown Creek)	73°56'	74°00'	74°09'	74°02'	73°48'	72°55'

Laboratory (1972) reported a highest concentration of 0.13 ppm dry weight of DDT, DDD and DDE in four sediment samples from the vicinity of the dredged material and sewage sludge dump sites in the Bight. Raytheon (1975 a,b) reported a greatest concentration of less than 0.05 ppm, dry weight for Toxaphene, while EPA (1976) reported a greatest concentration of less than 0.04 ppm dry weight of PCBs each in offshore continental sediments. A far more extensive series of data, on Bight Apex sediment and sewage sludge, for PCBs and for DDT, DDD and DDE are provided by West et al. (1977). Table 8 shows such data for samples of sewage sludge from four New York-New Jersey metropolitan area sewage treatment plants, while Fig. 4 shows total PCB distribution in sediments of the New York Bight Apex.

MacLeod (1978) finds large amounts of some organics in several portions of the New York Bight ecosystem (Table 9). Surface micro-layer samples taken in the Bight are quite low in aromatic hydro-carbons. On the other hand, the concentrations of aromatic hydro-carbons in Hudson-Raritan Estuary sediments, destined as dredged ma-terial, are very high.

The recent data on the New York Bight environment seem to indi-cate its deteriorating environmental quality, and/or the increasing contaminated quality of the waste materials inputs to the Bight. However, as the data comprise measurements of environmental param-eters that have not been closely examined before, it is not certain whether the deterioration has been more rapid in recent years or whether it has taken place slowly over the several decades of past rapid industrialization and urbanization.

Effects on the Biota

Results from numerous studies of the New York Bight allow defi-nition of some of the effects on the biota. In some instances, these effects can be tied indirectly to one or more types of the wastes, by spatial correlations of Bight quality to conditions of biota in field studies, by relating patterns of chemical observations to waste input locations, and by making use of waste volume and composition data.

The phytoplankton species assemblages of the inner Bight are modified greatly by processes related directly to the Hudson-Raritan Estuary, i.e., primarily natural factors and eutrophication. Zoo-plankton species composition, however, is more strongly influenced by oceanic processes (Malone, 1977).

The effects of wastes in the New York Bight have the greatest impact upon the benthic ecosystem. Benthic organisms in the Bight are important as commercial or recreational food resources, as inte-gral parts of marine food webs, as modifiers of the sedimentary envi-ronment, and as indicators of environmental perturbations and stress. Shellfish such as the common clam, Spisula solidissima; ocean quahog,

Arctica islandica; American lobster, Homarus americanus; and sea scallop, Placopecten magellanicus, are among the more important taken commercially in the New York-New Jersey metropolitan area. The hard-shelled clam or quahog, Mercenaria mercenaria is collected intensive-ly by both commercial and sport fishermen, but primarily from bays adjoining the Bight. Several species of crabs, including the can-croid crab, Cancer irroratus and the blue crab, Callinectes sapidus have more limited commercial importance (Pearce et al., 1978).

Most benthic invertebrates have no direct commercial or recrea-tional value but are of extreme importance in the food chain or web-of-life which culminates in demersal finfishes. The amphipods, crus-taceans, bivalve mollusks, and polychaetes are particularly signifi-cant as food of bottom-living fishes. Substantial impacts upon the benthic invertebrates of the already-severely-contaminated New York harbors were noted by Goode (1887) and Jacot (1920). These early impacts were aggravated by the dredging and several kinds of dumping which were practiced in the harbors then. More recent studies have documented that many of the benthic invertebrates normally found in temperate estuaries are no longer present in the Lower, Raritan, and Sandy Hook Bays (McGrath, 1974; cf Dean, 1975; Goode, 1887; Jacot, 1920; Pearce et al., 1978). The species no longer tolerant of the New York harbors are most often the crustacea and molluscs (Pearce et al., 1978) i.e., among the most important food organisms of demersal fishes (Edwards and Bowman, 1978).

These same tendencies toward low species diversity of benthic invertebrates and unusually few crustacea and mollusks are found in the most contaminated areas of the Bight (O'Connor, 1976; McDonough, 1976; Pearce et al., 1978) (see Fig. 5). These most contaminated areas are the dredged material dump site, Christiaensen Basin, and Hudson Shelf Valley (see Fig. 2).

A novel sampling strategy designed to assess effectively the in-fluences of wastes on the benthos of the Bight and compensate for the extreme small scale heterogeneity of sediments and biota has been outlined by Walker et al. (1979). They first define sampling strata in terms of sediment characteristics -- natural (e.g., median grain size) and people-dominated (e.g., trace metal concentration). Once strata are defined, the patterns of distribution of benthic species within and among strata are evaluated to assess implicitly how waste inputs have affected these benthic distributions. Sediment strata are intentionally made independent of geographic location. Those species having a significant, between-strata difference in abundance, with 95% confidence, are identified in Table 10. This extensive list of species is evidence of how much more distinct these strata are vice conventionally-defined geographic strata (e.g., cf Pearce et al., 1978).

The same data were analysed further by graphically comparing

biological responses to contaminants in two strata having similar
sediment grain size. The comparison here is between relatively uncontaminated sediments with low percentages of organic material and
low heavy metal concentrations vs waste-impacted sediments with high
percentages of organic material and high heavy metal levels. The
five species shown in Fig. 6 have mean abundances which are significantly different between the two strata. These species might be
worthwhile indicators of stressed vs unstressed environments of
similar sediment characteristics in the New York Bight Apex. For
each species, a significantly reduced abundance is evident in the

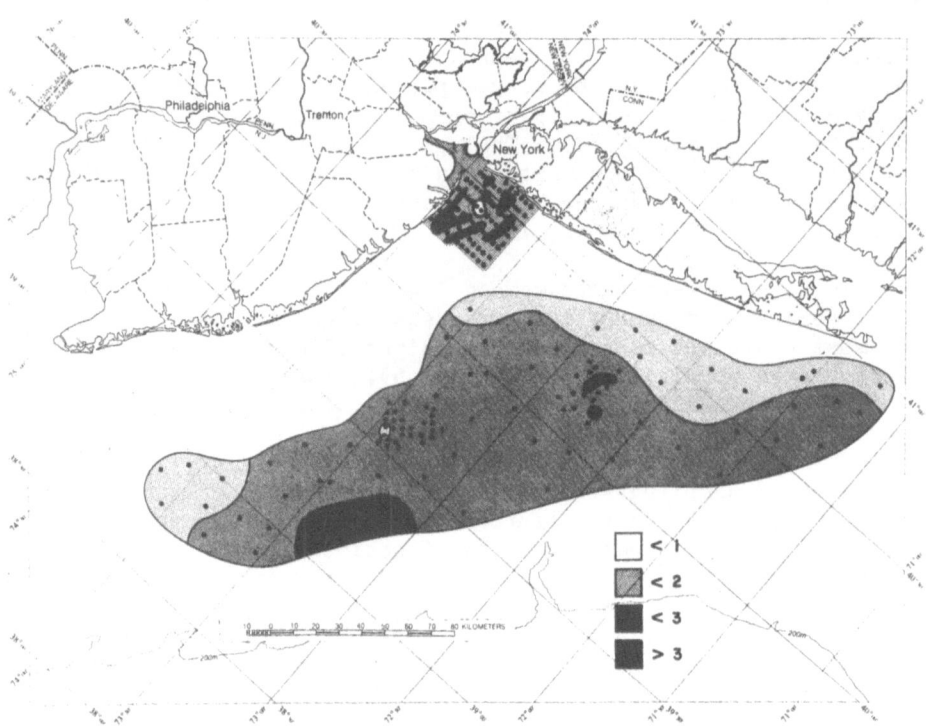

Fig. 5. Species diversity of benthic invertebrate macrofauna.
Smith-McIntyre grab locations are denoted by dots, with
two grabs per location in the Apex. The sampling density
(78 stations) in the bays south of New York City does not
permit indication of sampling locations there. (From
Pearce et al., 1978.)

stressed environment. These results may be used to help design
future monitoring efforts.

A circular area of 11-km radius, centered on the sewage sludge
dump site, was closed to shellfishing by the Food and Drug Administration (FDA) in 1970. This area, closed due to coliform bacterial
concentrations (Buelow et al., 1968), was expanded to the Long Island

Table 10. Species having significant, between-strata differences in
abundance, with 95% confidence, in the New York Bight.

Edwardsia sp.	Ninoe nigripes
Edwardsia elegans	Drilonepeis longa
Edwardsia sipunculdides	Stauronereis caecus
Cerianthus americanus	Haploscoloplos robustus
Rhynchocoela	Cossura longocirrata
Protodrilus symbioticus	Tharux annulosus
Harmothoe extenuata	Tharux acutus
Harmothoe imbricata	Owenia fusiformis
Lepidonotus squamatus	Asabellides oculata
Hartmania moorei	Polycirrus sp.
Sthenelais limicola	Pherusa affinis
Pholoe minuta	Potamilla neglecta
Eteone longa	Euchone rubrocincta
Autolytus cornutus	Hydrobia minuta
Nereis grayi	Crepidula plana
Nereis succinea	Unid bivalve #2
Aglaophamus circinata	Nucula proxima
Nephys bucera	Nucula delphinodonta
Nephtys incisa	Yoldia limatula
Glycera americana	Mytilus edulis
Glycera dibranchiata	Astarte castanea
Hemipodus sp.	Cerastoderma pinnulatum
Hemipodus armatus	Pitar morrhjana
Goniadella aracilis	Tellina agilis
Ophioglycera gigantea	Mulinia lateralis
Pisione remota	Copepoda
Ophelia denticulata	Heteromysis formosa
Mediomastus ambisetae	Leptocuma minor
Aricidea jeffreysii	Cirolana polita
Paraonis gracilis	Unciola inermis
Prionospio malmgreni	Unciola irrorata
Prionospio steenstrupi	Phoxocephalus holbolli
Spiophanes bombyx	Cancer borealis
Lumbrineris tenuis	Cancer irroratus
Lumbrineris fragilis	Phoronis architeta
Lumbrineris acuta	Asterias forbesii

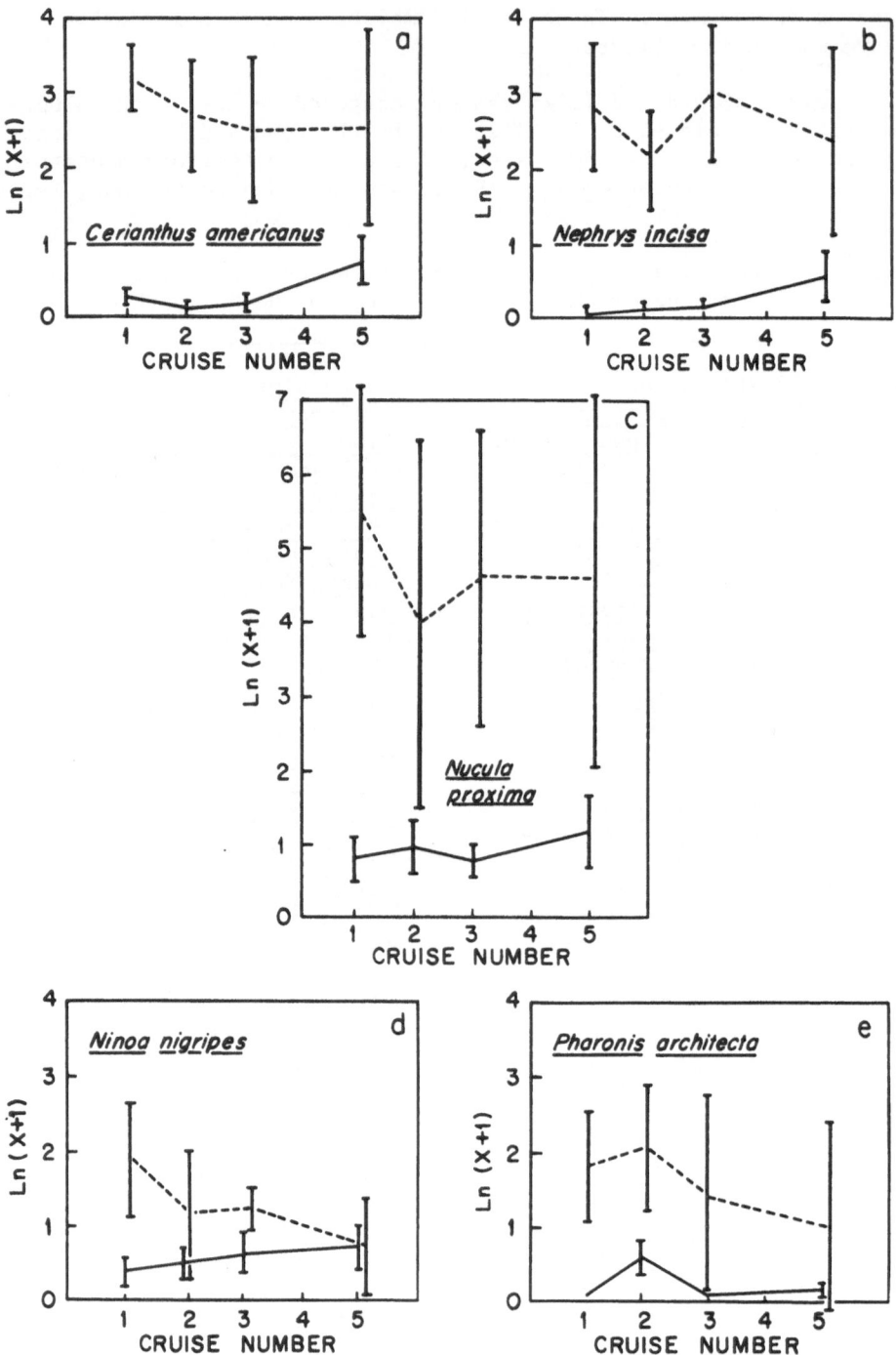

Fig. 6.   Species in the New York Bight with mean abundances
          significantly different between strata.

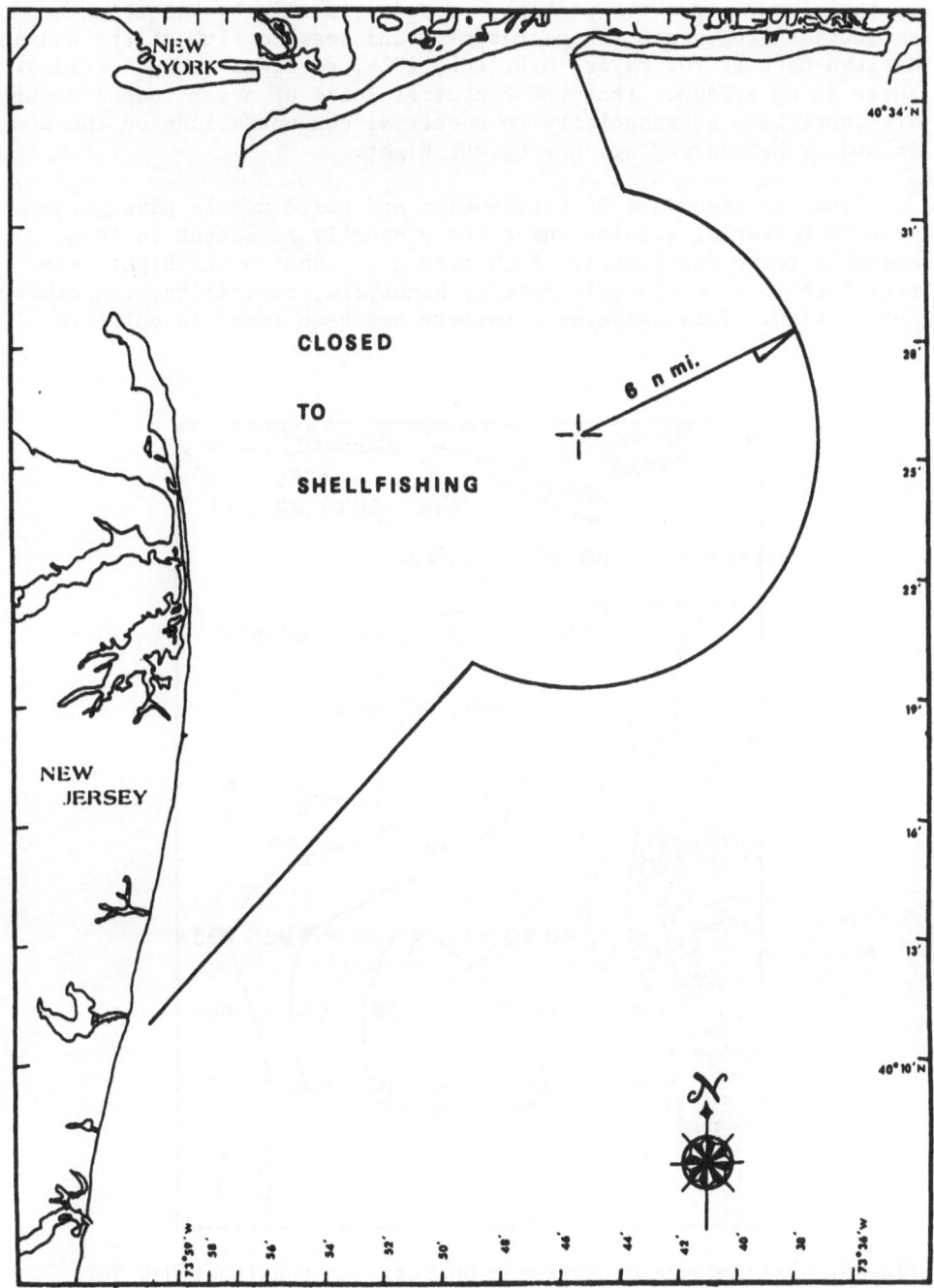

Fig. 7.   Shellfish closure area in New York Bight Apex.

and New Jersey shorelines in 1974 (see Fig. 7). The later extension
of the closure area was probably caused primarily by bacterial contamination from ocean sewage outfalls and seaward flow of the Hudson-
Raritan Estuary (G. Mayer, U.S. PHS, 1974, personal communication).
There is no evidence that the bacterial loads of ocean dumped materials contribute substantially to bacterial concentrations on the Long
Island or New Jersey beaches of the Bight.

The increased use of antibiotics and toxic metals have given
rise to bacterial strains which are unusually resistant to these
normally-toxic substances. Such strains, found in the Bight, are
resistant to several toxic metals, kanomycin, ampicillin, and other
antibiotics. This resistance pattern has been found in coliform

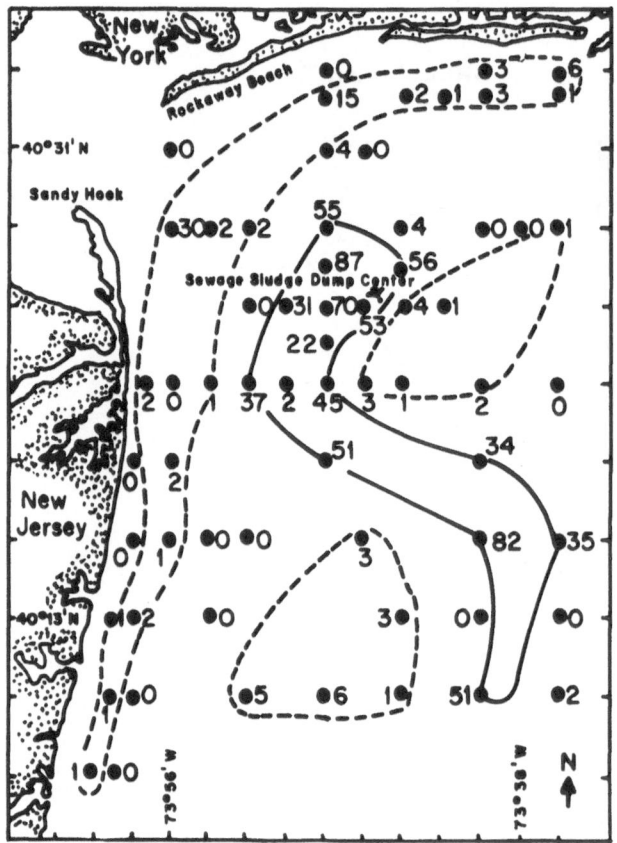

Fig. 8. Percentages of sediment bacteria in the inner New York
         Bight resistant to 20 mg Hg/ml (shown as figures beside
         each station). Approximate isopleths are drawn at 1 and
         30%. (From Timoney et al., 1978.)

bacteria (Koditschek and Guyre, 1974) and in Escherichia coli, Vibrio, and Bacillus (Timoney et al., 1978). These resistant bacteria are much more abundant at the dredged material and sewage sludge dump sites, and in the Christiaensen Basin. Their distributions correlate well with metals concentrations in the sediments (Timoney et al., 1978; see Fig. 8). There can be little doubt that these resistant bacteria are introduced via discharges and dumped dredged materials and sewage sludge. There is no evidence that these resistant forms pose as much of a health hazard as do those reaching man by terrestrial and airborne routes.

Two additional measurable contaminant effects upon fishes of the Bight are not exclusively due to, but are exacerbated by, ocean dumping: (1) contaminant accumulations in marine food organisms and their food webs, and (2) diseases of fish and shellfish.

The bioaccumulation of mercury, cadmium, and perhaps lead, has apparently not resulted in hazardous levels, but these metals are in sufficiently high concentration in food organisms and their food webs to be of concern (MESA, 1978). Dumped materials are the major sources of several toxic metals. Although the biological availability of the metals is unknown, losses of metals from the dump site to overlying waters must be viewed as potentially significant sources of some bioaccumulated metals.

Given the above-mentioned PCB concentrations in sediments at the dredged material and sewage sludge dump sites, these dumped materials probably contribute to the PCB contamination of striped bass, bluefish, and other biota of the Bight. However, scanty existing data and the widespread movements of most fishes preclude knowing the relative importance of each of the major PCB sources. It does appear that fishes from the Hudson River typically have higher PCB body burdens than those from the Bight, although many of the species analyzed travel to and from both environments (Hetling et al., 1978; MESA, 1978).

Some diseases of crustacea and finfish are much more prevalent in chemically impacted regions of the Bight than elsewhere. The exoskeletal "shell disease" of lobsters (Homarus americanus) and rock crabs (Cancer irroratus) appears to occur primarily in specimens on and near the benthic deposits of dredged material and sewage sludge (Sandy Hook Laboratory, 1972). The skeletal erosion occurs primarily on the joints of appendages where flocculant-contaminated sediments would be expected to accumulate. Equal numbers of crabs and lobsters from relatively uncontaminated areas were exposed, in aquaria, to organic deposits taken near the sewage sludge and dredged material dump sites, and to clean sand substrates by Young and Pearce (1975). Skeletal erosion appeared in all animals exposed to both contaminated sediments, but none of the controls developed any pathology. A similar shell disease of the shrimp, Crangon septemspinosa, was found

to be common in the inner Bight and Raritan and Sandy Hook Bays by
Gopalan and Young (1975).

The so-called "black-gill" disease of rock crabs and lobsters is
probably aggrevated by flocculent sediment. Continuing histolgical
analyses of over 1,000 lobster and crab gills indicate that a wide
variety of particles and microorganisms clog the gills. To date,
only circumstantial evidence points to dumped materials as significant causes of this disease.

The causes of fin rot disease in the Bight, primarily found in
winter flounder, are still uncertain. A study now being completed
hypothesizes that PCB body burdens are related to fin rot and to
changes in other tissues (M. Sherwood and others, Southern California
Coastal Water Research Project, 1978, personal communication). However, it seems probable that any of several chemical stresses may
conspire to cause fin rot in winter flounder (R. Murchelano, NEFC,
Oxford Laboratory, 1978, personal communication). The incidence of
fin rot in winter flounder and other species of the Bight was unusually low in 1979 (less than 1%), in marked contrast to an incidence of
16% in the Apex during the spring of 1973 (Ziskowski and Murchelano,
1975).

The genetic damage to eggs and early larvae of Atlantic mackerel
(Longwell, 1976) is apparently correlated with the concentrations of
several contaminants in surface waters and surface slicks (Longwell,
NEFC, Milford Laboratory, personal communication, 1978). However,
existing sampling levels are probably inadequate to ascertain the degree of influence of dumped materials.

Ocean Dumping in the New York Bight -- in Perspective

In the New York Bight, studies of the ecosystem have demonstrated diverse effects of the contaminants in the wastes reaching the
Bight. Identification of the specific causes of most of the observed
effects has not generally been possible, because of: the variability
of the many inputs; the delay times for effects to be seen on the
ecosystem; the "mixing-bowl nature" of the Bight where contaminants
from specific sources lose their identity; and the natural fluctuations of the ecosystem which can, erroneously, be attributed to
people-related factors.

Without tracers for particular wastes and their various physical
and chemical fractions it is possible to obtain only some knowledge
of the effects of the sum of all wastes reaching the Bight on the
quality of the sediment and water, and the condition of the biota.
In a number of instances, particular contaminants are known to have
considerable effects on organisms. Where those particular contaminants can be associated with specific waste sources, management can
reduce the level of input. If one or more of the specific waste

sources are identified as recurring culprits, then attention can be focused on such culprits. As regards the New York Bight, the leading culprit most often identified is not ocean dumping but the outflow from the Hudson-Raritan Estuary. Therefore, the waste inputs to the Hudson-Raritan Estuary must be reduced if it is desired to improve the quality of the Bight ecosystem. At the same time, the quality of the estuarine ecosystems will be improved. It is probable that reduction of the volumes of the other inputs, or improvement in the quality of the other wastes will not result in any observable improvement in the New York Bight ecosystem as long as the Estuary situation is not addressed.

It should be recognized that many individual waste inputs to the Estuary must be controlled to effectively reduce the input of contaminants from the Estuary to the Bight. In comparison, source control of ocean dumped wastes is more readily accomplishable.

The evidence is solid that ocean dumping, as it is presently carried out in the New York Bight, is adversely affecting the ecosystem. Other wastes reaching the Bight are also adversely impacting the ecosystem, probably to a greater degree than ocean dumping. The quality of the New York Bight or the Hudson-Raritan Estuary can only be improved significantly if effective, integrated, waste management is implemented to progressively reduce contaminant inputs.

EFFECTS OF OCEAN DUMPING ON OTHER MARINE ECOSYSTEMS

The New York Bight ecosystem analysis experience allows the design of a general framework for considering or approaching similar waste management problems in coastal marine ecosystems elsewhere.

In any area of concern, a two-part initial effort must be conducted to assess the effects of ocean dumping on the ecosystem. First, the quantity and characteristics, particularly contaminant concentrations, of the major wastes entering the ecosystem must be examined. The variability of the contaminant inputs should be determined on time scales of the order of the mixing and dispersion time scales of the disposal ecosystem. While it would be informative to know how much of particular contaminants in the wastes are potentially available to the biota, efforts to obtain such definition in the field are very time consuming, difficult, and costly. However, laboratory experiments on the effects of specific kinds of wastes (and not particular contaminants within them) on organisms from the ecosystem can be very useful.

The second part of the initial effort is an assessment of the general character of the ecosystem under consideration. Important matters are the bottom topography, as it relates to potential accumulation of dumped or other wastes reaching the ecosystem, the residence times for waters in appropriate portions of the ecosystem, and the general energy levels of the waters, which affect transport, dispersion, erosion, and deposition of waste material. The variances of these controlling factors are most important. General knowledge of the structure of the ecosystem is necessary. In some cases this knowledge will be available from measurements made in the area; in others, extrapolation of information from similar ecosystems may be warranted. Again, an important matter is the degree of variability of the character of the ecosystem.

Exhaustive study of the wastes and the ecosystem are not necessary in order to project some expected effects of waste materials on the biotic and abiotic portions of the ecosystem. In the case where the waste disposal is already occurring, minimal field efforts could confirm (or deny) such projected effects.

REFERENCES

Achrem, T. J. (1973) Ocean waste disposal in the New York Bight. Report No. 4460C1559, Interstate Electronics Corporation, Oceanics Division.

Buelow, R. W., B. H. Pringle, and J. L. Verber (1968) Preliminary investigation of waste disposal in the New York Bight. U. S. Department of Health, Education and Welfare, Northeast Marine Health Sciences Lab.

Callaway, R. J., A. M. Teeter, D. W. Browne, and G. R. Ditsworth (1976) Preliminary analysis of the dispersion of sewage sludge discharged from vessels to New York waters. In: Am. Soc. Limnol. Oceanogr. Spec. Symp., 2, M. G. Gross, editor, pp. 199-211.

Charnell, R. L. and G. A. Maul (1973) An oceanographic observation of New York Bight from ERTS-1. NOAA TR ERL 262-AOML 9.

Dean, D. (1975) Raritan Bay macrobenthos survey, 1957-1960. NMFS Data Report 99.

Drake, D. E. (1974) Suspended particulate matter in the New York Bight Apex: September-November 1973. NOAA TR ERL 318-MESA 1.

Edwards, R. L. and R. E. Bowman (1979) An estimate of the food consumed by continental shelf fishes in the region between New Jersey and Nova Scotia. In: Predator-Prey Systems in Fisheries Management, Sport Fishing Institute, Washington, DC, pp. 387-406.

EPA (1973) Ocean dumping in the New York Bight, facts and figures. U. S. Environmental Protection Agency, Region II, Surveillance and Analysis Division, July 1973.

EPA (1975) Ocean disposal in the New York Bight, Technical Briefing Report Number 2. U. S. Environmental Protection Agency, Region II, Surveillance and Analysis Division, April 1975.

EPA (1976) Environmental impact statement on the ocean dumping of

sewage sludge in the New York Bight. Draft. U. S. Environmental Protection Agency, Region II, February 1976.

Federal Register (1977) U. S. Environmental Protection Agency,
Ocean Dumping, Final Revision of Regulations and Criteria,
January 11, 1977, Part VI.

Freeland, G. L. and G. F. Merrill (1976) Decomposition and erosion in
the dredge spoil and other New York Bight dumping areas.
Proceedings of the Specialty Conference on Dredging and Its
Environmental Effects, ASCE, Mobile, Alabama, January 26-28,
1976.

Freeland, G. L., D. J. P. Swift, and R. A. Young (1979) Mud deposits
near the New York Bight dumpsites: Origin and behavior.
In: Ocean Dumping and Marine Pollution: Geological Aspects of
Waste Disposal, H. D. Palmer and M. G. Gross, editors, Dowden,
Hutchinson, and Ross, Stroudsburg, PA, pp. 73-95.

Garside, C. and T. C. Malone (1978) Monthly oxygen and carbon budgets
of the New York Bight Apex. Estuarine Coast. Mar. Sci., 6,
93-104.

Goode, G. B. (1887) The fisheries and fishery industries of the
United States, Section 2. A Geographical Review of the
Fisheries Industries and Fishing Communities for the Year 1880.
U. S. Government Printing Office.

Gopalan, U. K. and J. S. Young (1975) Incidence of shell disease in
shrimp in the New York Bight. Mar. Poll. Bull., 6, 149-153.

Gross, M. G. (1970) Analysis of dredged wastes, fly ash and waste
deposits, New York metropolitan region. Marine Sciences
Research Center Technical Report 7, State University of New
York, Stony Brook, New York.

Gross, M. G., J. A. Black, R. J. Kalin, J. R. Schramel, and R. N.
Smith (1971) Survey of marine waste deposits, New York
metropolitan region. Marine Sciences Research Center Technical
Report 8, State University of New York, Stony Brook, New York.

Hansen, D. V. (1977) Circulation. MESA New York Bight Atlas
Monograph 3, MESA New York Bight Project and New York Sea Grant
Institute, October 1977.

Hatcher, P. G. and L. E. Keister (1976) Carbohydrates and organic
carbon in New York Bight sediments as possible indicators of
sewage contamination. In: Am. Soc. Limnol. Oceanogr. Spec.
Symp., 2, M. G. Gross, editor, pp. 240-248.

Hatcher, P. G. and P. A. McGillivary (1979) Sewage contamination in
the New York Bight: Coprostanol as an indicator. Env. Sci. and
Tech., 13 (10): 1225-1229.

Hatcher, P. G., G. A. Berberian, A. Cantillo, P. A. McGillivary, P.
Hanson, and R. H. West (1978) Chemical and physical processes in
a dispersing sewage sludge plume. Report submitted to the
MESA New York Bight Project.

Hetling, L., E. Horn, and J. Tofflemire (1978) Summary of Hudson
River PCB study results. New York State Department of
Environmental Conservation, Technical Paper #51.

Jacot, A. (1920) On the marine mollusca of Staten Island, N.Y.

Nautilus, 33, 111-115, 1919-20.

Ketchum, B. H., A. C. Redfield, and J. C. Ayers (1951) The oceanography of the New York Bight. Pap. Phys. Oceanogr. Meteorol., 12.

Ketchum, B. H., C. S. Yentsch, and N. Corwin (1958) Some studies of the disposal of iron wastes at sea. Woods Hole Oceanographic Institution, Ref. 58-7.

Klein, L. A., M. Lang, N. Nash, and S. L. Kirschner (1974) Sources of metals in New York City wastewater. Department of Water Resources, City of New York, January 1974.

Koditschek, L. and P. Guyre (1974) Antimicrobial-resistant coliforms in New York Bight. Mar. Poll. Bull., 5, 71-74.

Lee, G. F. and R. A. Jones (1977) An assessment of the environmental significance of chemical contaminants present in dredged sediments dumped in the New York Bight. Report submitted to the New York District Corps of Engineers.

Longwell, A. C. (1976) Chromosome mutagenesis in developing mackerel eggs sampled from the New York Bight. In: Am. Soc. Limnol. Oceanogr. Spec. Symp., 2, M. G. Gross, editor, pp. 337-339.

MacLeod, W. D., Jr., L. S. Ramos, A. J. Friedman, D. G. Burrows, P. G. Prohaska, D. L. Fisher, and D. W. Brown (1981) Analysis of residual chlorinated hydrocarbons, aromatic hydrocarbons, and related compounds in selected sources, sinks, and biota of New York Bight. NOAA, Office of Marine Pollution Assessment, Tech. Memo., Rockville, MD (in press).

Malone, T. C. (1977) Plankton systematics and distribution. MESA New York Bight Atlas Monograph 13, MESA New York Bight Project and New York Sea Grant Institute, May 1977.

MESA (1975) Ocean dumping in the New York Bight. NOAA TR ERL 321-MESA 2.

MESA (1978) MESA New York Bight Project Annual Report for Fiscal Year 1977, October 1978.

McDonough, K. B. (1976) A benthic index of environmental quality for the New York Bight Apex and Raritan Bay. Master of Science Thesis, Marine Environmental Sciences Program, State University of New York, Stony Brook, New York.

McGrath, R. (1974) Benthic macrofaunal census of Raritan Bay - Preliminary results, pap. 24. In: Proc. Symp. Hudson River Ecol. (3rd) March 1973, Hudson River Environ. Soc., Inc.

Mueller, J. A., J. S. Jeris, A. R. Anderson, and C. F. Hughes (1976) Contaminant inputs to the New York Bight. NOAA TM ERL MESA-6.

Mytelka, A. I., L. P. Cagliostro, D. J. Deutsch, and C. A. Haupt (1973) Combined sewer overflow study for the Hudson River Conference, Environmental Protection Agency - Region II, EPA-R2-73-152, January 1973.

Nelsen, T. A. (1979) Suspended particulate matter in the New York Bight Apex: Observations from April 1974 through January 1975. NOAA, TM ERL MESA-42, 78 pp.

NOAA (1977) Long Island beach pollution: June 1976. MESA Special Report, February 1977.

O'Connor J. S. (1976) Contaminant effects on biota of the New York
    Bight. Proceedings 28th Annual Session Gulf and Caribbean
    Fisheries Inst.
Pararas-Carayannis, G. (1973) Ocean dumping in the New York Bight: An
    assessment of environmental studies. U. S. Army Corps of
    Engineers, Tech. Memo, No. 39, May 1973.
Parker, J. H., I. W. Duedall, H. B. O'Connors, Jr., and R. E.
    Wilson (1976) Raritan Bay as a source of ammonium and
    chlorophyll a for the New York Bight Apex. In: Am. Soc. Limnol.
    Oceanogr. Spec Symp., 2, M. G. Gross, editor, pp. 212-219.
Pearce, J., C. MacKenzie, J. Caracciolo, and L. Rogers (1978)
    Reconnaissance survey of the distribution and abundance of
    benthic organisms in the New York Bight Apex 5-14 June 1973.
    NOAA DR ERL MESA-41.
Proni, J. R., F. C. Newman, R. A. Young, D. Walter, R. Sellers, P. A.
    McGillivary, P. G. Hatcher, I. Duedall, H. Stanford, and C.
    Parker (1978) Observations of the intrusion into a stratified
    ocean of an artificial tracer and the concomitant generation of
    internal oscillations. Presentation at A.G.U. Spring Meeting,
    May 1977, Washington, D.C.
Raytheon (1975a) Cruise 1 data report, baseline survey - New York
    Bight, September 24 - October 5, 1974, Volumes 1-5.
Raytheon (1975b) Cruise 2 data report, baseline survey - New York
    Bight, April 11 - May 2, 1975, Volumes 1-6
Redfield, A. C. and L. A. Walford (1951) A study of the disposal of
    chemical waste at sea. National Academy of Sciences - National
    Research Council Publication 201.
Sandy Hook Laboratory (1972) The effects of waste disposal in the New
    York Bight. Final report submitted to the U. S. Army Corps of
    Engineers, National Marine Fisheries Center.
Segar, D. A. and G. A. Berberian (1976) Oxygen depletion in the New
    York Bight Apex: Causes and consequences. In: Am. Soc. Limnol.
    Oceanogr. Spec. Symp., 2, M. G. Gross, editor, pp. 220-239.
Segar, D. A. and A. Y. Cantillo (1976) Trace metals in the New York
    Bight. In: Am. Soc. Limnol. Oceanogr. Spec. Symp., 2, M. G.
    Gross, editor, pp. 171-198.
Swanson, R. L., H. M. Stanford, J. S. O'Connor, S. Chanesman, C. A.
    Parker, P. A. Eisen, and G. F. Mayer (1978) June 1976 pollution
    of Long Island beaches. Jour. Env. Eng. Div., ASCE, 104,1067-
    1085.
Swanson, R. L. and C. J. Sindermann, editors (1979) Oxygen Depletion
    and Associated Benthic Mortalities in New York Bight, 1976.
    U.S. Government Printing Office.
Swift, D., A. Cok, D. Drake, G. Freeland, W. Lavelle, T. McKinney, T.
    Nelson, R. Permenter, and W. Stubblefield (1974) Sedimentation
    in the New York Bight Apex and application to problems of waste
    disposal: An interim assessment. In: Geological Oceanography
    Section of Technical Background Relating to Offshore Dumping
    Assessment, Interim report submitted to the MESA New York Bight
    Project, August 1974.

Timoney, J. F., J. Port, J. Giles, and J. Spanier (1978) Heavy-metal
    and antibiotic resistance in the bacterial flora of sediments of
    New York Bight. Applied and Environmental Microbiology, 36,
    465-472.
Walker, H. A., S. B. Saila, and E. L. Anderson (1979) Exploring
    data structure of New York Bight benthic data using
    post-collection stratification of samples, and linear
    discriminant analysis for species composition comparisons.
    Estuarine Coastal Mar. Sci., 9, 101-120.
West, R. H., P. G. Hatcher, and D. K. Atwood (1977) Polychlorinated
    biphenyls and DDTs in sediments and sewage sludge of the New
    York Bight. Report submitted to the MESA New York Bight
    Project.
West, R. H. and P. G. Hatcher (1980) Polychlorinated biphenyls in
    sewage sludge and sediments of the New York Bight. Mar. Poll.
    Bull., 11, 126-129.
Wiebe, P. H., G. D. Grice, and E. Hoagland (1973) Acid-iron waste as
    a factor affecting the distribution and abundance of zooplankton
    in the New York Bight II. Spatial variations in the field and
    implications for monitoring studies. Estuarine Coast. Mar.
    Sci., 1, 51-64.
Young, J. S. and J. B. Pearce (1975) Shell disease in crabs and
    lobsters from New York Bight. Mar. Poll. Bull., 6, 101-105.
Ziskowski, J. and R. Murchelano (1975) Fin erosion in winter
    flounder. Mar. Poll. Bull., 6, 26-29.

II

PHYSICAL ASPECTS OF OCEAN DUMPING

PHYSICAL VARIABILITY AT AN EAST COAST

UNITED STATES OFFSHORE DUMPSITE

James J. Bisagni[1] and Dana R. Kester[2]

[1]NOAA/NMFS
South Ferry Road
Narragansett, RI 02882

[2]Graduate School of Oceanography
University of Rhode Island
Kingston, RI 02881

ABSTRACT

Three hydrographic stations conducted during a 1977 cruise to
Deep Water Dumpsite 106 inadvertently sampled an established anticyclonic Gulf Stream ring both prior to and after it interacted with
the Gulf Stream. Continuous STD data and discrete dissolved oxygen
measurements in the ring were analyzed using temperature-salinity
(T-S) and temperature-oxygen (T-O_2) diagrams. These data showed
good correlation with T-S and T-O_2 diagrams from Gulf Stream, Sargasso and Slope Waters obtained during a cyclonic Gulf Stream ring
study. Apparently, a new entrainment of Gulf Stream Water around the
ring occurred within the sampling period and was manifested by increased salinity and decreased levels of dissolved oxygen. Combined
satellite surveillance and at-sea measurement of temperature, salinity and oxygen provide an accurate method of describing this highly
dynamic and variable region in and around Deep Water Dumpsite 106.

INTRODUCTION

The metropolitan areas and industries of the Middle Atlantic
Bight have used ocean dumping for many years to dispose of a variety
of wastes (Anderson and Dewling, 1981). Deep Water Dumpsite 106 has
received increasing amounts of chemical wastes in recent years.

Since May 1974, NOAA has been collecting physical oceanographic data in and near DWD-106 through a series of baseline and experimental cruises. Data of this type collected during the baseline cruises of May 1974, July 1975 and February 1976 have been reported by Warsh (1975), Goulet and Hausknecht (1977) and Bisagni (1977) respectively. Additional experimental study cruises during June 1976, August 1976 and July 1977 have provided more information on the hydrographic features of this region.

THE WESTERN NORTH ATLANTIC NEAR DWD-106

Oceanographically, DWD-106 lies within the Slope Water region, i.e., between the Gulf Stream and the Shelf Water (Fig. 1). Separating the Shelf and Slope waters is a temperature and salinity frontal zone, usually termed the Shelf/Slope Front. This front has been observed and discussed by Bigelow (1933), Bigelow and Sears (1935), Miller (1950), Cresswell (1967) and Bowman and Weyl (1972).

The surface position of the Shelf/Slope Front generally lies near the 200 m isobath. Based on 30 years of historical data however, Wright (1976) found a close correlation between the near-bottom position of the Front and the 100 m isobath. He used the 10°C isotherm as both the surface and near-bottom Frontal criterion. Gunn (1978) reports that in a region near DWD-106, the Front (defined by satellite infrared imagery) occurs shoreward of the 200 m isobath by about 10 km (1973-77 mean). The Front is most apparent during mid-winter when both its thermal and salinity gradients are large (Beardsley and Flagg, 1976). The thermal gradient across the Front decreases during the spring and summer due to increased surface heating which also causes a seasonal thermocline to develop on the shelf (Ingham et al., 1977).

Exchange of water across this seasonal thermocline is small. The colder near-bottom water or "cold cell" has been described by Bigelow (1933), Ketchum and Corwin (1964) and Cresswell (1967) and generally persists until seasonal mixing occurs. Salinity increases in a seaward direction across the shelf with a marked increase in the vicinity of the Front. The 34.5°/oo isohaline usually denotes the center of the Front.

The mean position of the Front varies from day to day due to certain exchange processes across it. Exchange of waters across the Shelf/Slope Front can occur as Shelf Water "calving", entrainment of surface Shelf Water and Slope Water intrusions. "Calving" has been described by Cresswell (1967) as a process in which the offshore edge of the Shelf Water "cold cell" detaches and moves into the Slope Water as a discrete parcel. Entrainment of surface Shelf Water across the Front has been documented by Bisagni (1977) and Morgan and Bishop (1977). This process results in tongues of Shelf Water

Wastes
in the
Ocean

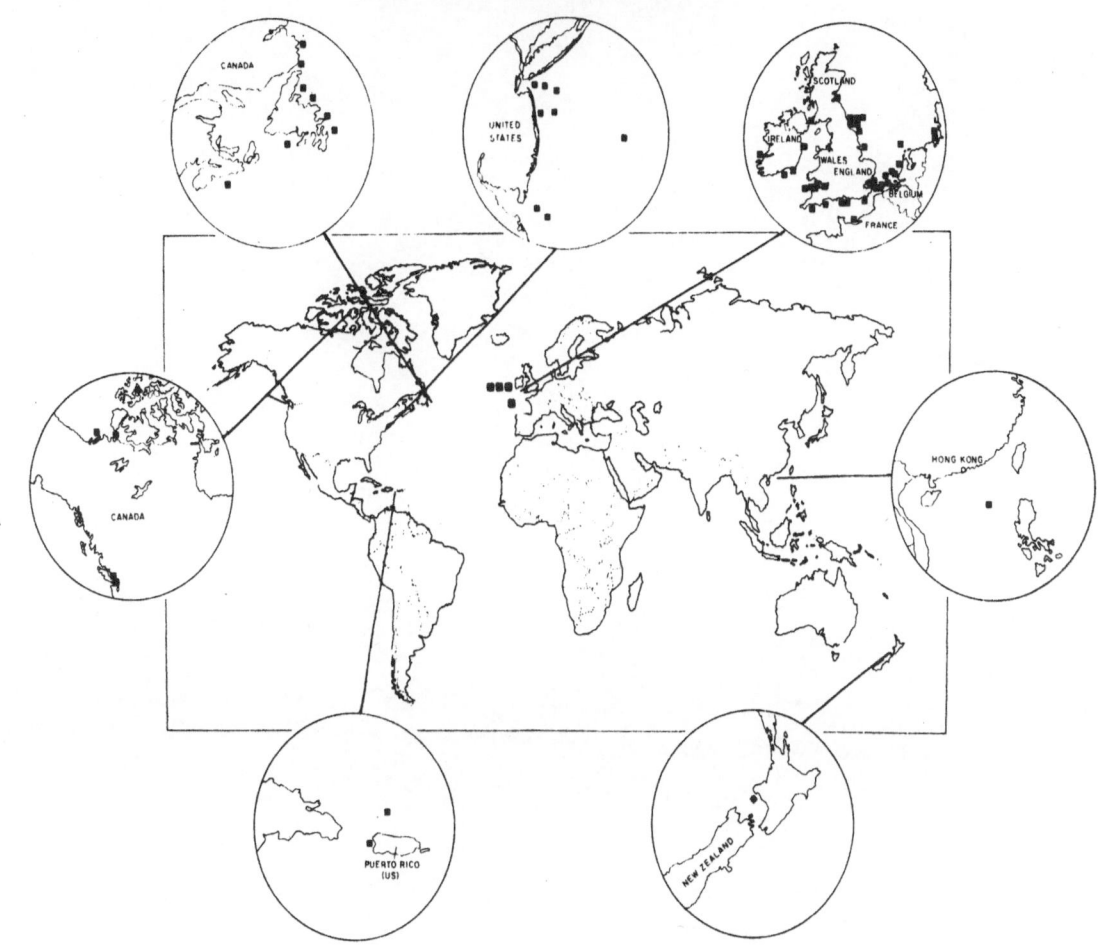

Volume 1: Industrial and Sewage Wastes
in the Ocean

Edited By: I.W. Duedall, B.H. Ketchum, P.K. Park, D.R. Kester

Wiley-Interscience

extending far into the Slope Water region due to the velocity shear
of an anticyclonic Gulf Stream eddy in close proximity to the Front.
Shoreward-moving Slope Water intrusions onto the shelf occur also
(Boicourt and Hacker, 1976). This process resulted from an offshore
movement of Shelf Water due to Ekman drift forced by strong southerly
winds. Any or all of these processes can modify the position of the
Shelf/Slope Front throughout the year.

A second frontal zone termed the "North Wall" lies generally
southeast of DWD-106 and separates the Slope Water region from the
Gulf Stream, (Fig. 1). Meandering of the North Wall has been revealed by many workers including Fuglister and Worthington (1951);

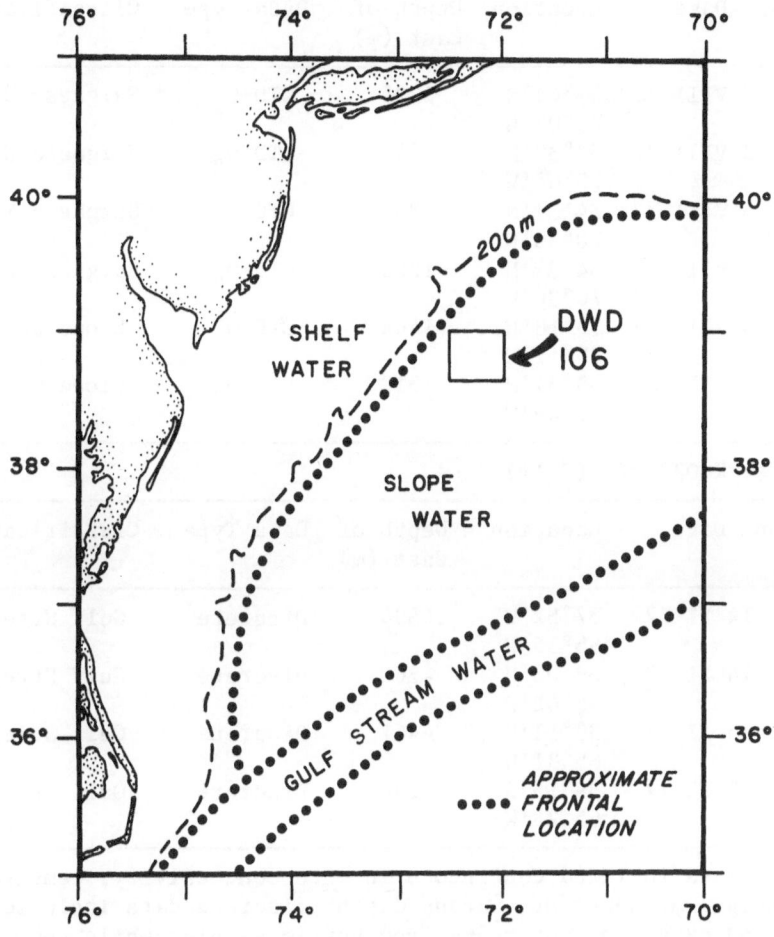

Fig. 1.   Schematic diagram of DWD-106 relative to the frontal
zones separating Shelf, Slope and Gulf Stream Waters.

Table 1. Hydrographic stations from the Cold-Core Gulf Stream Ring
          study used for comparison with data obtained at DWD-106.

Cruise KNORR 065 (Spring)

Station	Date	Location	Depth of Cast (m)	Data Type*	Classification
25	30 IV 77	39°21'N 69°43'W	627	CTD-O_2	Slope Water
26	30 IV 77	39°28'N 69°38'W	1127	CTD-O_2	Slope Water

Cruise ENDEAVOR 011 (Summer)

Station	Date	Location	Depth of Cast (m)	Data Type	Classification
6	2 VIII 77	33°41'N 73°07'W	1225	CTD-O_2	Sargasso Water
7	3 VIII 77	33°39'N 73°05'W	945	CTD-O_2	Sargasso Water
18	5 VIII 77	34°31'N 70°36'W	855	CTD-O_2	Sargasso Water
19	6 VIII 77	34°32'N 70°36'W	1263	CTD-O_2	Sargasso Water
37	5 VIII 77	38°18'N 69°11'W	1464	CTD-O_2	Slope Water
39	6 VIII 77	38°31'N 68°49'W	1512	CTD-O_2	Slope Water

Cruise KNORR 071 (Fall)

Station	Date	Location	Depth of Cast (m)	Data Type	Classification
25	14 XI 77	37°52'W 65°35'W	4594	Discrete	Gulf Stream
27	14 XI 77	39°03'N 65°42'W	4360	Discrete	Gulf Stream
28	15 XI 77	39°11'N 65°31'W	4461	Discrete	Gulf Stream
30	15 XI 77	39°15'N 65°26'W	4136	Discrete	Gulf Stream

*CTD-O_2 data included continuous in situ conductivity, temperature
and oxygen measurements versus depth; discrete data included salinity and oxygen measurements from bottle samples while temperature
data resulted from reversing thermometers.

Fuglister (1963) and Fuglister and Voorhis (1965), and since 1972 it has been evident in satellite infrared imagery. Extreme northward meandering of the North Wall has been shown to sometimes result in detachment of the meander from the Gulf Stream, forming an anticyclonic eddy (Saunders, 1971; Gotthardt, 1973; and Gotthardt and Potocsky, 1974). The movements of these eddies through the Slope Water region have been documented by Gotthardt and Potocsky (1974) while Bisagni (1976) discussed their presence and trajectories through the dumpsite. The eddies travel basically to the southwest or west at speeds which vary between 3 and 14 km per day. An important effect of the eddies is their importation of Gulf Stream and Sargasso Water into the Slope Water region. Mixing of these waters into the Slope Water may alter the corresponding Slope Water T-S characteristics (Bisagni, 1977).

The Slope Water region, including DWD-106, is a complex dynamic area of the western North Atlantic. The occurrence of oceanic fronts near DWD-106 along with their movements caused by various cross-frontal exchange processes may allow a variety of water types to be present at the site. This spatial variability at DWD-106 has been well documented with many synoptic vertical temperature, salinity and density sections, often revealing a complex, stratified water column. However, biological and chemical studies at the site have often required a simplified classification regarding water masses. This has been difficult to accomplish, especially for the near-surface portion of the water column. Empirically derived representations of water masses using T-S, T-O_2 and σ_t-O_2 diagrams have been used by many workers. Wright and Worthington (1970) developed a volumetric temperature and salinity census for the water masses contained in the 5 basins of the North Atlantic. However, the Slope Water region was not given detailed enough treatment in their census. Wright and Parker (1976) discussed volumetric T-S diagrams of the Middle Atlantic Bight and identified Slope Water in both their winter and summer diagram. Oxygen-density diagrams in the western North Atlantic were studied by Richards and Redfield (1955) and allowed discrimination between the Gulf Stream and the Sargasso Sea. This was based on the anomalously low oxygen content of the Gulf Stream. Lambert (1974) easily discriminated between a cyclonic Gulf Stream ring and the surrounding Sargasso Sea also on this basis. The use of T-S and T-O_2 diagrams in this paper will allow discrimination between water masses at DWD-106 and permit their classification.

METHODS

A two ship cruise operation was conducted between July 20-29, 1977 to determine the dispersal pattern, chemical interactions and biological effects of two waste dumps at DWD-106. Supplemental oceanographic measurements were made in the vicinity of the dumpsite and a control area aboard the FRS ALBATROSS IV and the OSS PEIRCE. The

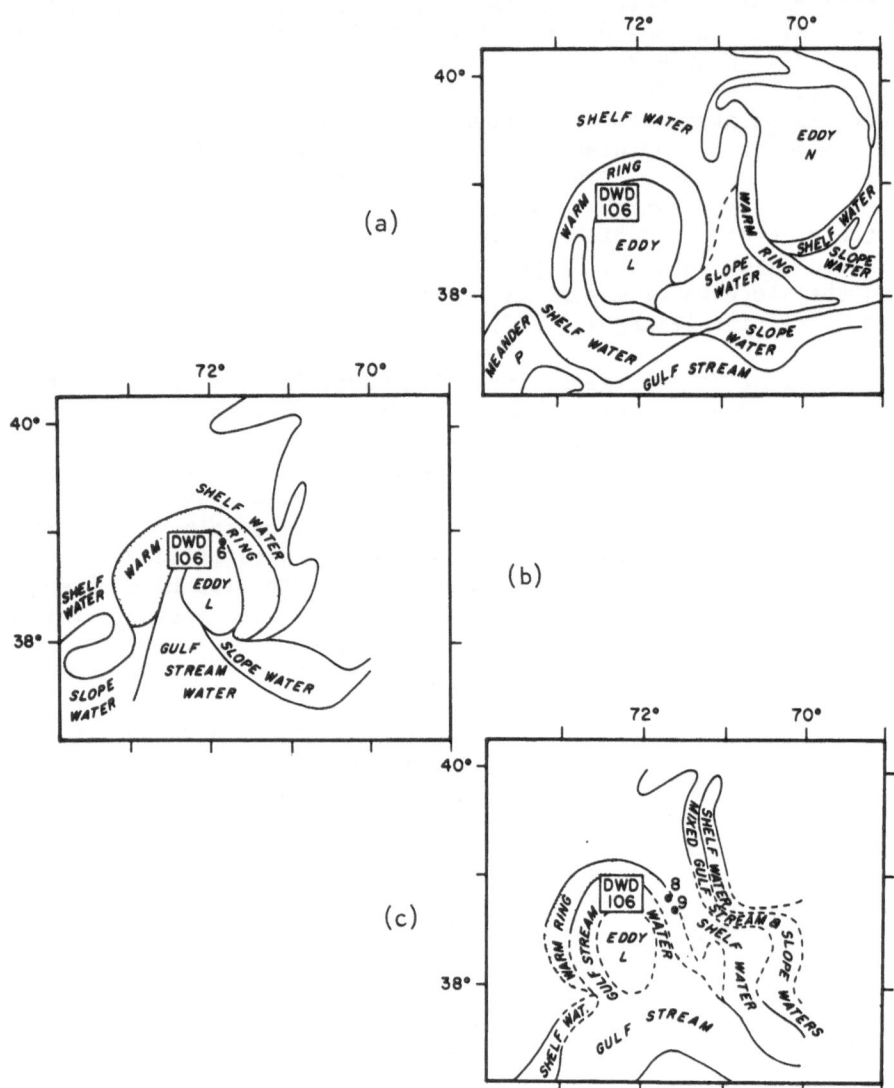

Fig. 2a, b, c.   NAVOCEANO Experimental Ocean Frontal Analysis of
                 VHRR Satellite Imagery. Water mass boundaries are
                 based on sea-surface temperature.
                     a.   July 20, 1977
                     b.   July 24, 1977
                     c.   July 27, 1977

surface hydrography was described both prior to and during the cruise by utilizing very high resolution radiometer (VHRR) imagery from the NOAA 5 satellite prepared daily by NAVOCEANO's Applications Research Division.

Hydrographic stations used a continuous profiling Plessey 9040 STD which produced analog traces of salinity and temperature versus depth. This was used in conjunction with a General Oceanics "rosette" sampler and 8 liter Niskin bottles with reversing thermometers. Salinity, temperature and depth were corrected using the discrete sample and thermometer data. Oxygen values were measured from discrete samples using the modified Winkler technique (whole bottle method). Three of the STD hydrographic stations conducted from ALBATROSS were taken to 1000 m depth or more, while the remainder were not taken deeper than 200 m.

A total of 12 hydrographic stations from 3 cruises which were part of a multidisciplinary study of cyclonic Gulf Stream rings were

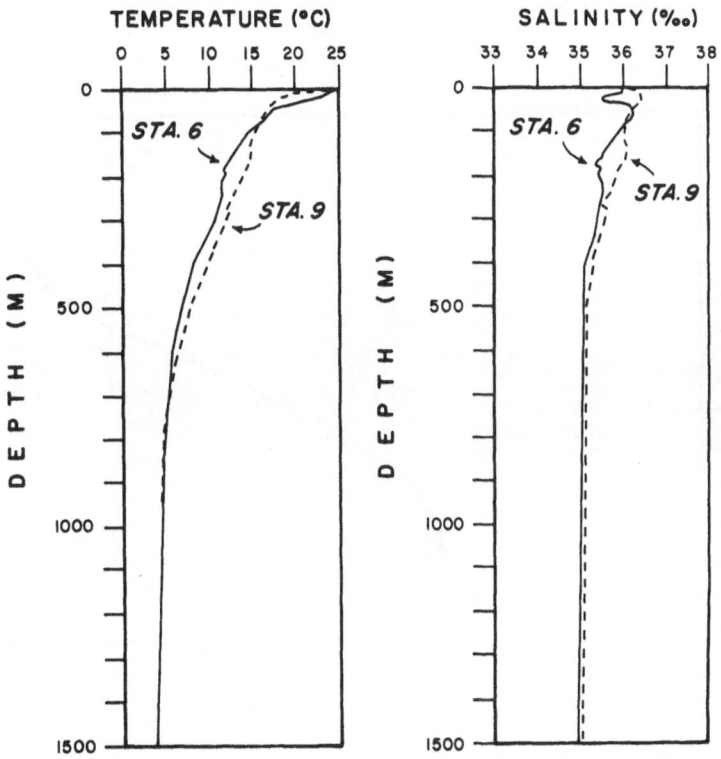

Fig. 3.   Temperature and salinity profiles for ALBATROSS IV
          stations 6 and 9.

selected to characterize Sargasso, Gulf Stream and Slope Waters
(Table 1). These stations were conducted during cruises KNORR 065,
ENDEAVOR 011 and KNORR 071 to specifically sample each of these water
types to provide comparison with stations in and near cyclonic Gulf
Stream rings in the western North Atlantic. Comparison of these data
with data from anticyclonic Gulf Stream ring stations 6, 8 and 9 from
the summer 1977 ALBATROSS IV cruise to DWD-106 may provide a way of
describing the complicated hydrography noted at the site.

    ALBATROSS IV 77-05 stations 6, 8 and 9 should be analyzed using
strictly summer data for comparison. However, the limited number of
Slope Water and Gulf Stream stations from the ENDEAVOR 011 summer effort necessitated the use of additional stations from KNORR 065
(spring) and 071 (fall). Seasonal differences between data from
KNORR 065 and ENDEAVOR 011 can be virtually eliminated from KNORR 065
Slope Water stations 25 and 26 by using only data from below the
depth of 13°C isotherm. This depth in the Slope Water was perceived
to be the limit of seasonal influence based upon a volumetric T-S diagram for the region and has been termed the "upper Slope Water ther-
mostad" by Wright and Parker (1976). Additional Gulf Stream stations
from KNORR 071 conducted in the fall of 1977 may be compared with the

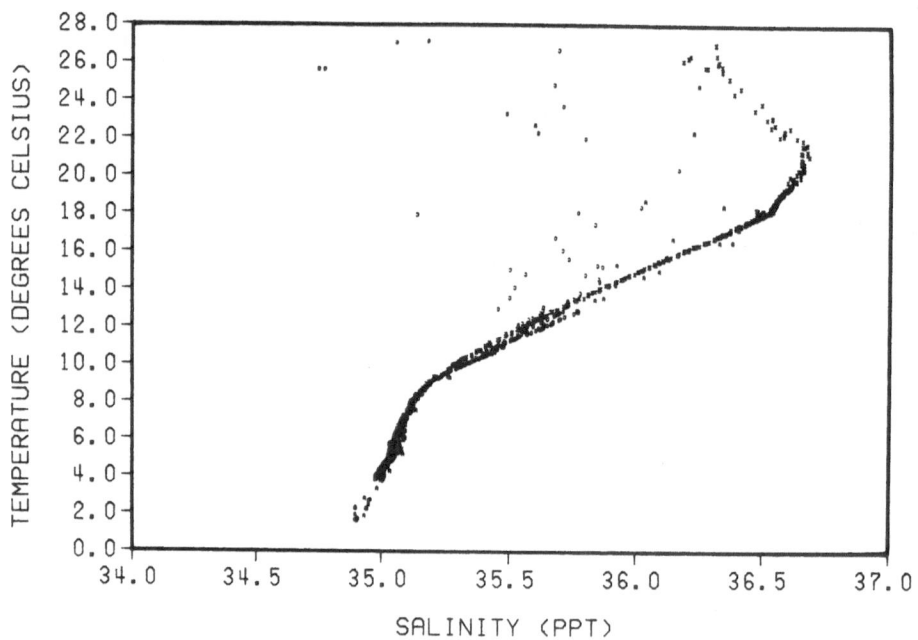

Fig. 4. Composite temperature-salinity (T-S) diagram of 12
         comparison stations from ENDEAVOR 011, KNORR 065 and
         KNORR 071.

summer ENDEAVOR effort because the Gulf Stream shows minimal sea
surface temperature variation from season to season. Also, complete
winter cooling had not yet occurred.

The locations of the 12 comparison hydrographic stations were
plotted on the NAVOCEANO interpretation of NOAA 5 satellite VHRR
imagery for the date each station was conducted. Each station plotted within the water type it was supposed to represent based on the
interpretation of the imagery. A total of 1249 temperature, salinity
and dissolved oxygen measurements (CTD-O_2 and discrete bottle and
thermometer data) from these stations were used in the analysis with
32% defining the Sargasso Sea Water, 4% defining Gulf Stream Water,
and 65% characterizing Slope Water.

RESULTS

Comparison between the NAVOCEANO interpretations of VHRR imagery
and ALBATROSS stations 6, 8 and 9 provides some information on the
hydrographic regime present at DWD-106. The imagery revealed a complex assemblage of waters including the Gulf Stream, a warm intrusion, and a portion of eddy L (ring L) (Figs. 2a-c). Eddy L moved

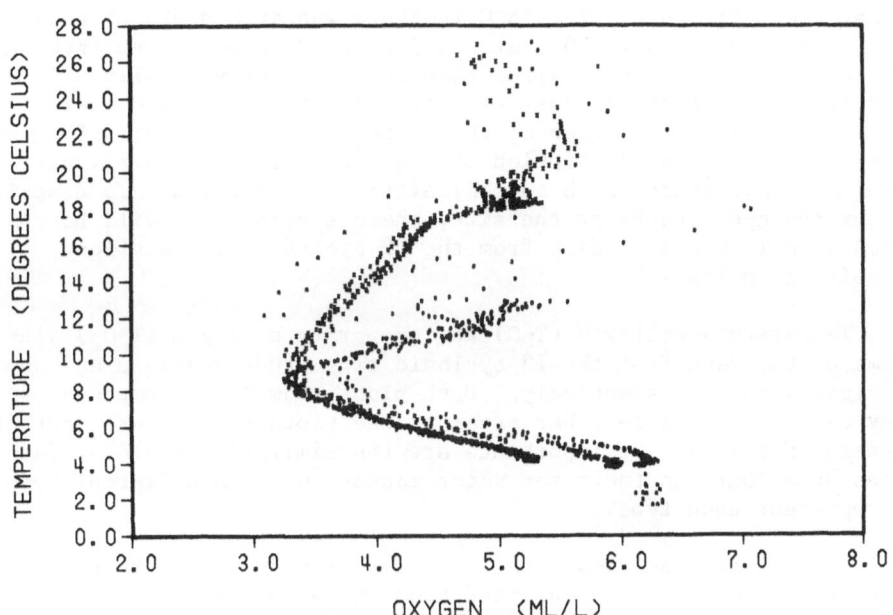

Fig. 5. Composite temperature-oxygen (T-O_2) diagram of 12
comparison stations from ENDEAVOR 011, KNORR 065 and
KNORR 071.

Table 2. Data summary for hydrographic stations 6, 8 and 9 ALBATROSS
          IV 77-05; all data were discrete observations.

Station	Date (GMT)	Time (GMT)	Location	Depth of Cast (m)
6	24 VII 77	1940	38°56'N, 71°50'W	1500
8	27 VII 77	1430	38°48'N, 71°41'W	20
9	27 VII 77	2145	38°42'N, 71°36'W	1500

approximately 10 nautical miles to the west between July 24 and 27
(Figs. 2b and c). Consequently, station 6 was conducted near the
edge of the eddy core, but stations 8 and 9 were in the warm ring
surrounding the core. Also, a closed entrainment of Gulf Stream
Water around the margin of eddy L is shown in Fig. 2c while in Fig.
2b the entrainment feature was not present. Subsurface temperature
and salinity profiles from ALBATROSS stations 6 and 9 (Fig. 3) show
variations which may be attributable to these changes in the position
of eddy L. Between 80 and 700 m, station 9 is approximately 2.5°C
warmer than station 6. The 15°C isotherm was at a depth of about 150
m at station 9 but only 90 m at station 6. Between the surface and
1500 m station 9 is more saline than station 6 at virtually all
depths. Water of 36 °/$_\circ$$_\circ$ salinity extended to almost 180 m at
station 9, while at station 6, this water lay in regions; 0-10 m and
40-80 m. However, a description of the water masses is difficult
based on temperature-depth and salinity-depth profiles. To describe
better the hydrography at the site, these station data will be compared with T, S and O_2 data from the 12 cyclonic ring stations
identified in Table 1.

Temperature-salinity (T-S) and temperature-oxygen (T-O_2) diagrams of the data from the 12 cyclonic ring study stations are shown
in Figs. 4 and 5 respectively. Both plots show limbs composed of
many data points while other areas of the plots show a more scattered
pattern of points. Of importance are the similarities and differences shown between the three water masses, using a different symbol
to represent each type.

The T-S diagram shows that for temperatures warmer than 12°C,
Slope Water can be differentiated from Sargasso Sea Water because of
the large difference in salinity between them; this is consistent
with a T-S diagram from a cyclonic ring reported by Lambert (1974).
For temperatures less than 12°C, Slope Water cannot be discerned from
Sargasso Water based on T-S analysis. However, when the T-O_2 relationships of Slope and Sargasso Waters are compared (Fig. 5) they
form separate limbs for temperatures down to 9°C. At the surface

dissolved oxygen values are similar (5.0 ml/l) due to air-sea exchange and mixing. At 9°C an oxygen minimum occurs (3.2 ml/l) where
the limbs join. The T-O_2 diagram for temperatures colder than 9°C
shows little differentiation between these water masses. Near surface Sargasso and Slope Waters (>19°C) can be discriminated using the
salinity difference while between 19°C and 9°C the greater dissolved
oxygen values of Slope Water are more easily discernable.

     Similarly, discrimination between Slope and Gulf Stream Waters
at temperatures above 12°C can be made using the difference in salinity. Between 9°C and 19°C, the strong oxygen depletion of Gulf
Stream Water relative to Slope water can be used to distinguish these
waters. At temperatures less than 9°C however, the variation in the
Slope Water T-O_2 data is too great to allow separation of the Gulf
Stream and Slope Waters. The T-S data likewise shows little difference between these waters.

     The T-S data shows little, if any, variation between Sargasso
and Gulf Stream Waters. However at temperatures greater than 12°C,
but less than 19°C, the T-O_2 diagram indicates that Gulf Stream

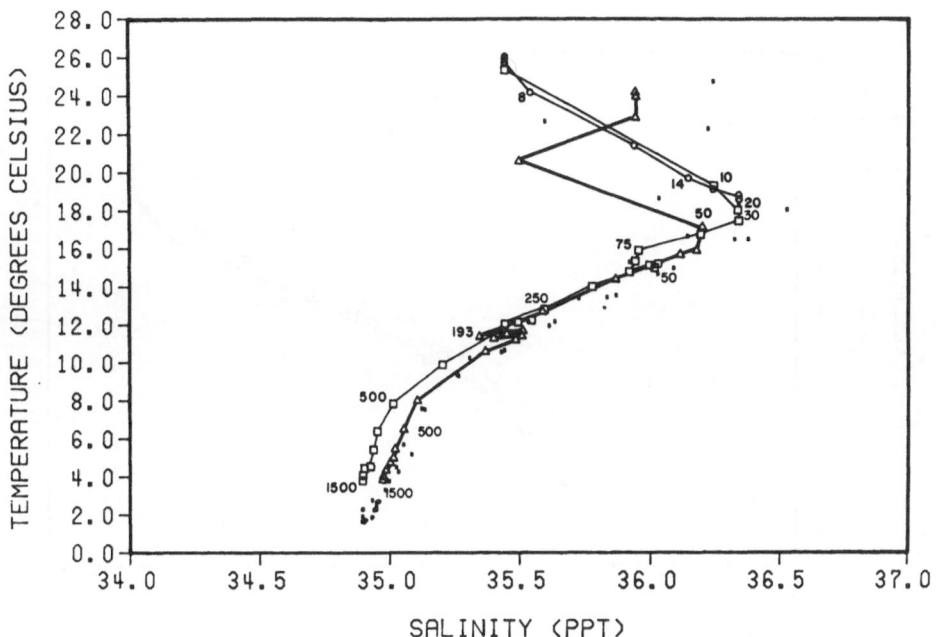

Fig. 6a.   Comparison of temperature-salinity correlations from
           ALBATROSS IV 77-05 stations 6 (triangles), 8 (circles)
           and 9 (squares) with Gulf Stream (a), Sargasso (b) and
           Slope Water (c) data of KNORR 071.

Water is depleted in oxygen relative to Sargasso Water. This depletion of approximately 0.5 ml/l is at least one order of magnitude greater than the uncertainty of both the <u>in situ</u> oxygen probe and the modified Winkler technique for discrete bottle samples. Oxygen depletion of up to 1.5 ml/l for Gulf Stream water relative to Sargasso Sea Water was measured at 3 crossings of the Gulf Stream in 1950 south of New England (Richards and Redfield, 1955).

ANALYSIS OF SUMMER ALBATROSS IV DATA FROM DWD-106

Table 2 summarizes the ALBATROSS IV data at DWD-106. Station 6, at the edge of the core of eddy L, appeared to be a mixture of Slope and Sargasso waters above 50 m depth based on the T-S correlation (Fig. 6a,b,c). From 50 to 250 m depth the T-S data indicate the presence of either Slope or Sargasso waters. The $T-O_2$ correlation however showed that between 50 and 100 m depth station 6 plotted close to the Sargasso curve (Fig. 7a,b,c). Between 193 and 250 m depth, the T-S correlations show a change. This change appeared to result from an intrusion of Slope Water at about those depths shown by the 250 m point on the $T-O_2$ correlation. Below 250 m at station

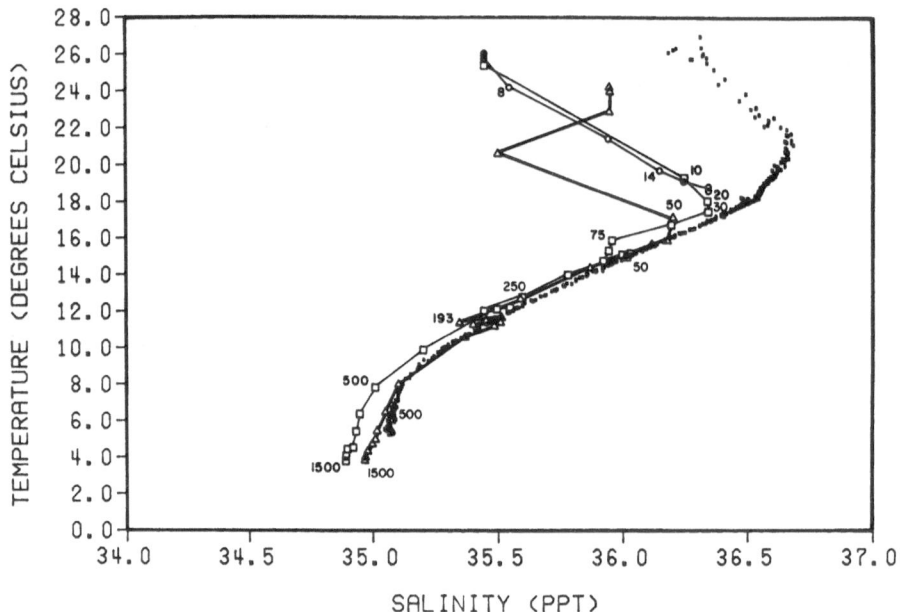

Fig. 6b. Comparison of temperature-salinity correlations from ALBATROSS IV 77-05 stations 6 (triangles), 8 (circles) and 9 (squares) with Gulf Stream (a), Sargasso (b) and Slope Water (c) data from ENDEAVOR 011.

6 the temperature was less than 9°C which precluded the distinction
of water masses.

The T-S correlation between the surface and 20 m depth at station 8 showed a large change. Based upon the comparison T-S data,
Slope Water occurred above 8 m at station 8, and became almost pure
Sargasso water at 20 m. Based upon the T-O_2 correlation however
the water below 8 m depth was depleted in oxygen relative to Sargasso
Sea water and plotted close to the Gulf Stream data. A possible explanation is that Slope Water overlay Gulf Stream water to a depth of
8-10 m. Below 14 m depth, water composed entirely of Gulf Stream
Water was present based on the high salinity T-S correlation and the
measured oxygen depletion.

A similar fluctuation also occurred in the T-S correlation of
the upper 20 m at station 9. Slope Water at the surface changed abruptly with depth to high salinity, low oxygen Gulf Stream at 20 m.
Oxygen was not sampled for the near surface level at station 9. However, dissolved oxygen at 10 m was depressed relative to Sargasso Sea
Water values. A sub-surface oxygen minimum of 3.57 $\pm$ 0.03 ml/1

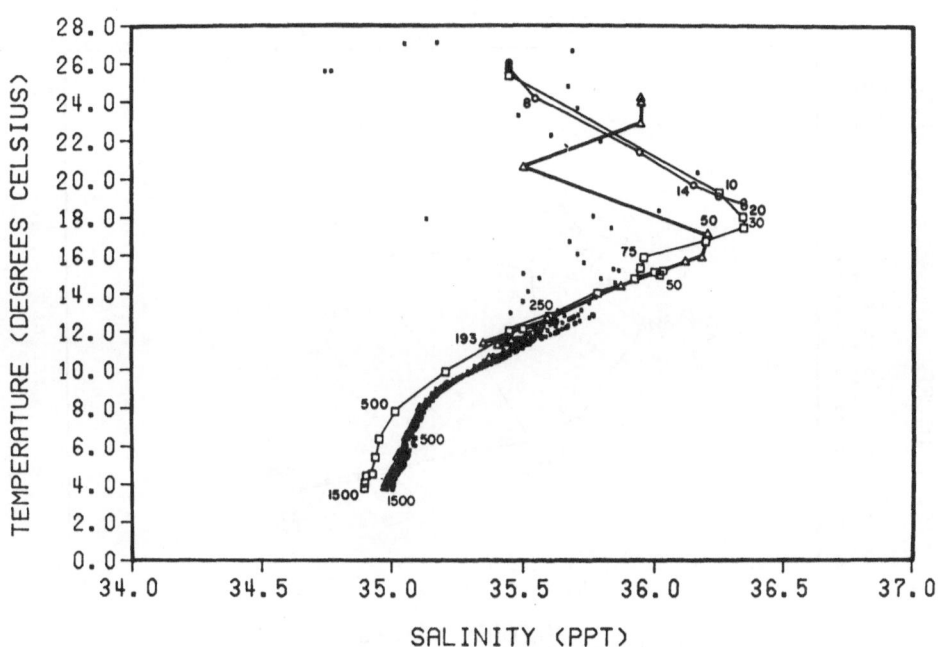

Fig. 6c.   Comparison of temperature-salinity correlations from
           ALBATROSS IV 77-05 stations 6 (triangles), 8 (circles)
           and 9 (squares) with Gulf Stream (a), Sargasso (b) and
           Slope Water (c) data from ENDEAVOR 011 and KNORR 065.

occurred at 30 m and corresponds to a salinity maximum of 36.5 $\pm$ 0.1 °/oo. These points provide an almost perfect fit to the T-O$_2$ correlation of Gulf Stream Water. At depths of 50 to 100 m this correlation shows a deviation towards the Sargasso Water curve. Between 150 and 1500 m the T-S correlation shows that station 9 is 0.1°/oo less saline than station 6 and also the comparison data. The T-O$_2$ correlation, however, allows station 9 to be discerned from station 6 to a depth of only 250 m because the deeper T-O$_2$ correlations for these stations lay below the depth of the deep oxygen minimum. This 0.1°/oo salinity difference of station 9 within the deeper waters is an artifact of an instrument scale setting which allowed a $\pm$ 0.1°/oo uncertainty. This uncertainty is least important near the surface, but becomes increasingly more important at depths where the normal variation in salinity is small.

SUMMARY AND CONCLUSIONS

The Slope Water region of the western North Atlantic is an area of strong oceanic frontal dynamics as shown by many workers. Deep Water Dumpsite 106 located within this region is subjected to these

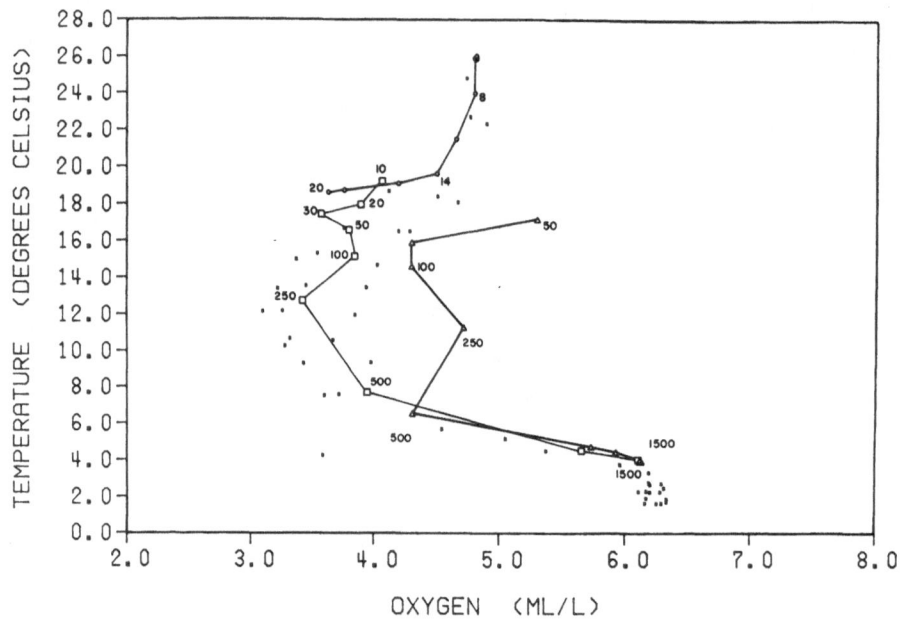

Fig. 7a. Comparison of temperature-oxygen correlations from ALBATROSS IV 77-05 stations 6 (triangles), 8 (circles) and 9 (squares) with Gulf Stream (a), Sargasso (b) and Slope Water (c) data from KNORR 071.

frontal movements and cross frontal exchange processes. Physical
oceanographic measurements conducted during baseline and experimental
cruises to the site have been collected since 1974. Detailed water
mass analysis techniques however were never adequately applied to the
data set because of lack of comparison data, especially above 200 m
depth.

     New data from a series of oceanographic cruises which were con-
ducted to study cyclonic Gulf Stream rings have been compared with
satellite VHRR sea-surface temperature interpretations. These data
were analyzed and yielded 12 stations, which were conducted in one of
3 water masses: Slope, Gulf Stream or Sargasso Waters. Composite T-S
and T-O_2 diagrams of these stations were analyzed. Internal comparisons between the 12 stations based on the T-O_2 and T-S diagrams
showed:
     1. The similarity of the Slope, Sargasso and Gulf Stream Water
T-O_2 correlations below 9°C (the depth of the deep oxygen minimum)
and above 19°C.
     2. The distinct T-O_2 correlation for each water mass between
9°C and 19°C.

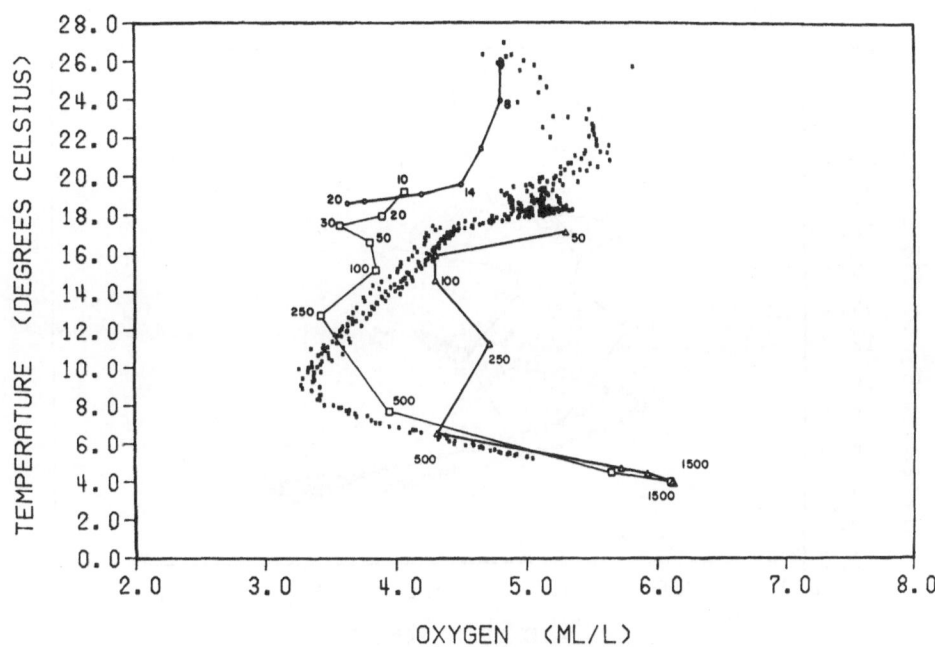

Fig. 7b. Comparison of temperature-oxygen correlations from
          ALBATROSS IV 77-05 stations 6 (triangles), 8 (circles)
          and 9 (squares) with Gulf Stream (a), Sargasso (b) and
          Slope Water (c) data from ENDEAVOR 011.

   3. The similarity of Slope, Sargasso, and Gulf Stream Water T-S
correlations below 12°C.
   4. The large differences between Slope and Sargasso Waters and
Slope and Gulf Stream Waters above 12°C using the T-S correlation.
   5. The similar T-S correlations of Sargasso Water and Gulf
Stream Water.

   The application of the above relationships to data from the July
1977 cruise at DWD-106 aboard the ALBATROSS IV showed differences in
the hydrographic regime near the site. Station 6 at the edge of the
anticyclonic ring L's core showed core characteristics by an absence
of Gulf Stream water and the presence of a 50 m thick layer of pure
Sargasso Water. An intrusion of Slope Water occurred at 250 m depth.
In contrast to station 6, stations 8 and 9 were conducted near ring
L's new Gulf Stream entrainment observed on July 27 in satellite
imagery. These stations showed almost pure Gulf Stream Water present
from about 20 m depth to the depth of the oxygen minimum. Based on
these data, an older but still energetic ring can re-entrain the Gulf
Stream after possibly losing its initial formational Gulf Stream
water. Above 10 m depth, stations 8 and 9 were composed of Slope
Water (based on the T-S correlation).

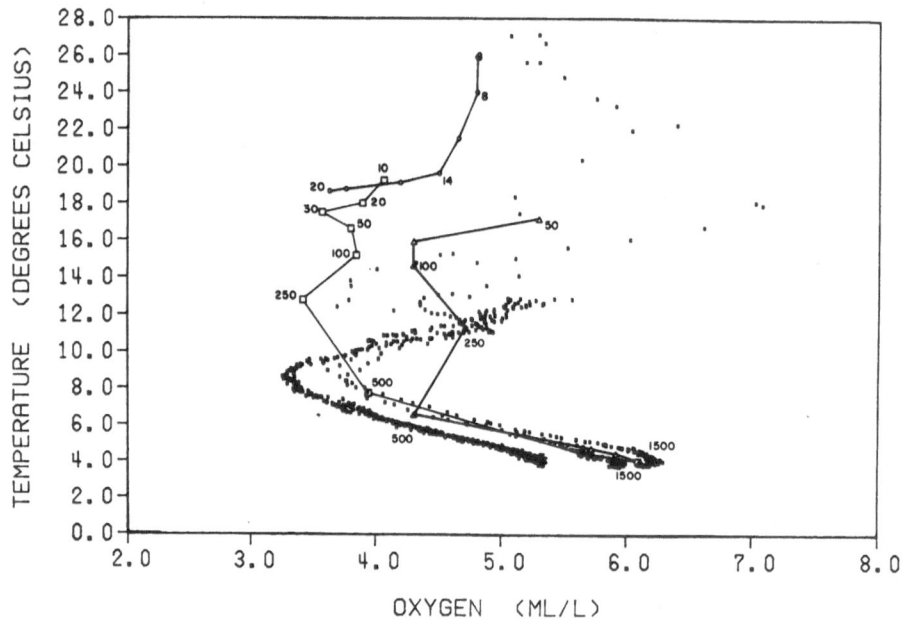

Fig. 7c. Comparison of temperature-oxygen correlations from
          ALBATROSS IV 77-05 stations 6 (triangles), 8 (circles)
          and 9 (squares) with Gulf Stream (a), Sargasso (b) and
          Slope Water (c) data from ENDEAVOR 011 and KNORR 065.

The extreme variation in the hydrography at DWD-106 shown by these data requires that in future work, continuous temperature, salinity and oxygen measurements (STDO) be made on a routine basis. Comparison of these measurements with more refined data from each individual water mass could produce detailed hydrographic representations of the site. The water-mass interactions discussed above could have an important effect on industrial waste plume movement and dispersion. More importantly, however, each water mass possesses a characteristic assemblage of biota which might be expected to behave differently in the presence of toxic waste, thus making water mass identification a necessary feature of future impact studies and monitoring programs.

ACKNOWLEDGEMENTS

We would like to thank Mr. Talbot E. Murray for his diligent work in providing the necessary corrections to the data sets and the plots using the University of Rhode Island computer facility. Special thanks and appreciation is due Mr. Rudolph Perchal of NAVOCE-ANO's Applications Research Division for his consistently high quality VHRR satellite imagery interpretations. Finally, we would like to extend thanks to the crew of the FRS ALBATROSS IV NOAA/NMFS for their efforts during the cruise and to S. Agronick who typed the manuscript.

REFERENCES

Anderson, P. W. and R. T. Dewling (1981) Industrial ocean dumping in EPA Region II - Regulatory Aspects. In: Ocean Dumping of Industrial Wastes, B. H. Ketchum, D. R. Kester, and P. K. Park, editors, Plenum Press, New York. This volume, pp. 25-37.

Beardsley, R. C. and C. N. Flagg (1976) The water structure, mean currents and Shelf-Water/Slope-Water Front on the New England Continental Shelf. Mem. Soc. Roy. Sci. de Liege, 6, X, 209-225.

Bigelow, H. B. (1933) Studies of the waters on the Continental Shelf, Cape Cod to Chesapeake Bay I the Cycle of Temperature. Pap. Phys. Oceanog. Met. V. II, n. 4, 103 pp.

Bigelow, H. B. and M. Sears (1935) Studies of the waters on the Continental Shelf, Cape Cod to Chesapeake Bay II Salinity. Pap. Phys. Oceanog. Met. V. IV, n. 1, 91 pp.

Bisagni, J. J. (1976) Passage of anticyclonic Gulf Stream eddies through Deepwater Dumpsite 106 during 1974 and 1975. NOAA Dumpsite Eval. Report 76-1, U. S. Dept. of Commerce Pub., 39 pp.

Bisagni, J. J. (1977) Physical oceanography of Deepwater Dumpsite 106, February-March 1976. In: Baseline Report of Environmental Conditions in Deepwater Dumpsite 106. NOAA Dumpsite Eval. Rept. 77-1, U. S. Dept. of Commerce Pub., 87-115.

Boicourt, W. C. and P. W. Hacker (1976) Circulation on the Atlantic
    Continental Shelf of the United States, Cape May to Cape
    Hatteras. Mem. Soc. Roy. Sci. de Liege, 6, X, 187-200.
Bowman, M. J. and P. K. Weyl (1972) Hydrographic study of the shelf
    and slope waters of New York Bight. State University of New
    York, Mar. Sci. Res. Center Tech. Rept. Series #16.
Cresswell, G. M. (1967) Quasi-synoptic monthly hydrography of the
    transition region between coastal and slope water south of Cape
    Cod, Massachusetts. Woods Hole Oceanographic Institution, Ref.
    67-35.
Fisher, A. (1973) Environmental guide to the Virginia Capes Operating
    Area. U.S. Naval Oceanographic Office, Special Publ. 211,
    58 pp.
Fuglister, F. C. (1963) Gulf Stream '60. Prog. Oceanogr., 1,
    265-283.
Fuglister, F. C. and A. D. Voorhis (1965) A new method of tracking
    the Gulf Stream. Limnol. Oceanogr., Suppl. 10, 115-124.
Fuglister, F. C. and L. V. Worthington (1951) Some results of a
    multiple ship survey of the Gulf Stream. Tellus, 3, 1-14.
Gotthardt, G. A. (1973) Observed formation of a Gulf Stream
    anticyclonic eddy. J. Phys. Oceanogr., 3, (2), 237-238.
Gotthardt, G. A. and G. J. Potocsky (1974) Life cycle of a Gulf
    Stream anticyclonic eddy observed from several oceanographic
    platforms. J. Phys. Oceanogr., 4, (1), 131-134.
Goulet, J. R., Jr. and K. A. Hausknecht (1977) Physical oceanography
    of Deepwater Dumpsite 106, Update: July 1975. In: Baseline
    Report of Environmental Conditions in Deepwater Dumpsite 106.
    NOAA Dumpsite Eval. Rept. 77-1, U.S. Dept. of Commerce Pub.,
    pp 55-86.
Gunn, J. T. (1978) Variation in the Shelf Water front position in
    1977 from Georges Bank to Cape Romain. Annales Biologique 34,
    36-39.
Ingham, M. C., J. J. Bisagni and D. Mizenko (1977) The general
    physical oceanography of Deepwater Dumpsite 106. In: Baseline
    Report of Environmental Conditions in Deepwater Dumpsite 106.
    NOAA Dumpsite Eval. Rept. 77-1, U. S. Dept. of Commerce Pub.,
    pp 29-54.
Ketchum, B. H. and N. Corwin (1964) The persistence of winter water
    on the continental shelf south of Long Island, N. Y. Limnol.
    Oceanogr., 9, (4), 467-475.
Lambert, R. B. (1974) Small-scale dissolved oxygen variations and
    the dynamics of Gulf Stream eddies. Deep-Sea Res., 21,
    529-546.
Miller, A. R. (1950) A study of the mixing processes over the edge
    of the continental shelf. J. Mar. Res., 9, (2), 145-159.
Morgan, C. W. and J. M. Bishop (1977) An example of Gulf Stream
    eddy-induced water exchange in the Mid-Atlantic Bight. J.
    Phys. Oceanogr., 7, (5), 472-479.
Richards, F. A. and A. C. Redfield (1955) Oxygen-density

relationships in the western North Atlantic. Deep-Sea Res., 2, 182-199.

Saunders, P. M. (1971) Anticyclonic eddies formed from shoreward meanders of the Gulf Stream. Deep-Sea Res., 18, 1207-1219.

Warsh, C. E. (1975) Physical oceanographic observations at Deepwater Dumpsite 106, May 1974. In: May 1974 Baseline Investigation of Deepwater Dumpsite 106. NOAA Dumpsite Eval. Rept. 75-1, U.S. Dept. of Commerce and U. S. Environmental Protection Agency Pub., pp. 141-188.

Wright, W. R. and L. V. Worthington (1970) The water masses of the North Atlantic Ocean -- a volumetric census of temperature and salinity. Folio 19 Serial Atlas of the Marine Environment, Amer. Geog. Soc. New York, 8 pp.

Wright, W. R. (1976) The limits of shelf water south of Cape Cod, 1941 to 1972. J. Mar. Res., 34, (1), 1-14.

Wright, W. R. and C. E. Parker (1976) A volumetric temperature/ salinity census for the Middle Atlantic Bight. Limnol. Oceanogr., 21, (4), 563-571.

AN ANALYSIS OF DUMPSITE DIFFUSION EXPERIMENTS

G. T. Csanady

Woods Hole Oceanographic Institution
Woods Hole, MA 02543

ABSTRACT

The oceanic dispersion of industrial waste barged to Deep Water
Dumpsite 106 is effected in the first instance by barge-bound eddies
("wake dispersion"), then by naturally occurring mean shear and turbulence. Wake dispersion is very efficient, producing initial dilution up to a factor of 10^4, and is controlled by the dimensions and
the forward speed of the barge. Subsequent dispersion by natural
oceanic processes is slow under stratified summer conditions, des-
cribable by an effective diffusivity of order 300 cm^2 sec^{-1} or a diffusion velocity in the neighborhood of 0.2 cm sec^{-1}, for the first
12 hours or so after release. A comparison of the observed data with
shear-dispersion theory shows that: (a) horizontal dispersion in the
upper 10-20 m of a stratified ocean is mainly due to mean shear-ver-
tical mixing interaction; (b) the low observed rates of dispersion
may be attributed to the fact that an early phase of shear diffusion
was in evidence. Extrapolation to diffusion times of the order of
several days should be possible using a constant diffusion velocity,
or an effective diffusivity increasing in direct proportion to time.

INTRODUCTION

Several diffusion experiments have been carried out during the
last few years at Deep Water Dumpsite 106 in an attempt to determine
the rate at which the various industrial wastes discharged in that
area are diluted with sea water (Frye and Williams, 1977a, b; Kohn
and Rowe, 1976; Orr and Hess, 1977; Bisagni, 1977). The method of
discharge is from a moving barge, the waste being released directly
into the wake. As one might expect, there is vigorous mixing in the

wake of the barge and the waste is effectively dispersed in a considerable volume of seawater in a matter of a minute or two. Although
some wastes are 10% or so denser than seawater, the initial ("wake")
dispersion is so effective that the mixed waste does not sink, remaining in the top 10-15 m of the water column at least for some
hours after discharge (under summer conditions, when the mixed layer
is of this depth).

Of interest is the exact degree of dilution accomplished in the
wake, and the rate of subsequent dilution by oceanic processes. From
a practical point of view the key question is, whether (or how) it is
possible to extrapolate the dilution observed in the experiments referred to above, which were carried on only for some tens of hours
after release. How long a time is necessary to achieve a total dilution by, say, a factor of 10^7, which would presumably render such
waste innocuous? The intent of the present study is at least to put
some bounds on estimates of this kind, following an analysis of the
available evidence and a discussion of the key physical processes
involved in the dispersal of barged waste.

In some of the experiments referred to above the dilution was
inferred from the concentration history of a tracer (Rhodamine B dye)
added to the waste in the barge for this purpose. In other experiments the iron content of the waste was used as a tracer of opportun-
ity, or the particulate phase ("floc") formed on the entry of the
waste into seawater was traced acoustically. A considerable amount
of qualitative and quantitative evidence has been accumulated in this
manner on waste dilution over approximately the first 24 hours after
release.

It is difficult to follow a waste patch or plume in the open
ocean and not all encounters with the waste field several hours after
release necessarily yield quantitative data on the minimum dilution
achieved by that time: there is always a possibility that the sampling vessel misses the most concentrated part of the waste field.
As pointed out by Okubo (1971) in an analysis of many oceanic diffusion experiments, only such data on cross-plume or patch concentration profiles can be used for quantitative diffusion estimates which
satisfy at least a rough mass balance. If a cross section of a diffusing plume contains most of the tracer which was released per unit
length of the initially generated line-source cloud, one can be reasonably sure that the thickest part of the plume was sampled.

Of the experiments referred to above only two (Frye and Williams, 1977a, b) have been documented to the point that mass balances
can be checked to a reasonable approximation. Only a fraction of the
waste plume transects satisfy the mass balance criteria. Dilution
estimates based on other transects may not reflect the minimum dilution likely to have been accomplished by that time.

The few solid data points which remain demonstrate the slowness of dispersion by oceanic processes under stratified conditions, yielding a dilution following wake-mixing of only something less than a factor of 10 in one day. On the other hand, the mixing in the wake is found to be very efficient, yielding dilution by up to a factor of 10^4, or 1 1/2-2 orders of magnitude better than initial mixing over a typical, well designed oceanic sewage outfall.

A number of years ago, Ketchum and Ford (1952) reported some observations on the dispersion of barged industrial waste in shallow water, from which they have calculated horizontal mixing coefficients. No mass balances were possible, and Ketchum and Ford duly note the difficulty of interpreting the scatter in their observations. Only in one of the three experiments reported do Ketchum and Ford consider concentration measurements to have been at all reliable at dispersion times greater than 1/2 hour. In this one experiment, a dilution of about 50 has been reached at 3 1/2 hours after release. There is no assurance that this was in fact representative of the minimum dilution reached at that time.

WAKE DISPERSION

As the waste is pumped into the turbulent wake of the barge, vigorous mixing takes place. The characteristic velocity of wake turbulence is a high fraction of the barge's forward velocity (which is typically 3 m sec^{-1}). Turbulence length scale is proportional to the barge's dimensions, mainly width (13.2 m in the case of the barge used on the dump described in the EG&G Grasselli report, Frye and Williams, 1977a). A rule of thumb for calculating wake width a short distance downstream is about 3 times the width of the object causing the wake: this gives 40 m for the barge just referred to. The observations included data on wake width a few minutes after dump: 25 and 32 m were noted on the first two transects. Fig. 1 illustrates schematically the wake mixing process.

Given the vigorous turbulence existing in the wake, it is unlikely that the density of the waste in the range encountered (1.0

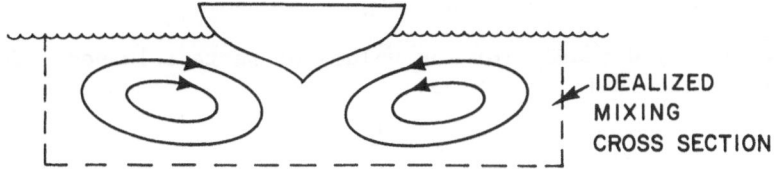

Fig. 1. Schematic illustration of wake mixing behind barge; trailing vortices only decay a long distance behind the barge (order 100 h, if h is initial mixing depth).

to 1.15 g cm^{-3}) plays a significant role in wake dispersion. Dilution by a factor of 1000 and more was quickly reached in all experiments, so that the effective gravity of the mixture dropped to about 0.1 cm sec^{-2}. The initial mixing depth was about 10 m (roughly 3 times the barge draft). If the eddying motions are assigned a characteristic velocity of 50 cm sec^{-1}, the Froude number $v/(\varepsilon gh)^{1/2}$ based on this velocity, the effective gravity and mixing depth just quoted, was 5, meaning that the inertial forces of the eddying motion greatly outweighed buoyancy forces. This explains the lack of any significant influences on initial dispersion by the density of the waste.

Some of the waste during the EG&G Grasselli dump was tagged with Rhodamine B dye. Allowing for losses due to reactions between seawater, waste and dye, an estimated 130 kg of dye was laid down along a wake 7.9 km long. Distributed over a 30 m wide by 10 m deep cross section, the concentration of the dye works out at 55 ppb (10^{-9} g cm^{-3}). The observed maximum concentration at the surface 3 minutes after the dump was 60.5 ppb, or at 5 m depth 16 minutes after the dump, 43 ppb. It seems therefore that a satisfactory estimate of wake dilution may be obtained by distributing the tracer over a 300 m^2 wake, or a wake 2 1/2 times wider than the beam and 3 times deeper than the draft of the vessel.

Along with the dye, 500 m^3 of waste were released in the course of the dumping operation. On the reasonable assumption that this waste was mixed with seawater in the same proportion as the dye, one calculates the wake dilution as

$$D_w = \frac{AL}{W} = 4740 \qquad (1)$$

say 5000. Here A is cross sectional area of the wake

$$A = b\,h \cong 8\,b_b\,h_b \qquad (2)$$

with b the width, h the depth of the wake, b_b, h_b the beam and draft of the barge. L is wake length, or if the barge speed is u, the time period of the dump t_d,

$$L = ut_d \qquad (3)$$

In Eq. (1) W stands for the volume of waste released. Sometimes the dumping rate R is prescribed:

$$R = \frac{W}{t_d} \qquad (4)$$

In terms of the dumping rate, the waste dilution is $D_w = Au/R$ (from 1, 3 and 4). Note, however, that controlling the dumping rate

alone does not determine the wake dilution, which remains relatively
low for a slow barge.

Wake dilution of barged waste is thus seen to be an efficient
process in comparison with the initial dilution of sewage from a submarine outfall, where an initial dilution of 100 is considered good
practice (Rawn et al., 1960). A not unreasonably long barge track
ensures a wake dilution by a factor of 5000 or more.

The other waste tracking experiments referred to above, although
generally not as well documented as Frye and Williams (1977a), confirm the above results regarding wake dispersion. Wake width has
been reported from visual estimates to range from 20 to 50 m. Concentration measurements in cross sections of the wake close to the
barge showed similar widths. The initial mixing depth is less well
documented, but it is certainly true to say that the waste-seawater
mixture rapidly came to occupy the available mixing depth of 10-15 m
under conditions of summer stratification. This is at least indirect
evidence that wake mixing penetrates to a depth of the same order.
Where a satisfactory mass balance is possible, Eq. (1) was more or
less confirmed. In the EG&G Edgemoor waste study (Frye and Williams,
1977b), iron concentration was observed. The wake concentration according to (1) should have been 10 mg l^{-1}, the maximum observed
concentration was 7 mg l^{-1}. However, only a few spot samples were
obtained from the wake, rather a long time (order 1 hour) after release.

From a practical standpoint one may note that A in Eq. (1) is
fixed by the dimensions of a vessel. Wake dilution from a vessel of
given size is controlled by the waste laid down per unit length of
barge track W/L. For fixed barge speed (and only then) this is
proportional to the dumping rate R.

OCEANIC DISPERSION

Subsequent to the vigorous mixing in the wake of the barge
further dispersion is controlled by oceanic processes, as the waste
cloud spreads downward to occupy the available mixing depth, and as
it grows horizontally from an initial width of some 30 m to some
hundreds of meters within about 24 hours. Under conditions of summer
stratification the wake mixing depth and surface mixed layer depth
are (fortuitously) closely similar so that vertical mixing contributes little to further dilution. Any decrease in peak concentration
under these conditions is then a result of a one-dimensional diffusion process in the cross-plume direction, which results in a relatively slow further decrease in waste concentration. Physically, uncontaminated water is only entrained into the plume at its lateral
boundaries, but not at its bottom. However, when the surface mixed
layer is much deeper than the wake mixing depth (as it certainly is

on a stormy winter day), a two-dimensional mixing process should operate for a period following such dispersion, resulting in relatively
rapid dilution until some deeper diffusion floor stops further vertical penetration.

    The dispersion experiments referred to above were all carried
out under strongly stratified (summer) conditions and provide information on horizontal diffusion under these circumstances. In analyzing the results of these experiments it is essential to establish
mass balance for any given transect. This is necessary if one wishes
to identify the maximum observed concentration with the peak concentration in the plume. When a transect contains, say, an order of
magnitude less tracer than released per unit plume length, the
chances are that the peak concentration has not been observed.

    In the case of the EG&G Grasselli observations it is possible to
estimate mass balances from Table 3-4 of Frye and Williams (1977a),
although the accuracy of these estimates is low for two reasons. One
is that vertical resolution is nonexistent, the main plume having
been traversed usually only at a single level. In addition, the orientation of the plume axis was assumed to be invariant and cross
plume distances were calibrated on this basis. These may have been
considerably in error, but only in the direction of overestimating
plume width. Fig. 2 shows the peak concentrations observed on the
various plume crossings in this experiment.

    In this light it may be stated that the main plume was missed at
the surface and at level 1 more or less immediately upon release. At
level 2 the main plume was apparently encountered 2 1/2-3 hours after
release (transects 12-15), but this was presumably at the lower
fringe of the cloud where concentrations were less than the peak,
perhaps by a factor of 2.

    At level 1 the main plume was again encountered 7-8 hours after
release (transects 26-28). Even if one assumes negligible vertical
diffusion following the initial wake-mixing, the observed plume width
is at these transects about 3 times too wide (mass balance being more
than satisfied). One possibility is that the plume "pancaked" out,
fortuitously, at the level of the uppermost instrument (5 m). A more
likely explanation is that the wake width was overestimated, on account of the fixed orientation assumption. In any event, the maximum
concentrations at these transects were about 16 ppb, only a factor of
3.5 times less than following wake dispersion. The much lower maximum concentrations observed at this level at earlier transects were
clearly not the peak concentrations in the plume. Eleven hours after
release a maximum concentration of about 10 ppb was observed: this
may not have been a true peak, however, the drop between 9 and 10
hours having been too abrupt. The repeatedly observed maxima between
7-11 hours, of some 15-16 ppb, leave little doubt that the true peak
concentration in the plume was at least this great.

Somewhat similar considerations may be applied to the EG&G Edgemoor study, where iron content was used as a tracer, although the
data are even cruder. At transect 18 (4 hours after release) a plume
of about 100 m width was in evidence, with a peak concentration of 5
mg/l or a factor of two less than after wake mixing. Assuming absence of vertical mixing, mass balance may be confirmed at this transect. The next transect, taken 20 minutes later shows a maximum concentration of only 0.75 mg 1^{-1}, much too abrupt a drop to attribute
to mixing, even if mass balance did not rule this value out

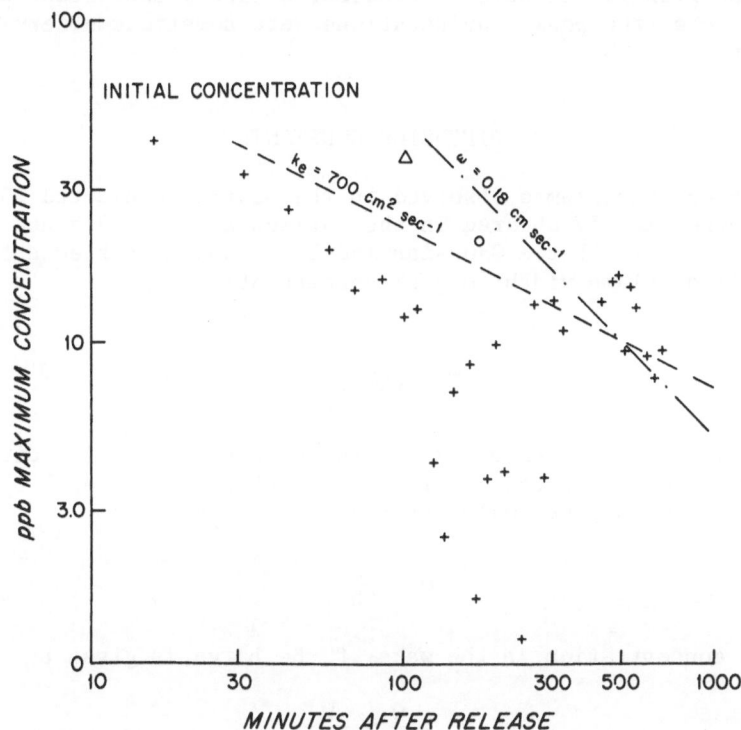

Fig. 2. Peak concentration at level 1 (5 m) in successive plume
cross sections observed by Frye and Williams (1977a)
(crosses); triangle: peak concentration at 100 min after
release estimated from observed plume width (assuming
uniform concentration in wake); circle: peak observed at
level 2 (14.5 m); broken line: constant diffusivity model
fitted to 10 ppb at 500 min diffusion time; chain-dotted
line: constant diffusion velocity model fitted to same
point. Many of the concentration maxima observed at
level 1 were clearly not true peak concentrations at
those diffusion times.

completely: the width of the plume remains more or less the same. The true peak concentration must be assumed to have been missed in this and later transects.

Mass balances in the other experiments cannot be verified, and in view of the experience with the EG&G studies, their results on dilution cannot be accepted as necessarily representing minimum dilution at any given time. However, minimum dilution estimates on a number of observed occasions roughly agree with the estimates made from the better documented experiments. On other occasions much higher dilution ratios have been calculated, more or less randomly interwoven with low dilution estimates, entirely consistent with the idea that the true peak concentrations were sometimes observed, sometimes not.

DIFFUSION CONSTANTS

Although the plumes observed in the better documented EG&G studies were clearly sheared in the horizontal, they did not show much skewness, so that a Gaussian model reasonably reflects the relationship of plume width to peak concentration, c_m:

$$c_m = \frac{Q}{\sqrt{2\pi}\ \sigma_y} \tag{5}$$

where σ_y is the standard deviation in the cross-plume (y) direction and Q is the waste released per unit length and unit depth of the plume, i.e., with previously defined quantities:

$$Q = \frac{W}{Lh} \tag{6}$$

The concentration in the wake of the barge is given by

$$c_w = \frac{Q}{b} = \frac{W}{hLb} \tag{7}$$

The mixing depth h may be taken to be constant for the duration of these experiments, the upper thermocline acting as an efficient diffusion floor, at a depth about equal to wake mixing depth. The observable "width" of the plume will be taken to be the distance between the points at the edges where the concentration drops to 10% of the center concentration c_m. In a Gaussian plume this observable width is 4.3 σ_y. From (5) and (7) the dilution due to oceanic mixing is then:

$$\frac{c_w}{c_m} = \frac{\sqrt{2\pi}\ \sigma_y}{b} \tag{8}$$

As noted before, the wake width b was 30 m.

The only reliable point on the concentration versus time curve of the Grasselli study was observed at about 480 minutes after release, when the concentration was 16 ppb or 3.5 times lower than in the wake. Substituting c_w/c_m = 3.5 into (8) one finds

$$\sigma_y = 42\ m \qquad (480\ min) \tag{9}$$

The observable width of the plume should have been about 200 m, instead of which the study reports 600 m. As noted before, however, this was a projection of the observed width to the original waste plume normal and did not necessarily reflect the shortest distance across the plume.

In the EG&G Edgemoor study a dilution factor c_w/c_m of about 2 was observed 3 hours after release. From (8) one calculates

$$\sigma_y = 24\ m \qquad (180\ min) \tag{10}$$

The observed plume width was about 100 m, which agrees with the usual 4.3 times standard deviation rule.

Owing to the scatter and unreliability of most of the maximum concentration results the time history $\sigma_y(t)$ cannot be deduced from either study with any kind of confidence. However, standard models of constant effective diffusivity or diffusion velocity (cf. below) may be applied to estimate key diffusion constants. If a constant effective diffusivity model applies, the standard deviation σ_y should grow as

$$\sigma_y = \sqrt{2\ K_e\ t} \tag{11}$$

where K_e is effective diffusivity. In a constant diffusion velocity model, on the other hand,

$$\sigma_y = \omega t \tag{12}$$

where ω is the diffusion velocity. In coastal waters, in experiments lasting for a period on the order of 10 hours, typical values of K_e have been from 200 to 2000 $cm^2\ sec^{-1}$, for ω from 0.2 to 2.0 $cm\ sec^{-1}$ (Csanady, 1966, 1973; Okubo, 1971; Pritchard and Okubo, 1969).

Given the empirical results expressed by (9) and (10) for the waste dilution studies, it is possible to determine K_e and ω for the Grasselli and Edgemoor studies.

The results are:

	Grasselli	Edgemoor
K_e, cm^2 sec^{-1}	700	300
ω, cm sec^{-1}	0.15	0.24

These constants are on the low end of the range found in coastal waters and show that oceanic mixing under summer conditions is a slow and inefficient process. To illustrate this conclusion, the dilution rates specified by K_e = constant or ω = constant will be extrapolated to an ocean dilution c_w/c_m of 10^3. Given a wake dilution of 10^4, this would correspond to total dilution by 10^7, which is presumably sufficient to render the waste innocuous. With the aid of (8) and (11) or (12) it is possible to calculate how long it would take to achieve such dilution, given that the oceanic mixing process continues to be characterized by: (1) K_e = 300 cm^2 sec^{-1} = constant; (2) ω = 0.2 cm sec^{-1} = constant.

The results of this calculation are:

Time t_c for oceanic dilution by a factor of 10^3, given

(1)	(2)
K_e = 300 cm^2 sec^{-1}	ω = 0.2 cm sec^{-1}
t_c = 75 years	t_c = 69 days

The extrapolation based on constant K_e is particularly unrealistic because any plume would be caught up in some major oceanographic feature (such as a warm core ring) within months, even forgetting more vigorous winter mixing. Nevertheless, these calculations suggest longevity of the plumes in summer. Their ultimate fate might well be significantly influenced by such large eddies in the slope region as warm core rings.

Physically, the difference between the "diffusivity" and "diffusion velocity" models is briefly that K_e = const implies constant step-length in the random walk of dispersing particles, while ω = const implies steps increasing in length in linear proportion to time. The increase in step-length is, in turn, due to the enlistment of greater velocity differences to distort the diffusing cloud as it grows in a region of non-uniform flow. These effects are discussed in greater detail in the following section.

Ketchum and Ford (1952) have also used a Gaussian model (Eq. 5

here) to infer a horizontal mixing coefficient (effective diffusivity, K_e, in the present terminology) from their observations. They have not distinguished initial (wake) dispersion from oceanic dispersion, so that most of the diffusivity values quoted in their paper include the effects of barge-bound eddies. At long diffusion times this does not distort K_e significantly, however, and a comparison with the present analysis should be possible. Unfortunately, there is only one data point at a diffusion time greater than 1 hour. The diffusivity inferred from this by Ketchum and Ford is 2700 cm^2 sec^{-1}, although the accuracy of this estimate cannot be assessed. This value of the diffusivity is not too different from what has been reported at similar coastal sites (order 50 m depth) by other investigators.

PHYSICS OF SHEAR DIFFUSION

It is not unreasonable to suppose that a waste plume released in summer may remain subject to the same slow one-dimensional diffusion process which characterizes its first 24 hours or so in the ocean (discussed above) for a period of the order of a month. It becomes important then to decide which of the two models introduced above is

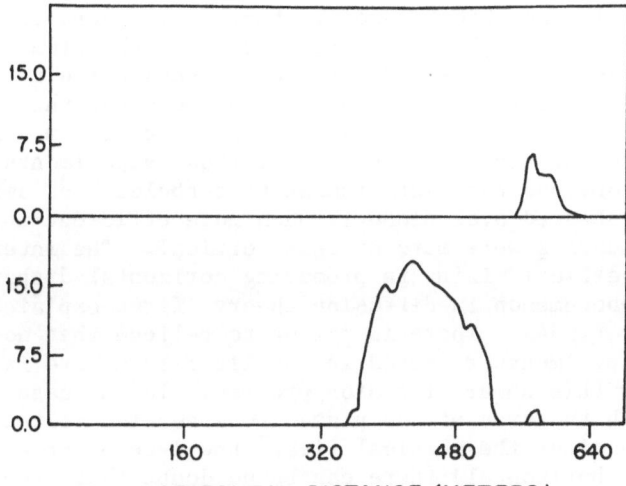

Fig. 3. Dye concentration, ppb, at 5 m (top) and 14.5 m (bottom) depth observed 145 min after release. Width of plume at 14.5 m is about 100-120 m; this is also approximately the offset of apparent plume centers between the two levels. Note, however, that true peak at 5 m was probably missed. The plume centers at these two levels are being displaced at a relative velocity of 1-2 cm sec^{-1}.

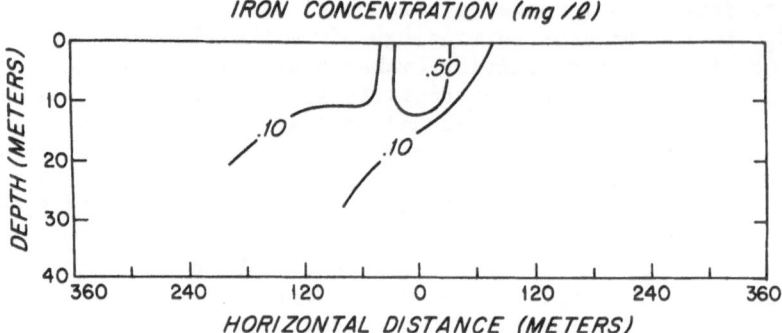

Fig. 4.   Contoured cross section of waste plume 152 min after
          release (Frye and Williams, 1977b). Vertical scale is
          exaggerated 6 times.

the more realistic in making extrapolations similar to those above.
This requires a closer look at the basic physics of the problem.

In both EG&G studies the waste plume was sheared to a spectacu-
lar extent. In the EG&G Grasselli study, 8-10 hours after release
the plume at level 1 (5 m) was displaced from the plume at level 2
(14.5 m) by several kilometers, the exact separation being unknown.
In the EG&G Edgemoor study, 4 hours after release, the plume center
at levels 2 and 3 (9.5 and 15 m) were offset by 240 m. Figs. 3 and
illustrate skewness at early stages of these experiments. It is
intuitively obvious that entrainment by turbulence of ambient fluid
into such a sheared-over plume is much more efficient than if the
lateral boundaries were more or less vertical. The interaction of
shear with vertical mixing in promoting horizontal dispersion is a
well known phenomenon in diffusion theory, first explained clearly b
G. I. Taylor (1954). There is reason to believe that most oceanic
diffusion experiments reported in the literature have in fact been
dominated by this shear diffusion process. In the case of the EG&G
studies, with the axis of the plume cross section tilted something
like 100:1 against the vertical (i.e., the isoconcentration lines
being nearly horizontal) there can be no doubt that such was the
case, as the following discussion should make clear.

The properties of shear-diffusion may be best understood by
focusing on the behavior of concentration-moments in shear flow, fol
lowing the work of Aris (1956) and Saffman (1962). Given horizontal
mean flow sheared in the vertical, $u(z)$, $v(z)$, and turbulence parame
terized by eddy diffusivities K_x, K_y and K_z, the diffusion equation
for a conservative substance may be written

$$\frac{\partial c}{\partial t} + u \frac{\partial c}{\partial x} + v \frac{\partial c}{\partial y} = \frac{\partial}{\partial x} \left(K_x \frac{\partial c}{\partial x} \right) +$$

$$+ \frac{\partial}{\partial y} \left(K_y \frac{\partial c}{\partial y} \right) + \frac{\partial}{\partial z} \left(K_z \frac{\partial c}{\partial z} \right) \tag{13}$$

where c is mean concentration.

Expressions are sought for various <u>moments</u> of the concentration field in horizontal planes. The zeroth moment or total amount of diffusing material per unit depth is

$$q = \iint c \, dx \, dy \tag{14}$$

Typical first and second moments are

$$m_1 = \iint x \, c \, dx \, dy$$

$$s_{11} = \iint x^2 \, c \, dx \, dy \tag{15}$$

The first moments define the position of the center of gravity of a diffusing cloud, hence they characterize its <u>trajectory</u>, while the second moments yield the cloud size, hence their time history shows how the cloud <u>spreads</u>. The relationships of the above moments to trajectory and spread are:

$$\ell_x = \frac{m_1}{q}$$

$$\sigma_x^2 = \frac{s_{11}}{q} - \ell_x^2 \tag{16}$$

where ℓ_x is the x-coordinate of the center of gravity (at a given level z and a given time t) and σ_x is standard deviation along x, also $\sigma_x(z,t)$.

Differential equations describing the behavior of various moments in function of time and depth may be derived from (13). For the three moments of (14) and (15) they are

$$\frac{\partial q}{\partial t} - \frac{\partial}{\partial z} \left(K_z \frac{\partial q}{\partial z} \right) = 0$$

$$\frac{\partial m_1}{\partial t} - \frac{\partial}{\partial z} \left(K_z \frac{\partial m_1}{\partial z} \right) = qu \tag{17}$$

$$\frac{\partial s_{11}}{\partial t} - \frac{\partial}{\partial z} \left(K_z \frac{\partial s_{11}}{\partial z} \right) = 2K_x q + 2m_1 u$$

These are all one-dimensional diffusion equations, some with
source terms on the right. If, as in the oceanic mixed layer, the
vertical spread of the material is limited by a diffusion floor and a
ceiling, a simple asymptotic solution of (17), valid at a long enough
time after release, is

$$q = \frac{W}{h} = \text{constant} \tag{18}$$

or a uniform distribution over the available depth. Here W is the
total amount of waste released and h is mixed layer depth. Corresponding asymptotic solutions for m_1 and s_{11} are now easily written
down, for details see e.g., Saffman[1] (1962). These depend on the distribution in the vertical of the velocity and of the horizontal eddy
diffusivity. However, they are always of such a form that the center
of gravity and the standard deviation can be written as follows:

$$\ell_x = Ut + \ell_x^*(z)$$

$$\tag{19}$$

$$\sigma_x^2 = 2D_x t + \sigma_x^*(z)$$

where U is depth-average velocity over the mixed layer and D_x is an
effective "shear diffusivity" which depends on the velocity differences across the mixed layer and on the vertical diffusivity. In
engineering applications of shear diffusion theory attention is usually focused on the magnitude of the asymptotic shear diffusivity
D_x (see e.g., Fischer, 1973) because the time-dependent terms in
(19) dominate the time-independent distributions ℓ_x^* and $\sigma_x^*(z)$
at long diffusion times. However, at least the ℓ_x^* distribution is
quite illuminating, if for no other reason, to diagnose the presence
of shear diffusion.

A very simple model of shear flow is a linear variation of
velocity over the mixed layer depth, from 0 at the bottom to 2U at
the surface (U = depth average velocity), accompanied by a constant
vertical diffusivity:

$$u = 2U \left(a + \frac{z}{h}\right)$$

$$K_z = K = \text{constant} \tag{20}$$

$$(0 > z > -h)$$

This simple model yields the following asymptotic quantities
characterizing shear diffusion:

$$D_x = K_x + \frac{U^2 h^2}{30K}$$

$$\ell_x^* = h \left[\frac{Uh}{12K} \left(1 - 6 \frac{z^2}{h^2} - 4 \frac{z^3}{h^3} \right) \right] \tag{21}$$

$$\sigma_x^{*2} = h^2 \frac{U^2 h^2}{360K^2} \left[\frac{z^2}{h^2} \left(1 + \frac{z}{h} \right) \left(8 \frac{z^3}{h^3} + 16 \frac{z^2}{h^2} - \frac{z}{h} + 9 \right) \right]$$

For a mixed layer well stirred by the application of wind stress τ_0 at the surface a crude formula for the vertical eddy diffusivity is

$$K = \frac{u_* h}{20} \tag{22}$$

where $u^* = (\tau_0 / \rho)^{1/2}$ is the friction velocity (ρ = water density). The nondimensional parameter Uh/K, which largely determines the magnitudes of the various quantities in Eq. (21), is then equal to 20 U/u_*. A typical value of the U/u_* ratio under these conditions is about 10, which yields

$$D_x \cong K_x + 1300\ K$$

$$\ell_x^* \text{ between } \pm 16\ h$$

$$\sigma_x^* \text{ between 0 and 13 } h$$

The position of the center of gravity of a diffusing cloud at different depths is illustrated in Fig. 5, with the vertical scale exaggerated in the ratio 50:3, given these typical figures. The whole pattern of course translates with the mean velocity U.

Physically, the important points are:
(1) The large magnitude of the asymptotic shear diffusivity D_x. In a well-stirred mixed layer, the diffusivity K_x due to eddying motion is at most one order of magnitude greater than the vertical diffusivity K. The typical magnitude of u_* is 1 cm sec^{-1}, which for a 10 m mixed layer corresponds by Eq. (22) to K = 50 cm sec^{-1}. The shear diffusivity with the above typical Uh/K ratio is then some 7×10^4 cm^2 sec^{-1}.
(2) The flatness of the ℓ_x^*-distribution, showing the cloud center to be sheared over between top and bottom of the mixed layer horizontally by a distance of some 33 times the depth of that layer, even under these well-mixed conditions.

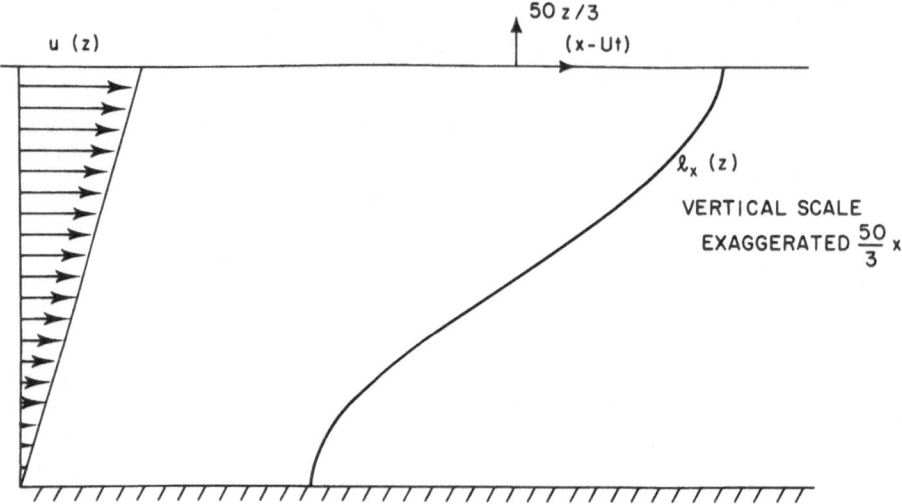

Fig. 5. Distribution of diffusing cloud center of gravity $\ell_x(z)$
in shear flow, at different levels, relative to a moving
center (which moves with the depth-average velocity U).
Velocity profile is linear, indicated on left. Note that
vertical scale is exaggerated, for typical conditions, by
a factor of 50/3.

        (3) Similarly large variations of cloud size. Since the
minimum cloud size may be expected to be at least as large as these
variations, by the time the asymptotic stage is reached, one immediately suspects that the time required to reach this stage is quite
long.

        When the mixed layer is not subject to strong stirring by wind
stress, the vertical eddy diffusivity may be much lower than the
above quoted values. However, fairly large variations of the horizontal velocity in the vertical direction may remain, so that the
Uh/K ratio may be higher then the value of 200 used above. This
would lead to a further increase of D_x, more horizontal shearing of
the cloud centerline, and larger variations in spread.

        The precise magnitude of the constants in the above calculations
of course depends on the simple model of velocity distribution adopted. However, very similar results are obtained for more realistic
velocity profiles. One should also note that although the above
model was written down for one coordinate direction, similar results
hold for the other horizontal coordinate direction, if the velocity
component in that direction also varies in the vertical. This is
generally the case in the surface mixed layer, where the typical

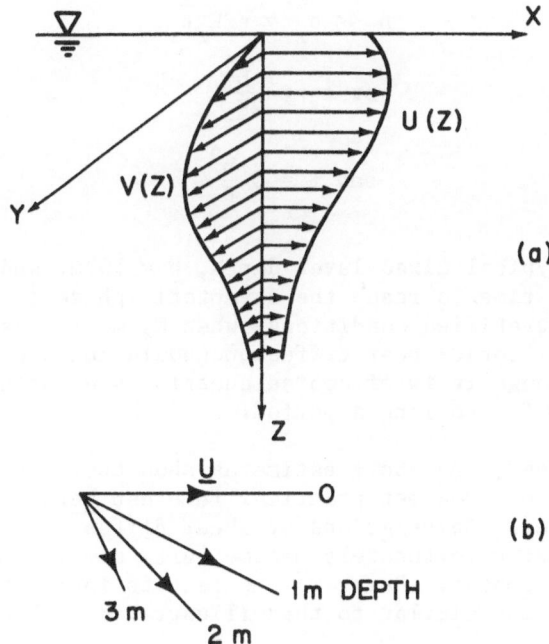

Fig. 6. Schematic picture of skewed shear flow in near surface Ekman layer.

velocity distribution is as illustrated in Fig. 6. In the dispersion of barged waste the direction of importance for diffusion is across the serially released clouds which form a continuous plume. The above results may then be thought to apply to effective cross-plume diffusion, the shearing-over of the cloud centers manifesting itself in a sheared-over plume cross-section. As pointed out before, the EG&G observations showed that the plume axis was sheared over to about the extent predicted by the simple theoretical model, providing strong evidence for the importance of shear diffusion in the observed plumes.

EARLY BEHAVIOR IN SHEAR DIFFUSION

The asymptotic stage of shear diffusion may be regarded as an equilibrium between the distorting effect of shear and the equalizing effect of vertical mixing. Model calculations of the early behavior of shear diffusion show that this occurs when

$$h \ll \sigma_z = \sqrt{2K_z t}$$

$$\text{or } t \gg \frac{h^2}{2K_z}$$ (23)

$$\text{say } t = 5 \frac{h^2}{K_z}$$

For the typical mixed layer depth, $h = 10$ m, and if $K_z = 50$ cm^2 sec^{-1}, the time to reach the asymptotic phase is 10^5 sec (≈ 1 day). Under stratified conditions, when K_z may be as low as 1 cm^2 sec^{-1}, asymptotic shear diffusion should not occur for about 2 months, although it is of course questionable whether K_z can be quite this low for so long a period.

Nevertheless, the above estimates show that the behavior of a waste cloud is of greatest practical interest <u>before</u> the asymptotic stage is reached. Calculations of shear diffusion, even using the simplest models, unfortunately become quite complex between initial and asymptotic phases. However, the results invariably show a history of cloud size similar to that illustrated in Fig. 7, from

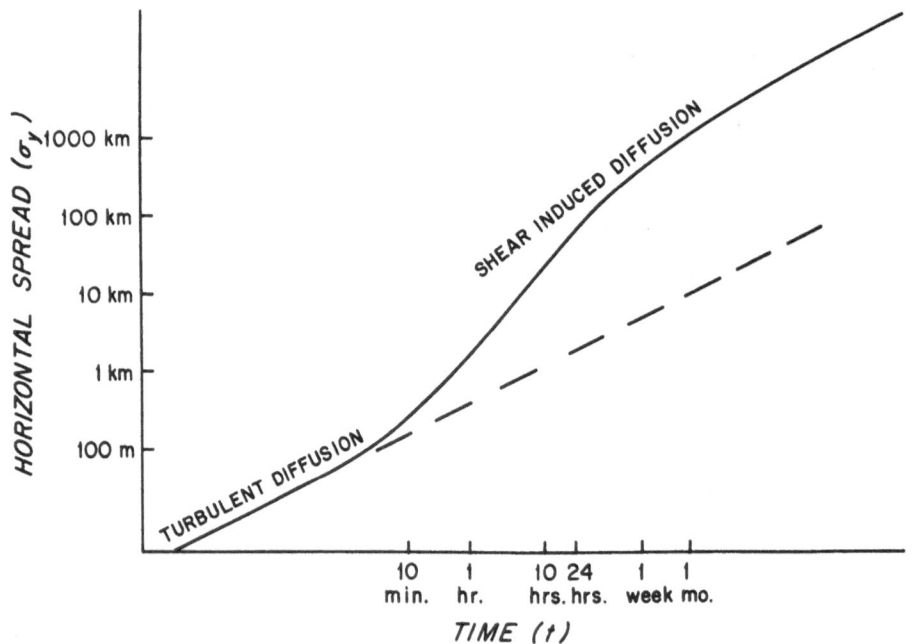

Fig. 7. Cloud spread versus diffusion time calculated for an
         Ekman layer shear diffusion model, from Csanady (1969).

Csanady (1969). The velocity distribution in this case was the
skewed shear flow of an Ekman layer with constant eddy viscosity, and
σ_y is cross-plume standard deviation.

Cloud size in an "intermediate" phase in such calculations is
found to grow as t^p with $p = 1.1$ to 1.2, most usually $p = 1.15$. A
tolerably good fit is a simple linear relationship,

$$\sigma_y = \omega t \tag{24}$$

where ω is a "diffusion velocity". This relationship may be taken to
apply approximately from wake mixing to the asymptotic shear diffusion phase. Very crudely, the diffusion velocity is of the order of
$U/4$, if U is the characteristic velocity difference across the mixed
layer (in the coordinate direction in which diffusion is of interest,
i.e., across the plume for a barge wake). Fairly strong winds sometimes give rise to vigorous cross-plume flow, say $U = 10$ cm sec^{-1},
which yields $\omega = 2.5$ cm sec^{-1}. On the other hand, under quiescent
conditions U is only of the order of 1 cm sec^{-1} and $\omega = 0.25$ cm
sec^{-1} or so.

CONCLUSION

Evaluating the two EG&G studies in terms of the above theoretical framework, we note first that current shear was clearly strong
enough to produce the large observed displacements of the plume
center between different levels. Hence it may be legitimately supposed that horizontal spread was mainly due to cross-plume shear
diffusion.

The inferred effective diffusivities of 300–700 cm^2 sec^{-1} were
orders of magnitude too low to be asymptotic shear diffusivities.
The period of observations for which reliable dilution data were
available was 8 hours or less. Under the more or less quiescent
conditions of the observations a low value of vertical diffusivity
may be hypothesized, which leads to the conclusion that shear diffusion was a long way from reaching the asymptotic phase. This is consistent with the inference of a low effective diffusivity. The value
of a diffusion velocity, ω, appropriate to quiescent conditions is of
the order of 0.25 cm sec^{-1}, which is about what was observed.

In extrapolating the observations, one may suppose that the "intermediate" phase would last for many more days, provided that conditions remained quiescent. In other words, calculations with ω = constant remain realistic for a period of quiescence. However, a storm
would change these conditions and accelerate diffusion materially,
especially a storm which brought cold, dry continental air in contact
with warm water. The vigorous thermal convection resulting in the
latter case would lead to much more rapid mixing indeed. One is led

to speculate that catastrophic events of this kind (storms or entrainment by warm core rings) govern the ultimate fate of barged
waste, following quiescent periods characterized by sluggish diffusion.

Two points may be made here in regard to the conduct of diffusion experiments designed to elucidate the behavior of barged waste.
One is that concentration measurements without some form of mass
balance are virtually useless. At the very least, a reasonable
estimate of plume width should be obtained along with the value of
apparent maximum concentration at a given cross section. This of
course begs the question of plume-axis definition, sometimes a very
difficult one to answer.

The second point is that in order to understand observed variations in the apparent efficiency of horizontal mixing it is necessary
to describe the structure of the flow, i.e., the variation of the
horizontal velocity vector with depth. A resolution on the order of
1 m separation between adjacent velocity determinations is needed at
least close to the surface and near the top of a pycnocline, with
perhaps 3-5 m being acceptable elsewhere in the mixed layer. The
practical difficulties in conducting such observations in deep water,
far from shore are non-trivial.

ACKNOWLEDGEMENT

The work described herein has been supported by the Ocean Dumping Program, National Ocean Survey, NOAA. T.P. O'Connor kindly supplied the reports on the various empirical studies used in the analysis.

REFERENCES

Aris, R. (1956) On the dispersion of solute in a fluid flowing
        through a tube. Proc. Roy. Soc. A 235, 67-77.
Bisagni, J. J. (1977) The physical oceanography and experimental
        studies at Deepwater Dumpsite 106 during June 1976.
        Unpublished manuscript, National Marine Fisheries Service,
        Atlantic Environmental Group.
Csanady, G. T. (1966) Accelerated diffusion in the skewed shear
        flow of lake currents. J. Geophys. Res., 71, 411-420.
Csanady, G. T. (1969) Diffusion in an Ekman layer. J. Atmos.
        Sci., 26, 414-426.
Csanady, G. T. (1973) Turbulent Diffusion in the Environment. D.
        Reidel Publishing Co., Boston, 248 pp.
Fischer, H. B. (1973) Longitudinal dispersion and turbulent
        mixing in open-channel flow. Ann. Rev. Fluid Mech., 5,
        59-78.

Frye, D. and G. Williams (1977a) Measurements of the dispersion
    of barged waste near 38°50'N latitude and 72°15'W longitude
    at the "106" dump site. Unpublished manuscript, EG&G
    Environmental Consultants, Waltham, MA.
Frye, D. and G. Williams (1977b) Measurements of the dispersion of
    barged waste near 38°33'N latitude and 74°20'W longitude.
  Unpublished manuscript, EG&G Environmental Consultants,
    Waltham, MA.
Ketchum, B. H. and W. L. Ford (1952) Rate of dispersion in the
    wake of a barge at sea. Trans. AGU, 33, (5), 680–684.
Kohn, D. and G. T. Rowe (1976) Dispersion of Two Liquid
    Industrial Wastes Dumped at Deep Water Dumpsite 106, Off the
    Coast of New Jersey, USA. Final Report, DWD 106 Large Scale
    Dumping Study, NOAA, Rockville, MD.
Okubo, A. (1971) Oceanic diffusion diagrams. Deep-Sea Res., 18,
    789.
Orr, M. H. and F. R. Hess (1978) Acoustic monitoring of industrial
    chemical waste released at deep water dump site 106. J.
    Geophys. Res., 83, 6145–6154.
Pritchard, D. W. and A. Okubo (1969) Summary of Our Present Know-
    ledge of the Physical Processes of Mixing in the Ocean and
    Coastal Waters, and a Set of Practical Guidelines for the
    Application of Existing Diffusion Equations in the
    Preparation of Nuclear Safety Evaluations of the Use of
    Nuclear Power Sources in the Sea. Chesapeake Bay Institute,
    The Johns Hopkins University, Report No. NYO-3109-40 (U.S.
    Atomic Energy Commission) Reference 69-1, September, 1969.
Rawn, A. M., F. R. Bowerman and N. H. Brooks (1960) Diffusers for
    disposal of sewage in sea water. J. Sanit. Div. ASCE, 86,
    SA2, 65-105.
Saffman, P. G. (1962) The effect of wind shear on horizontal
    spread from an instantaneous ground source. Quart. J. Roy.
    Meteor. Soc., 88, 382-393.
Taylor, G. I. (1954) The dispersion of matter in turbulent flow
    through a pipe. Proc. Roy. Soc. A 223, 446-467.

HORIZONTAL DIFFUSION IN OCEAN DUMPING EXPERIMENTS

Takashi Ichiye, Masamichi Inoue and Michael Carnes

Department of Oceanography
Texas A & M University
College Station, TX 77843

ABSTRACT

During two experiments monitoring ocean dumping operations in the Gulf of Mexico and off Arecibo, Puerto Rico, a number of drifters were tracked with a ship-borne radar and aerial photographs of the plumes were analyzed by use of the results of drifter experiments in the Gulf of Mexico. Horizontal eddy diffusivity was determined from increases of the variances of the drifter positions with time and were calculated with Okubo's method (1976). The variances of drifter positions did not increase with time when the wind speed exceeded critical values off Puerto Rico. It was speculated that the wind rows caused the drifter convergence during the Puerto Rico experiments. The speculation is partially confirmed by analyzing the aerial photos of the plumes which show striations along the direction of the prevailing winds.

INTRODUCTION

In order to determine the fate of pollutants discharged into the ocean, information on the mean current and its fluctuations is essential. The velocity fluctuations are random but obey the Navier-Stokes equations. These are called turbulence or turbulent flows and their statistical properties have been studied for more than forty years (Monin and Yaglom, 1975; Lipmann, 1979). However, the difference between turbulence and mean flow is relative and depends on scales of time and space over which the mean flow is determined. Statistical properties of turbulence have been determined by use of a finite number of experiments on a hypothesis of ergodicity postulating that averages of all possible quantities should converge to the

131

same mean values as the averaging time increases.

One outstanding effect of turbulence is its ability to transfer or mix momentum and matter at rates several orders of magnitude greater than those due to molecular diffusion. This effect is summarily called eddy diffusion and parameterized by eddy diffusivity which depends on characteristics of the flow and thus is not constant as is molecular viscosity or diffusivity. In the atmosphere and ocean where a horizontal dimension is far greater than a vertical dimension, turbulence and thus horizontal eddy diffusivity is orders of magnitude greater in horizontal directions than in the vertical.

Both vertical and horizontal diffusion are important for dispersion of pollutants. However, the vertical diffusion in the upper ocean layer is mainly dependent on the wind stress, wave conditions and density stratification of the water and has been parameterized in terms of these quantities (Defant, 1961; Huang, 1979). On the other hand, horizontal diffusion depends on the mean current and its gradients, topography, and large scale density distributions and also covers enormous ranges in time and space scales. There is no parameterization of horizontal diffusivity in relation to the mean state of the flow except its dependence on a horizontal length scale of the motion as the four-thirds scale law (Monin and Yaglom, 1975). Therefore it is necessary to conduct experiments to determine this coefficient in specific situations.

There are two types of description of the current field, Eulerian and Lagrangian. Thus turbulent motion also can be described in these ways. In practice, however, the Eulerian method is very difficult to employ for determining the mean flow and turbulence which are responsible for transporting and dispersing plumes ocean dumping, because a number of current meters should be deployed for a prolonged period of time in the area where the plumes are to be discharged. Although the Lagrangian measurements are difficult in analysis and interpretation, they are simpler to make in the field and more relevant to transport and dispersion of the plumes. For this reason, our study is limited to the Lagrangian method. Our field work is devoted to monitoring the movement of a number of drifers at two dump sites. Further, in order to study micro-structures of a plume, aerial photographs of the plume are analyzed.

HORIZONTAL DIFFUSIVITY DETERMINED FROM THE EXPERIMENT IN THE GULF OF MEXICO

The experiment was carried out on 25 and 26 July, 1977, at the Western Gulf Dump Site about 200 km south of Galveston during the experiment monitoring an ocean dumping operation. Sixteen drifters equipped with radar reflectors were released from the Coast Guard Cutter ACUSHNET along the east-west direction with intervals of one

to three hundred meters and were tracked for 19 hours by photograph-
ing the radar scope of the NOAA R/V RESEARCHER at intervals of fifteen to twenty minutes (Ichiye et al., 1978). The positions of the
drifters were determined for each photograph by means of a magnifying
scale relative to the ship's position which was determined by Loran
C.

     Eight drifters were drogued with current crosses at 5 m and five
with crosses at 10 m. For each cluster the variances of the drifter
positions σ_X^2 and σ_Y^2 along the major and minor axes of the scatter
ellipse were computed and the direction of the major axes from the
eastward direction was determined from each photograph.

     Change of the direction of the major axis of the scatter ellipse
of 5 and 10 m drifters is plotted in Fig. 1. The trajectories of the
centroid and the major and minor axes of the scatter ellipse are
plotted in Fig. 2 for the 5 m drifters. The corresponding figure for
the 10 m drifters is almost similar to this figure and thus not shown
here. The mean velocity is 17.6 and 20.0 cm/sec in speed and 112°
and 115° from the north in direction for the 5 m and 10 m drifters,
respectively. The major axis of the scatter ellipse deviated clockwise from the mean direction by 18° or 21° in the initial hour for
the 5 m or 10 m drifters, respectively, but counterclockwise by less
than 3° from the mean direction at both depths. The drifter

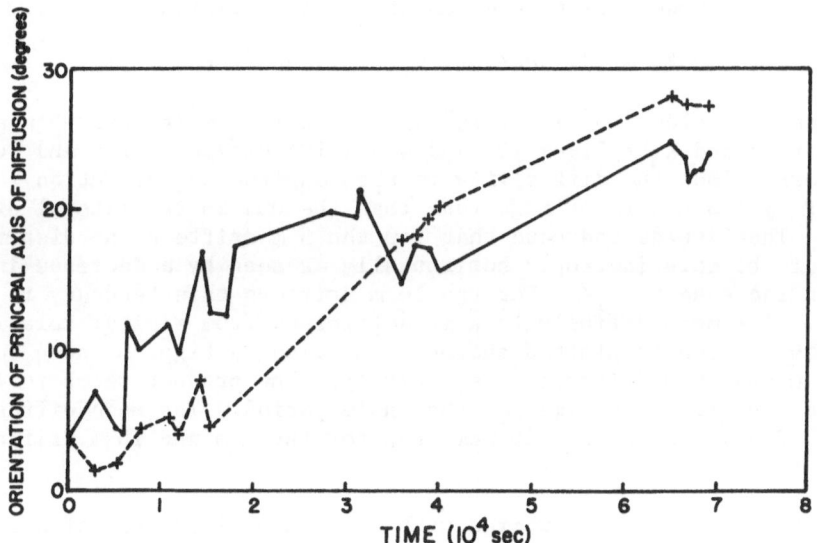

Fig. 1. Time series of direction of the major axis of the scatter
         ellipse. Direction is measured positive clockwise from
         east. Solid and dashed lines are for 5m and 10m drifters,
         respectively. Time origin is at 0112Z, 26 July, 1977.

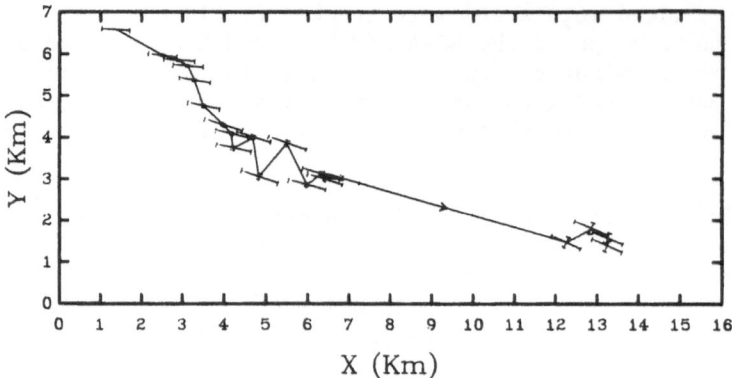

Fig. 2. Centroid trajectory and major and minor axes of 5m drifters.
The X- and Y-axis correspond to east-west and north-south,
respectively. Major and minor axes measure one standard
deviation.

dispersion seems to be biased along the mean current direction.

The variances σ_X^2 and $\sigma_Y^2 \cdot$ of drifter positions in the longitudinal direction (major axis of a scatter elipse) X and lateral direction
Y (minor axis) are plotted against time in Figs. 3 and 4 for the 5 m
and 10 m drifters, respectively. The horizontal diffusivity K_X and
K_Y in X- and Y-directions are computed from relations

$$K_X = 1/2 \, d\sigma_X^2/dt, \quad K_Y = 1/2 \, d\sigma_Y^2/dt \qquad (1)$$

The result yields K_X = 1.4 x 10^4 and 8.0 x 10^4 cm^2/sec for 5 m and 10
m drifters and K_Y = 3.9 x 10^2 and 4.3 x 10^2 cm^2/sec for 5 and 10 m
drifters. Thus the diffusivity in the longitudinal direction is
larger by two orders of magnitude than the one in the lateral direction. The figures indicate that for the 5 m drifters the dispersion
tends to be more isotropic horizontally as seen by a decrease in σ_X^2
and an increase in σ_Y^2. For the 10 m drifters this tendency is not
clear. The mean diffusivity $\bar{K}$ is determined from similar relations
from $\sigma_X\sigma_Y$ which is plotted against time also in Figs. 3 and 4 for
the 5 m and 10 m drifters, respectively. The product seems to increase linearly with time for the whole period. The mean diffusivity
$\bar{K}$ is 3.2 x 10^3 and 5.2 x 10^3 cm^2/sec for the 5 m and 10 m drifters,
respectively.

These values are compared with the synthetic result of oceanic
diffusion from dye experiments by Okubo (1971). The variance of dye
patches σ_{rc}^2 was plotted against time by Okubo (1971) (referred to
as 0-1 curve) from various sources. This variance can be determined
from

$$\sigma_{rc}^2 = 2\sigma_X\sigma_Y \qquad (2)$$

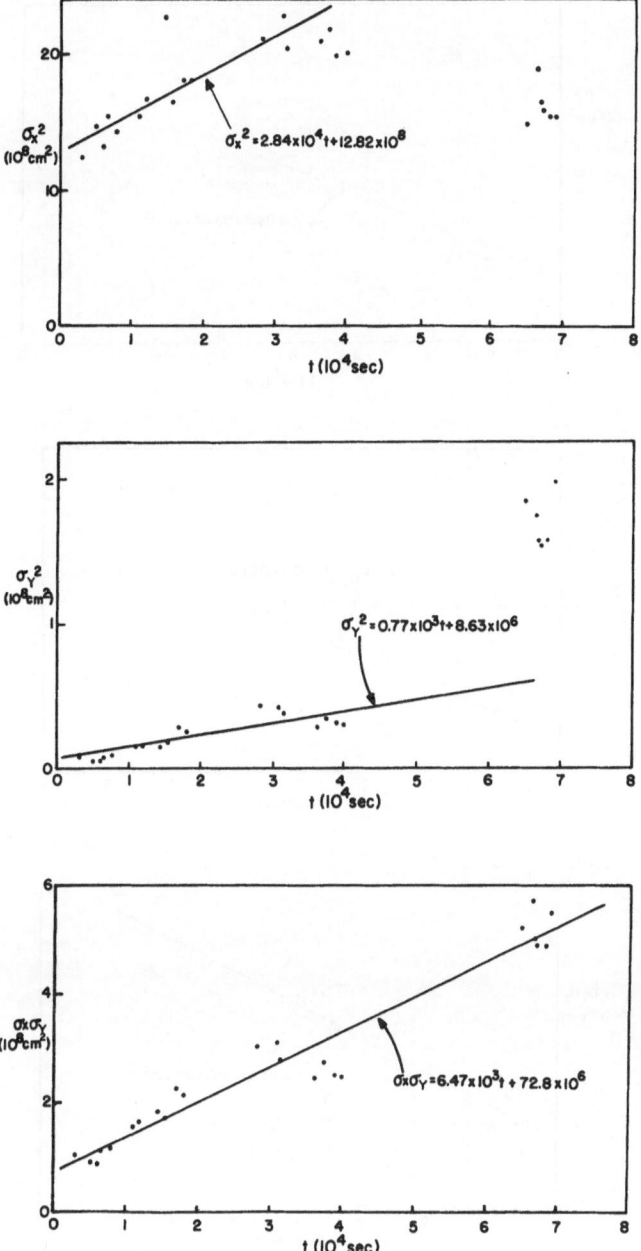

Fig. 3. Variances versus time for 5m drifters. In the linear regression calculation, data points for $t > 5 \times 10^4$ sec
are not included for σ_X^2 and σ_Y^2. Time origin is at
0112Z, 26 July, 1977.

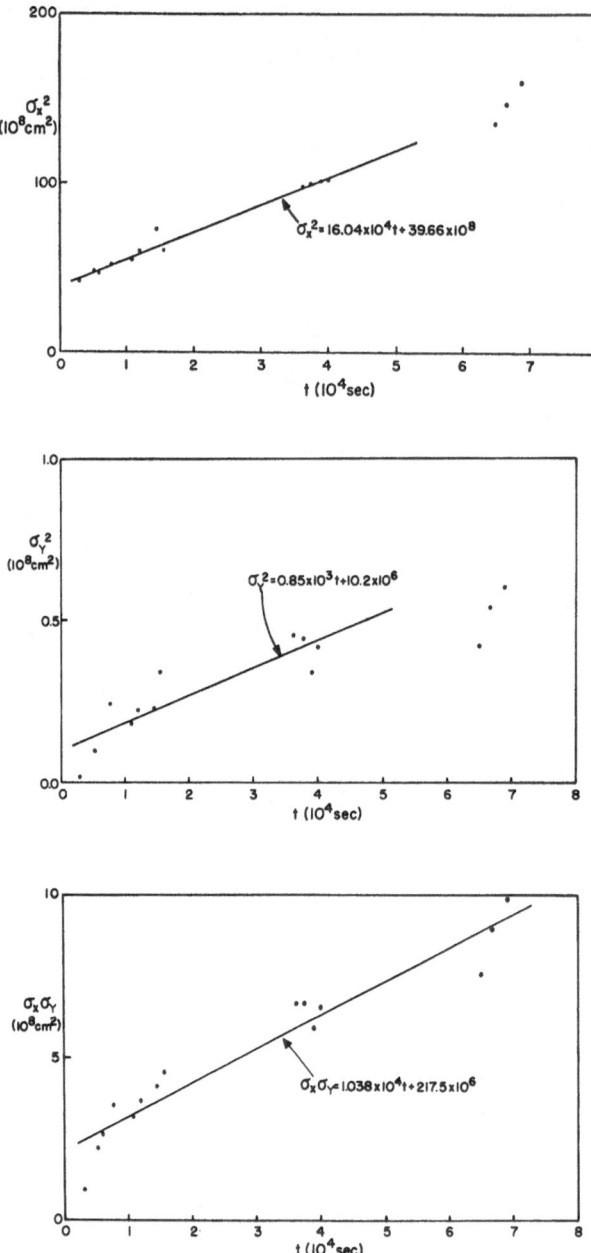

Fig. 4.   Variance versus time from 10m drifters.   In the linear
regression calculation, data points for $t > 5 \times 10^4$ sec
are not included for σ_X^2 and σ_Y^2.   Time origin is at
0112Z, 26 July, 1977.

by use of the present data, though our data show non-isotropicity.

Okubo's model is based on diffusion from a point source. In the present experiment the observed initial values of σ_X^2, σ_Y^2 are not zero. The time to reach the observed initial values is calculated from

$$\sigma^2 = 0.0108 \ t^{2.34} \tag{3}$$

based on Okubo's formula of dependency of σ^2_{rc} on time. This relation yields $t_0 = 2.13 \times 10^4$ and 3.40×10^4 sec for the 5 m and 10 m drifters, respectively for $\sigma_0^2 = 1.43 \times 10^8$ cm (for 5 m drifters) and 4.35×10^8 cm^2 (for 10 m drifters).

The curves of $2\sigma_X\sigma_Y$ versus time from Fig. 3 and Fig. 4 are replotted in Fig. 5 with the time origin replaced with t_0 instead of zero for 5 m and 10 m drifters (denoted as A-5 and A-10, respectively). The O-1 curve and its range of scatter are plotted for comparison. The values of $2 \ (\sigma_X\sigma_Y - \sigma_{Xo}\sigma_{Yo})$ calculated from the data are also plotted, where σ_{Xo} and σ_{Yo} are initial values of σ_X and σ_Y, respectively. The curves denoted as B-5 and B-10 in Fig. 5 represent the mean curves of these values which are linearly proportional to time with its origin at the beginning of the experiment.

It is clear that curves A-5 and A-10 from our data are within the scatter range of Okubo's (1971) data but the slopes of these curves are less steep than that of his mean curve. The drifters were subjected to the initial deployment pattern in the early part of the experiment and influenced by turbulent diffusion later. The observed points and curves B-5 and B-10 are almost linearly proportional to time and also they reach within the scatter range of Okubo's curve after a period of 10^4 and 3×10^4 sec. These two facts suggest that Okubo's curve is based on the data with a wide range of σ_0^2 and that the diffusion with a finite initial value of σ_0^2 can be explained as the Taylor diffusion.

Horizontal diffusivities obtained from σ_X^2 and σ_Y^2 are also compared with Okubo's curve which plots the diffusivity against the scale of diffusion . The latter can be defined as either the initial value or the time averaged value of $3\sigma_{rc} = 3(2\sigma_X\sigma_Y)^{1/2}$ for $\bar{K}$ and of $3(2)^{1/2}\sigma_X$ or $3(2)^{1/2}\sigma_Y$ for K_X or K_Y, respectively. The time averaged scale of diffusion can be obtained from

$$\bar{\ell}_X, \bar{\ell}_Y, \bar{\ell} = 2^{3/2}(at)^{-1} \left[(at + b)^{3/2} - b^{3/2}\right] \tag{4}$$

for K_X, K_Y or $\bar{K}$, respectively. In this equation, t is the duration of time which σ_X^2, σ_Y^2 and $\sigma_X\sigma_Y$ can be expressed by a linear function of time, t, as

$$\sigma_X^2, \sigma_Y^2, \sigma_X\sigma_Y = at + b \tag{5}$$

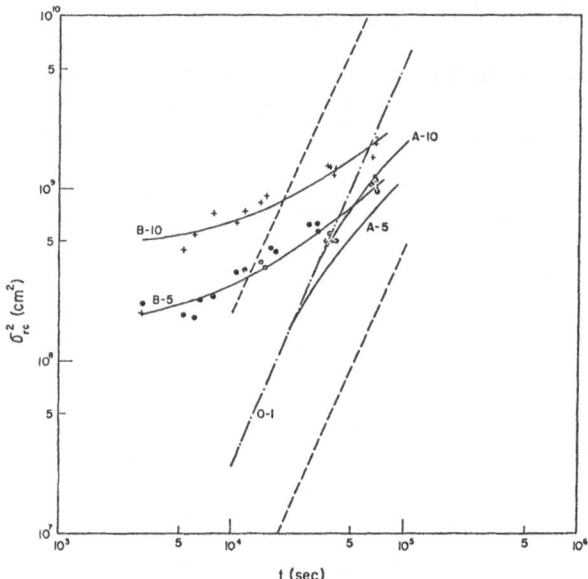

Fig. 5.   Comparison of $2\sigma_X\sigma_Y$ versus time t with Okubo's curve of
          σ_{rc}^2 denoted as 0-1 with dashed lines indicating the
          scatter range. A and B curves are the mean curves with time
          origin at the extrapolated value and at start of the experi-
          ment, respectively, and 5 and 10 denote 5 m and 10 m drift-
          ers, respectively. Full circles and crosses are $\sigma_X\sigma_Y$ –
          $\sigma_{Xo}\sigma_{Yo}$ for 5 m and 10 m drifters, respectively.

The value of t is taken as 4.0×10^4 and 6.9×10^4 sec for $\bar{l}_X$, $\bar{l}_Y$,
and for $\bar{l}$, respectively. The eddy diffusivities and the scale of
diffusion are listed in Table 1.

     In Fig. 6 the values of K_X, K_Y, K are plotted against l_X, l_Y,
together with the curve of Okubo (1971) on the relationship between
the apparent horizontal eddy diffusivity (K_a) and the scale of
diffusion (l). The latter is given by an equation (Okubo, 1971).

$$K_a = 0.0103l^{1.15} \tag{6}$$

     This figure indicates that the observed data become closer to
Okubo's curve for $\bar{l}$ and l_X than for l_o and l_{Xo}. It again indicates
that K_Y is smaller by a factor of 2 than the value predicted by
Okubo's curve using $\bar{l}_Y$ as the scale of diffusion. It is noted that
K_X and K_Y are larger or smaller than the values predicted by Okubo's
curve regardless of the definition of the values of l, whereas K is
larger or smaller than Okubo's curve according to the definition of
l. The longitudinal diffusivity determined by the present experiment

Table 1. Apparent eddy diffusivity and scale of turbulence.

ν (eddy diffusivity)	Depth	
λ (turbulence scale)	5 m	10 m
K_X	1.42×10^4	8.02×10^4
ℓ_{Xo}	1.52×10^5	2.67×10^5
$\bar{\ell}_X$	1.82×10^5	3.56×10^5
K_Y	3.9×10^2	4.3×10^2
ℓ_{Yo}	1.25×10^4	1.35×10^4
$\bar{\ell}_Y$	2.04×10^4	2.17×10^4
$\bar{K}$ (cm^2/sec)	3.24×10^3	5.19×10^3
ℓ_o (cm)	3.62×10^4	6.26×10^4
$\bar{\ell}$ (cm)	7.10×10^4	1.00×10^5

is larger than the curve determined by Okubo (1971) by averaging the results of all the previous experiment whereas the lateral diffusivity is smaller than the value predicted by his curve.

It is generally recognized that the turbulence in nature is not isotropic even in the horizontal direction and is more intense in the direction of the mean current than in the direction lateral to the mean current except in a zone of strong shear (Csanady, 1973; Tennekes and Lumley, 1972). Therefore it is speculated that the longitudinal or lateral eddy diffusivity determined by individual experiments may be respectively larger or smaller than the values determined by averaging over different experiments.

LAGRANGIAN VELOCITY GRADIENTS

Okubo and his group (1976, 1976a, and 1976b) developed statistical methods to determine the mean values of the Lagrangian velocity gradients using positions of a number of current followers

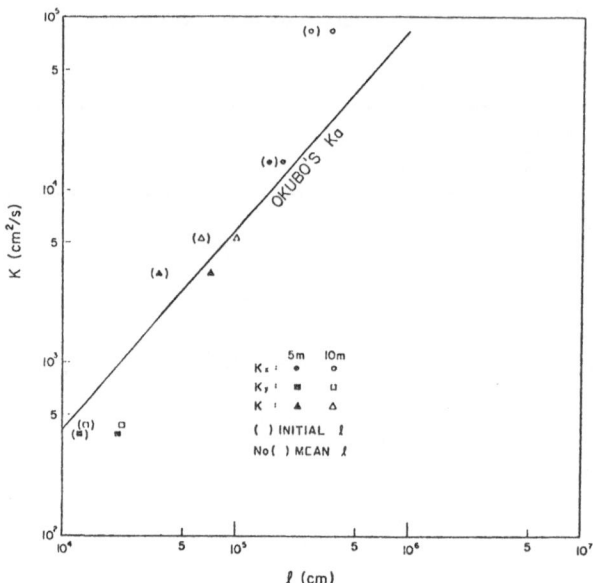

Fig. 6.   Comparison of eddy diffusivity with the curve in Fig. 2 of
          Okubo (1971). Data points without the parenthesis indicate
          that the time averaged value is used for ℓ.

(drifters), and applied the methods to about six sets of data based
on drifter released near the shores of Lake Erie. In each set of experiments, some 50 to 90 drifters were tracked by aerial photography
at 5 to 10 minute intervals.

      Our experiment did not employ as many drifters as were used by
Okubo and his group. However, the area of our experiment is far from
the coast and has depths of nearly 1000 m. Therefore our data are
free from complications caused by bottom topography and coastal effects. In fact, each cluster of either 5 m and 10 m drifters can be
grouped with a single ellipse for at least 11 hours after the release, in contrast to the clusters of the Lake Erie experiment. The
result of these experiments indicate that each cluster was separated
into smaller sub-clusters within several hours of the release.

      Since the number of 10 m drifters is too small to be statis-
tically meaningful, the analysis is limited to the 5 m drifters. The
regression analysis used follows the method of Okubo and Ebbesmeyer
(1976) instead of Okubo et al. (1976a) because our data sets had
small numbers and thus were not smoothed. The analysis of the kth
photograph denoted as time k produced mean velocities $\bar{u}(k)$ and $\bar{v}(k)$,
horizontal divergence $\gamma(k)$ ($=\bar{u}(k)_x + \bar{v}(k)_y$), relative vorticity

$\eta(k)$ $(=\overline{v}(k)_x - \overline{u}(k)_y)$, stretching deformation rate $\alpha(k)$ $(=\overline{u}(k)_x - \overline{v}(k)_y)$ and shearing deformation rate $h(k)$ $(=\overline{v}(k)_x + \overline{u}(k)_y)$, where the x and y denote partial differentiation. The mean velocities represent the mean Lagrangian velocity of the center of the drifters at time k.

The values of $\overline{u}(k)$ and $\overline{v}(k)$ are plotted against time in Fig. 7. The mean speed determined from two sets of drifters positions at time intervals shorter than about 15 minutes (10^3 sec) becomes abnormally large because of errors in positioning. Therefore, these sets of data were discarded. Fig. 7 indicates that the magnitude of the mean current is the order of 0.5 m/sec directed mainly towards east-southeast to south-southeast similar to the major axis of the scatter diagram ellipse.

The divergence, relative vorticity, stretching rate, and shearing rate are plotted against time in Fig. 8. The divergence is almost constant with time with positive values of the order of 10^{-4} sec^{-1} and the stretching rate has negative values of the order of 10^{-3} to 10^{-4} sec^{-1}. The relative voriticity and shearing rate change their signs as well as their magnitudes more irregularly.

In contrast to the results by Okubo et al. (1976a), the fluctuations of vorticity and shearing rate from the present data are so irregular that there is no consistent picture. However, the results of Okubo's group are smoothed with time and their original values show large fluctuations. More interesting is that the divergence and stretching rates show more systematic change in the present data.

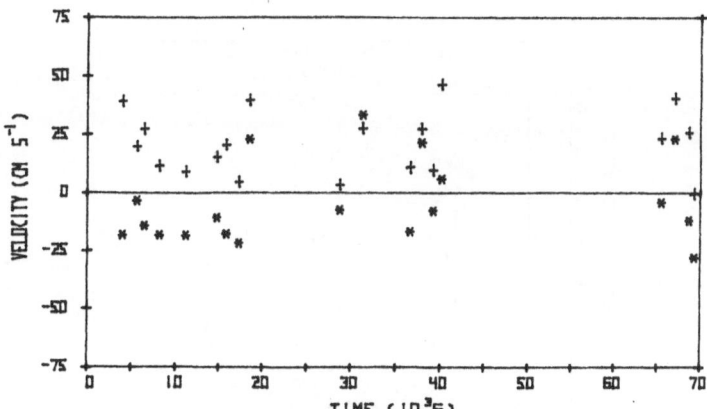

Fig. 7. Centroid velocity versus time. Crosses are east-west velocity component (east positive). Asterisks are north-south velocity component (north positive). Time origin is at 0112Z, 26 July, 1977.

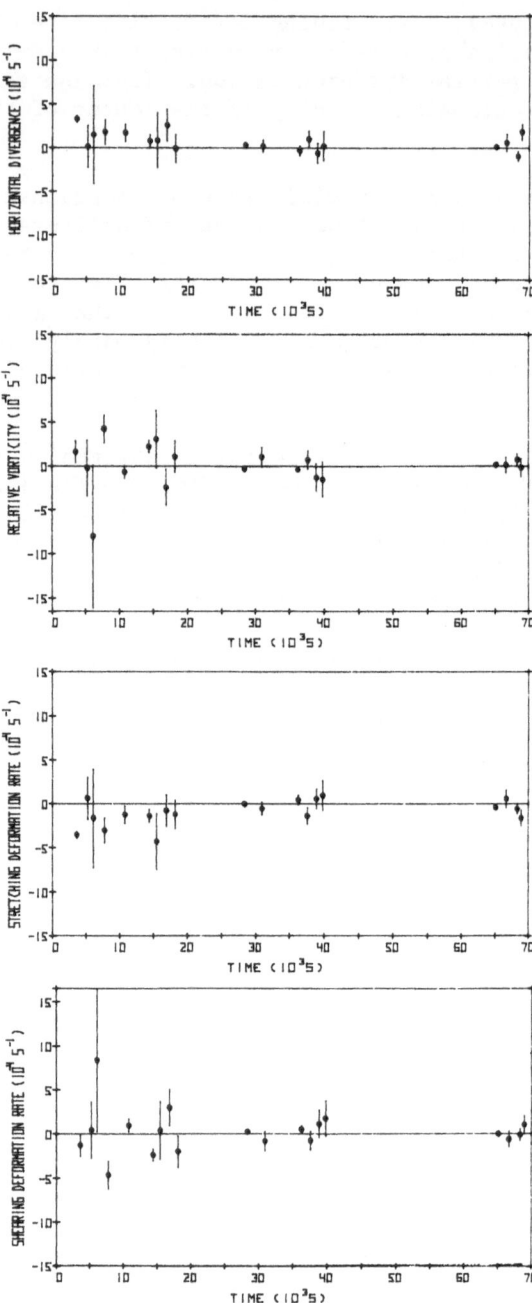

Fig. 8.   Time series of Lagrangian deformation rates.   Bars indicate
          95% confidence interval.   Time origin is at 0112Z, 26 July,
          1977.

The Lagrangian deformation parameters can be used to detemine characteristics of horizontal dipersion of drifters around the centroid according to classification of flow singularities by Okubo (1970). The stability parameter ($S = \alpha^2 + h^2 - \eta^2$) of each photograph is plotted against horizontal divergence in Fig. 9. It is noted that the range of S is larger by one order of magnitude than that obtained by Okubo et al. (1976b) from drogue experiments in Lake Erie. This may be due to larger magnitudes of the Lagrangian deformation parameters in the present experiment, in which the parameters are calculated without smoothing.

Although the points are scattered, the general tendency was that "saddle" trajectories were predominant during the initial phase within 6 hours after release and that there were a few "outward nodal" trajectories but almost no "outward spiral" trajectory. The trajectories may deviate from a starting point for the three unstable categories "saddle," "outward nodal" and "outward spiral" but the former two categories do not provide quick far-reaching deviations as the last one. Ten hours after release the two stable categories appeared because of convergence ($\gamma < 0$). These features of trajectory characteristics might have contributed to the observed small horizontal dispersion of the waste plume in this particular experiment.

The eddy diffusivity determined from σ_x^2 and σ_y^2 is independent of time and may be used to predict the diffusion over the period of about 10 hours. However, the eddy diffusivity may vary with time as well as in space. The momentary eddy diffusivities are defined as contributions from random motion which are due to deviations of predicted positions of the drifters from the Lagrangian deformations

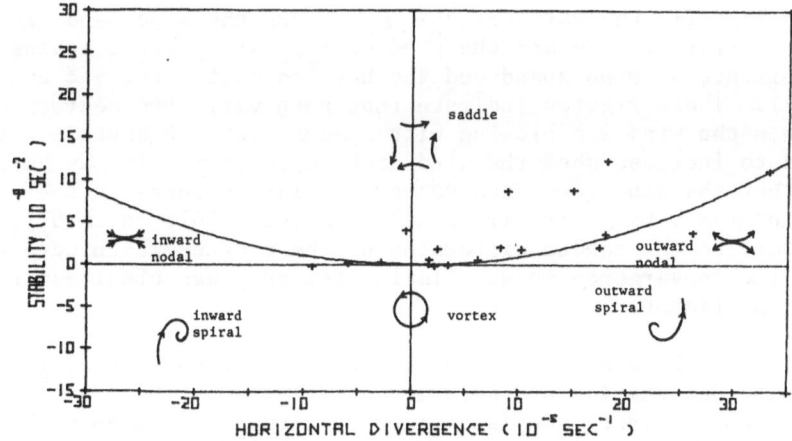

Fig. 9. Singularity diagram. Crosses indicate data points. Curved line represents division between saddle and nodal singularities.

(Okubo and Ebbesmeyer, 1976). Components of momentary eddy diffusivities M_X and M_Y and their mean value M are defined as

$$M_X = \sigma_X \sigma_u, \ M_Y = \sigma_Y \sigma_v, \ M = (M_X M_Y)^{1/2} \qquad (7)$$

in the X- and Y-directions, where σ_u and σ_v are standard deviations
of the velocity components u and v directed along the major and minor
axes, respectively. The results calculated for the 5 m drifters are
shown in Fig. 10. Both M_X and M_Y are larger by an order of magnitude
than the averaged eddy diffusivities determined before. The definition (7) of momentary eddy diffusivities is based on the mixing
length theory which postulates that M is proportional to the product
of the mixing length and magnitude of the turbulence velocity $(\ell\epsilon)^{1/2}$
where ϵ is the energy dissipation. In relation (7) σ_X or σ_Y is considered as the mixing length and σ_u or σ_v is considered as the turbulent velocity. However, the proportionality coefficient may be of
the order 0.1 rather than 1 as argued by Ozmidov (1960). Then the
discrepancy of the order of magnitude in the eddy diffusivities may
thus be partially resolved.

DRIFTER EXPERIMENTS OFF ARECIBO, PUERTO RICO

Two sets of drifter experiments were carried out about 40 nautical miles north of Arecibo, Puerto Rico on February 6 and 9, 1978 on
board the R/V KNORR. All the current crosses were at 5 m depth. The
drifters were released in an elliptical shape rather than on a line
as in the Gulf of Mexico. In the first experiment five drifters were
tracked by the same method as in the Gulf of Mexico from about 1800
on February 6 until 0930 on February 7.

In Fig. 11, the variance σ_X^2, σ_Y^2 along the major and minor axes
of the scatter ellipse and the product $\sigma_X\sigma_Y$ are plotted against time
with comments on wind speed and the bar indicating the 95% confidence
interval. These figures indicate that both variances decreased with
time when the wind was blowing with speeds above 10 knots but they
started to increase when the wind dropped to calm. It may be speculated that the wind generated rather regular currents in the upper
layer, probably Langmuir circulation (Ichiye, 1967; Pollard, 1977),
thus reducing the random dispersion of the drogues which were caught
within the convergence zone. Similar tendency was observed in the
second experiment.

Fig. 12 shows horizontal divergence, relative vorticity,
stretching deformation and shearing deformation. The magnitude of
the horizontal divergence is smaller by a factor of 3 than that by
the Gulf of Mexico experiment. However, the values were negative
when the wind was blowing and the variances σ_X^2 and σ_Y^2 decrease,
whereas they became positive when it was calm and the variances
started to increase. This suggests some correlation between the

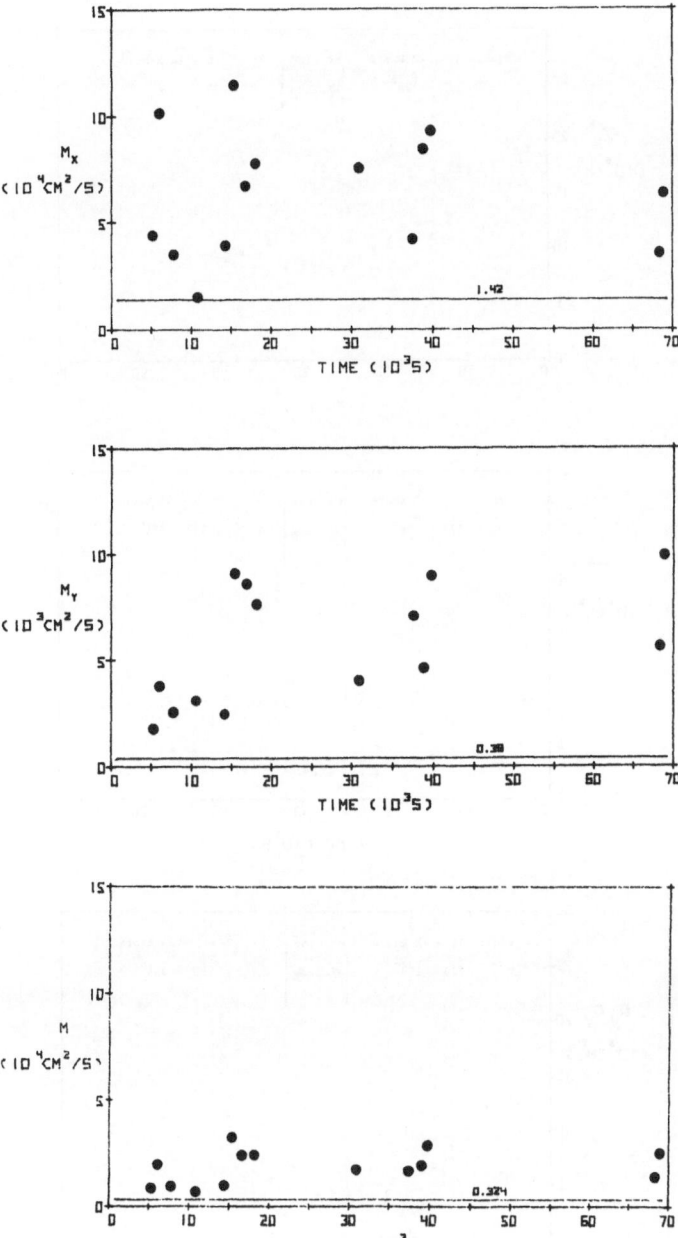

Fig. 10.   Time series of momentary eddy diffusivities (full circles).
Horizontal lines represent eddy diffusivities obtained by
the linear regression shown in Fig. 3.   Time origin is at
0112Z, 26 July, 1977.

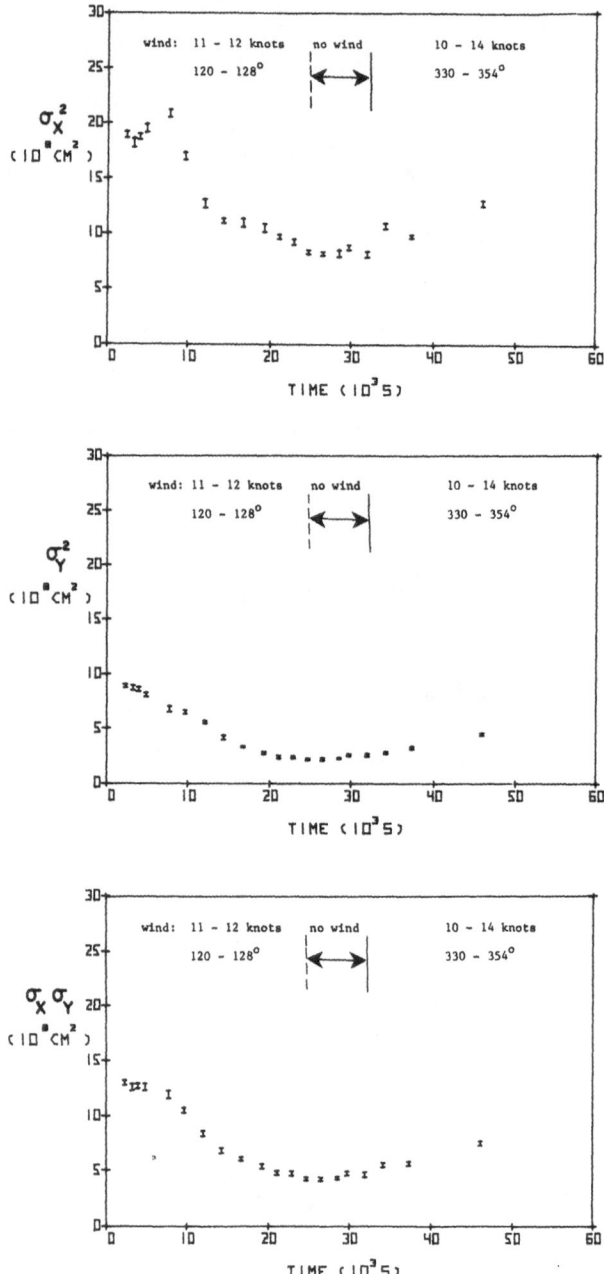

Fig. 11. Variances versus time. Time origin is at 2038 LT, 6 February., 1978. Bars indicate 95% confidence interval.

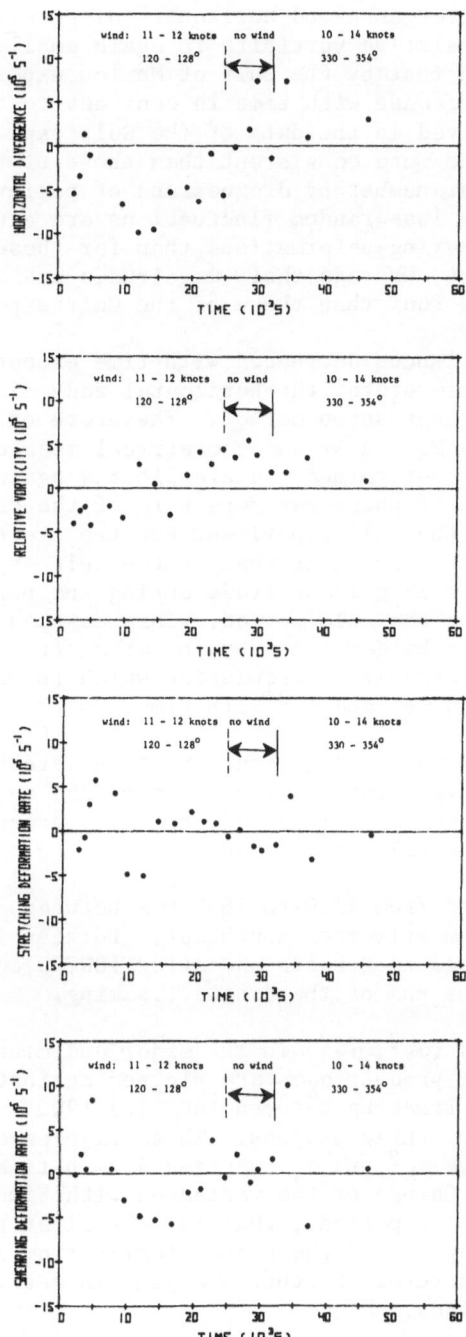

Fig. 12.   Time series of Lagrangian deformation rates (full circles). Time origin is at 2038LT, 6 February, 1978.

Lagrangian mean divergence and horizontal dispersion due to turbulence. The mean relative vorticity is again smaller in magnitude by a factor of 5 than that by the Gulf of Mexico experiment and further it shows steady increase with time in contrast to the rather random fluctuations observed in the data of the Gulf experiment. The present data seem to be more consistent than those of the Gulf experiment in spite of smaller number of drogues and of poorer quality of navigational fix. The less random fluctuations are also observed for stretching and shearing deformations than for those determined from the Gulf experiment, whereas their magnitudes are again smaller by a factor of three to four than those of the Gulf experiment.

Since the variances decreased with time except during a short interval of the calm state, the horizontal eddy diffusivity cannot be determined by the regression method. Therefore only the momentary eddy diffusivities M_X and M_Y and geometrical mean of the two M defined by (7) are determined and are plotted against time in Fig. 13. The magnitude of these parameters is of the order of 10^4 cm^2 sec^{-1} as those of the Gulf experiment but the scattering is much less in the present experiment than in the Gulf experiment. Further both M_X and M_Y show larger magnitude during the period of wind blowing than during the calm period. This may confirm speculation that turbulence may be generated by the wind stress in spite of the generation of the organized circulation which is manifested in decrease of variances σ_X^2 and σ_Y^2 with time.

In the second monitoring, four drifters were tracked with the radarscope about every thirty minutes from 1255 on February 9. At about 1800 three drogues were added and seven drogues were tracked at fifteen minute intervals until 2300.

The wind speed from 1300 to 1600 was between 3 to 6.5 knots and its direction was mostly from southeast. Between 1600 and 1800 the wind speed decreased to 2 knots and after 1800 a complete calm state continued until the end of the drogue tracking.

The variances (σ_X^2, σ_Y^2) in the major and minor axis of scatter ellipses and their product $\sigma_X \sigma_Y$ are plotted against time in Fig. 14. There is a gap in tracking between 1600 and 1900 because of the second launching of three drogues. These figures clearly indicate that both variances σ_X^2 and σ_Y^2 decreased with time when the calm state prevailed. Change of the variances with time was completely different for the two periods, that is, the first period from 1255 to 1600 with four drogues and the second period from 1900 to 2300 with seven drogues. Therefore further analysis is carried out separately for these two periods.

The mean values of Lagrangian deformation rates for the first period are plotted against time in Plate A of Fig. 15. Magnitudes of the four rates are of the same order as those of the first

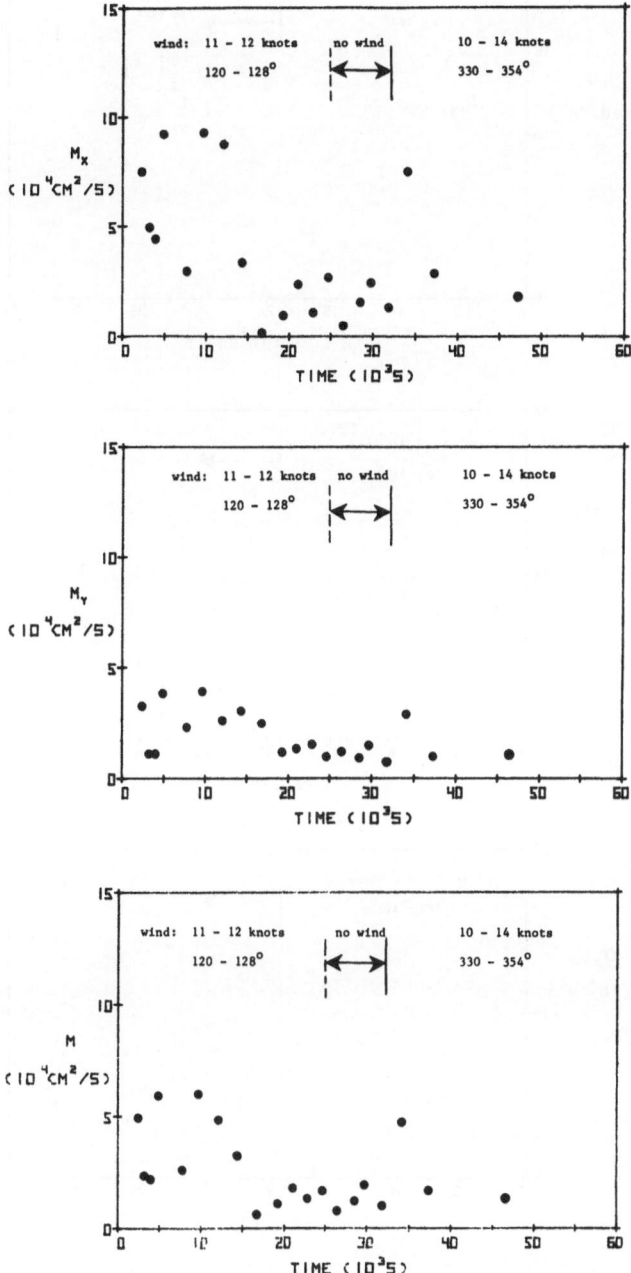

Fig. 13.   Time series of momentary eddy diffusivities (full circles).
           Time origin is at 2038LT, 6 February, 1978.

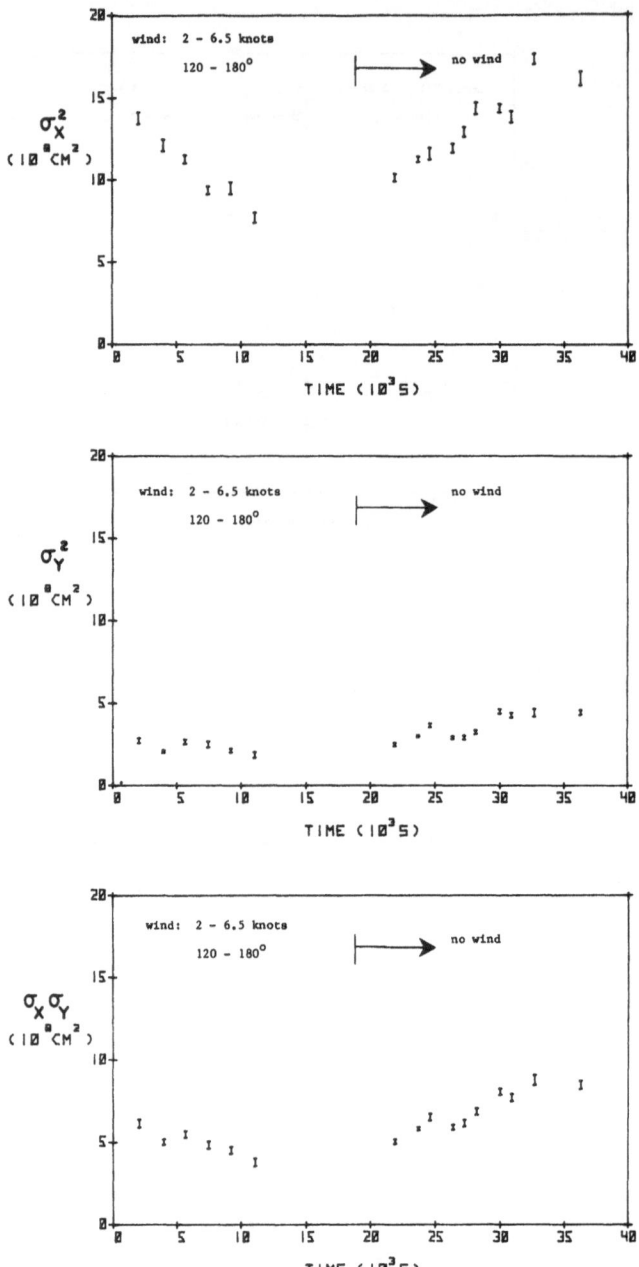

Fig. 14.   Variances versus time.   Time origin is at 1255LT, 9 Febru-
           ary, 1978.   Bars indicate 95% confidence interval.   Data
           points for time < 15 x 10³ sec are with 4 drifters and
           those for time > 20 x 10³ sec are with 7 drifters.

experiment. Also rather large negative values of divergence correspond to the decrease of variances with time. The relative vorticity
was always negative during this period and its fluctuation with time
is less than other deformation rates and also less than the first
experiment.

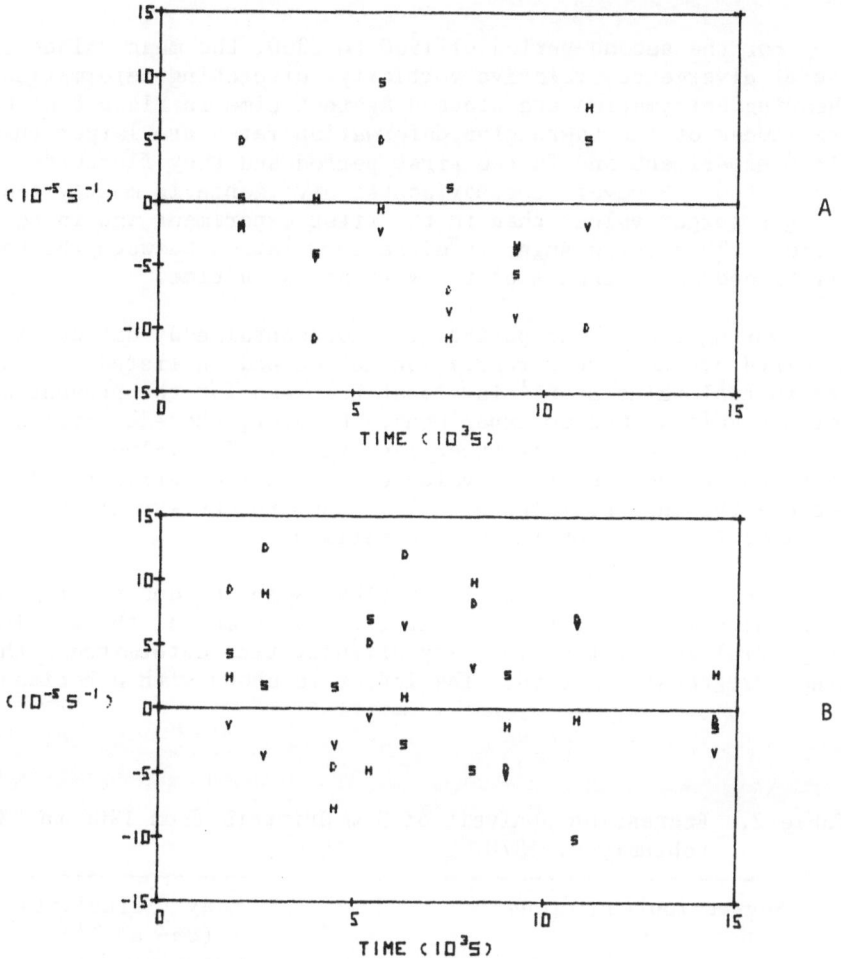

Fig. 15. Time series of Lagrangian deformation rates. D, V, S and H
denote horizontal divergence, relative vorticity, stretching deformation rate and shearing deformation rate, respectively. Plate A is for the first period with 4 drifters.
Time origin is at 1255LT, 9 February, 1978. Plate B is for
the second period with 7 drifters. Time origin is at
1900LT. 9 February, 1978.

Since the regression method cannot be applied to calculate the eddy diffusivity, only the instantaneous eddy diffusivity M_X and M_Y and their geometric mean are determined and plotted against time in Fig. 16. It is noted that the magnitude of the longitudinal diffusivity M_X is less than that of the lateral diffusivity M_Y. The mean diffusivity M is of the same order of magnitude as in the first experiment but it is almost stationary with time in contrast to the first experiment.

For the second period of 1900 to 2300, the mean values of horizontal divergence, relative vorticity, stretching deformation and shearing deformation are plotted against time in Plate B of Fig. 15. Magnitudes of the Lagrangian deformation rates are larger than in the first experiment and in the first period and they fluctuate more widely too. However, the horizontal divergence is mostly positive and has larger values than in the first experiment and in the first period. This again suggests close correlation between the horizontal divergence and increase of the variance with time.

During the second period, the horizontal eddy diffusivity is determined by the linear regression method and is listed in Table 2. The initial value of σ_X^2 is almost the same for the present case as for the Gulf of Mexico experiment. However, the eddy diffusivity K_X is almost twice the Gulf experiment value. The value of K_Y is almost 18 times the corresponding value of the Gulf experiment. This is because the initial value of σ_Y^2 in the present experiment is about 30 times the value of the Gulf experiment.

The momentary eddy diffusivities M_X and M_Y and their geometric mean value M are plotted in Fig. 17. As expected, these values are in general larger than the eddy diffusivities determined with the linear regression method. The latter is shown with a horizontal

Table 2. Regression analysis of 5 m drifters from 1900 to 2300 of February 9, 1978

Regression equations	Eddy Diffusivity (cm^2 sec^{-1})
$\sigma_X^2 = 4.58 \times 10^4 t + 10.55 \times 10^8$	$K_X = 2.29 \times 10^4$
$\sigma_Y^2 = 1.35 \times 10^4 t + 2.79 \times 10^8$	$K_Y = 6.8 \times 10^3$
$\sigma_X \sigma_Y = 2.50 \times 10^4 t + 5.40 \times 10^8$	$\bar{K} = 1.25 \times .04$

(Variances in cm^2, t in sec)

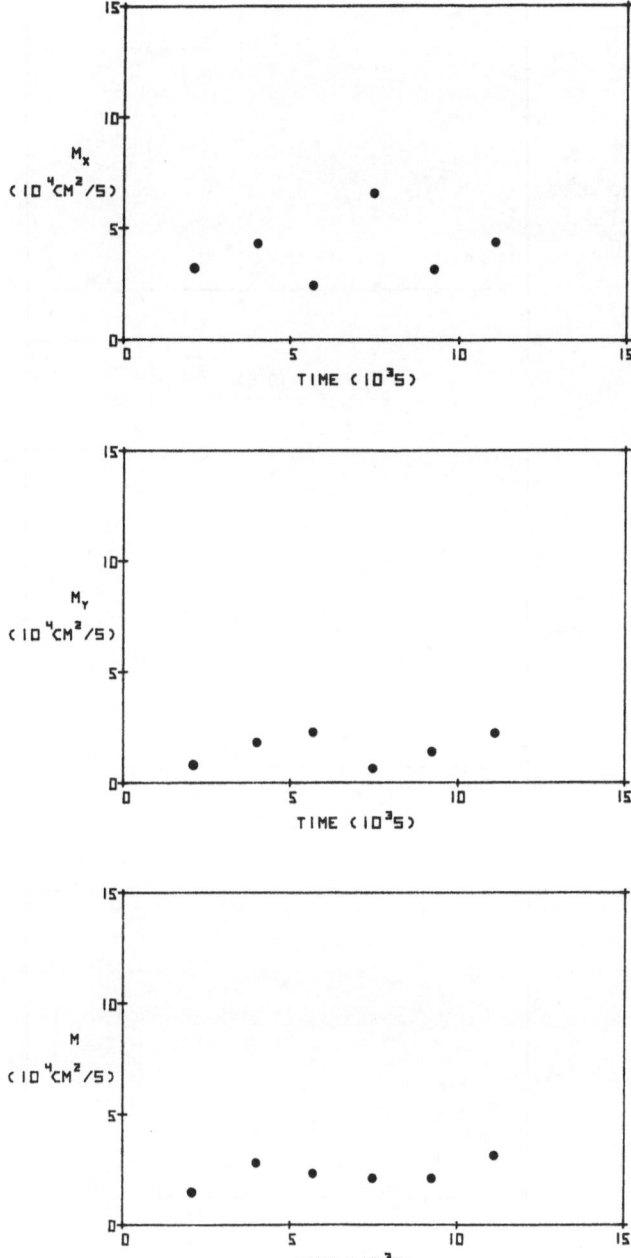

Fig. 16.   Time series of momentary eddy diffusivities (full circles)
            for the first period. Time origin is at 1255LT, 9
            February, 1978.

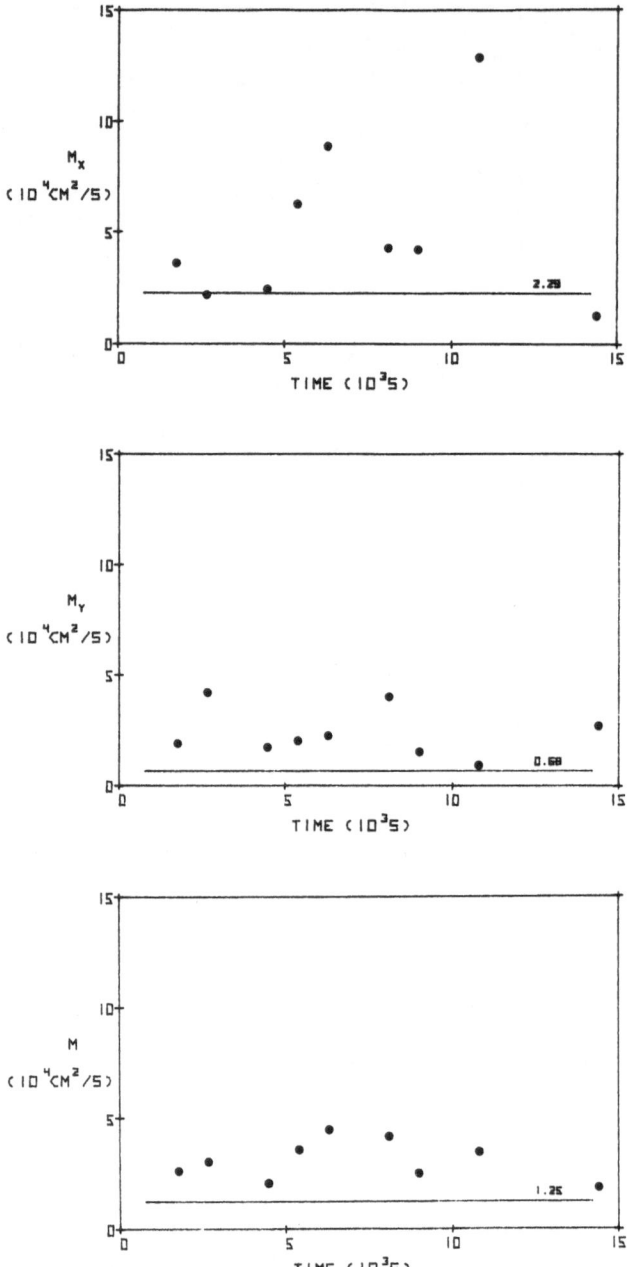

Fig. 17. Time series of momentary eddy diffusivities (full circles)
          for the second period. Time origin is at 1900LT, 9 Febru-
          ary, 1978. Horizontal lines indicate eddy diffusivities
          obtained by the linear regression listed in Table 3.

line.   The magnitude M_X fluctuates with time more widely than that of M_Y and M. This may be due to the difference in definition rather than in difference in physical situation. In fact, the definition for M_X and M_Y from equation 7 should include a numerical constant c on the right side. Table 3 lists eddy diffusivities determined from linear regression and momenary diffusivities and the mean and standard deviation of c calculated by the ratio of the former to the latter for data off Puerto Rico and in the Gulf of Mexico. The mean values of c thus estimated is in the range of 0.1 to 0.7. Ozmidov (1960) conjectured that c is of the order of 0.1. The present examples agree with his conjecture except c_X in the Puerto Rico experiment.

Table 3. Comparison between eddy diffusivity determined from linear regression (K_X, K_Y, $\bar{K}$) and momentary eddy diffusivity (M_X, M_Y, M)

A				B			
K_X	K_Y	$\bar{K}$		K_X	K_Y	$\bar{K}$ (x 10^4 cm^2/sec)	
2.29	0.68	1.25		1.42	0.04	0.32	
M_X	M_Y	M		M_X	M_Y	M (x 10^4 cm^2/sec)	
5.14	2.38	3.11	(Mean)	6.34	0.57	1.84	(Mean)
3.69	1.10	0.90	(SD)	2.92	0.28	0.79	(SD)
c_X	c_Y	c		c_X	c_Y	c	
0.69	0.35	0.44	(Mean)	0.30	0.09	0.22	(Mean)
0.49	0.17	0.13	(SD)	0.21	0.05	0.11	(SD)

A: Based on data off Puerto Rico 1900 to 2300 of Feb. 9, 1978. Number of momentary diffusivity is nine.

B: Based on data in the Gulf of Mexico. Number of momentary diffusivity is fourteen.

SD: Standard deviation

c_X, c_Y, C: Factor in $M_X = c_X \sigma_X \sigma_u$ etc. instead of (6)

ANALYSIS OF AERIAL PHOTOGRAPHS

Aerial photographs of the chemical waste plume were taken for
several hours from a NASA aircraft in the Gulf dumping experiment on
July 25, 1977. These photographs indicate the waste plume, though
not mixed with dye, is brownish in color and also show that striations began to develop perpendicular to the plume axis within one
hour after being dumped. Selected aerial photographs were digitized
by use of a microphotodensitometer at Johnson Space Center. The
scanning "window" is 25 μ wide in the direction of the plume axis and
125μ long across the plume. The instrument measures the opacity of
the photonegative while scanning it at a preselected aperture and
scan interval. Plate A of Fig. 18 indicates a portion of the digitized plume density distribution with arbitrary scale for

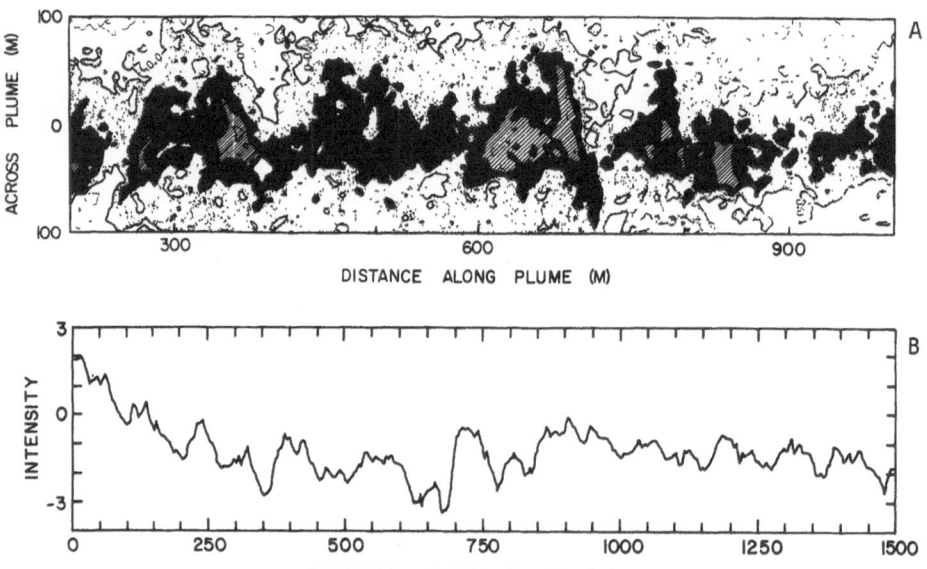

Fig. 18A.   Photo—density distribution (determined from an aerial pho-
            tograph) for the chemical waste plume in the Gulf dumping
            experiment on 25 July, 1977. Density is indicated with
            different shades from the lowest value in blank and the
            highest value in hatching as determined with photo—densi-
            tometry (Courtesy of NASA Johnson Flight Center, Photo-
            densitometry Division).

Fig. 18B.   Photo—density averaged across the plume of Fig. 18A versus
            distance along the plume axis in an arbitrary scale.

photodensity. Though the wind force was about 2 in Beaufort scale, the striations started to develop normal to the plume axis indicated in this figure.

In order to show the density change along the axis, plate B of Fig. 18 indicates the density averaged across the plume against the distance along the plume axis. This figure shows that the averaged density fluctuates with approximate wave length of about 160 m. The power spectrum of the averaged density is computed by use of FFT and by the Maximum Entropy Method (Smylie et al., 1973) and is plotted in Fig. 19. Both spectra show the distinguished peak near the wave number 0.06 cycle/meter or wavelength of 160 m as expected from Fig. 18. The secondary peaks at wave lengths of 81 m and 51 m are due to the

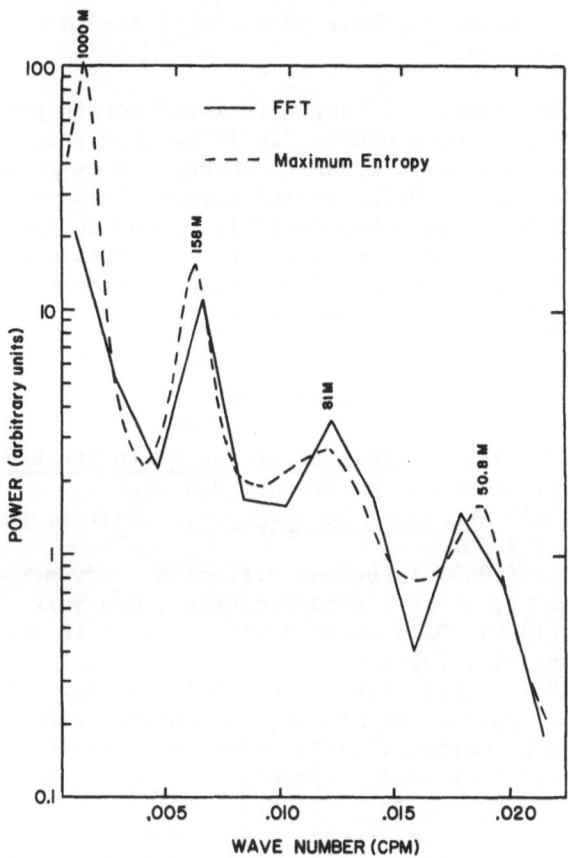

Fig. 19.   Power spectrum of averaged density for the chemical waste plume in the Gulf dumping experiment on 25 July, 1977. The data are from Fig. 18B. The wave number is in the direction of the plume axis.

overtones of the fundamental peak.

Those spectra seem to confirm that the plume is not a continuous
jet as assumed in dispersion of smoke (Frenkiel, 1973) or in discharged mining waste (Ichiye and Carnes, 1977) but consists of striations. The striations were apparently caused by the wind, since they
were almost lined up along the wind.

CONCLUSION

The works described here represent an initial assessment of
waste dispersion. The results demonstrate that drifters can be
tracked easily with a ship-borne radar to provide data for determining horizontal diffusivity and Lagrangian deformation rates. The
waste plumes are not always turbulent jets as described by simple
models or by analogies but have more complicated structures with
striations or streaks.

Although horizontal diffusion is important in predicting behavior and fate of the waste plumes, it is necessary to obtain information on vertical mixing in order to assess the environmental effects
of the waste disposal. Drifters and aerial photos do not provide
data for vertical mixing. One feasible method to assess vertical
mixing is application of fluorometry to the artificially dyed waste
plumes when other hydrographic data such as velocity and density
profiles are collected simultaneously.

REFERENCES

Csanady, G. T., (1973) Turbulent Diffusion In The Environment. D.
    Reidel Publishing Co. Dordrecht, 248 pp.
Defant, A., (1961) Physical Oceanography. Vol. I, Pergamon Press,
    New York, 729 pp.
Frenkiel, F. N., (1953) Turbulent diffusion. Advances in Applied
    Mechanics. 3, 61-107, Academic Press, New York.
Huang, N. E., (1979) On surface drift currents in the ocean. J.
    Fluid Mech., 91, 191-208.
Ichiye, T., (1967) Upper ocean boundary-layer flow determined by dye
    diffusion. Physics of Fluids, Supplement, 270-277.
Ichiye, T., and M. Carnes, (1977) Modeling of sediment dispersion
    during deep ocean mining operations. 9th Ocean Technology
    Conference Proceedings in Houston, 421-432.
Ichiye, T., C. Mungall, M. Inoue, and D. Horne, (1978) Gulf of Mexico
    dispersion calculations. Texas A&M University Research
    Foundation, Technical Report, Ref. 78-10-T, 49 pp.
Liepmann, H. W., (1979) The rise and fall of ideas in turbulence.
    American Scientist, 67, No. 2, 221-228.
Monin, A. S., and A. M. Yaglom, (1971) Statistical Fluid Mechanics.

Vol. I, MIT Press, Mass., 769 pp.

Okubo, A., (1971) Ocean diffusion diagrams. Deep-Sea Res., 18,
     789-802.

Okubo, A., and C. C. Ebbesmeyer, (1976) Determination of vorticity,
     divergence, and deformation from analysis of drogue observa-
     tions. Deep-Sea Res., 23, 349-352.

Okubo, A., C. C. Ebbesmeyer, and J. M. Helseth, (1976a) Determination
     of Lagrangian deformation from analysis of current followers.
     J. Phys. Oceanogr., 6, 524-527.

Okubo, A., C. C. Ebbesmeyer, J. M. Helseth, and A. S. Robbins,
     (1976b) Reanalysis of the Great Lakes drogue studies data. SUNY
     Marine Research Center, Spec. Rept. 2, Ref. 76-2, 85 pp.

Ozmidov, R. V., (1960) On the rate of dissipation of turbulent energy
     in sea currents and on the dimensionless universal constant in
     the "4/3-power law." (English Translation, American Geophysical
     Union) Bull. of Acad. Sci. USSR Geophysics 8, 821-823.

Pollard, R. T., (1976) Observations and theories of Langmuir circula-
     tions and their role in near surface mixing. Deep-Sea Res., Sir
     George Deacon Anniversary Suppl., 235-251.

Smylie, D. E., G. K. C. Clarke, and T. J. Ulrych, (1973) Analysis of
     irregularities in the earth's rotation. Methods in
     Computational Physics, 13, 391-430, Academic Press, New York.

Tennekes, H. and J. L. Lumley, (1972) A First Course in Turbulence.
     The MIT Press, Cambridge, Mass., 300 pp.

DISPERSION OF PARTICULATES IN THE OCEAN STUDIED ACOUSTICALLY:

THE IMPORTANCE OF GRADIENT SURFACES IN THE OCEAN

John R. Proni and Donald V. Hansen

Environmental Research Laboratory
National Oceanic and Atmospheric Administration
15 Rickenbacker Causeway
Miami, FL 33149

ABSTRACT

Knowledge of the rate of dispersion of particulate matter dumped in the upper ocean and of the processes governing that dispersion is fundamental to an understanding of the effects of dumping on the ocean environment. Accurate chemical dilution rates of dumped material cannot be determined without a knowledge of the spacial distribution of ocean dumped material and the way in which that distribution changes with time. Acoustical methods offer one approach to studying ocean dispersion of dumped material. Acoustical results presented herein show some of the complicated effects which gradients in oceanic parameters, such as temperature and density, produce in the dispersion of particulate matter dumped in the ocean.

INTRODUCTION

The subject of this paper is the dispersion of particulate matter in the ocean with special emphasis being given to acoustic methods for studying that dispersion. Special emphasis is also given to particulate matter introduced into the ocean by man such as in ocean dumping. The concept that biota accumulate at density gradient surfaces will be re-examined from the point of view of ocean dumping. The hypothesis we will consider is that dumped particulates accumulate at least in the short term (e.g., order of weeks) and possibly in the long term (e.g., months, years) on the same density surfaces as natural immobile particulates. Obviously, this hypothesis cannot be true for all classes of materials which contain some particulates achieving or nearly achieving neutral buoyancy within, say, the upper

100 m of the water column. This hypothesis has been considered in
discussions dealing with oceanic plutonium (Bowen, 1968) and polychlorinated biphenyls (Harvey, 1976). Recent evidence has indicated
the importance of gradient surfaces in particulate dispersion (Proni
et al., 1975; Proni et al., 1977; Orr and Hess, 1978).

     The existence of gradient surfaces within the upper 100 m of the
water column can significantly alter the horizontal and vertical dispersion of dumped material from that in a homogeneous, turbulent
ocean. Little oceanic field data exist on the influence of gradient
surfaces on dump material dispersion. A larger amount of work has
been done on the effects of density stratification, in the atmoshere,
upon dispersion (Csanady, 1967). One of the advantages of oceanic
acoustics is its ability to delineate the topography of gradient surfaces prior to dumping and its ability to study the effects of the
gradient surfaces on the space-time dispersion of dumped material
after dumping.

     There are several examples in the literature of coincidence of
biota concentrations with gradient surfaces (Pieper, 1977). Of most
interest in the ocean dumping context is the accumulation of passive,
immobile biota along easily measurable gradient surfaces such as
those of temperature, salinity and density. In the case of mobile
scatterers other effects such as light levels might come into play,
or food-web dynamics thus potentially reducing the correlation between gradients and accumulation of particulates.

     In the following section several examples of acoustically ob-
served gradient surfaces will be presented. The first example shows
gradient surfaces with no dumping having occurred. Examples two and
three come from the sewage sludge dumpsite in the New York Bight.

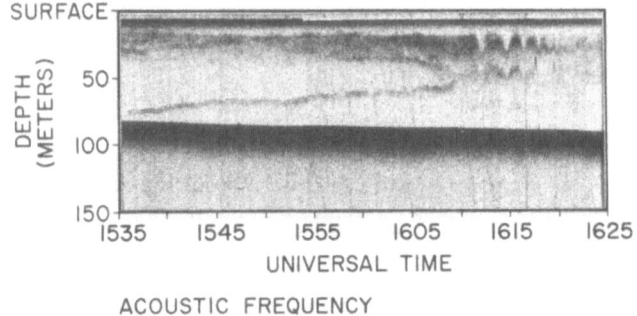

ACOUSTIC FREQUENCY
20 KHz

Fig. 1. Real-time acoustic data gathered while moving in a
         general southeasterly direction at 2 m/sec aboard the RV
         KELEZ. Shown in this print are the shelf water/slope
         water mass boundary and an internal wave wavepacket.

EXAMPLES OF GRADIENT SURFACES

The first example is the New England shelf-slope water mass
front in the vicinity of latitude 40°25'N, longitude 71°W, as it appeared on June 22, 1976 (Fig. 1; Newman et al., 1977). These data
were gathered by a ship using a 20 kHz acoustic echo-sounder while
moving seaward away from the coast. The upper and lower boundaries
of the shelf-slope front are clearly visible; also present in the
figure is an internal bore (Proni and Apel, 1975), which is moving
towards the front. In this instance, three gradient surfaces are
present. They are (a) the shelf-slope water mass boundary (both the
upper and lower boundaries are visible), (b) the horizontal surface
in the depth range of 15 to 20 m and (c) the horizontal surface at 50
m depth located seaward of the shelf-slope front. Patchiness of particulates appears to be present in the water column because of the
presence of the internal bore. A particulate sampler, or plankton
sampler, or chlorophyll a sampler, towed at 20 m depth and capable of
50 m or so horizontal resolution probably would confirm the existence
of patchiness of particulates. The boundaries of the shelf water/
slope water front are known to be regions of high gradients of oceanic parameters such as temperature and salinity. Newman et al. (1977)
present temperature profile data corresponding to the acoustic data
shown in Fig. 1.

An interesting comparison of the influence of slight temperature
and density gradients in the water column upon particulate dispersion

SLUDGE TRACKING EXPERIMENT I

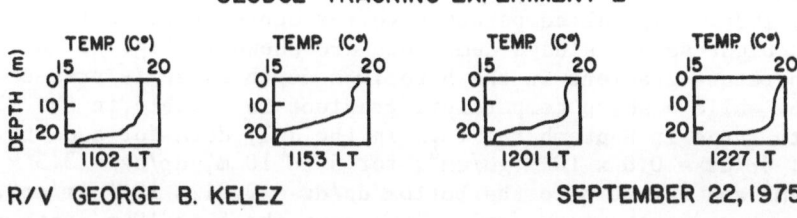

R/V GEORGE B. KELEZ   SEPTEMBER 22,1975

SLUDGE TRACKING EXPERIMENT II

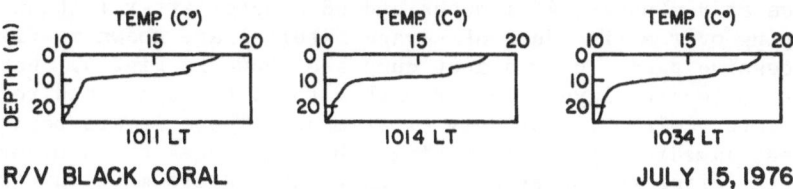

R/V BLACK CORAL   JULY 15, 1976

Fig. 2. Typical temperature profiles in the N. Y. Bight Sewage
          Sludge dump area for September 1975 and July 1976.

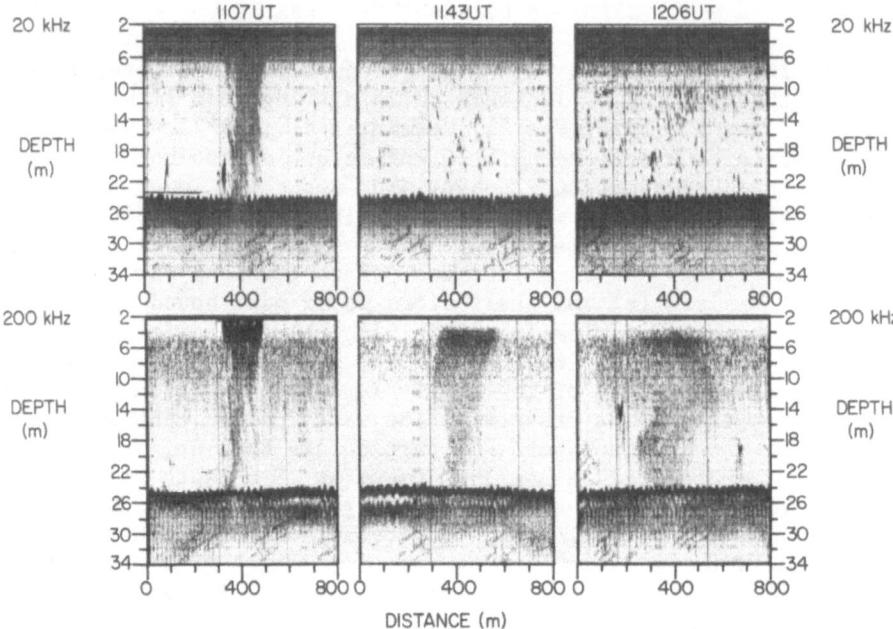

Fig. 3.   Acoustic data gathered during three successive passes over
          a line dump commercing at 1058 U.T. in September 1975.

can be made by utilizing acoustic data from two dumping experiments
carried out in the New York Bight, one in September 1975 and one in
July 1976.   Typical temperature versus depth profiles from the New
York Bight sewage sludge dump zone are shown in Fig. 2.   A strong
temperature gradient in the 5 to 10 m depth range is present in July
1976, while a sharp temperature gradient is visible in the 15 to 22 m
depth range in September 1975.   In the July data for a depth of 0 to
5 m, $d\rho/dz = 0.8 \times 10^{-6}$ gm/cm^4, for 5 to 10 m, $d\rho/dz = 3.3 \times 10^6$
gm/cm^4 and for 10 m to the bottom $d\rho/dz = 0.01 \times 10^{-6}$ gm/cm^4, where
ρ = water density in gm/cm^3.   Thus over the 5 to 10 m depth range a
total density increment of approximately 1.5×10^{-3} gm/cm^3 occurs.
Acoustic data gathered in the September and July experiments are
shown in Figs. 3 and 4 respectively.   Three successive passes taking
place at 9 minutes, 45 minutes and 68 minutes after initiation of
dumping over a line dump of sewage material are shown in Fig. 3. Two
successive passes over a spot dump are shown in Fig. 4.   Note that
dark reflecting layers appear within the 5 to 10 m depth region of
the water column in the July experiment.   These layers begin to
appear immediately after dumping (the first pass shown occurred 16
minutes after the onset of dumping).   The layers move outward from
the principal body of the dump at a speed of about 50 cm/sec,
decreasing with distance from the dump (Proni et al., 1978).

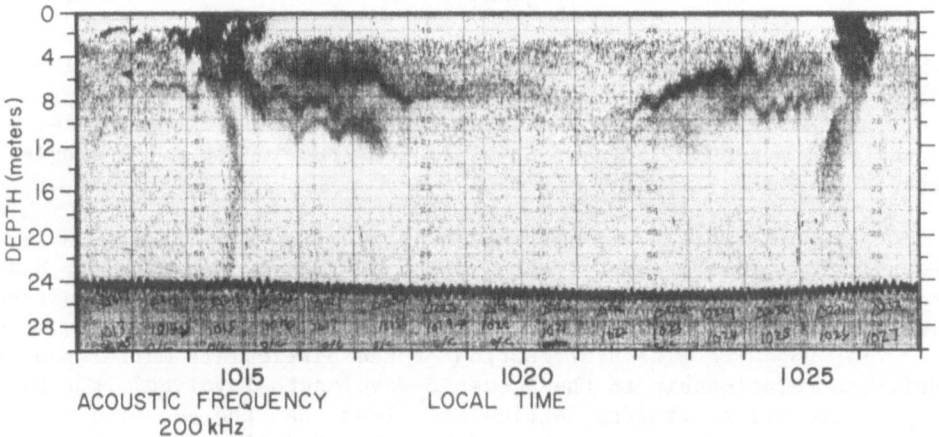

Fig. 4. Two successive passes over a spot dump in July 1976 are
         shown. Clearly indicated are dark reflecting layers of
         particulate matter in the 5-10 m depth range.

     In the July dump the sewage sludge dispersed horizontally in the
5 to 10 m depth range at a speed far in excess of any to be expected
in dispersion in a homogeneous ocean. The material shallower than
the 5 m depth and deeper than the 10 m depth does in fact disperse at
a rate in agreement with traditional estimates (Proni et al., 1978b).
This fast dispersion of dump material in the 5 to 10 m depth range
continued (with diminishing speed) for a period in excess of 3 hours.

ACOUSTICAL CONSIDERATIONS

     The fundamental measurable quantity available for interpre-
tation is the received back-scattered, acoustical signal intensity I.
In a dumping experiment, I consists of two basic components, I_D and
I_B, where I_D denotes the received signal intensity from the dump
material itself, and I_B the received signal intensity from the
background particulate material -- i.e., material present in the
water column prior to the dumping event.

     Now in general $I_B = I_B (t,\bar{x})$ and $I_D = I_D (t,\bar{x})$ where $\bar{x} =$
x,y,z. If we let $I (t,\bar{x})$ denote the received signal intensity for a
given $(t,\bar{x})$ then the fundamental acoustical quantity of interest for
ocean dumping dispersion measurements, as far as the solid phase of
the dump material is concerned, is $I_D = I - I_B$.

     The basic assumption made in utilizing I_D for concentration
measurements is that:

$$I_D = CN_D \qquad\qquad\qquad (1)$$

where N_D is the number of particles per unit volume under consideration and C is a proportionality constant. Several complicating factors are contained within C. These include particle shape, size, density, compressability and physical construction (e.g., a single entity or a floc).

In a wide variety of circumstances equation (1) is reasonable. However, no rigorous experimental proof of this proportionality has been made for material dumped in the ocean or for background particulate material in the ocean. Several factors must be considered:
   i) Does the size distribution of the particulate matter bear a definite relationship to the acoustic wavelength involved? For example, are all scattering particulates less than the acoustic wavelength λ? For 20 kHz sound, λ = 7.5 cm and for 200 kHz sound λ = .75 cm. In these two frequencies the particulate matter in dump studies is generally smaller than the wavelength, thus avoiding certain difficulties in applying equation (1).
   ii) Does the size distribution or particle characteristics (compressibility, density) change in going from scattering volume to scattering volume? The more the size distribution changes from volume to volume in space the less accurate equation (1) is likely to be.

There are other questions regarding equation (1) which may be asked, such as: Is the density of the dumped material sufficiently high for multiple acoustic scattering to occur? Are there scatters other than dumped particulates present in the dump such as bubbles or flocs which may contribute to the scattered acoustic signal? The answer to these questions must be examined on a case by case basis.

In using I_D for concentration estimation it is not necessary that:

$$I_B = CN_B \qquad\qquad\qquad (2)$$

I_B is an experimentally measured quantity which is subtracted from I. However, as the notion of gradient surfaces in the ocean is discussed, equation (2) may need to be invoked.

DISPERSION OF PHARMACEUTICAL WASTES

An example of dumping in the deep ocean in which gradient surfaces may or may not be of importance in both the short and long term dispersion of particulates is provided by pharmaceutical waste dumping 40–60 km north of San Juan, Puerto Rico. Data from two pharmaceutical dumping events, one occurring in February 1978 and one in October 1978 are presented. The pharmaceutical material contains a

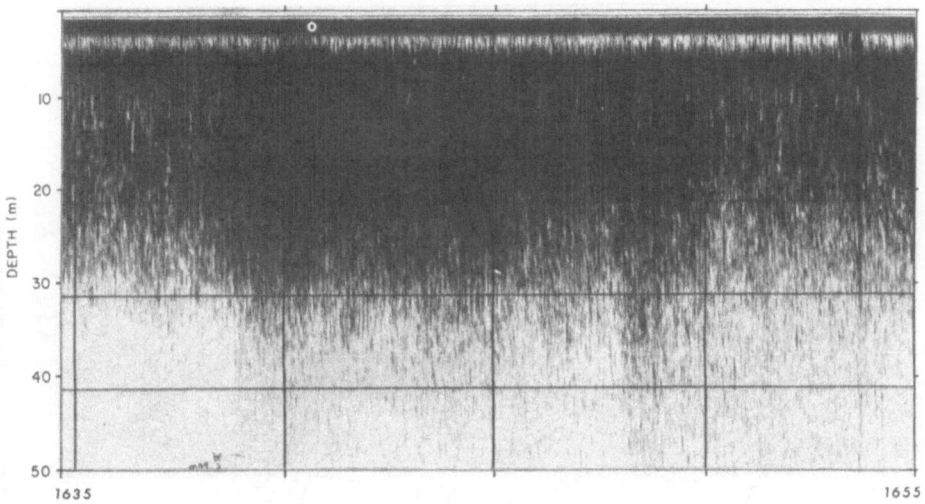

Fig. 5.   Acoustic data (real-time) obtained during a pharma-
          ceutical waste dump north of San Juan, Puerto Rico in
          February 1978.

small amount of solids, less than 2%, and is dumped from a moving
barge.

During the February dump the oceanic water column was well-
mixed. The uppermost detectable temperature feature was the thermocline at about 50 m. Fig. 5 shows some graphical 200 kHz acoustic
data obtained in passing over the dump material about 1 hour after
the dumping event occurred. In order to begin to quantitatively in-
terpret this record in terms of concentration of material, we must
turn to plots of I_D (Fig. 6) for the time interval 1637 (marked as
0.00 on the bottom of the plot) to 1653 U.T. The data displayed in
Fig. 5 have had I_B $(t,\bar{x})$ removed and have also been compensated for
geometrical spreading effects. The background data I_B are removed
in the following way. Prior to the dump event an acoustical background survey of the dump area is carried out and a map of I_B $(t,\bar{x})$
is obtained for the period prior to the dump event. After the dump
event occurs, further data on I_B $(t,\bar{x})$ is obtained when the research ship is clearly out of the dump zone. In this way the I_B
map is continually updated. After the dump event occurs maps of I
$(t,\bar{x})$ are obtained. The data in Fig. 6 results from subtracting
corresponding map points of I_B $(t,\bar{x})$ from I $(t,\bar{x})$, so that plots of
I_D $(t,\bar{x})$ result. From this plot we see that indeed most of the
material remains in the upper 50 m of the water column and that in
fact the peak of the concentration of material occurs in the 20 to 40
m depth range. We note that three distinct concentration peaks are
visible in Fig. 6, all at approximately 30 m depth; these peaks have

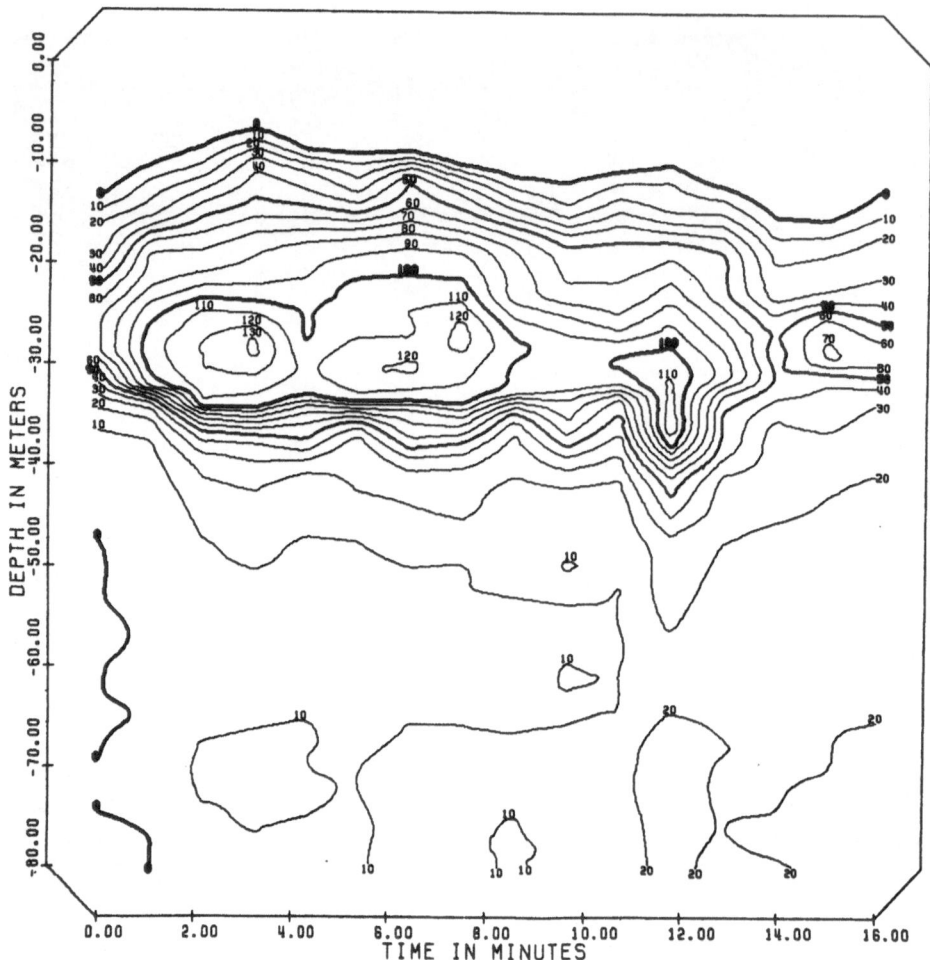

Fig. 6. Computer processed acoustic data corresponding to the
         real-time data shown in Fig. 5. Three concentration
         peaks are visible.

values of 130 acoustic scattering units, 120 acoustic scattering
units and 110 acoustic scattering units respectively.

     In terms of dispersion calculations the data in Fig. 6 are one
realization of a statistical process, i.e., the oceanic dispersion of
dumped particulates. It is clear that the acoustic data provide a
far more complete realization of a dumping event than any other known
method, including fluorometry (depending of course on sufficient particulates being present in the dump material), by obtaining more
passes over the dump materials, values of the horizontal dispersion

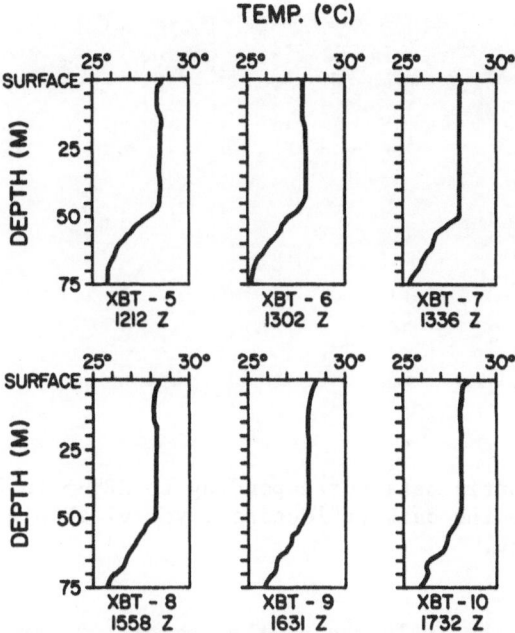

Fig. 7. A series of temperature profiles made during the
         October pharmaceutical waste dump showing a temperature
         gradient within 15 m of the ocean surface.

coefficients may be obtained (Proni et al., 1978c). In terms of
chemical analysis and chemical sampling the data portrayed in Fig. 6
provide a means of separating the effects of spacial gradients in concentration and true dilution effects. For example, if one were attempting to measure the rate of dilution at the peak of concentration
within the water column (at 130 acoustic units), closing a water sampling bottle at a depth of 24 m instead of 29 m results in an error
of 30% for the dilution calculation.

    Let us now return to the question of gradient surfaces by exam-
ining the dumping experiment which occurred in October 1978. Prior
to and during the October dumping experiment a substantial quantity
of rain fell and winds exceeded 20 knots. These conditions undoubt-
edly contributed to the presence of a layer of colder water which
formed in the upper 15 m of the water column. The bottom of this
layer of colder water was marked by a vertical temperature gradient
which clearly separated the cold layer from the well-mixed layer below it. A series of expendable bathythermograph traces made during
the October cruise are shown in Fig. 7. The bottom depth of the cold
water varies from location to location and the depth of the thermocline is about 45 meters.

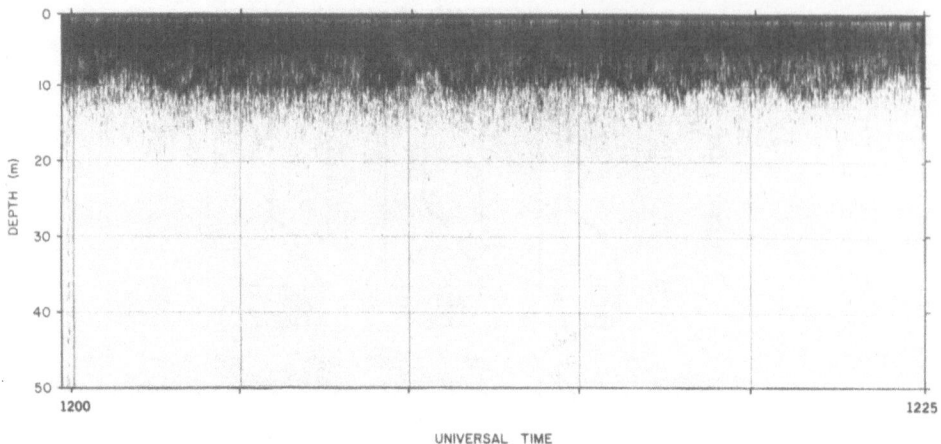

Fig. 8.   Acoustic data corresponding to XBT-5 in Fig. 7.
          Note the dark reflecting layer visible at about 12 m
          depth.

     To examine the influence of the temperature gradient at the bot-
tom of the cold water pool upon the dispersing pharmaceutical material consider XBT-5 in Fig. 7.   The cold pool temperature gradient begins at about 11 m depth and extends to about 14 m depth.   This corresponds to the dark reflecting layer visible at about 9 m depth in
Fig. 8.   (One must add 2 to 3 meters to the depth shown in Fig. 8 to
correct for the depth of the acoustic transducer below the surface.)
In MXBT-6 the cold pool temperature gradient begins at about 15 m
depth.   The dark reflecting layer shown in Fig. 8 now occurs at about
15 m depth (Fig. 9).   Finally, in XBT-9 the cold pool temperature
gradient has essentially disappeared as has the cold layer itself;
instead the surface temperature is now warmer than the mixed layer
temperature.   Fig. 10 shows the acoustic record corresponding to
XBT-9.   The dark reflecting layer has essentially disappeared and the
overall appearance of the acoustic data is similar to February 1978
data.

     The "overall" dispersive behavior of the February and October
1978 pharmaceutical dumps were very similar: in both cases the dump
material generally proceeded in a west-northwesterly direction, in
accordance with the prevailing currents (Hansen, 1974) in the dump
area.   Both dumps had a sharply defined eastern edge but a generally
diffuse western boundary.   A conjecture which has evolved out of the
Puerto Rican dump experiments is that pharmaceutical material from
successive dumps remain near the ocean's surface (within 50 m) and
that the dumps eventually "merge" with one another forming a rather
broad carpet of material extending west-northwesterly from the dump
zone.

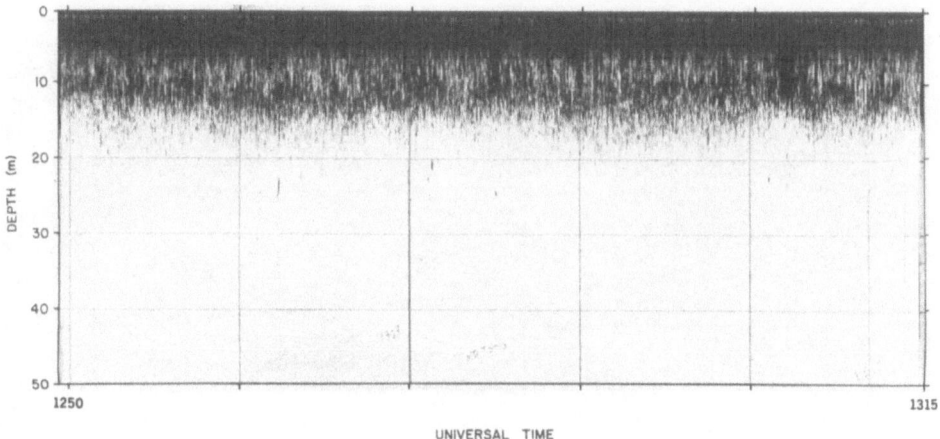

DEPTH (m)

UNIVERSAL TIME

Fig. 9. Acoustic data corresponding to XBT-6 in Fig. 7. Note the
dark reflecting layer visible at about 15 m depth.

        In the Puerto Rican experiments two examples of preferred sur-
faces appeared. The first is the thermocline itself, located at
about 50 m depth. In neither of the experiments did the thermocline
play a major role in particulate dispersion over the lifetime of the
field experiment. Nevertheless enhanced dump material concentrations
were detected on the thermocline, suggesting its potential importance
for longer term dispersion studies. The second is the cold pool temperature gradient which did play a role in short term particulate
dispersion in that (a) enhanced concentrations of dump particulate
matter were detected within the cold pool temperature gradient, (b)
the vertical dispersion of the dump material was limited to some degree by the cold pool temperature gradient and, (c) the importance of
near-surface dispersion mechanisms were given a role of increased importance since the dump material removed was held within 15 m of the
surface by the cold pool temperature gradient.

                        CONCLUSIONS AND SUMMARY

        In this paper several examples of gradient surfaces within the
ocean have been presented. From the point of view of ocean dumping
the existence of such surfaces is important for at least two reasons;
the first is that such surfaces can substantially influence the dispersion of material dumped within the ocean, the second is that over
the short term and possibly over the long term gradient surfaces act
as zones of accumulation of dump material. Such gradient surfaces
may also act as boundary regions for portions of the water column
moving with differing horizontal velocities -- i.e., a vertical
current shear exists across the gradient surface. These surfaces

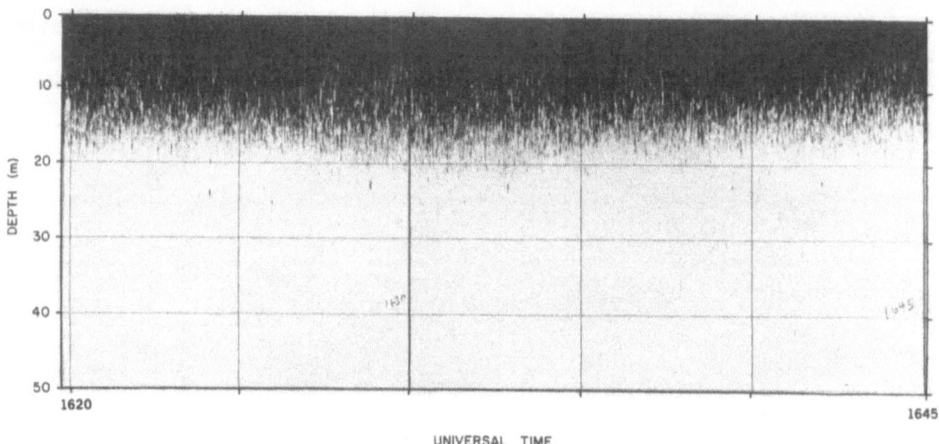

UNIVERSAL TIME

Fig. 10.   Acoustic data corresponding to XBT-9 in Fig. 7.   Note the
lack of a dark reflecting layer.

also served as a zone of accumulation of dumped sewage material.

Some effects of transient gradient surfaces (or at least thought
to be transient) have been illustrated through the use of pharmaceutical waste acoustic data. A study carried out by Orr and Hess
(1977) at Deep Water Dumpsite 106, off New York City reveals many
dispersive characteristics similar to those observed in the October
1978 Puerto Rican pharmaceutical waste study. They also find that
particulates which are near neutral buoyancy are subject to dynamic
effects involving gradient surfaces.

In summary, caution must be exercised in treating the ocean as a
spatially homogeneous well mixed layer bounded by a seasonal thermocline when considering the dispersion of dumped material in the
ocean. The existence and influence of gradient surfaces should be
ascertained in a dumping experiment. Regions of relatively sharp
gradients are likely places to sample for dumping related contaminants. Such sampling in gradients should be done both before and
after the dumping process begins.

REFERENCES

Bowen, V. T., V. E. Noshkin, H. L. Wolchok and T. T. Sugihara (1968)
    Fallout strontium-90 in Atlantic Ocean Surface Waters.
    Health and Safety Laboratory Quart. Report NYO-2174-77, pp.
    1-2-I-64.
Csanady, G. T. (1973) Turbulent Diffusion in the Environment. D.
    Reidel Publishing Co., Dordrecht, Holland.

Hansen, D. V. (1974) A.I.A.A. Conference on Free Drifting Buoys NASA CP-2003, Hampton, VA, pp. 175-192.

Harvey, G. R. and W. G. Steinhauer (1976) Transport pathways of polychlorinated biphenyls in Atlantic Water. Journal of Marine Research, 34, (4), 561-575.

Newman, F., J. R. Proni, D. J. Walter and H. M. Byrne (1977) Acoustic imaging of the New England Shelf-Slope Water mass interfaces. Nature, 269, (5631), 790-791.

Orr, M. H. and E. R. Hess (1978) Acoustic monitoring of industrial chemical waste released at Deep Water Dump Site 106. Journal of Geophysical Research, 83, (C12), 6145-6154.

Pieper, R. E. (1977) Some comparisons between oceanographic measurements and high-frequency scattering of underwater sound. In: Oceanic Sound Scattering Prediction. Neil R. Anderson and Bernard J. Zahuranec, editors.

Proni, J. R., D. C. Rona, C. A. Lauter and R. L. Sellers (1975) Acoustic observations of suspended particulate matter in the ocean. Nature, 254, (5499), 413-415.

Proni, J. R., F. C. Newman, R. Young, D. Walter, I. Duedall, H. Stanford and H. Parker (1980) On the observation of the intrusion into a stratified ocean of dumped sewage sludge and concomitant generation of internal oscillations. Submitted: Deep-Sea Research

Proni, J. R., F. C. Newman, F. Ostapoff and D. J. Walter (1978) Vertical particulate spires or walls within the Florida Current and near the Antilles Current. Nature, 276, 360-362.

APPLICATION OF REMOTE SENSING TO MONITORING AND STUDYING

DISPERSION IN OCEAN DUMPING

Robert W. Johnson and Craig W. Ohlhorst

NASA Langley Research Center
Hampton, VA 23665

ABSTRACT

Experiments conducted in the coastal waters of the United States indicate that plumes resulting from ocean dumping of sewage sludge and industrial wastes have distinguishable spectral characteristics. Remotely sensed wide area synoptic coverage provides information on these pollution features that is not readily available from other sources. Results indicate that qualitative analysis techniques may be used for the location, identification, and mapping of plumes resulting from ocean dumping of waste materials. An in-scene background elimination technique was developed that "normalizes" atmospheric and other environmental effects, thereby potentially providing a means of plume identification that is independent of the specific scene and the multispectral scanner used. Application of this technique to data from several experiments demonstrates that plumes resulting from sewage sludge and several industrial wastes have distinctive spectral characteristics over a range of environmental conditions and for two multispectral scanners flown on aircraft at altitudes of 3.0 and 19.7 kilometers.

In addition to qualitative analyses that used the in-scene background elimination, quantitative analysis techniques were applied to remotely sensed data collected over sewage sludge and industrial waste plumes. Results indicate that quantitative relationships exist between remotely sensed data and suspended solids in sewage sludge plumes and iron concentrations in acid-iron waste plumes. Calibrated regression equations from the multiple regression analyses were applied to map quantitative distributions of those parameters in remotely sensed scenes. These quantitative values and/or radiance differences also have been used to study the temporal dispersion of plumes and surface water movement in the dump areas.

INTRODUCTION

Remotely sensed wide area synoptic data provides information on ocean dumping that is not readily available by other means. These remotely sensed data may be interpreted by two approaches. The first approach is qualitative and has been used to map features, such as river plumes, without concurrent sea-truth measurements. Photographic and multispectral scanner imagery or digital data have been used in conjunction with other available data, following photoanalysis techniques. The second approach is quantitative analyses in which sea-truth collected at a limited number of locations is used to calibrate remotely sensed digital data and to extend the results to the remotely sensed scene. Results of the quantitative analyses, in the form of calibrated regression equations, have been used to develop maps showing quantitative distributions of one or more water quality parameters, such as suspended solids or chlorophyll a.

Qualitative results have been reported by Scherz et al. (1973) in which aircraft photographs were used to locate and trace suspended sediment plumes and study their dispersion characteristics; and by Klemas et al. (1978) in which satellite (LANDSAT and SKYLAB) multi-spectral scanner imagery and/or photographs were used to study suspended sediment distributions, acid-iron waste plume persistence, and other charateristics of coastal zone pollution features. Quantitative results have been reported in which Yarger et al. (1973), Williamson and Grabau (1973), Klemas et al. (1974), Rogers et al. (1975), Johnson and Bahn (1977), and Johnson (1978) applied classification or regression techniques to calibrate satellite (LANDSAT) and aircraft multispectral scanner data and to map distributions of water quality parameters in inland and estuarine systems. Results from the analyses of spacecraft data (LANDSAT) indicate certain features may be monitored (Bowker and Witte, 1977). However, present satellite coverage of the coastal zones precludes the use of these systems for routine monitoring of ocean dumping. Distributions of parameters were for surface waters (about top 1 meter) in these relatively turbid lake and coastal waters (Whitlock et al., 1978). The regions, parameters monitored, and remote sensors for these prior studies are summarized in Table 1.

More recently, joint National Aeronautics and Space Administration (NASA)/National Oceanic and Atmospheric Administration (NOAA) experiments have been conducted at the Environmental Protection Agency (EPA) designated dump areas in the United States coastal zones to determine the applicability of aircraft remote sensing systems to (1) locate, identify, and map plumes resulting from ocean dumping of sewage sludge and industrial wastes, and (2) evaluate previously developed quantitative analysis techniques that may be used for studying dispersion of materials in these plumes. These investigations include development of multispectral analysis techniques that may be used to identify ocean dumped materials over a range of environmental

Table 1. Previous studies using remotely sensed data.

Sensor*	Platform	Region	Parameter	Reference
Camera	Aircraft	Lake Superior	Suspended Sediment	Scherz et al. (1973)
Camera MSS	Satellite	Delaware Bay Mid-Atlantic Coastal Zone	Suspended Sediment Acid-iron Waste	Klemas et al. (1978)
MSS	Satellite	Kansas	Suspended Sediment	Yarger et al. (1973)
MSS	Satellite	Chesapeake Bay	Suspended Sediment	Williamson and Grabau (1973)
MSS	Satellite	Delaware Bay Mid-Atlantic Coastal Zone	Suspended Sediment Acid-iron Waste	Klemas et al. (1974)
MSS	Satellite	Lake Huron	River Plume	Rogers et al. (1975)
M2S OCS	Aircraft	James River, VA New York Bight	Suspended Sediment Chlorophyll a	Johnson and Bahn (1977) Johnson (1978)
MSS	Satellite	Chesapeake Bay Mid-Atlantic Coastal Zone	Suspended Sediment Chlorophyll a Acid-iron Waste	Bowker and Witte (1977)

*MSS - Multispectral Scanner (e.g., LANDSAT); M2S - Modular Multispectral Scanner;
OCS - Ocean Color Sensor.

conditions (Johnson, 1977a). In these relatively clear ocean waters,
remote sensors probably can measure materials down to Secchi disc
depths in the most transparent spectral region (e.g., blue-green;
400-500 nanometers) with decreased sensitivity in other spectral
ranges due to water absorption (Williams, 1970). In addition to the
remote sensing monitoring objectives, physical oceanographers have
used the remotely sensed data to evaluate plume dispersion and to
study water movement characteristics in the ocean dump areas. This
report will summarize results from completed experiments and discuss
development of a technology for use in a routine monitoring system,
based on remote sensing techniques.

                              EXPERIMENTS

Description

     Remotely sensed data have been collected by NASA in conjunction
with sea-truth measurements by NOAA, the United States Coast Guard
and cooperating universities in experiments over nearshore and deep
water dumpsites (DWD) in the United States' Coastal Zone (Fig. 1).
Table 2 summarizes a number of remote sensing experiments over these
dumpsites. Experiment 3, conducted on September 22, 1975, in the New
York Bight, will be used as an example to discuss data collection,
remotely sensed data format and preprocessing, and analysis approaches. Results of the remote sensing aspects on this experiment have
been reported by Johnson et al. (1977a).

     In experiment 3, plumes resulting from dumping sewage sludge by
two methods were monitored. The first dumping method was a "line"
dump in which the sewage sludge was dispersed over a 3 to 4 kilometer
(km) track in a 30 to 45 minute period. The second dumping method
was a "spot" dump in which the sewage sludge was discharged in one
location in about 5 minutes.

Sea Truth

     Sea-truth measurements were made in the line dump plume by col-
lecting surface water samples while the NOAA ship KELEZ zigzagged in
and out of the plume. Ten samples were taken, five in the plume and
five in water adjacent to the plume. These samples were analyzed in
the laboratory for suspended solids (SS) and chlorophyll a concentrations. Locations of the sea-truth stations were determined from a
plot of the ship's path during the sampling period, overlaid on the
essentially instantaneous location of the plume from the remote sensors and taking into account the apparent plume drift. Sea-truth
stations and the ship's path are shown in Fig. 2. The SS and chlorophyll a concentrations determined for the 10 sea-truth stations are
listed in Table 3, along with the corresponding radiance values obtained by the remote sensor, which will be described in the following

Table 2. Summary of experiments included in this work; all experiments used photographic cameras; experiment 1 used the OCS sensors and the remaining experiments used the M2S sensor.

Experiment no. & date	Location (See Fig. 1)	Material	Quantity m^3	Dump type	Reference
1. 13 Apr 75	New York Bight (1)	Acid-Iron	3200	12 Km line	Johnson (1977b)
2. 28 Aug 75	Off Cape Henlopen (2)	Acid-Iron	3785	45 Km line	Lewis and Collins (1977) Ohlhorst (1978)
3. 22 Sept 75	New York Bight (1)	Sewage Sludge Sewage Sludge	1812 1456	4 Km line Spot	Johnson et al. (1977a)
4. 15 Jul 76	New York Bight (1)	Sewage Sludge Acid-Iron	2890 3200	Spot 12 Km line	Johnson (1977a) Johnson et al. (1979)
5. 7 Oct 76	Off Cape Henlopen (2)	Sewage Sludge Acid-Iron	3785 3785	45 Km line 45 Km line	Johnson et al. (1977b)
6. 25 Jul 77	Galveston (3)	Petrochemical	2460	46 Km line	See Text
7. 4-6 Feb 78	Puerto Rico (5)	Pharmaceutical	2460	36 Km line	See Text
8. 24 Jul 78	Site 106 (4)	DuPont Grasselli	2170	45 Km line	See Text

Table 3. Sea truth and remotely sensed measurements in the New York Bight on September 22, 1975.

| Station | Sea Truth | | Remotely sensed radiances, mw/cm² · ster · μm M2S Band | | | | |
	Suspended solids mg/l	Chlorophyll a mg/m³	2	3	4	7	9
1	1.53	1.0	3.0	3.05	2.22	0.93	0.37
2	1.26	1.0	2.88	2.81	1.98	0.75	0.30
3	27.00	1.5	3.04	2.96	2.07	1.08	0.49
4	1.60	1.2	2.59	2.99	2.16	0.85	0.34
5	30.10	1.9	2.99	2.92	2.02	1.04	0.46
6	1.70	1.1	2.96	2.99	2.16	0.88	0.34
7	32.20	1.8	3.10	3.07	2.16	1.16	0.53
8	1.11	1.2	3.12	3.14	2.24	0.95	0.39
9	13.26	1.5	3.26	3.24	2.29	1.20	0.49
10	13.36	1.5	3.30	3.30	2.30	1.05	0.41
SPOT*	12.0**	—	3.30	3.45	2.51	1.36	0.52

*Average radiances near SUNY ship ONRUST.
**Interpolated value.

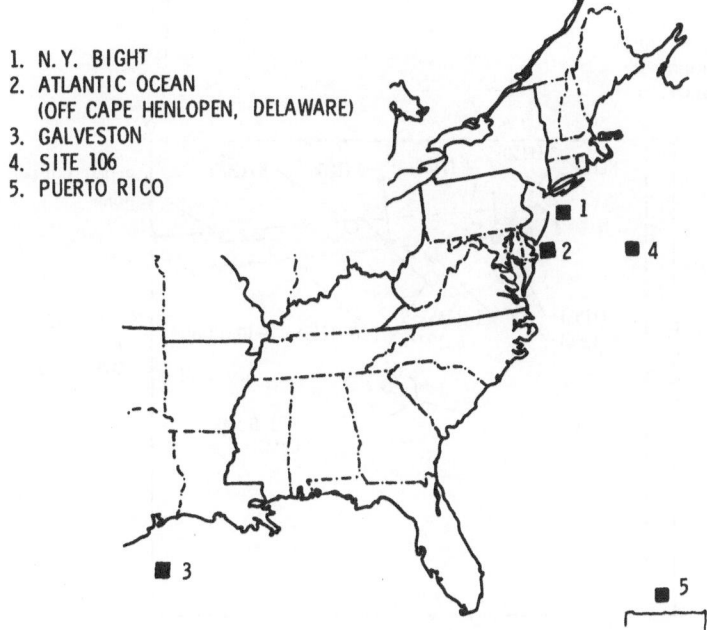

1. N.Y. BIGHT
2. ATLANTIC OCEAN
   (OFF CAPE HENLOPEN, DELAWARE)
3. GALVESTON
4. SITE 106
5. PUERTO RICO

Fig. 1. Ocean dumpsites in the U.S. coastal zone.

section. Additional samples were taken from the State University of
New York (SUNY) ship ONRUST and an interpolated value for SS corresponding to the time of the remote sensing data collection is listed
in Table 3.

Remote Sensing

    Remote sensors that measure electromagnetic radiation in the
visible and near-infrared (NIR) wavelengths were flown over the test
site. The NASA NP-3A aircraft flew at a nominal altitude of 3.0 km
(10,000 ft.) and a speed of 444 km/h (240 knots). Two onboard remote
sensor systems collected data for these analyses: An 11-band Modular
Multispectral Scanner (M2S), which consisted of 10 bands in the visible and NIR spectral range and 1 thermal band; and a Zeiss mapping
camera. Spectral and spatial characteristics of the M2S scanner and
mapping camera are listed in Table 4. Because of instrumentation
problems, the M2S unit recorded usable data only in bands 2, 3, 4, 7,
and 9 in this particular experiment.

    Digital data from the M2S scanner were recorded inflight on mag-
netic tape in a high density format. Inflight calibration also was
provided for each line of data. Screening imagery in conjunction
with the mapping camera products were used to locate areas of

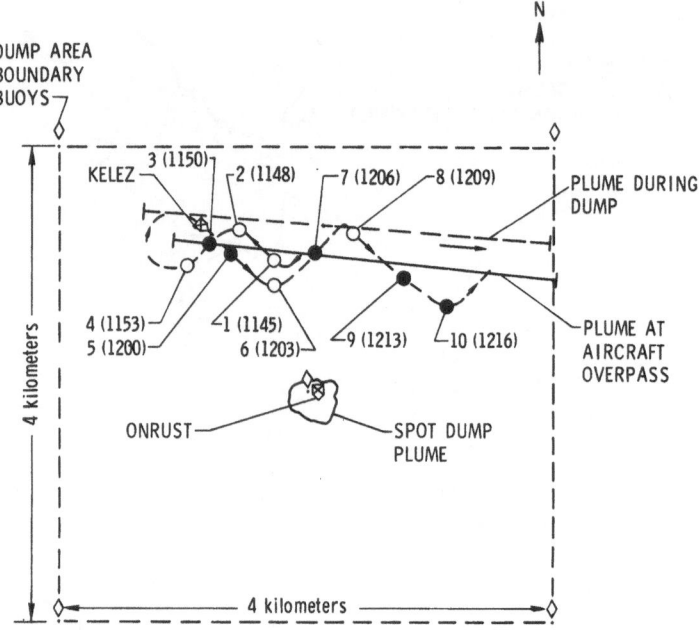

Fig. 2.   Plumes from line and spot dumping of sewage sludge in the
          New York Bight on September 22, 1975 (about 1 hour after
          dumping). NOAA ship KELEZ path during water sampling and
          approximate location and time of surface water samples
          are indicated (solid circles are samples taken in the
          plume and open circles outside the plume). SUNY ship
          ONRUST is shown in the spot dump plume (Taken from
          Johnson et al., 1977a).

interest in the scanner data. Digital data (measured in counts) in
these areas were transferred to computer compatible tapes (CCT) with
a typical format of nine-track, 800 bits per inch. Inflight calibration provided information to convert instrument count data to average
radiances in each band.

     Fig. 3 is a photograph taken at 11:59 a.m. (EDT) (about 1 hour
after the dumping events) showing the plumes of the morning line and
spot dumps. The KELEZ is shown near the line dump plume and the ON-
RUST is shown in the spot dump plume. Radiance measurements and aerial photographs were also taken during the afternoon and are discussed by Johnson et al. (1977a).

Preprocessing of Remotely Sensed Data

     Data preprocessing included locating the sample points in the

Table 4. Spectral and spatial characteristics of remote sensors used in the New York Bight on September 22, 1975.

I. Modular Multispectral Scanner (M2S)

Band	Band Width
1	380 – 440 nm
2*	440 – 490 nm
3*	495 – 535 nm
4*	540 – 580 nm
5	580 – 620 nm
6	620 – 660 nm
7*	660 – 700 nm
8	700 – 740 nm
9*	760 – 860 nm
10	970 – 1,060 nm
Thermal	8000 – 13,000 nm
Scan width, m	8500
Resolution, m	8

*Radiance data are availble in these bands on September 22, 1975.

II. Zeiss Mapping Camera

Spectral Range	Film	Resolution, m	Foot Print, m
300 – 700 nm	2402 (B&W)	1.3	4550 x 4550

imagery and determining radiance values in the M2S bands to be compared to the sea-truth measurements obtained from the KELEZ and ONRUST. In the analysis, representative radiance values were determined by taking a 7 by 7 pixel field centered at the best estimate of the sea-truth measurement. This pixel field was determined empirically as the minimum size that compensated for uncontrollable spectral and spatial errors. The average count in the 7 by 7 field was multiplied by a radiance conversion value obtained from calibrating the inflight system to known radiance sources. Evaluation of inflight calibration data indicates less than 2 percent, or negligible, instrument drift. Average radiances in individual band ranges have the units $mW/cm^2 \cdot sr \cdot \mu m$, or an average radiance per unit bandwidth. Representative radiance values in each of the bands for the sea-truth stations are listed in Table 3. The time interval between remote sensor overpass and sea-truth measurements was less than 20 minutes.

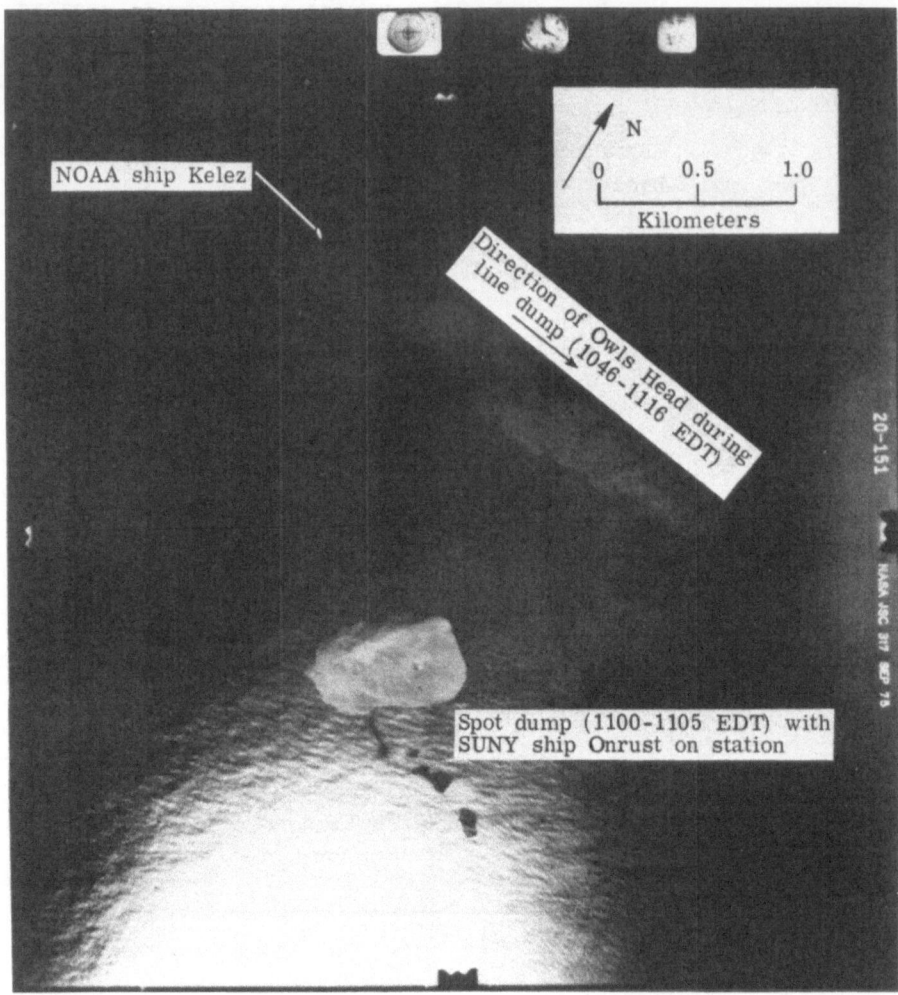

Fig. 3.   Photograph taken at 11:59 EDT over sewage sludge dump
          area.   NOAA and SUNY monitoring ships are shown.   OWLS
          HEAD was barge that performed line dump (see Fig. 2).
          (Taken from Johnson et al., 1977a).

DATA ANALYSIS

Qualitative Analysis

     Qualitative analysis of pollution features includes location,
identification, and mapping their extent.   In general, pollution
features have radiance levels different from those of the background

water in one or more spectral regions. In almost all cases, plumes
in the coastal zone have higher radiances (than the background water)
that are due to suspended materials in the plumes. These differences
have been observed in photographic and digital remotely sensed data.
Identification of pollution features without concurrent sea-truth
measurements requires consideration of atmospheric as well as pollutant spectral responses. One method of "normalizing" environmental
effects (e.g., atmospheric, sun angle and background) between scenes
is to use an in-scene background elimination (Johnson, 1977a). This
approach, used in the results reported here, was to determine ratios
of plume radiances to background ocean water for the same remotely
sensed scene in each of the multispectral scanner bands. These ratio
values as a function of wavelength indicate distinctive characteristics that may be used for identification of plumes resulting from
ocean dumping of industrial waste and sewage sludge. After the
plumes have been located and identified, they may be mapped using
radiance differences by density slicing or classification techniques.
Also, multiple data collections by the remote sensors may be used to
study dispersion patterns and plume persistence.

Quantitative Analysis

     The quantitative analysis used is based on multiple regression
techniques. Water parameters (e.g., suspended solids) are the dependent variables and radiances in each of the multispectral scanner
bands are the independent variables. Specifically, a stepwise regression analysis is applied in which only statistically significant
spectral bands are included in regression equations that relate remotely sensed data to parameters in the water column. These calibrated regression equations have been used to map distribution of
water quality parameters, such as suspended sediment and chlorophyll
a (Johnson and Bahn, 1977).

                              RESULTS

     Qualitative analyses of the remotely sensed data collected under
moderate to calm environmental conditions indicate that normalized
(to ocean water) spectral characteristics may be used to distinguish
the plumes of several waste materials. Results will be presented
from the nearshore experiments (1-5, see Table 2) in the New York
Bight and off Cape Henlopen, Delaware, where sewage sludge and acid-
iron waste plumes were monitored; and from the DWD experiments (6-7,
see Table 2) at Galveston and Puerto Rico where petrochemical and
pharmaceutical wastes, respectively, were dumped. Data from experiment 8, which monitored dumping of DuPont Grasselli waste at DWD
106, are being analyzed.

     Typical spectral characteristics of materials (sewage sludge and
acid-iron) monitored in the nearshore experiments are shown in Fig 4.

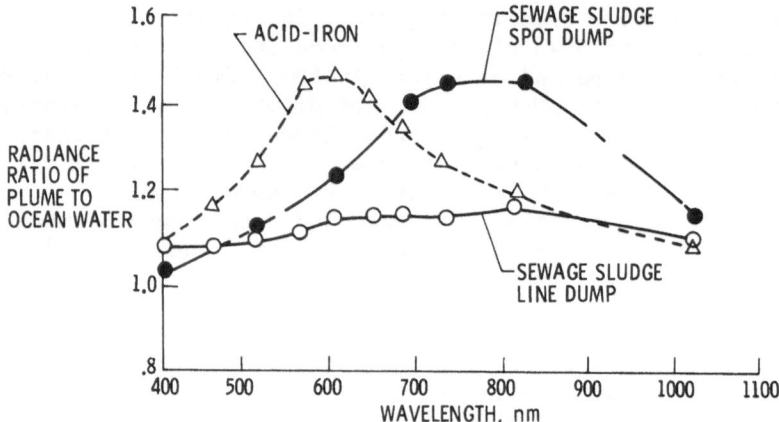

Fig. 4.   Spectral characteristics of sewage sludge and acid-iron
          waste plume from near-shore experiments.

The acid-iron industrial waste has a pronounced radiance ratio peak
at about 600 nanometers (nm) wavelength.   In contrast, the sewage
sludge plume has a relatively gradual increase in normalized radiance
ratio values with a maximum in the 700 to 800 nm range.   Spectral
characteristics curves shown in Fig. 4 are for plumes about 30 minutes after the dumping events.   In general, both sewage sludge and
acid-iron waste dump plumes undergo some initial spectral (color)
variations for the first 15-20 minutes then radiances decrease with
no significant spectral shifts as the plume disperses.   Sewage sludge
plumes usually disappear from surface waters within 4-8 hours while
acid-iron waste plumes have been observed for several days after
dumping (Bowker and Witte, 1977; Klemas et al. 1977).

     Sewage sludge is the only ocean dumped material that has used
both the "line" and "spot" dump methods; all other waste dumps we
have observed have been from moving barges by the "line" method.
Dispersed spot dumps are spectrally similar to line dumps (Johnson
et al., 1979).   Acid-iron waste materials dumped in the New York
Bight and in the Atlantic Ocean off Cape Henlopen have similar constituents (i.e., high concentrations of iron compounds, ferrous sulfate or ferric chloride, and their associated acid).   Laboratory and
theoretical analyses indicate a resultant suspended solid containing
ferric hydroxide when either material is buffered in sea water (Lewis
and Collins, 1977).   Results of our experiments indicate that spectral characteristics are similar for plumes resulting from disposal
of either of these acid-iron wastes.

     Spectral characteristics of petrochemical and pharmaceutical
industrial wastes dumped at the Galveston and Puerto Rico deep water

sites, respectively, are shown in Fig. 5. These materials have quite
different spectral characteristics as indicated by moderate broadband
increased radiance ratio values from the petrochemical waste material
and the absorbing (e.g., radiance ratio values les than 1.0) characteristics of the pharmaceutical waste plume. Also shown for reference is a spectral characteristic curve for acid-iron waste. The
DuPont Edgemoor plant waste material is now dumped at DWD 106 (instead of in the Atlantic Ocean off Cape Henlopen, Delaware). This
curve probably will be applicable at DWD 106 since experiments conducted under a variety of environmental conditions indicate that
spectral variations of background ocean water are not a significant
factor when the background elimination technique is used for identification of this spectrally distinctive plume.

Persistences of the petrochemical and pharmaceutical waste
plumes were about 4-8 hours based on visual observations from the
monitoring aircraft. It should be noted that dumping at the Galveston DWD has been discontinued (at least temporarily); however, spectral information have been included for comparison with other geographic areas.

Two additional industrial waste materials are dumped frequently
at DWD 106. These are DuPont (Grasselli) and American Cyanamid
wastes. From laboratory evaluations, it appears that these materials
have different spectral characteristics from other materials we have

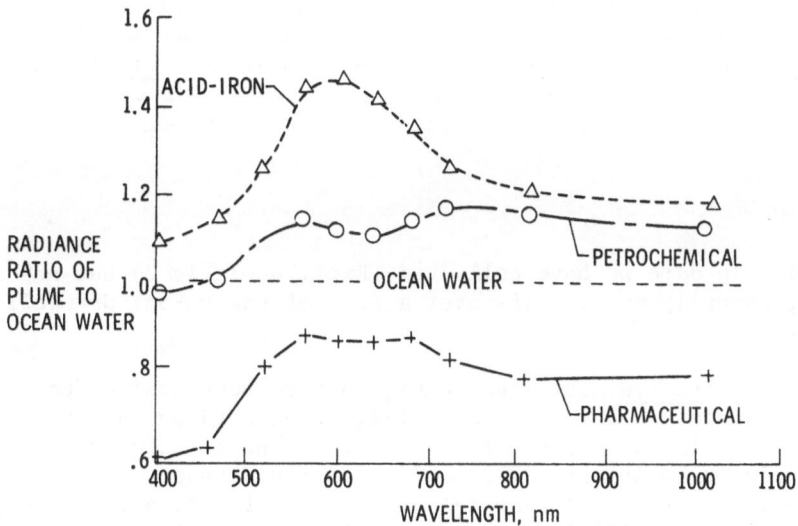

Fig. 5.   Spectral characteristics of petrochemical and pharma-
          ceutical waste plumes from Galveston and Puerto Rico
          DWD. Acid-iron curve is the same as Fig. 4 (see text).

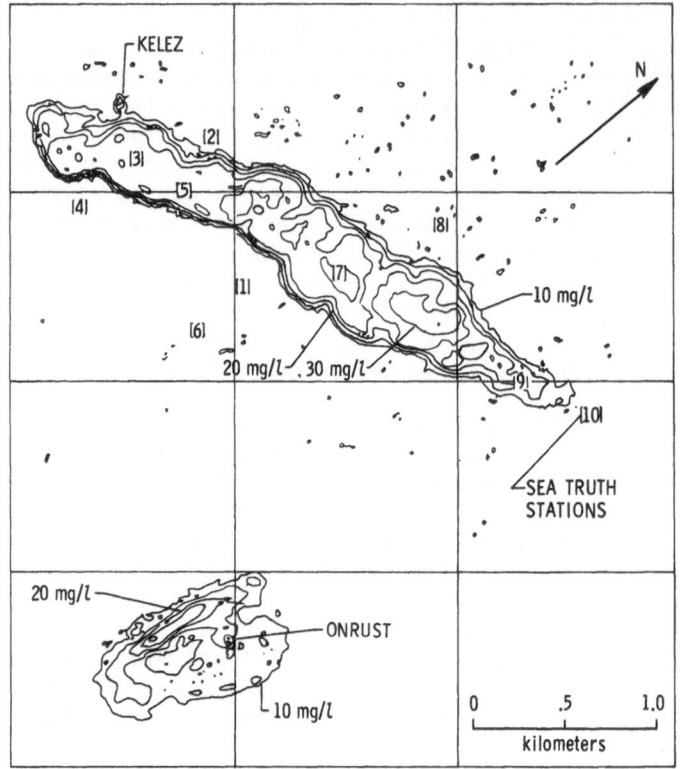

Fig. 6.   Quantitative distribution of suspended solids in sewage
          sludge plumes in the New York Bight on September 22, 1975
          (see Figs. 2 and 3). (Taken from Johnson et al., 1977a).

tested. To date we have collected (Expt. 8, Table 2) but not analyzed, remotely sensed data over a dump of the DuPont Grasselli
waste.

     Quantitative analysis techniques have been applied to suspended
solids concentrations in sewage sludge plumes (Johnson, 1977a; Johnson et al., 1977a) and to iron concentrations in acid-iron industrial
waste plumes (Ohlhorst, 1978). In addition, these techniques have
been used to study the temporal dispersion of a sewage sludge plume
(Johnson et al., 1979). An example of the quantitative results from
experiment 3 (Table 2) conducted on September 22, 1975, is shown in
Fig. 6 where suspended solids concentration distributions in the sewage sludge plumes are mapped.

CONCLUDING REMARKS

Results of remote sensing field experiments conducted at several ocean dumping sites indicate that plumes resulting from dumping of four waste materials have distinctive spectral characteristics. In nearshore regions acid-iron (industrial) waste plumes may be distinguished from those of sewage sludge. At the deep water sites, petrochemical waste (dumped at the Galveston DWD) is spectrally different from pharmaceutical wastes (dumped at the Puerto Rico DWD). Additional measurements are required for the materials dumped at DWD 106. However, available results indicate that spectral responses, when combined with geographic information, may be used to provide a technology base for monitoring ocean dumping from aircraft. Current satellite systems have limitations due to coverage frequency and spectral and spatial resolution.

Multiple regression quantitative analysis techniques have been applied to study spatial and temporal concentration distributions of suspended solids associated with plumes from dumping sewage sludge and acid-iron waste. Maps of these distributions will be useful for determining dispersion coefficients and studying surface water movement in the dump areas. Acid-iron waste plumes should be of particular interest since these materials may be observed in surface waters for up to several days.

Development of remote sensing monitoring techniques should consider factors such as plume persistence since most materials are not visible in surface waters for more than 4 to 8 hours after dumping. The exception is acid-iron waste as noted above. In addition, moving waste dumping from one location to another may be important. For example, the acid-iron wste previously dumped off Cape Henlopen is now dumped at DWD 106. In this case, it appears that the distinctive spectral characteristics shown in Fig. 4 should be applicable at Site 106, since ocean water spectral characteristics do not appear to be a significant factor when the background elimination technique is used. However, this may not be true for other materials dumped at other locations.

REFERENCES

Bowker, D. E. and W. G. Witte (1977) The use of LANDSAT for monitoring water parameters in the coastal zone. In: Proceedings of the AIAA Joint Conference on Satellite Applications of Marine Applications, American Institute of Aeronautics and Astronautics, New Orleans, pp. 193-198.

Johnson, R. W. (1977a) Multispectral analysis of ocean dumped materials. In: Proceedings of the Eleventh International Symposium on Remote Sensing of Environment, Ann Arbor, Michigan, pp. 1619-1625.

Johnson, R. W. (1977b) Mapping the Hudson River plume and an acid
    waste plume by remote sensing in the New York Bight Apex,
    April 1975.   In: Results from the National Aeronautics and
    Space Administration Remote Sensing Experiments in the New
    York Bight - April 7-17, 1975.   J. B. Hall, Jr. and A. O.
    Pearson, compilers.   NASA TM X-74032 Langley Research Center,
    Hampton, Virginia, pp. 106-129.

Johnson, R. W. (1978) Mapping of chlorophyll a distributions in
    coastal zones by remote sensing. Photogrammetric Engineering
    and Remote Sensing, 44, (5), 617-624.

Johnson, R. W. and G. S. Bahn (1977) Quantitative analysis of
    aircraft multispectral-scanner data and mapping of water-
    quality parameters in the James River in Virginia.   NASA TP
    1021, 31 pp.

Johnson, R. W., I. W. Duedall, R. M. Glasgow, J. R. Proni, and T. A.
    Nelsen (1977a) Quantitative mapping of suspended solids in
    wastewater sludge plumes in the New York Bight Apex. Journal
    Water Pollution Control Federation, 49, (10), 2063-2073.

Johnson, R. W., R. M. Glasgow, I. W. Duedall, and J. R. Proni (1979)
    Monitoring the temporal dispersion of a sewage sludge plume
    in the New York Bight by remote sensing. Photogrammetric
    Engineering and Remote Sensing. 45, (6), 763-768.

Johnson, R. W., C. W. Ohlhorst, and J. W. Usry (1977b) Location,
    identification and mapping of sewage sludge and acid waste in
    the Atlantic coastal zone.   In: Proceedings of the Fourth
    Joint Conference on Sensing of Environmental Pollutants,
    American Chemical Society, Washington, D. C., pp. 644-647.

Klemas, V., D. Bartlett, W. Philpott, R. Rogers, and L. Reed
    (1978) SKYLAB/EREP Applications to ecological, geological and
    oceanographic investigations of Delaware Bay. NASA CR
    144910, 67 pp.

Klemas, V., G. R. Davis, and R. D. Henry (1977) Satellite and
    current drogue studies of ocean-disposed waste drift. Journal
    Water Pollution Control Federation, 49, (5), 757-763.

Klemas, V., M. Otley, W. Philpott, C. Wethe, and R. Rogers (1974)
    Correlation of coastal water turbidity and circulation with
    ERTS-1 and SKYLAB imagery.   In: Proceedings of the Ninth
    International Symposium on Remote Sensing of Environment,
    University of Michigan, Ann Arbor, Michigan, pp. 1289-1317.

Lewis, B. W. and V. G. Collins (1977) Remotely sensed and laboratory
    spectral signatures of an ocean-dumped acid waste. NASA TN
    D-8466, 36 pp.

Ohlhorst, C. W. (1978) Quantitative mapping by remote sensing of an
    ocean acid-waste dump. NASA TP 1275, 23 pp.

Rogers, R. H., N. J. Shah, J. B. McKeon, C. Wilson, L. Reed, V. E.
    Smith, and A. Thomas (1975) Application of LANDSAT to the
    surveillance and control of eutrophication in Saginaw Bay.
    Proceedings of the Tenth International Symposium on Remote
    Sensing of Environment, Ann Arbor, Mighigan, pp. 437-446.

Scherz, J. P., J. F. Van Domelen, and S. A. Klooster (1973) Aerial
    and satellite photography. A valuable tool for water quality
    investigations. In: Proceedings of the American Society of
    Photogrammetry Fall Convention, Lake Buena Vista, Florida,
    pp. 883-905.
Whitlock, C. H., W. G. Witte, J. W. Usry, and E. Z. Gurganus (1978)
    Penetration depth at green wavelengths in turbid waters.
    Photogrammetric Engineering and Remote Sensing, 44, (11),
    November 1978, 1405-1410.
Williams, J. (1970) Optical Properties of the Sea. United States
    Naval Institute, Annapolis, Maryland, 123 pp.
Williamson, A. N. and W. E. Grabau (1973) Sediment concentration
    mapping in tidal estuaries. In: Third Earth Resources
    Technology Satellite-I Symposium, Washington, D. C. NASA
    SP-351, pp. 1347-1386.
Yarger, H. L., J. R. McCauley, G. W. James, and L. M. Magnuson (1973)
    Water turbidity detection using ERTS-1 imagery. In:
    Symposium of Significant Results Obtained from the Earth
    Resources Technology Satellite-I, GSFC, New Carrollton,
    Maryland. NASA SP-327, pp. 651-658.

REMOTE SENSING OF OCEAN-DUMPED WASTE

DRIFT AND DISPERSION

Vytautas Klemas and William D. Philpot

College of Marine Studies
University of Delaware
Newark, DE   19711

ABSTRACT

The drift and dispersion of sixteen acid waste plumes 64 km off the Delaware coast were investigated using Landsat imagery, current drogues and ship data. The waste plumes imaged by Landsat were found to be drifting at average rates from 0.59 km hr^{-1} to 3.39 km hr^{-1} into the southwest quadrant. The plumes seemed to remain above the thermocline which was observed to form from June through August at depths ranging from 13 m to 24 m. During the remainder of the year the ocean at the test site was not stratified, permitting wastes to mix throughout the water column.

The magnitudes of plume drift velocities were compatible with the drift velocities of current drogues released over a 12-month period at the surface, at mid-depth and near the bottom. However, during the stratified warm months, more drogues tended to move in the north-northeast direction, while during the non-stratified winter months a southwest direction was preferred.

Rapid waste movement toward shore occurs primarily during storms, particularly northeasters. During such storms, however, the plumes are rapidly dispersed and diluted. The plume width was observed to increase at a rate of about 1.5 cm sec^{-1} during calm sea conditions, yet attain spread rates in excess of 4 cm sec^{-1} on windy days. These results indicate that by the time a waste plume would reach shore, dilution would be at least one million to one.

INTRODUCTION

The continuing use of the continental shelf for waste disposal and the extraction of oil and other resources may pose an environmental threat to the shelf regions. The offshore-onshore transport rates of pollutants, sediments, and nutrients strongly influence the ecology of the coastal zone. In order to keep the environmental impact within acceptable levels, it is important to understand the circulation and exchange processes on the shelf. Satellites can make synoptic observations of certain large scale processes indicative of pollution transport on the continental shelf. The objective of this study was to determine from LANDSAT imagery the drift and dispersion of plumes generated by the dumping of acid waste and to compare the satellite results with drogue and ship data.

WASTE COMPOSITION AND DISPOSAL

The barged wastes come from the manufacture of titanium dioxide pigment. The specially designed barge, described by Fader (1972), is 82 m long, 18 m wide and 5.5 m deep, and has a total capacity of about 3800 m^3 (1 million gallons). Radio-controlled signals from a towing tug release the wastes from the unmanned barge in a designated disposal area.

Wastes flow from the barge by gravity at a controlled rate into a disposal area recommended by the U. S. Interior Department's Bureau of Commercial Fisheries (now the National Marine Fisheries Service, Department of Commerce). The waste disposal area encompasses a rectangle of 9.3 km by 14.8 km centered approximately 73 km southeast of Cape Henlopen, Delaware. The area is bounded by 38°30' and 38°35' north latitude, and 74°15' and 74°25' west longitude. The sea throughout the area is between 38 m and 48 m deep. The barge is towed to sea approximately two to three times a week. Initially the discharge time was about 60 minutes at a speed of 11 km hr^{-1} (6 knots), weather and other conditions permitting. Since May 1974 a bow tie dump pattern was adopted with a discharge time of about 5 hours at a speed of 15 km hr^{-1} (8 knots). The waste originally was 17 to 23 percent acid (expressed as H_2SO_4) and 4 to 10 percent ferrous sulfate (Falk, 1974). During 1975 the composition of the waste was changed to a solution of 10 percent acid (expressed as HCl) and 4 percent iron as iron chloride salts (EPA 1975). Similar wastes have been disposed at sea for 28 years in the New York Bight (Peschiera et al., 1968).

A limited amount of information on the waste disposal site is given by Hydrographic Office bathymetric charts and in papers by Bumpus (1965, 1969, 1973), Ketchum (1953) and Myers (1974). Their work included releases of bottom drifters and surface floats. Bumpus (1965) observed a line parallel to the shore at approximately 55 to

64 m depth where bottom drifters released inside of this line move
shoreward and drifters released outside of the line move offshore.
This line lies approximately 69 km off the coast of Delaware. Myers
found that isothermal conditions prevailed throughout the water column during winter and spring months until mid-April, when a distinct
thermocline formed between 18 and 23 m depth.

A study conducted by Falk (1974) at the time when the waste was
disposed in a 1-hour period included Eulerian type current measurements with current meters. They found that the ocean currents were
southwesterly at the bottom. In the summer, the stratified surface
waters moved north or northwesterly. The general movement of the
waste field was to the southwest, except during late spring and summer when the wastes above the thermocline moved to the northwest.

SATELLITE OBSERVATIONS

The frequency of the dumping made it possible for LANDSAT satellites and aircraft to observe the waste plumes in various stages of
dispersion ranging from minutes to days after dump completion. Sixteen satellite images taken from October 1972 to March 1976 were

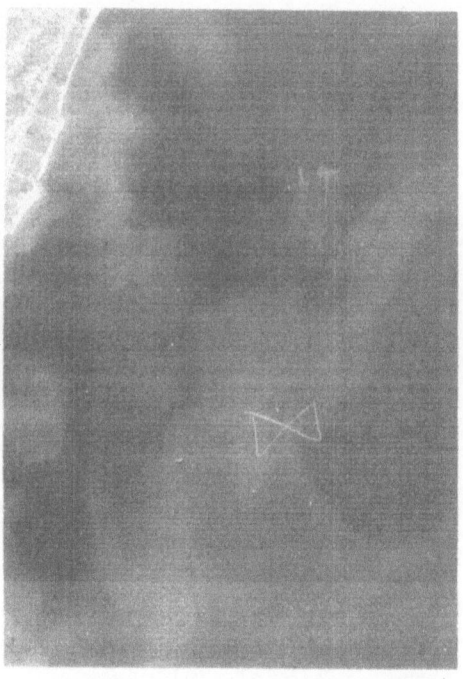

Fig. 1.   Acid-iron plume visible in LANDSAT imagery on August
           28, 1975 (during dump).

found which show water discoloration in the general vicinity of the
waste dumpsite. The spectral characteristics and position of the
discoloration, the dump pattern and the time difference between the
dump and photograph gave strong indications that the discolorations
are the waste plume (Fig 1).

The LANDSAT imagery was produced by a four-channel multispectral scanner (MSS) having the following bands:

Band 4	500-600 nanometers
Band 5	600-700 nanometers
Band 6	700-800 nanometers
Band 7	800-1100 nanometers

From an altitude of 920 km, each satellite frame covered an area
of 185 km by 185 km and had a resolution of about 80 m. The imagery
was recorded on 9-track 800 bpi (bits per inch) magnetic tapes, from
which 70 mm transparencies and 230 mm prints were reconstructed. Before visual interpretation, some of the imagery was enhanced optically. Maps and print-outs of the waste plumes were prepared by

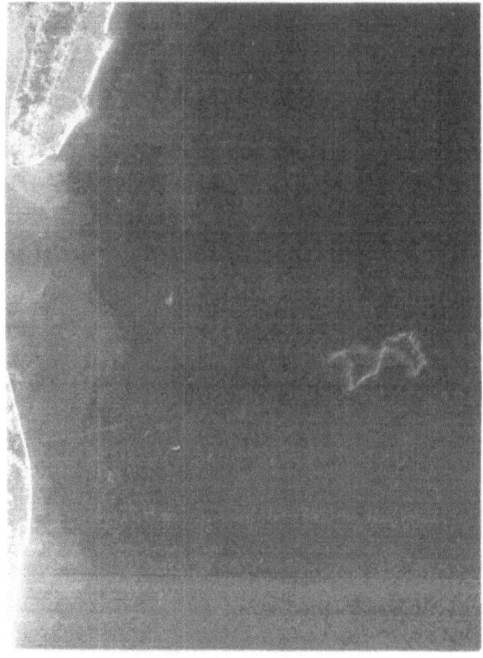

Fig. 2.   Acid-iron waste plume visible in LANDSAT imagery on
           February 24, 1976 (6 hours after dump).

computer analysis of digital tapes and by direct photo-interpretation
of the transparencies.

There is considerable shearing and dispersion of the waste plume
during a six hour period (Fig 2). Spectrometric measurements indicate that upon combining with seawater, the waste develops a strong
reflectance peak in the 550 to 600 nanometer region, resulting in a
stronger contrast in the LANDSAT Band 4 than the other bands. This
spectral appearance seems to be caused by the formation of a sparse
but optically persistent suspended ferric floc.

MULTISPECTRAL ANALYSIS AND ENHANCEMENT OF LANDSAT IMAGERY

Much of what is unique about our approach is an indirect result
of the use of LANDSAT/MSS digital data. LANDSAT was not designed for
observations in water. The gain is low making the dynamic range of
the sensor very limited. The four spectral channels were selected
for land use applications and are hardly ideal for water observations. Yet, there is a surprising amount of information in the
LANDSAT imagery. LANDSAT data has been used to map sediment distribution patterns (Klemas et al., 1977), to observe the occurrence of
estuarine fronts (Klemas and Polis, 1977) and to observe the occurrence of internal waves (Apel, 1974), to cite only a few of the many
papers in which LANDSAT imagery has been used in sensing of water.

Satellite images in which the acid iron waste plume were partly
masked by clouds or other substances were enhanced using multispectral eigenvector analysis techniques. The use of spectral reflectance characteristics to identify substances in the water has at-
tracted considerable attention in the past several years (McCluney,
1974; Mueller, 1976; Gordon et al., 1975; Hovis, 1977; Gordon, 1978
and Plass et al., 1978). Our approach is most similar to that suggested by Mueller (1976) -- eigenvector (principal component) analysis. Eigenvector analysis has been described by a number of investigators including Mueller (1976) and Simmonds (1963). Perhaps
the best and most complete presentation can be found in Morrison
(1976) and the reader is referred to this text for a detailed discussion of the technique.

One major reason for using eigenvector analysis is that it allows the reduction of significant variates with minimal loss of information. With LANDSAT/MSS data there are only four spectral bands
and therefore only four variables to begin with and the analysis will
rarely reduce this number by more than one. However, the eigenvectors can also provide an efficient representation of variations in
water color which can be readily adapted to an automatic classification process.

To illustrate the technique one can imagine a body of water,

part of which is clear, part of which is heavily sediment-laden and
part of which contains pollution of some sort. A LANDSAT image of
the area would show the clear water as relatively dark while both the
sediment and pollutant would appear relatively bright. The sediment
might show up brighter than the pollutant in Band 5 (600 to 700 nanometers) and the reverse might be true in Band 4 (500 to 600 nanometers). If one were to plot the radiances observed in both bands for
each picture element (pixel) the result would appear as in Fig. 3.
The origin represents the clear water pixels and the two lobes represent sediment pixels ($\bar{B}$) and pollutant pixels ($\bar{A}$). Clear water has a
particular spectral signature and addition of any material to the
water will produce a deviation from the clear water signature. If
the deviation is in different directions for two materials then they
will be distinguishable to some extent. Vectors A and B represent
the first eigenvectors associated with each material. The simplest
measure of the spectral separability of the two materials is the
angular separation of these two vectors. The eigenvector analysis
also provides measures of the dispersion of the data about the axis
of the first eigenvectors. This procedure is covered in considerable
detail by Klemas et al. (1978).

   We will consider only the angular separation of the eigenvectors
as a measure of spectral separability. Table 1 shows the results of
analyzing six different coastal Delaware LANDSAT scenes for sediment,
ice, clouds and an acid-iron industrial waste. There is some

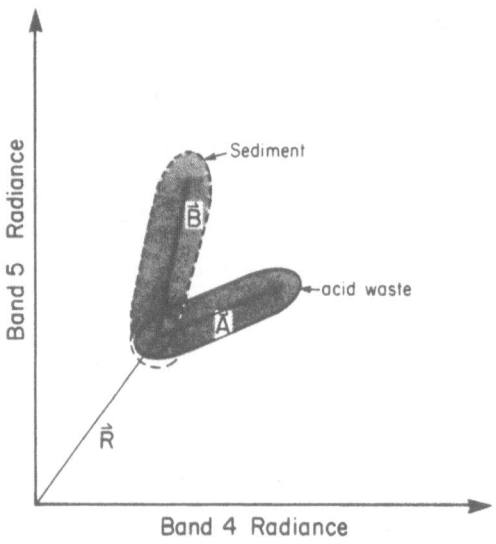

Fig. 3.   Diagram of the geometry of the eigenvector analysis. The
           origin has been placed at the position of the "clear" water
           standard.

Table 1. Angular separation (in degrees) between primary eigenvectors.

		ACID–IRON WASTE							SEDIMENT				CLOUDS				ICE
		1	2	3	4	5	6	7	8	9	10	11	12	13	14	15	16
		24 FEB 76	19 JAN 76	21 OCT 75	19 AUG 75	17 NOV 75	15 MAR 74	AVERAGE	24 FEB 76 (NORTH)	24 FEB 76 (SOUTH)	19 JAN 76	AVERAGE	19 JAN 76	19 AUG 76	15 MAR 74	AVERAGE	19 JAN 76
ACID	24 FEB 76 — 1	—															
	19 JAN 76 — 2	14.4	—														
	21 OCT 75 — 3	3.9	13.7	—													
	19 AUG 75 — 4	12.7	6.2	10.5	—												
	17 NOV 75 — 5	4.4	13.6	7.8	13.8	—											
	15 MAR 74 — 6	6.3	10.7	8.6	11.6	3.3	—										
	AVERAGE — 7	8.5	11.1	8.6	10.3	8.8	7.9	—									
SEDIMENT	24 FEB 76 — 8	36.1	22.9	33.9	23.7	36.4	33.5	31.5	—								
	24 FEB 76 — 9	37.9	24.8	35.7	25.4	38.3	35.4	33.3	2.0	—							
	19 JAN 76 — 10	34.2	22.6	31.4	21.8	35.2	32.7	30.1	6.1	6.5	—						
	AVERAGE — 11	36.1	23.5	33.7	23.6	36.6	33.8	31.8	3.6	3.9	5.1	—					
CLOUDS	19 JAN 76 — 12	47.2	33.8	45.9	35.7	46.1	42.9	42.2	19.8	18.9	24.1	21.0	—				
	19 AUG 76 — 13	31.6	18.5	30.8	21.3	30.2	26.9	27.0	16.9	17.8	20.2	18.3	16.2	—			
	15 MAR 74 — 14	39.4	25.7	38.1	27.8	38.4	35.2	34.5	13.5	13.4	18.0	15.1	8.3	9.2	—		
	AVERAGE — 15	39.8	26.7	38.7	28.8	38.7	35.5	35.0	16.9	16.9	20.9	18.3	10.5	10.8	7.2	—	
ICE	19 JAN 76 — 16	42.6	28.6	41.1	30.6	41.9	38.7	37.6	11.2	10.5	16.5	13.0	9.4	14.3	5.9	10.4	—
	AVERAGE																

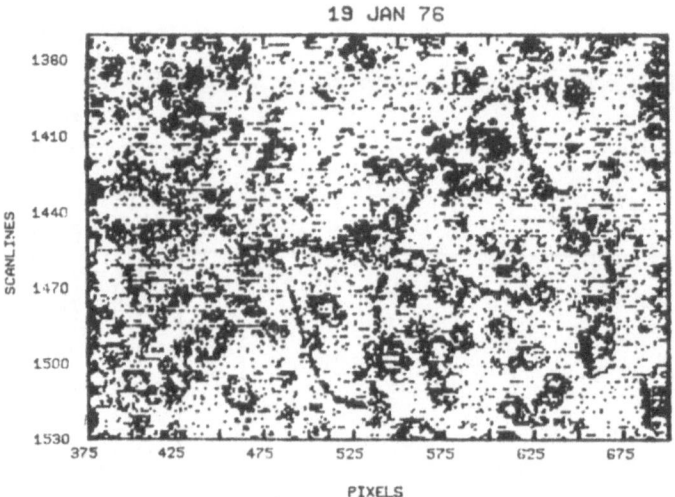

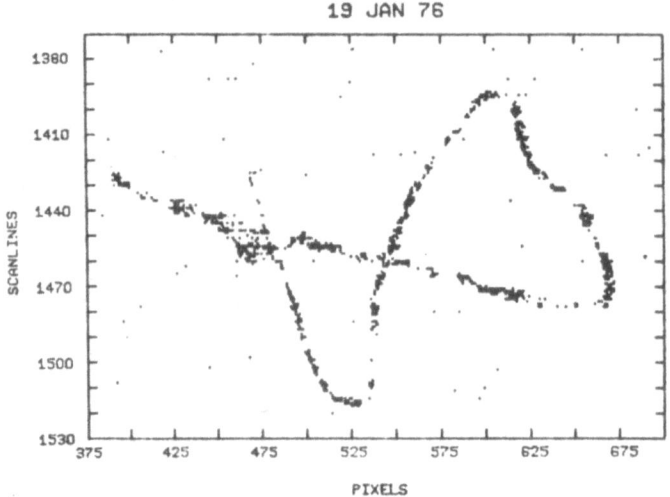

Fig. 4.   Enhancement of the iron-acid waste plume of 19 January
           1976, against a cloud background.

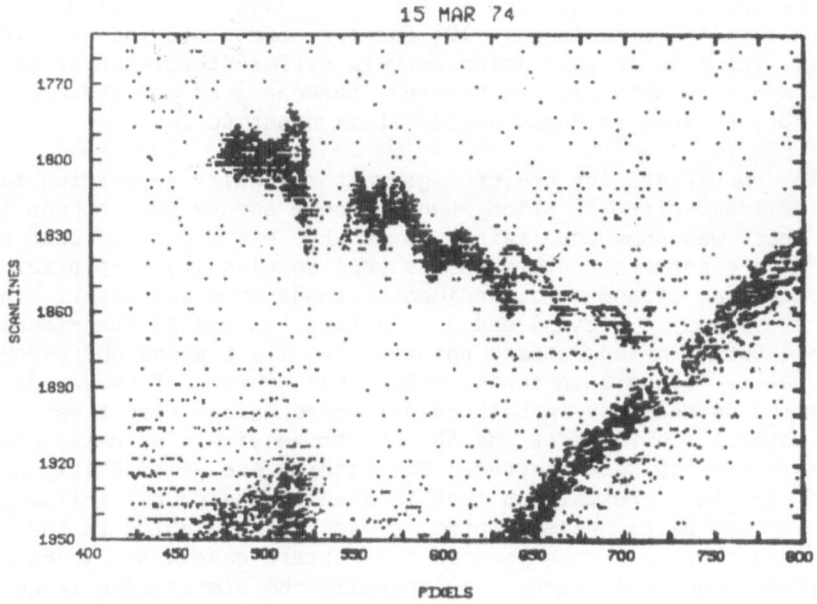

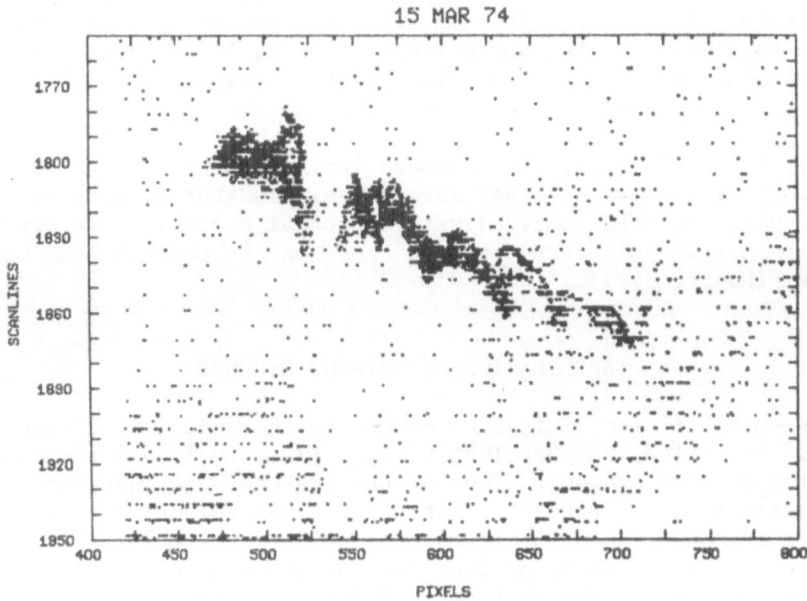

Fig. 5.   Enhancement of the iron-acid waste plume of 15 March 1974, against a cloud background.

dispersion among vectors identifying each material resulting solely
from the use of data acquired on different days. This dispersion
amounts to ~6° for sediment, ~10° for the acid waste and ~10° for
clouds. The angular separation between different substances is
always significantly higher, however, between acid and sediment it is
about 35°, between acid and clouds it is about 40°.

    To demonstrate the use of eigenvector angular separation in
terms of image classification, two of these scenes were chosen in
which there was some uncertainty as to what was cloud and what was
acid. The eigenvector analysis was used to classify each pixel in
both scenes as either acid, sediment, clouds or clear water. The results are shown in Figs. 4 and 5. In Figs. 4A and 5A the clouds and
acid are both plotted as dark points. The light areas correspond to
clear water. (Less than twenty points out of tens of thousands were
classified as sediment in both cases; these points were treated as
clear water.) In Figs. 4B and 5B only those pixels actually classified as acid-iron were plotted. The results are particularly strik-
ing in Fig. 4B. The pattern that is seen is the course followed by
the acid-iron barge while dumping. There is some noise in the background and there are some gaps in the pattern caused by clouds directly over the dump track, but generally the distinction is quite
good. The results illustrated in Fig. 5B are still good although
much more noisy. The waste had been in the water for more than 6
hours as compared to a fresh dump in the earlier example. Because of
dispersion and settling, the signal is not as clear, as is apparent
in the relatively higher noise level. Still, it is clear that the
acid was dumped in a straight line track on this day and that the
other linear feature in the scene was a cloud.

    It is likely that this approach can be extended with the LANDSAT
data to include several other substances, and that considerably better results could be achieved using spectral channels more appropriate for analysis of water, such as those on the Coastal Zone Color
Scanner (Hovis, 1977).

APPLICATION OF CURRENT DROGUES

    A total of nine cruises were made to the acid-iron disposal site
during the period from May 1975 to June 1976, to launch radio-signal-
emitting current drogues. Four of the cruises were made when a summer thermocline was present. The other five cruises were made during
the winter months when isothermal conditions existed in this area.
During each cruise three of four current drogues were deployed to
measure currents at the surface, mid-depth or above the thermocline,
and below the thermocline. Also weather and sea conditions were
noted during each cruise and a temperature profile was made to determine the presence and depth of the thermocline.

The drogues used are expendable units each consisting of an 0.6
m plastic pipe containing the electronics. Buoyancy is provided by a
pair of floatation chambers so attached that when properly ballast-
ed, the antenna portion of the drogue projects 1.9 m above a still
sea datum plane. The transmitter broadcasts less than 100 milliwatts
of power in the 2- to 6-megahertz band. When using the sea surface
as a ground plane and transmitting via ground wave, ranges in excess
of 160 km have been obtained. The drogue's operating life is select-
able from 1 to 6 weeks through choice of battery type and control of
transmitter power (Klemas et al., 1974).

Current sensing is achieved by the use of a current trap, such
as a biplane, which is suspended at a controllable depth beneath the
drogue hull. The current intercept area is isotropic and ranged from
0.9 m^2 for surface drogues to 1.5 m^2 for sub-surface drogues. A
ballast weight is attached to the bottom of the current trap in order
to provide a righting movement about the system's metacenter to resist heelover under strong wind and wave conditions. Since the buoy
portion of the drogue is nearly awash and has only a thin radio antenna protruding above the water surface, wind drag on the drogues
usually was not significant. A ratio in excess of 20:1 was maintained between the projected area of the buoy exposed to surface currents
and of the current trap exposed to currents at its depth. Careful
hydrodynamic design of the buoy hull structure further reduces the
effect of surface currents.

Position finding is accomplished by triangulation from two (or
more) radio direction-finding (DF) stations located on-shore near the
water's edge. The loop antenna used on the DF sets is highly direc-
tional and permits an accurate audio null of the received signal.
Ideally the baseline between DF stations should be such that the
bearing lines from the stations to each drogue intersect as close to
a right angle as possible. At an average distance of 65 km a standard deviation in bearing angle to a drogue of about 1 degree was obtained. To maintain this accuracy, each DF set must be recalibrated
on a moored drogue before each series of measurements. Best results
were obtained for tracking ranges of about 16 to 97 km. However, the
triangulation method gives large position errors for drogues moving
to the outer portion of the shelf.

DISCUSSION OF RESULTS

As shown in Table 2 and Fig. 6, the maximum range of measurable
wastes estimated by Falk (1974) as being about 18.5 km from the discharge point was substantiated by the satellite imagery, because only
1 out of 16 plumes was observed to be significantly beyond this
range. This was also the only plume observed by the satellite as
being 32 km from shore with two others about 40 km from shore, and
all remaining thirteen plumes more than 48 km from shore.

Waste plume drift speeds derived from LANDSAT imagery ranged from 0.46 km hr^{-1} (0.25 knot) to 2.74 km hr^{-1} (1.48 knots), with an average of 1.1 km hr^{-1} (0.59 knot). Plume drift speeds were calculated by dividing the distance between plume centroid and the center of the waste dumpsite by the time between dump and satellite overpass. The dump time was defined as the midpoint between waste dump initiation and completion. Barge captain's logs were consulted to validate this procedure. The authors' observations confirm that most of the waste dumps are centered on the proper site.

A total of 35 current drogues were released at the waste disposal site, during the time period from May 1975 through June 1976. The drogues were successfully tracked from 1 to 13 days, with the

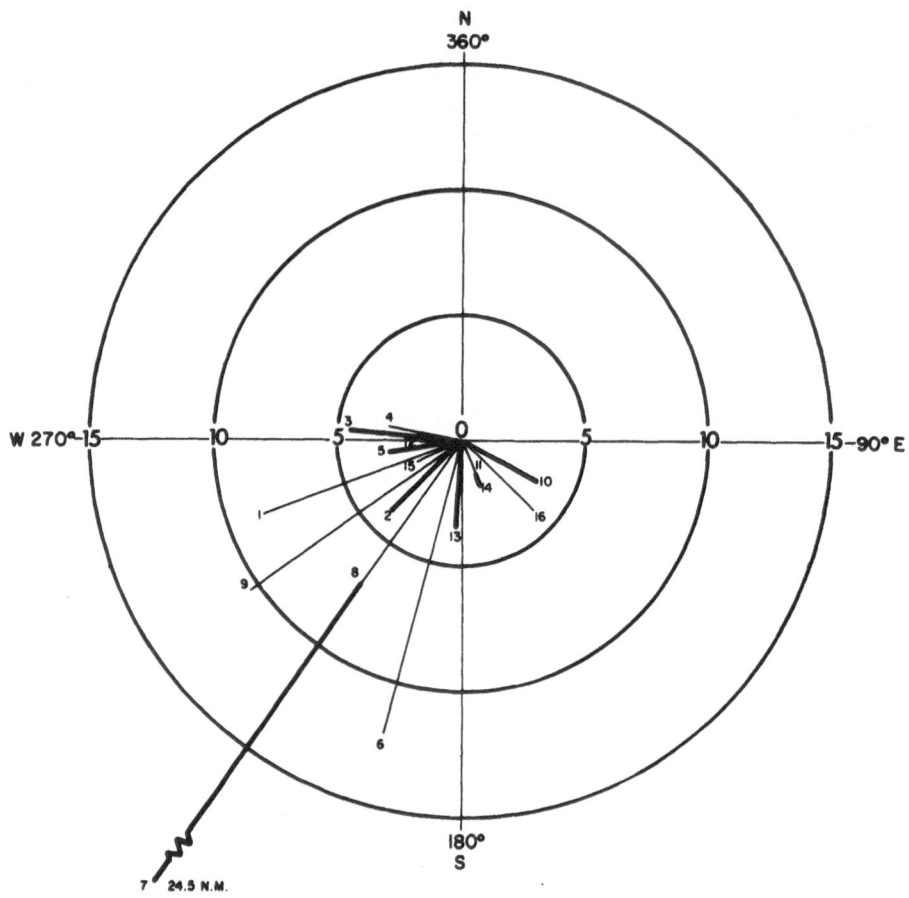

Fig. 6.   Distance and direction from the center of the dumpsite
to the centroid of each imaged plume. (statute miles).

Table 2. Waste plume characteristics derived from LANDSAT imagery.

Date	Hours After Dump	Lateral Extent (km)	Plume Axis Orientation	Distance Between Centroids (km)	Drift Vector Orientation	Average Drift Velocity kmhr⁻¹	(knots)
1) 10/10/72	9.63	18.5	155°	15.7	250°	1.63	(0.88)
2) 10/27/72	14.13	9.3	260°	7.4	225°	0.52	(0.28)
3) 01/25/73	3.05	9.3	120°	8.3	275°	2.74	(1.48)
4) 04/07/73	6.63	14.8	110°	5.6	280°	0.83	(0.45)
5) 05/13/73	0.00	1.9	105°	5.6	260°	0.00	(0.00)
6) 10/22/73	29.50	26.9	225°	22.2	195°	0.76	(0.41)
7) 10/23/73	53.60	25.9	200°	45.4	215°	0.85	(0.46)
8) 04/20/74	13.78	13.0	145°	13.0	215°	0.94	(0.51)
9) 05/26/74	24.33	13.9	145°	19.4	235°	0.89	(0.48)
10) 11/04/74	14.00	13.0	240°	6.5	110°	0.52	(0.28)
11) 08/19/75	0.00	13.9	120°	2.8	160°	0.00	(0.00)
12) 08/28/75	0.00	11.1	220°	2.8	270°	0.00	(0.00)
13) 10/21/75	6.58	14.8	250°	11.9	188°	1.80	(0.97)
14) 11/17/75	5.00	14.8	120°	3.7	160°	0.74	(0.40)
15) 01/19/76	3.00	12.0	245°	3.7	245°	1.24	(0.67)
16) 02/24/76	6.00	18.5	105°	7.4	135°	1.24	(0.67)
AVERAGE		14.5		9.8		1.09	(0.59)

average drift speeds of surface drogues ranged from 0.10 km hr^{-1} (0.05 knot) to 3.52 km hr^{-1} (1.9 knots), with a combined average of 0.85 km hr^{-1} (0.46 knot). The near-bottom drogues moved at average speeds of 0.11 km hr^{-1} (0.06 knot) to 2.22 km hr^{-1} (1.2 knots), with a combined average of 0.67 km hr^{-1} (0.36 knot). However, since only several near-bottom drogues survived for more than 2 days, near-bottom results may not represent "typical" conditions.

Most of the current drogues, which were part of another study, were not released or being tracked during satellite over-passes. However, the average magnitudes of waste plume and drogue drift velocities are compatible with each other and with net current velocities measured by Falk et al. (1974), including their maximum recorded net drift speed of 14 km d^{-1} (0.44 knot). Drogues were being released at the time of the October 21, 1975 pass of LANDSAT over an acid waste plume (Table 2). During the first 6 hours after dump the waste plume had drifted south (188° azimuth) at a velocity of 1.80 km hr^{-1} (0.97 knot). During the same period the near-surface drogue moved south-southeast (175° azimuth) at 1.96 km hr^{-1} (1.06 knots) and the 20 m depth drogue drifted to the south-southwest (190° azimuth) at 1.57 km hr^{-1} (0.85 knot). Since the waste plume disperses during the first few hours over intermediate depths, its drift direction and velocity lie between those of the two drogues. Twenty-four hours later the same plume was observed by aircraft at a distance of about 10.3 km and an azimuth direction of 166° from its location in the satellite image. The slowing of the plume drift velocity to 0.87 km hr^{-1} (0.47 knot) and change in direction indicates that, similar to the drogues, a northeasterly component was temporarily superimposed on the south-ward movement. The cause of this northeast current was most likely a steady 37 km hr^{-1} (20 knots) wind from the west which developed during that same period.

During the stratified warm months, more drogues tended to move in the north-northeast direction while during the non-stratified winter months a southwest direction prevailed. These results do not conflict with previous studies which found that there was a mean flow to the southwest, with stratified surface waters moving in the northerly or northwesterly direction during the summer. Identical drogues released at equal depths generally followed similar paths. However, drogues released at different depths frequently traveled along different paths, and at different speeds, indicating the presence of current shear.

Most rapid movement of the drogues at all depths occurred during a severe northeaster storm with drogue speeds in excess of 3.33 km hr^{-1} (1.8 knots). The circulation process at the waste dumpsite appears to be highly storm-dominated with an increase of water transport occurring during storms, particularly northeasters. This conclusion is in agreement with results obtained by other investigators in the Middle Atlantic Bight (Beardsley et al., 1974).

As shown in Fig. 7, a distinct summer thermocline was observed from June through August 1975 at depths ranging from 13 m to 24 m. In 1976, the first observation of a thermocline again occurred in June. The strongest thermocline was observed on August 19, 1975, having a change of temperature from 23° to 8°C between depths of 13 m and 20 m, respectively. In comparison, Myers (1974) observed the formation of a thermocline at the same site during April 1973 at depths between 18 m and 23 m. Ocean stratification conditions influence waste dispersion. The wastes do not reach the ocean bottom when a thermocline is present. They are distributed from top to bottom when the ocean is isothermal (Falk et al., 1974).

The spatial and temporal resolution of the satellite imagery was not sufficient to provide precise data on waste plume dispersion. However, a visual estimate of plume width was obtained from satellite imagery and plotted as a function of time after the dump (Fig. 8).

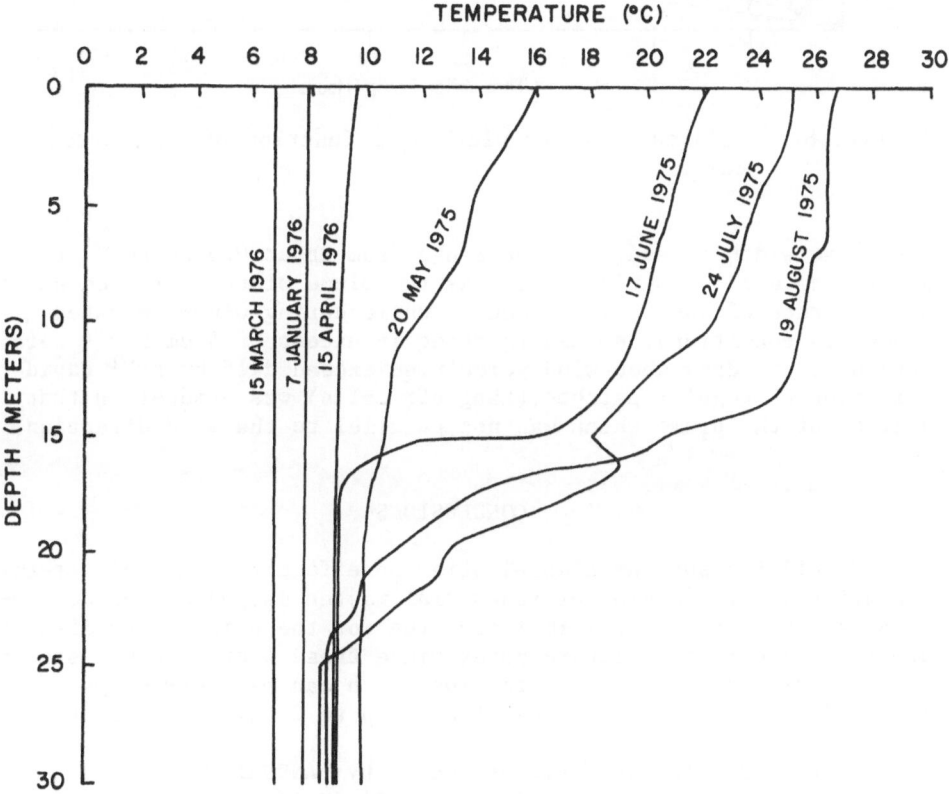

Fig. 7.   Temperature profiles obtained with an expendable bath-
           thermograph showing water stratification and formation
           of a thermocline during summer months.

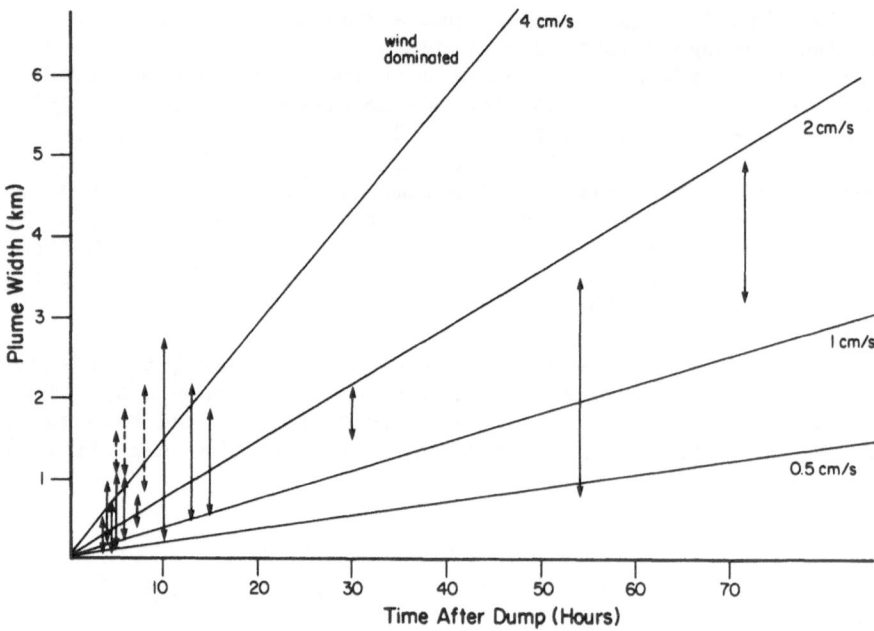

Fig. 8.   Acid waste plume width as a function of time after
           dump.

The plume width spreading rates range from about 0.5 cm sec^{-1} to
about 6 cm sec^{-1}. During calm seas the plume width increased at an
average rate of about 1.5 cm sec^{-1}, while during wind-dominated,
rough sea conditions, spreading rates in excess of 4 cm sec^{-1} were
attained. On days when wind velocities exceeded 15 km hr^{-1} rapid
formation of regular patches (Langmuir cells) was evident in that
section of the plume which was not parallel to the wind direction.

                              CONCLUSIONS

     Satellites such as LANDSAT offer an effective means of assessing
the drift and dispersion of industrial wastes dumped on the continental shelf. This is particularly true for the acid wastes disposal
about 64 km off the Delaware coast since these wastes form a sparse
but optically persistent ferric floc which can be observed by
LANDSAT's multispectral scanner band 4 up to 2 days after dump.

     Most of the 16 waste plumes imaged by LANDSAT were found to be
drifting at average rates of 0.59 km hr^{-1} (0.28 knot) to 3.39 km
hr^{-1} (1.83 knots) into the southwest quadrant. The plumes seemed
to remain above the thermocline which was observed to form from June
through August at depths ranging from 13 m to 24 m. During the

remainder of the year, the ocean at the test site was not stratified, permitting wastes to mix throughout the water column to the bottom.

The magnitudes of plume drift velocities were compatible with the drift velocities of current drogues released over a 12-month period at the surface, at mid-depth and near the bottom. However, during the stratified warm months, more drogues tended to move in the north-northeast direction, while during the non-stratified winter months a southwest direction was preferred. Drogues released at different depths frequently traveled along different paths and at different speeds, indicating the presence of current shear.

Rapid waste movement toward shore occurs primarily during storms, particularly northeasters. During such storms, however, the plumes are rapidly dispersed and diluted. The plume width was observed to increase at a rate of about 1.5 cm sec^{-1} during calm sea conditions, yet attain spread rates in excess of 4 cm sec^{-1} on windy days. These results are in agreement with Falk's (1974) estimate of plume dispersion shown in Fig. 9, which indicates that by the time a waste plume moves 37 km from the dumpsite, dilution is at least about one million to one. (See also Csanady; Hatcher et al.; and Kester et al., this volume.)

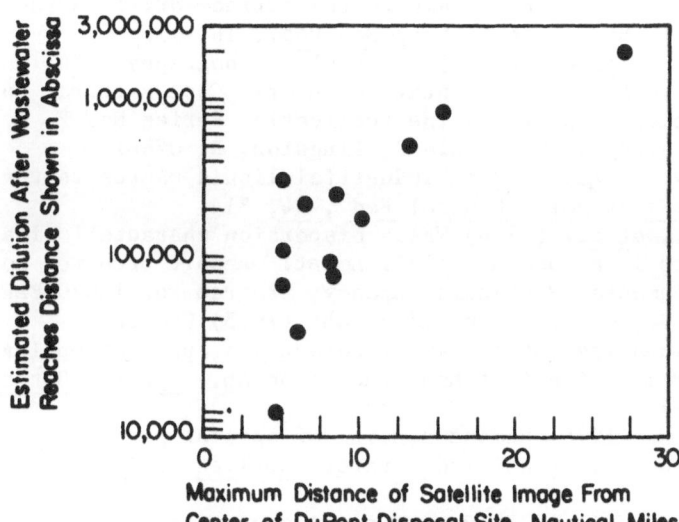

Fig. 9.   Estimated dilution of waste plume as a function of distance from dumpsite at the time of satellite overpass.

ACKNOWLEDGEMENTS

The research projects described in this article were partly funded through NASA, Contract Number NAS5-20983; and a contract from E. I. DuPont de Nemours & Company.

REFERENCES

Apel, J. R., R. V. Charnell and R. J. Blackwell (1974) Ocean internal waves off the North American and African coasts from ERTS-1-Proc. Ninth Intern. Symp. on Rem. Sens. of Environment, Ann Arbor, Mich. 14-19 April 1974.

Beardsley, R. C. and B. Butman (1974) Circulation on New England continental shelves: Response to strong winter storms. Geo-Phys. Res. Letters, 1, 181.

Boicourt, W. C. and P. W. Hacker (1976) Circulation on the Atlantic continental shelf of the United States, Cape May to Cape Hatteras. Memoires Societe Royale des Sciences de Liege, 6th Series, 10, 187.

Bumpus, D. F. (1965) Residual drift along the bottom of the continental shelf in the Middle Atlantic Bight area. Limnol. Oceanogr., 10, R50.

Bumpus, D. F. and L. M. Lauzier (1965) Surface circulation on the continental shelf off eastern North America between Newfoundland and Florida. American Geographical Society, New York, Folio 7, Serial Atlas of the Marine Environment.

Bumpus, D. F. (1969) Reversals in the surface drift in the Middle Atlantic Bight area. Deep-Sea Res., 16, 17.

Bumpus, D. F., et al. (1973) Physical oceanography. In: Coastal and Offshore Environmental Inventory, Cape Hatteras to Nantucket Shoals. Marine Publication Series No. 2, University of Rhode Island, Kingston, RI 02881.

Fader, S. W. (1972) Barging industrial liquid wastes to the sea. Jour. Water Poll. Control Fed., 44, 314.

Falk, L. L., et al. (1974) Waste dispersion characteristics and effects in an oceanic environment. Report prepared for U. S. Environmental Protection Agency, Program No. 12020 EAW.

Gordon, H. R., O. B. Brown and Jacobs (1975) Computer relationships between the inherent and apparent optical properties of a flat homogeneous ocean. Applied Optics, 14, 417-427.

Gordon, H. R. (1978) Remote sensing of optical properties in continuously stratified waters. Applied Optics, 17, (12), 1893-1897.

Hovis, W. A., Jr. (1977) Remote sensing of water pollution. Proc. 11th Internatl. Symp. on Remote Sensing of Environment, Ann Arbor, Michigan, 361-362.

Ketchum, B. H. (1953) Preliminary evaluation of the coastal water off Delaware Bay for the disposal of industrial wastes.

Unpublished manuscript, Reference No. 55-31, Woods Hole
Oceanographic Institution, Woods Hole, MA.

Klemas, V., C. Davis, J. Lackie, W. Whale and G. Tornatore
(1977) Satellite, aircraft and drogue studies of coastal
currents and pollutants. IEEE Transactions on Geoscience
Electronics, 2, (2), 119-126.

Klemas, V. and D. F. Polis (1977) Remote sensing of estuarine
fronts and their effects on pollutants. Photogram. Eng. and
Remote Sensing, 43, (5), 599-612.

Klemas, V., D. S. Bartlett and W. Philpot (1978) Remote sensing
of coastal environment and resources. Proc. Coastal Mapping
Symp., Am. Soc. Photogrammetry, August 14-16.

Morrison, D. F. (1976) Multivariate Statistical Methods.
McGraw-Hill, New York, 2nd edition.

Mueller, J. L. (1976) Ocean color spectra measured off the Oregon
coast: Characteristic vectors. Applied Optics, 15, 395-402.

Myers, T. D. (1974) An observation of rapid thermocline formation
in the Middle-Atlantic Bight. Estuarine and Coastal Marine
Science, 2, 74.

Peschiera, L., F. Freiherr II (1968) Disposal of titaniumn
pigment process wastes. Jour. Water Poll. Control Fed., 41,
127.

Plass, G. N., P. J. Humphreys and G. W. Katawar (1978) Color of the
ocean. Applied Optics, 17, (9), 1432-1446.

Simmonds, J. L. (1963) Application of characteristic vector
analysis to photographic and optical response data. J. Opt.
Soc. Amer., 55, 968-974.

U. S. Environmental Protection Agency (1975) Interim Ocean
Dumping Permit DE006. Issued to E.I. DuPont de Nemours and
Company, Inc. by EPA Region III, Philadelphia, PA.

III

CHEMICAL ASPECTS OF OCEAN DUMPING

TRANSITION AND HEAVY METALS ASSOCIATED WITH ACID-IRON WASTE DISPOSAL AT DEEP WATER DUMPSITE 106

Dana R. Kester, Richard C. Hittinger,
and Prithviraj Mukherji

Graduate School of Oceanography
University of Rhode Island
Kingston, R.I. 02881

ABSTRACT

Liquid acid-iron waste has been discharged into the surface waters at Deep Water Dumpsite 106 in 4 x 10^6 liter quantities since 1976. Upon mixing with seawater hydrous ferric oxide precipitates and can scavenge potentially toxic metals (Cu, Cd, and Pb) and organic substances from the waste plume and the seawater. A series of water samples were collected at various times from 0.5 to 27 hours after a dump of acid-iron waste using a non-contaminating pumping system and acoustic backscattering to locate the waste plume in the mixed layer (upper 20 m). The samples were analyzed for total and particulate Fe, Cu, Cd, and Pb, as well as for pH, total suspended matter, phosphate, silicate, temperature and salinity. Because of the high Fe concentrations in the waste (0.5 molar), the total Fe concentration in the waste plume provides a good index of the dilution of the waste. During the first 27 hours after the dump the dilution occurred as a two stage process. There was rapid initial dilution by 10^4 in the first 0.5 hr followed by slower dilution to 10^5 after 27 hr. Correlation analysis of the metal concentrations in the waste plume showed that Fe and Pb disperse in a closely coupled fashion related to particulate matter. The Cd and Cu disperse at a slower rate than Fe and Pb and they are less associated with the particulate phase.

INTRODUCTION

Deep Water Dumpsite 106 (DWD-106) has been a location for the regulated discharge of a variety of industrial chemical wastes for more than a decade. Ocean disposal of similar wastes on the continental shelf off the coasts of New York, New Jersey, and Delaware has been studied extensively, but prior to the NOAA Ocean Dumping Program little attention had been given to the effect of waste disposal at deep ocean dumpsites. Several studies at the shallow water (depth less than 100 m) continental shelf dumpsites indicate that components of the wastes become associated with the bottom sediments and benthic organisms (Gross, 1976; Pesch et al., 1977; Vaccaro et al., 1972). However it is likely that these wastes will be more highly dispersed at deepwater dumpsites before reaching the seafloor at depths greater than 2000 m. Consequently, the most likely impact, if any, of wastes at deepwater dumpsites is probably associated with planktonic rather than benthic organisms.

Our chemical studies related to the problem of industrial waste disposal at DWD-106 have focused on three objectives. First we would like to establish the extent to which waste disposal has increased the concentration of selected metals (Fe, Cu, Cd, and Pb) at DWD-106. Second we would like to identify the chemical reactions which occur when these wastes mix with seawater. And third we would like to determine the chemical form of these metals at DWD-106. Information on these three aspects of the chemistry of industrial waste disposal in the sea are required to assess the biological consequences of ocean dumping these wastes.

The principal chemical properties of the acid-iron waste dumped at DWD-106 are compared with those of seawater in Table 1. When this waste mixes with seawater the acid is neutralized and the iron precipitates forming hydrous ferric oxide. This precipitate is known to be an effective chemical scavenger of organic substances and of other metals which could be toxic to marine organisms. One of the processes which is important is the extent to which metals are concentrated into the particulate phase formed by the iron. It is possible that particulate metals could be taken up by filter-feeding planktonic organisms thereby providing rapid concentration and transfer of metals into the marine food web.

METHOD OF SAMPLING AND ANALYSIS

Studies were carried out in July 1977 aboard R/V ALBATROSS IV to determine the effect of an acid-iron waste disposal on the distribution of selected metals at DWD-106. Samples were collected for total metal analyses, for particulate metals, and for total suspended matter. The pH of the seawater samples was measured onboard the ship.

The metals included in this study were Fe, Cu, Cd, and Pb.

Water samples were collected by two methods designated as pump and rosette. The rosette consisted of up to 12 standard 8-liter Niskin bottles mounted on a rosette sampler which was used with an in situ salinity, temperature, depth (STD) measuring system. The pumping system was designed and built at the Woods Hole Oceanographic Institution by M. Orr and C. Winget; it was an all-Teflon system with

Table 1. Comparison of the chemical properties of an E.I. duPont deNemours and Co. waste and seawater in which it has been discharged. (Based on a 10 September 1976 EPA, Region III, Discharge Permit Application.)

Property	Edgemoor Waste	Seawater
Density (g/ml)	1.15	1.02
Total Dissolved Solutes (g/l)	240	35
Total Suspended Solids (g/l)	1.5	0.001
pH	0.01	8.2
Alkalinity (eq./l)	-4.50	0.002
Chloride (molar)	5.36	0.55
Sulfate (molar)	0.0083	0.029
Hydrogen ion (molar)	0.98	6×10^{-9}
Iron (molar)	0.90	$< 1 \times 10^{-8}$
Titanium (molar)	0.063	$< 1 \times 10^{-8}$
Aluminum (molar)	0.063	$< 1 \times 10^{-7}$
Manganese (molar)	0.031	$< 1 \times 10^{-8}$
Vanadium (molar)	0.004	$< 4 \times 10^{-8}$
Chromium (molar)	0.004	$< 2 \times 10^{-8}$
Cadmium (molar)	$< 7 \times 10^{-8}$	2×10^{-11}
Copper (molar)	3.9×10^{-5}	2×10^{-9}
Lead (molar)	0.0002	1.4×10^{-10}
Total organic carbon (mg/l)	960	1

Note: The data in this table for the waste are not internally consistent. Charge équivalents of cations and anions are not equal and the alkalinity is not fully accounted for by the hydrogen ions and iron. The pH may be inaccurate if it was measured by an instrument which does not permit pH < 0. Consequently, the hydrogen ion molarity which appears to be derived from pH may be in error. If the alkalinity is accepted as being correct the hydrogen ion concentration may be calculated by difference. The Fe contributes -2.70 eq/l and other metals (Al, Mn) provide about -0.2 eq/l. Thus H^+ could be about 1.6 eq/l in which case the pH would be -0.2.

six sampling ports spaced at one meter intervals. Attempts were made
to position the sampling ports at depths of maximum acoustic backscattering. Six samples could be collected simultaneously at a rate
of approximately 5 liters per minute. Samples for total metal analyses were drawn into precleaned and preweighed 250 ml linear polyethylene centrifuge bottles directly from the pump or as soon as possible after the STD-rosette were brought aboard. If more than a half-
hour delay was necessary, the 8-liter Niskin bottles were shaken
thoroughly before drawing either total metal or particulate metal
samples. The centrifuge bottles were returned to sealed plastic bags
for storage.

Particulate samples were collected on acid cleaned, preweighed,
0.4 μm Nuclepore filters. These filters were washed in a laminar
flow clean bench 5 times with ultraclean distilled deionized water
within hours of collection. Upon return to the laboratory, the filters were folded and placed in a precleaned polyethylene vial; 1.00
ml of 3 N ultrapure HNO_3 was added to cover the filter. The vial
was then sonicated in an ultrasonic cleaner to release the adsorbed
and physically trapped metal. The acid was diluted as necessary and
analyzed using the same graphite furnace atomic absorption analysis
as is used for the total metal analyses.

Total suspended matter data were obtained from the difference in
weight before and after filtration through the 0.4 μm Nuclepore filters used for particulate metal analyses.

The pH measurements were made with a glass electrode onboard
ship. Unfiltered samples were kept closed and were measured immediately after collection.

The analytical method that we used for total (dissolved + chemically active particulate) iron, copper, cadmium, and lead is precon-
centration from seawater followed by graphite furnace atomic absorption spectroscopy (AAS). In this work we modified the preconcentra-
tion technique of Boyle and Edmond (1977) so that sampling, coprecipitation with cobalt-APDC (ammonium pyrrolidine dithiocarbonate),
nitric acid dissolution, and AAS analysis could all be carried out
with a single 250 ml centrifuge bottle, thereby reducing the possibility of contamination during sample processing. Immediately after
returning to our laboratory the 250 ml centrifuge bottles were re-
weighed to determine the mass of seawater in the total metal sample.
The samples were acidified to pH 3.0 in a laminar flow clean bench
and allowed to sit for 48 hours. A solution of $CoCl_2$ was added to
the acidified sample followed by the addition of an APDC solution.
The flocculant precipitate which formed extracted the dissolved
metals from the acidified seawater. After at least 30 minutes the
bottles were centrifuged at an acceleration of 3000 g for 15-20
minutes. The supernatant was aspirated from the metal-APDC precipitate and HNO_3 plus distilled water were added to provide about 5 ml

Table 2. Analytical characteristics of the total metal analyses used
          in this work.

Metal	Detection (ng/kg)	Blank (ng/kg)	Efficiency (%)	Precision for a range (ng/kg)
Cd	1	$< 1 \pm 1$	90 ± 1	± 2 for 6-20
Cu	5	16 ± 10	99 ± 3	± 20 for 100-400
Pb	5	15 ± 19	97 ± 2	± 20 for 50-200
Fe	30	230 ± 36	92 ± 1	± 60 for 400-1400

of 3 N HNO_3 which dissolved the precipitate and which also would
desorb any metals from the container walls or from particulate matter
that may not have been removed during the acidification at pH 3.0 and
in the presence of APDC. The bottles were reweighed to determine the
concentration factor for the extraction. The sample was analyzed by
injection into a graphite furnace for AAS.

     With this method the sample comes into contact with only one
container for storage, processing, and analysis. Contamination is
minimized (1) by reducing the sample handling at sea, (2) by opening
the sample container only in a laminar flow clean bench except for
the initial drawing at sea and injection into the graphite furnace,
and (3) by using the least possible quantities and types of reagents
(e.g., buffers and organic solvents are not required).

     Several factors must be considered in the use of this method.
There is some carry-over of sea-salts in the final 3 N HNO_3 solution, the possible matrix effect of which is accounted for by preparing standard solutions in previously extracted seawater and by using
a deuterium arc background corrector in the AAS analysis. Table 2
summarizes the analytical characteristics of this method. The detection limit is defined as two times the instrumental noise. The blank
was determined by re-extracting seawater samples and it accounts for
metals added with the reagents and during handling. The efficiency
of the extraction was obtained from the yield of radioactive tracers
(^{64}Cu, ^{109}Cd, ^{210}Pb, ^{59}Fe) which were added to a series of seawater
samples. The precision is based on triplicate analyses of about 20
rosette samples collected during the field study. The precision is
given as the variability among the triplicates for samples with the
indicated range of concentration. The consideration of precision is
useful for interpreting variations found in the metal concentration
data. The high iron blank is contributed primarily by the reagents
and appears to be reasonably consistent within a data set. In re-
porting the values of total metal concentrations we have ignored the

Table 3.   Summary of total metal concentrations in the control ring
           based on the pump sampler and Niskin samplers. The mean,
           standard deviation (σ), and number of samples (n) are
           reported.

Pump Sampler (16-21 m)

Metal	ng/kg		
	Mean	σ	n
Fe	230	60	6
Cu	110	30	6
Cd	2	0.5	6
Pb	110	40	6

Niskin Sampler (20-1000 m)

Metal	ng/kg		
	Mean	σ	n
Fe	670	140	10
Cu	170	64	10
Cd†	13	6	5
Pb*	276	34	5

† Upper 100 m only due to systematic increase in deep
waters.

* Upper 100 m only due to decrease in deep waters.

Cd blank and we have subtracted the 16 ng/kg Cu blank. We have not
corrected the measured Fe and Pb for blank values because we are
still assessing the source of these blanks. Thus, the reported
values for these two metals represent the oceanic concentration plus
our analytical blank. These blank values are insignificant relative
to the observed post-dump concentrations, but they may be large compared to natural oceanic concentrations (Settle and Patterson, 1980).

METAL CONCENTRATIONS PRIOR TO THE ACID-IRON DUMP

     A set of pump and rosette samples were taken in a control region
which was hydrographically similar to the dumpsite for comparison
with post-dump samples. Table 3 lists the results of total metal
analyses (mean and standard deviations) for pumped samples at 16-21 m
and rosette samples collected from 20-1000 m. The pump sampler

Table 4. Results of particulate metal analyses for the control ring.

Pump Sampler - Depths 16-21 m

Metal	(ng/kg)			% of Total Metal		
	Mean	σ	n	Mean	σ	n
Fe	250	130	6	82%	34%	6
Cu	48	22	6	32%	19%	6
Cd	0.15	0.06	6	7%	3%	6
Pb	11	7	6	12%	10%	6

Niskin Sampler - Depths below 20-1000 m

Metal	(ng/kg)			% of Total Metal		
	Mean	σ	n	Mean	σ	n
Fe	470	400	9	60%	50%	8
Cu	28	16	9	17%	8%	8
Cd	0.40	0.26	9	3.2%	4.1%	8
Pb	12	7	9	5.0%	2.5%	3

yielded lower concentrations than the Niskin sampler. This conclusion appears to be valid even though different portions of the water column were sampled by the two systems. The 20 m Niskin sample gave values close to the mean for all the Niskin sampled metals reported in Table 3 and these values are significantly different from the 16-21 m pump samples.

The pump data represent some of the lowest concentrations for iron in seawater which have been reported. Earlier analytical data for iron reported concentrations 10-50 times greater than those found at this control station. These concentrations, however, may not represent the true value because they are comparable to the analytical blank.

During this study a warm core Gulf Stream ring was present at DWD-106. These rings occasionally transport Gulf Stream and Sargasso Sea waters through the dumpsite which is normally occupied by Slope Water with some Shelf Water intrusions. Consequently the hydrography of this region can be complex and highly variable (Bisagni and Kester, 1981) and the background chemical and biological characteristics may range from Shelf Water to Slope Water to Sargasso Sea characteristics. The control samples in this work were obtained from a warm core ring approximately 160 km east of the ring at DWD-106. The total

suspended matter in the upper 80 m measured at the control station
was quite low as expected for Sargasso Sea Water. The average of 9
samples was 130 µg/l with a standard deviation of 47 µg/l. Analyses
of this particulate matter are summarized in Table 4. For particulate Fe and Cd the Niskin data show greater variability (larger σ)
than the pump data. This result could be a consequence of the Niskin
data set sampling from 20-1000 m where variations may be greater than
the 16-21 m sampled by the pump. The data indicate that a large portion of the total Fe is in the particulate phase whereas most of the
total Cd is dissolved. These results are consistent with our general
understanding of the marine geochemistry of these two metals. The Pb
results are somewhat surprising in that data on ^{210}Pb indicate a high
affinity of this metal for surfaces of particles (Spencer et al.,
1978), whereas our data indicate that only a small portion of the Pb
is in the particulate phase.

CHEMICAL DISTRIBUTIONS AFTER THE ACID-IRON DUMP

Samples taken following the July 26, 1977 dump of an acidic
waste with extremely high iron content showed significant alterations
in pH; total and particulate Fe, Cu, Cd, and Pb; total suspended matter (TSM); and total organic carbon (TOC). The most dramatic signal
was that for iron. Immediately following the dump, samples more than
2×10^4 times background total iron were collected. Every sample
collected in the top 21 m after the dump was more than 2 σ above the
mean total iron in the control area, including samples collected more
than 27 hours following the dump.

These iron data for a series of stations following the dump are
illustrated by vertical plots in Fig. 1. The sampling was generally inadequate to define the true vertical distribution of the iron,
but the data indicate large increases in total Fe in the upper 10-20
m for at least 27 hours after a dump. Even though these data are
highly dependent on the limited sampling their magnitude provides an
estimate of the dispersion of the waste. A sample of the acid-iron
waste was analyzed to determine the initial concentrations of metals
(Table 5).

Table 5. Analyses of a sample of Edgemoor acid-iron waste dumped
          at DWD-106 on July 26, 1977.

Metal	Concentration (mg/kg)	Ratio to Fe
Fe	3.1×10^4	1
Cd	$< 8 \times 10^{-3}$	$< 2.6 \times 10^{-7}$
Pb	67	2.2×10^{-3}
Cu	2.5	8.1×10^{-5}

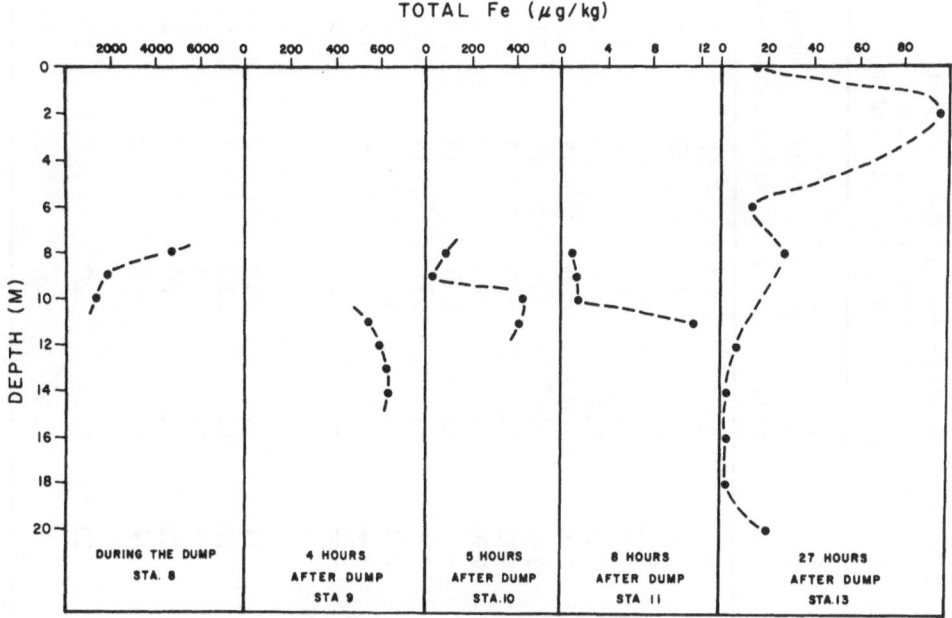

Fig. 1. Vertical plots of T(Fe) in the upper 20 m at various times after the acid-iron dump.

Fig. 2 shows a plot of the observed total Fe concentrations (on a logarithmic scale) sampled at various times following the dump. Due to the extremely small portion of the water column which was sampled at each time and the very large vertical gradients (Fig. 1) the results are highly dependent on the positioning of the sampler vertically to within a few meters (or tens of centimeters) and laterally relative to the main body of the waste plume. Thus, the values after 8 hours could be low relative to the 27 hour values due to sampling outside of the maximum intensity of the plume. Two conclusions are indicated by these data. In the first 10 hours after a dump there is rapid 10^6 dilution of the waste from its initial concentration. However, after this rapid dilution there is a persistence of elevated total iron concentrations up to 400 times the natural values.

DISCUSSION

In order to investigate the chemical relationships among the parameters measured after the acid-iron dump the data were rank ordered according to total Fe without regard to depth or time after the dump. Table 6 is a rank order listing of T(Fe) > 10 µg/kg plus the control values. Table 7 shows the matrix of correlation

Table 6. Rank order listing by total Fe concentration of observations after the acid-iron test dump with T(Fe) > 10 µg/kg plus the control analyses. The data include values for total suspended matter (TSM), particulate Fe, Cu, Pb, Cd, and total Fe, Cu, Cd, and Pb.

Hours After Dump	Sta.	Depth (m)	T(Fe) µg/kg	TSM µg/kg	Part. Fe µg/kg	Part. Cu ng/kg	Part. Pb ng/kg	Part. Cd ng/kg	T(Cu) ng/kg	T(Cd) ng/kg	T(Pb) ng/kg
0.5	8	8	4700	11,900	980	1400	—	2.0	710	52	6970
0.5	8	9	1830	5,192	1150	446	1250	5	470	33	1350
0.5	8	10	1300	3,465	892	830	950	3	470	23	880
4.0	9	14	630	2,130	420	444	707	1.2	380	14	640
4.0	9	13	625	1,450	280	208	501	1.9	260	26	490
4.0	9	12	594	1,410	250	196	440	1.8	430	43	610
4.0	9	11	546	980	190	183	349	1.3	180	13	370
5.0	10	10	420	1,320	310	264	483	1.6	450	18	440
5.0	10	11	410	990	109	73	280	1.0	200	7	80
27.0	13	2	95	260	25	27	—	0.4	430	18	190
27.0	13	8	27	130	6	22	—	0.5	270	19	1330
5.0	10	9	20	280	52	348	201	1.0	420	30	220
27.0	13	6	12	240	12	81	—	0.3	110	18	130
8.0	11	11	12	88	2	33	—	—	630	11	180
Control Samples Prior to Dump	16		0.3	—	0.3	40	11	0.2	150	2	170
	17		0.2	—	0.2	71	10	0.2	110	2	150
	18		0.2	30	0.1	18	5	0.1	70	2	120
	19		0.1	180	0.4	65	23	0.1	70	2	80
	20		0.3	140	0.2	27	14	0.1	140	3	70
Control Mean			0.22	117	0.24	44	13	0.14	120	2.2	120
Control σ			0.08	78	0.11	23	7	0.06	40	0.4	40

Table 7. Matrix of correlation coefficients for pairs of parameters listed in Table 6. The values of r ≥ 0.8 have been underlined for emphasis.

	T(Fe)	Part. Fe	Part. Cu	Part. Pb	Part. Cd	TSM	T(Cu)	T(Cd)	T(Pb)
T(Fe)	1.00								
Part. Fe	0.82	1.00							
Part. Cu	0.92	0.82	1.00						
Part. Pb	0.99	0.74	0.92	1.00					
Part. Cd	0.98	0.69	0.89	0.99	1.00				
TSM	0.99	0.84	0.93	0.98	0.97	1.00			
T(Cu)	0.74	0.74	0.79	0.73	0.70	0.75	1.00		
T(Cd)	0.72	0.65	0.70	0.70	0.70	0.72	0.85	1.00	
T(Pb)	0.94	0.62	0.85	0.97	0.98	0.94	0.67	0.68	1.00

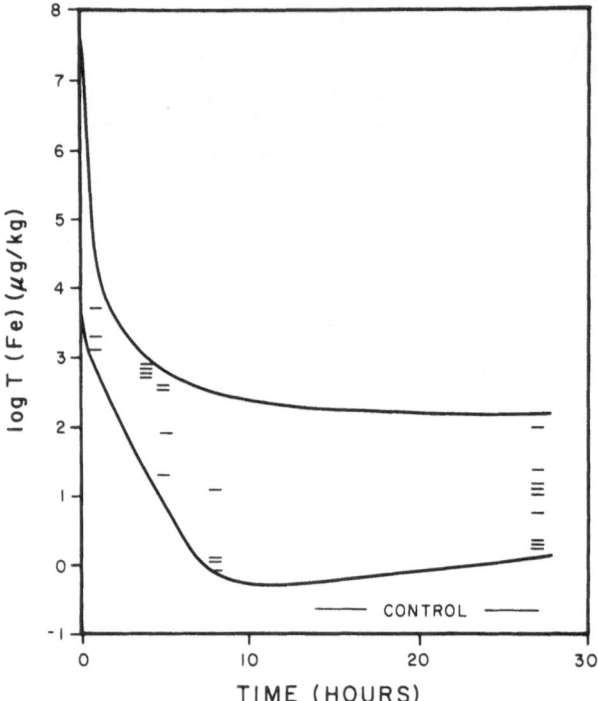

Fig. 2. Variation with time after the acid-iron test dump of the
         observed T(Fe).

coefficients for all pairs of these 9 parameters. Total suspended
matter and T(Fe) are very closely coupled and a scatter diagram of
these two parameters is shown in Fig. 3. The equation describing
this relationship is:

$$\text{TSM} = (2.54 \pm 0.09)\ \text{T(Fe)} + (124 \pm 48)\ \mu g/kg \qquad (1)$$

The plus and minus values are the standard deviation of the slope and
the intercept of the least squares linear regression between TSM and
T(Fe). Iron is a principal component of the TSM after the dump, comprising about 40% by weight of the TSM. We may assume that the iron
is primarily $FeOOH \cdot nH_2O$. The slope in equation (1) then yields the
following formula for the principal solid phase: $FeOOH \cdot 3H_2O$ which
would have a ratio of TSM/T(Fe) = 2.55.

     Examining the other correlation coefficients one finds that
there are basically two sets of coupled parameters in these data.
Total Fe, TSM, particulate Cu, Pb, and Cd, and total Pb all co-vary
with significant correlation coefficients. Total Cu and Cd vary more
independently from the other parameters and they show a moderate

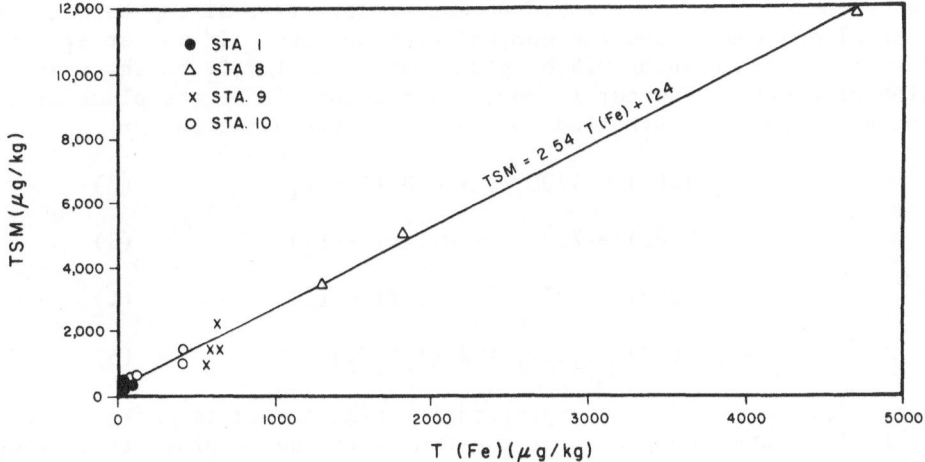

Fig. 3. Correlation of T(Fe) and TSM after the acid-iron test dump.

relation between each other. TSM and T(Fe) are clearly indicators of at least one phase of the acid-iron waste. It may seem surprising that particulate Fe is not more closely coupled to the TSM and the other particulate metal results. The value of particualte Fe obtained during the dump at 8 m depth is anomalously low compared to the TSM and the T(Fe). If this value is eliminated from the data set particulate Fe then shows high correlations with TSM (r = 0.99), particulate Pb (r = 0.97), particulate Cd (r = 0.96), and particulate Cu (r = 0.83). We consider that this analysis of particulate Fe was probably subject to large errors due to the extremely high concentration of Fe, and perhaps due to the colloidal nature of $FeOOH \cdot nH_2O$ soon after formation allowing much of it to pass through the 0.4 μm filter. It is possible that the more independent variations of T(Cu) and T(Cd) relative to the lead, iron and particulate parameters are related to the fact that these two metals have relatively low concentration in the acid-iron waste and their marine chemistry is more associated with the solution phase than with particulate phases. These characteristics are influenced by the tendency for Cu to form stable inorganic complexes with $CO_3{}^{2-}$ and OH^- as well as with organic substances and for Cd to form stable chloride complexes.

The following picture emerges from these data. In the waste plume 0.5 hr after dumping, concentrations of TSM=12,000 μg/kg, T(Fe) = 4700 μg/kg, T(Cd) = 52 ng/kg, T(Cu) = 0.71 μg/kg, and T(Pb) = 7.0 μg/kg existed in the waste-seawater mixture. Subsequent sampling represents dilutions of this mixture by the control station: TSM = 120 μg/kg, T(Fe) = 0.2 μg/kg, T(Cd) = 2 ng/kg, T(Cu) = 0.1 μg/kg, and T(Pb) = 0.1 μg/kg. If this dilution process occurred without any chemical differentiation among the metals and between the particulate

and solution phases, we would expect conservative mixing between the
initial end member and the control station data. If we let f_i represent the fraction of 0.5 hr plume water and $1 - f_i$ be the fraction of control seawater in subsequent dilutions of the plume we can
calculate the concentrations of metals in the dilution as:

$$T(Fe) = 4700\ f_i + 0.2\ (1 - f_i) \qquad (2)$$

$$T(Pb) = 7.0\ f_i + 0.1\ (1 - f_i) \qquad (3)$$

$$T(Cu) = 0.7\ f_i + 0.1\ (1 - f_i) \qquad (4)$$

$$T(Cd) = 52\ f_i + 2\ (1 - f_i) \qquad (5)$$

where Fe, Pb, and Cu are expressed in $\mu g/kg$ and Cd is ng/kg. Since
T(Fe) is a major component of the TSM of the waste plume it provides
a good index of the amount of waste in a mixture. Thus from equation
(2):

$$f_i = \frac{T(Fe)_{obs} - 0.2}{4700 - 0.2} \qquad (6)$$

where $T(Fe)_{obs}$ is the observed T(Fe) in a mixture. For the data
listed in Table 6 $T(Fe)_{obs} > 10\ \mu g/kg$ so that equation (6) may be
approximated by:

$$f_i = \frac{T(Fe)_{obs}}{4700} \qquad (7)$$

Substituting equation (7) into (3), (4), and (5) and rearranging
yields expressions which may be used to calculate the expected
T(Pb), T(Cu), and T(Cd) based on T(Fe) assuming conservative dilution of the waste plume:

$$T(Pb)_{calc} = \frac{T(Fe)_{obs}}{4700} \times 6.9 + 0.1 \qquad (8)$$

$$T(Cu)_{calc} = \frac{T(Fe)_{obs}}{4700} \times 0.6 + 0.1 \qquad (9)$$

$$T(Cd)_{calc} = \frac{T(Fe)_{obs}}{4700} \times 50 + 2 \qquad (10)$$

    Fig. 4 shows an application of these expressions for T(Pb). Due
to the large range in values a log-log plot has been used. The

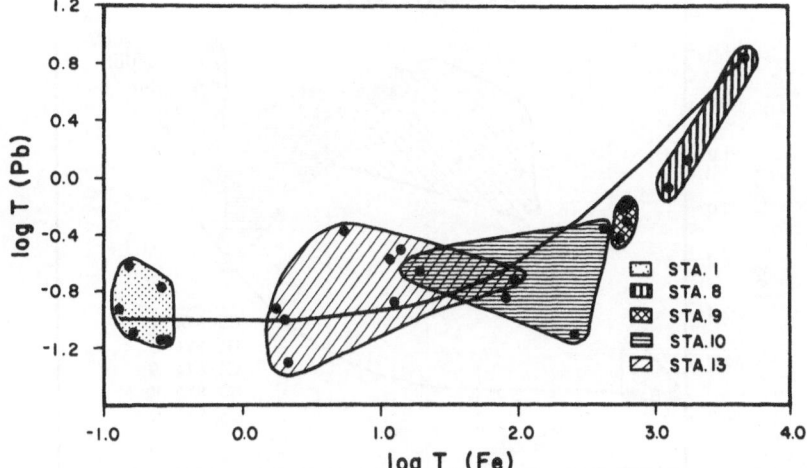

Fig. 4. Variation in T(Pb) and T(Fe) after the dump with the
         observations at each time (station) enclosed by textured
         regions. The curve is calculated from eq. (8).

textured zones encompass the data from each station or time of sampling. The solid curve is a plot of equation (8) on the log-log
axes. The total Pb data follow this relationship very closely and we
take this observation to be evidence that lead follows the particulate phase (the iron) in the waste plume. Based on the particulate
Pb data for stations 8, 9, and 10 essentially all the T(Pb) is particulate Pb in the waste plume. If lead were being scavenged from
the seawater and concentrated in the particles beyond that contributed by the waste the data would trend above the solid curve.

    Fig. 5 shows a similar plot for T(Cu) and T(Cd). When inter-
preting these results three factors must be kept in mind. First, the
sampling was directed by the acoustic scattering in order to track
the particulate phase as much as possible. If there was a chemical
fractionation of a metal between the particulate and the soluble
phases in the plume such that the metal remained soluble it would be
under-sampled in the data set. Second, changes among stations, 8, 9,
10, and 13 include two important factors: dilution as the plume disperses and equilibration or aging of the solid phase with time if the
rates of chemical exchange processes between the particles and the
seawater are slow compared to the rate of dilution. Third, if the
hydrous ferric oxide particles fall through the water by gravitation-
al settling the T(Fe) will decrease more quickly than metals in the
soluble phase.

    The copper data trend systematically above the initial plume
conservative mixing curve with control seawater. If one were to use

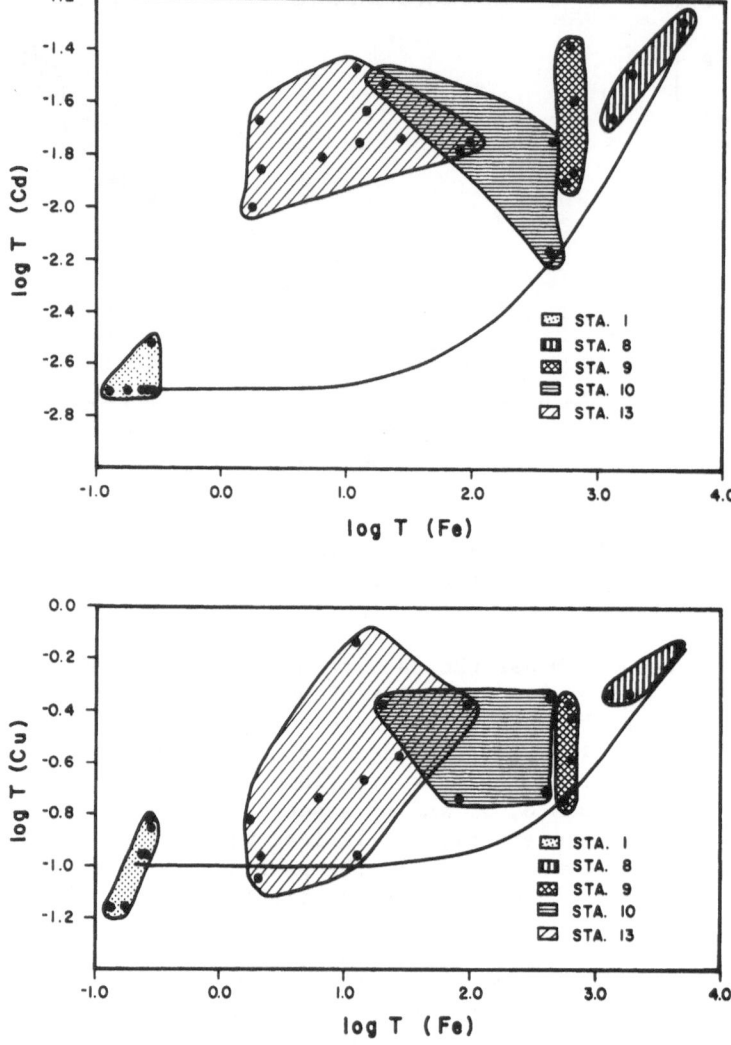

Fig. 5.   Variation in T(Cd) and T(Cu) with T(Fe) after the dump.   The
curve for T(Cd) vs T(Fe) is calculated from eq. (10) and for
T(Cu) vs T(Fe) from eq. (9).

as the two end members the composition of initial waste and the control seawater the conservative mixing would follow the same curve as
shown in Fig. 5.   Therefore, we conclude that the samples at stations
10 and 13 have about three times more Cu than is expected based on
this composition of the waste and conservative mixing of the plume.
Comparison of particulate and total metal analyses show that about
half of the copper is associated with the particulate phase.

The Cd data in Fig. 5 are even more dramatic than Cu in their departure from the conservative mixing curve. Even though the T(Fe) is being diluted 1000 fold between stations 8 and 13 the T(Cd) is diminished by only 3 fold. Only a few percent of the Cd is in the particulate phase. The relationship between Cd and Fe shown in Fig. 5 could occur if one considers that Cd is mainly in the soluble phase whereas Fe is mainly particulate. If a significant portion of the Fe is being removed by settling and the Cd is remaining in the upper waters in the waste plume one would see a persistence of Cd concentrations and a decrease in Fe.

The results shown in Figs. 4 and 5 are most consistent with the concepts that after the dump of acid-iron waste there are two main factors responsible for the distribution of the waste: dilution with seawater and settling of particles. Cd is affected primarily by dilution whereas TSM, T(Fe), T(Pb), and the particulate metals are affected both by dilution and settling. Cu being about 50% in the soluble phase is intermediate between Cd and Pb in its variation with Fe at various times after the dump. The sampling after the dump was too limited to provide a mapping of the plume or information on the fate of particles which may have settled beneath 20 meters within 30 hours.

This analysis of the results leaves one major unanswered question. Where did the high concentrations of Cd observed after the acid-iron dump ($\geq$ 18 ng/kg) originate? If we take the T(Fe) of the 0.5 hr 8 m sample to be an index of the amount of waste in that sample, we estimate from Table 5 that 0.15 grams of waste mixed with 1 kilogram of seawater would yield 4700 μg/kg of T(Fe). A mixture of 0.15 g of waste in 1 kg of seawater would yield T(Pb) = 10 μg/kg, T(Cu) = 0.5 μg/kg, and T(Cd) < 3.2 ng/kg. The observed values at 0.5 hr 8 m are T(Pb) = 7 μg/kg, T(Cu) = 0.7 μg/kg and T(Cd) = 52 ng/kg. The agreement with Pb and Cu is acceptable, but we can not reconcile the amount of Cd found in the samples with our measurement of the amount of Cd in the waste. The Cd analyses of the waste (Table 5) should be investigated further before the discrepancy in the field analyses can be resolved.

In future studies of waste plumes in the ocean there is a need to utilize continuous sampling and nearly real-time analyses. While considerable information can be obtained from the discrete sampling used in this work, the interpretation of the results is limited by the adequacy of the sampling. With continuous sampling and analysis of at least selected parameters, it should be possible to establish the maximum concentrations and the vertical distribution of waste with greater reliability than with discrete sampling.

ACKNOWLEDGEMENTS

    This research was supported by the NOAA Ocean Dumping Program
under grant NA79AA-D-00033. The non-contaminating pumping system developed by M. Orr and C. Winget and the information provided by their
acoustic backscattering measurements during sampling were extremely
valuable in conducting this work.

REFERENCES

Bisagni, J. J. and D. R. Kester (1981) Physical variability at an
    east coast United States offshore dumpsite. In: Ocean Dumping
    of Industrial Wastes, B. H. Ketchum, D. R. Kester, and P. K.
    Park, editors, Plenum Press, New York. This volume, pp. 89-107.
Boyle, E. A. and J. M. Edmond (1977) Determination of copper,
    nickel, and cadmium in sea water by APDC chelate coprecipitation
    and flameless atomic absorption spectrometry. Analytica Chimica
    Acta, 91, 189-197.
Gross, M. Grant (1976) Waste disposal. MESA New York Bight
    Atlas Monograph, 26, New York Sea Grant Institute, Albany,
    N.Y., 32 pp.
Pesch, G., B. Reynolds, and P. Rogerson (1977) Trace metals in
    scallops from within and around two ocean disposal sites.
    Marine Pollution Bulletin, 8, 224-228.
Settle, D. M. and C. C. Patterson (1980) Lead in albacore: Guide to
    lead pollution in Americans. Science, 207, 1167-1176.
Spencer, D. W., P. G. Brewer, A. Fleer, S. Honjo, S. Krishnaswami,
    and Y. Nozaki (1978) Chemical fluxes from a sediment trap
    experiment in the deep Sargasso Sea. Journal of Marine
    Research, 36 (3): 493-523.
Vaccaro, R. F., G. D. Grice, G. T. Rowe, and P.H. Wiebe (1972)
    Acid-iron waste disposal and the summer distribution of standing
    crops in the New York Bight. Water Research, 6, 231-256.

WASTE MATERIAL BEHAVIOR AND INORGANIC GEOCHEMISTRY

AT THE PUERTO RICO WASTE DUMPSITE

B. J. Presley, J. Scott Schofield, and John Trefry

Department of Oceanography
Texas A & M University
College Station, TX 77843

ABSTRACT

Samples of a pharmaceutical waste were taken prior to dumping at a site over the Puerto Rico Trench and analyzed for the cations Ag, Al, As, Ca, Cd, Co, Cu, Fe, Hg, K, Li, Mg, Mn, Mo, Na, Ni, Pb, Sb, Sr and Zn. Iron and Mn were found at concentrations of 170 ppm and 5.2 ppm respectively with the other heavy metals present at below 1 ppm. Of the remaining cations, the most abundant were: Ca at 290 ppm, K at 340 ppm, Mg at 33 ppm, and Na at 9,900 ppm.

The major anions were Cl^- at 8,000 ppm and SO_4^{2-} at 13,500 ppm. Total phosphorus was 105 ppm and PO_4^{3-}-P was present at 53 ppm. The quantity of organic material, pH and physical characteristics were also determined. Analyses of surface water samples, after dumping, for Cd, Cu, Fe, Hg and Ni indicated only Fe levels were distinctly affected by dumping. The highest value noted in the study was 30 ppb, 30 minutes after a dump. The results suggest rapid depletion of surface Fe and indicate a plume surface half-life of about 2.4 minutes.

Suspected dump-induced iron contamination of zooplankton was seen five hours after one dump but no effects were noted after 24 hours. The toxic trace metals were thus generally present in low concentrations initially and rapidly dispersed in the water after dumping.

INTRODUCTION

The Puerto Rico chemical waste dumpsite, which is located 42 miles north of Arecibo, has been used since its establishment in 1972 primarily as a disposal site for pharmaceutical wastes. Currently seven different companies dump a total of about one-half million tons of material per year at the site. Additional information about this dumpsite is provided by Anderson and Dewling (1981).

We participated in an integrated study of the dumpsite which included a February 1978 sampling cruise involving a number of different research groups. The goals of our part of the study were to: (1) analyze the waters of the site for selected chemical constituents to establish baseline levels for reference purposes; (2) monitor the effects of a dump on water quality and on the chemical composition of plankton; (3) characterize the dumped waste with respect to selected inorganic constituents which could be either toxic or useful as waste tracers.

SAMPLING AND FIELD WORK

Immediately before the February 1978 cruise, samples of waste material were collected from each of the individual production plants and a composite sample of mixed waste was taken as the dumping barge was being loaded. Over a two week period, the R/V KNORR was used to collect water, suspended material and plankton samples before, during and after two dumps in the designated dumpsite. These samples have been analyzed in our laboratory for a number of potentially toxic metals (Cd, Cu, Fe, Hg, Ni and Pb) to see how the waste disposal affects metal concentrations at the dumpsite and how the metals are distributed between the dissolved and particulate phases. In addition, the waste has been analyzed for a large number of elements to characterize it more fully.

Numerous precautions were taken to minimize trace metal contamination during sample collection, storage and analysis. Most of the water samples were taken at least 1 km from the research vessel by hand filling 1 liter conventional polyethylene bottles from the bow of a slowly moving rubber boat. The bottles had been cleaned by completely filling them with concentrated HCl and heating to 60°C for several days. They were repeatedly rinsed with distilled deionized water in the laboratory and individually packed in plastic "zip-lock" bags and then in air-tight plastic boxes. Immediately after filling at sea, the sample bottles were returned to the bags and boxes for transport and storage. A few water samples were taken from a 5 liter "Go-Flo" Nisken bottle which was suspended from a nylon rope and lowered and raised from the ship with the aid of a small power capstan of our own design.

Typically, 8-10 bottles were collected from just beneath the surface for each sample, and 5-6 of these were filtered immediately upon return to the ship from each trip in the rubber boat. Filtering was done through 47 mm diameter 0.4 μm pore size Nuclepore filters held in an acid washed, all glass Millipore apparatus. Filtering and all sample manipulation was done inside a class 100 laminar flow clean hood which was located in a special restricted-access clean van on the ship. The filters had been acid washed, rinsed, freeze dried, weighed to the nearest microgram and stored in individual plastic containers before the cruise. After filtering, each filter was rinsed with distilled deionized water and returned to its plastic container.

At least 1 liter bottle of each sample was kept unfiltered and at least 1 liter of filtered water was saved. These were immediately acidified with 0.5 ml of concentrated "Ultrex" HNO_3 which lowered the pH to about 2.2. The samples were stored at room temperature in their bottles, bags and boxes until analyzed 2 to 4 months after collection.

Zooplankton were collected with a custom made 202 μm net which had no metal parts. The net was allowed to trail behind a drifting rubber boat on a nylon line. The plastic cod end of the net was removed and placed in a plastic bag after about 30 minutes of towing. Plankton were washed from the cod end into a plastic jar in the clean van aboard ship immediately after each trip. Small samples resulted from this technique, but they were believed to be relatively free of contamination from sampling and handling. Conventional plankton samplers grossly contaminate the sample with a number of metals.

LABORATORY METHODS

About two months after the cruise, a 100 ml aliquot of each sample was transferred to a plastic gas stripping chamber with a fritted glass bottom. Stannous chloride was added to reduce Hg ions to the metallic state, and Hg was then rapidly stripped from the chamber by a flow of nitrogen gas. The flowing nitrogen carried Hg vapor through a dryer and into a Laboratory Data Control model 1235 UV monitor for analysis. Known spikes of Hg were added to stripped samples to prepare a standard curve for Hg in the samples. The detection limit of the system was 0.5 ng of Hg, thus the concentration detection limit was 5 ng/l (5 parts per trillion). Precision at the 20 ng/l level was estimated to be $\pm$ 5 ng (1 sigma).

A second set of 100 ml aliquots of the samples was coprecipitated by a modified Boyle and Edmond (1977) technique using 50 μl of 1,000 ppm cobalt as a carrier and 0.2 ml of 2% aqueous ammonium pyrrolidine dithiocarbamate (Eastman) as a chelating agent for trace metals. Ten minutes after reagent addition, the aliquots were vacuum

filtered using a 50 cc disposable syringe body attached to a Millipore polypropylene 13 mm swinnex-type filter holder containing a 0.4
µm Nuclepore polycarbonate filter. The filters were then leached in
0.1 ml of concentrated Ultrex HNO_3 in a polystyrene snap-cap vial
for at least ten minutes, after which 2.0 ml of 0.1N Ultrex HNO_3 were
added. All work was done inside a class 100 clean air hood. Final
analysis was by flameless atomic absorption spectrometry (AAS) using
a Perkin-Elmer 306 and HGA-2100 instrument with deuterium background
corrector.

Recovery of trace metals by the coprecipitation procedure was
checked by spiking 100 ml aliquots of samples with approximately the
amounts of metals in the average sample. Recoveries at the low concentrations and low pH used were as follows: Cd 60%, Cu 80%, Ni 100%,
Pb 80%, Fe 50%. Precisions are estimated to be: Cd $\pm$ 100%, Cu $\pm$ 20%,
Ni $\pm$ 40%, Pb $\pm$ 100%, Fe $\pm$ 50%.

Filtered and unfiltered water samples were analyzed, giving information on the partitioning of metals between the dissolved and
particulate phases. A better indication can be gained, however, from
the direct analysis of the particulates on filters processed on board
ship. Upon return to the lab, the filters in their plastic storage
containers were freeze dried and returned to the weighing room for
humidity equilibration. They were then reweighed to the nearest
microgram. From the weight gain and the volume of water filtered
(usually 5 - 6 liters) a total suspended material (TSM) concentration
was calculated. After weighing, the filters were carefully folded
with teflon forceps and were put into teflon centrifuge tubes. High
purity acids (1.5 ml HNO_3 + 0.5 ml HF) were added, the tubes were
closed and placed in a boiling water bath for 1.5 hours. After cooling, 3 ml of distilled-deionized water were added to the tubes and
the resulting mixture was transferred to polystyrene vials. Analysis
was by flameless AAS. Precision for the suspended matter analyses,
based on previous similar analyses in our laboratory, was estimated
to be $\pm$ 20%.

Zooplankton samples were transferred to petri dishes in a clean
room and inspected under a disecting microscope. No foreign matter
was seen. The samples were washed sparingly with distilled water and
drained on a nylon screen. They were then weighed, freeze dried, re-
weighed, and transferred to Pyrex electrolytic beakers. Five ml of
high purity concentrated HNO_3 + 1 ml of concentrated $HClO_4$ were
added and the mixture was refluxed using a tight fitting watch cover
for two hours. The watch covers were raised and the nitric acid
evaporated off. Refluxing with $HClO_4$ continued for two hours before the covers were removed and it was evaporated off. The contents
were washed into pre-weighed centrifuge tubes using about 10 ml of
0.1 N nitric acid. The final volume was determined by reweighing and
the final solution was analyzed by flameless AAS. Precision for the
zooplankton analyses, based on several hundred previous similar

Table 1.   Chemical and physical characterization of composite waste
           liquor sample discharged at the Puerto Rico Dumpsite,
           February 1978.

Solids		Anions (mg/l)	
dissolved 34 g/l		Cl^-	8,000
particulate 0.49 g/l		SO_4^{2-}	13,500
loss on ignition 8 g/l		PO_4^{3-}-P	53
		Total-P	105

Lipid Soluble Matter	pH	Specific Gravity
at pH 6.98 and pH 0.7	6.98	1.025
< 2 ml/100 ml		

Metal Levels   (mg/l)

	Technique	Wet Basis	Dry Basis
Ag	FL	< 0.02	< 0.45
Al	F	< 30	< 680
As	FL	< 0.15	< 3.40
Ca	F	290	6,500
Cd	FL	0.0006	0.014
Co	F	< 1.0	< 23
Cr	F	< 0.5	< 11
Cu	FL	0.1	2.3ppm
Fe	F	170	3,900
Hg	FL,CV	0.004	0.09
K	F	340	7,700
Li	F	< 0.5	< 11
Mg	F	33	750
Mn	F	5.2	120
Mo	F	< 1	< 23
Na	F	9,900	230,000
Ni	F	0.3	7.0
Pb	FL	.09	2.0
Sb	FL	< 0.1	< 2.3ppm
Sr	F	< 5	< 110
Zn	F	< 0.5	< 11

F = Flame AAS
FL = Flameless AAS
CV = Cold Vapor

analyses in our laboratory are estimated to be $\pm$ 10% except for lead which was $\pm$ 25%.

Individual and composite wastes were analyzed for a number of metals (Tables 1 and 2) by either flame or flameless AAS. For the flame work, the waste was simply made 1 N with HNO₃, then further diluted as necessary for the various metal analyses. For flameless AAS, the waste was mixed 1 to 1 with HNO₃ and boiled for one hour before being diluted 10 to 1 or more for the individual analyses. No replicate analyses were done, but because the matrix in the final solution was so low in salts, precisions should be similar to those for standards, i.e., within a few percent of the values given.

An aliquot of the composite waste was evaporated to dryness in the presence of HNO_3 and $HClO_4$, then redissolved in dilute HNO^3. Phosphate was determined on this aliquot by the Strickland and Parsons (1968) method and sulfate by precipitating and weighing BaSO . Phosphate was also determined by the same procedure on an undigested aliquot of waste.

Total dissolved solids were determined by filtering an aliquot of waste through a 0.4 µm filter (which was weighed to determine suspended solids) and slowly evaporating the filtrate to dryness. The residue was then heated at 400°C for 48 hours to determine loss on ignition.

Table 2. Concentration (mg/l) of selected metals in individual and composite waste liquor samples discharged at the Puerto Rico Dumpsite, February 1978

Source	Ca	Fe	K	Mg	Mn	Na	Pb
American Cyanamid	100	1.7	57	16	0.3	12,000	< 0.6
Bristol	110	15	39	18	0.4	3,200	< 0.6
Merck	27	960	9.5	14	1.0	17,000	0.9
Pfizer	44	1.4	16	7	< 0.3	6,500	< 0.6
Shering	7	6.2	390	37	0.5	382	< 0.6
Upjohn	360	3.8	410	40	13	2,150	< 0.6
Composite	290	170	340	33	5.2	9,900	0.09

In all cases Al < 30, Li < 0.05, and Sr < 5 mg/l.

The percent lipid matter was determined about six months after collection by agitating 50 ml of waste and 50 ml of CCl$_4$ in a stoppered graduated cylinder and checking the positions of the liquid-liquid and air-liquid interfaces.

RESULTS AND DISCUSSION

The composite waste sample was a dark brown, noxious-smelling liquid containing very fine grained particles which did not settle upon prolonged standing. The waste is primarily from fermentation and extraction processes used in pharmaceutical manufacturing and thus contains a mixture of organic compounds. Many of the organics are volatile, and the waste composition changes with time (see Atlas et al., 1981 and Hatcher et al., 1981 this volume).

Our analyses (Table 1) show the waste to be somewhat similar to seawater in total dissolved solid content. Compared to most industrial waste, it has a low suspended solid concentration, but it filters extremely slowly due to the very fine grain size of the particles. Of the heavy metals, only Fe and Mn were in high concentration, whereas such toxic metals as Cd, Pb and Hg were present in extremely low concentrations. Iron, at 170 ppm, is enriched by a factor of 10 to 10 over seawater values and it flocculates upon mixing of the waste with seawater.

Analysis of the individual wastes (Table 2) show Merck and Upjohn to be the chief sources of Fe and Mn, respectively. Based on these metals, Merck appears to contribute about 18% of the total sample volume and Upjohn about 40%, in agreement with total discharge data given by the companies. No combination of the individual wastes exactly matches the composite, however, showing that a chemical segregation occurs in the holding tank as the wastes mix. Sulfate and chloride are the most abundant anions in the waste, but a significant amount of phosphate is also found, about half of which is bound to organic matter. An approximate carbon analysis of the bulk waste by a carbon analyzer gave about 8 g/l, in good agreement with the loss on ignition at 400°C.

About one half of the iron in the composite waste was retained on a 0.4 m filter, but upon mixing with seawater almost all of the Fe becomes particulate. A settling rate experiment was performed in the lab by mixing a spiked composite waste sample 1 to 50 with seawater (Fig. 1). The upper 3 cm of the water column were sampled for several days and showed a very slow decline in the levels of all the metals to 50% of the original value in about 100-200 hours, indicating a very slow settling rate for the particulates.

Despite the low toxic metal levels in the composite waste there was concern that iron flocculation might scavenge metals from the

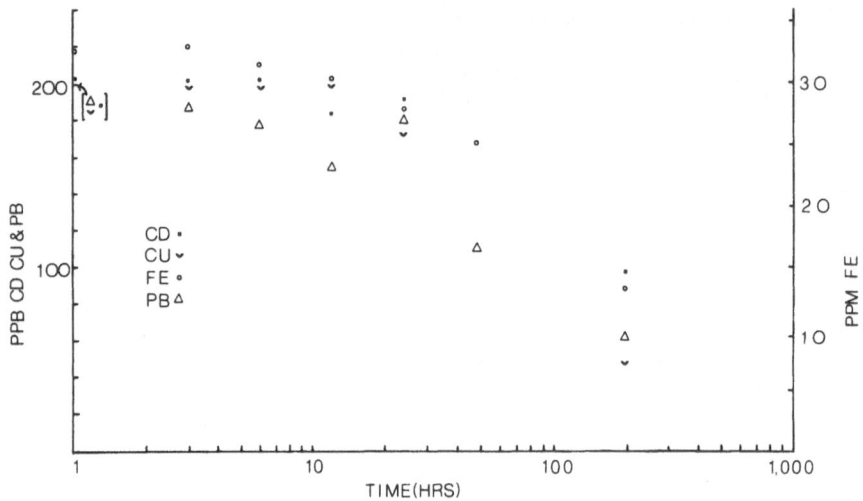

Fig. 1. Surface (upper 3 cm of a 30 cm water column) total metal
         levels composite waste settling experiment.

waste or from seawater, thereby concentrating them and making them
more available to organisms. Our seawater analyses, however, as
shown in Table 3, indicate low amounts of the potentially toxic
metals Cd, Cu, Hg, Ni and Pb, with little difference before and after
the waste dump. Furthermore, there was negligible difference between
filtered and unfiltered samples, showing that particulates contribute
little of these metals. It appears that despite our elaborate precautions to avoid contamination, the samples were contaminated with
Cd and perhaps other metals during filtering. However, this does not
detract from the conclusion that dumping seems to have negligible effect on the Cd, Cu, Hg, Ni and Pb chemistry at the dumpsite.

     The observed iron levels are similar to those reported by Bewers
et al. (1974, 1976) for central Atlantic and Gulf of St. Lawrence
waters respectively, however, they exceed those reported by Kester et
al. (1981) for Deep Water Dumpsite 106. Table 3 shows concentrations
increasing by at least an order of magnitude after the 9 II 78 dump.
Samples were not taken after the 6 II 78 dump until the iron had dispersed.

     Fig. 2 shows the rapid decline in post dump iron levels. The
data indicate a half-life for surface iron of about two hours. However, when the original Fe concentration in the waste is used with
the two 30 minute post dump 2 seawater Fe values, a plume half-life
of 2.2 to 2.4 minutes is calculated assuming an exponential decrease.
Even the latter estimates are maxima since they are based on sam-
plings of the most obviously persistent plume patches. This

Table 3. Seawater metal concentrations (ng/l) at the Puerto Rico Dumpsite, February 1978.

Date	Station	Time (hrs pre/post dump)	Depth* (m)	Filtered	Cd	Cu	Fe	Hg	Ni	Pb
4 II 78	1	34.8 Predump 1	10	No	3	200	1,200	20	270	113
4 II 78	1	34.8 Predump 1	110	No	15	280	830	10	230	935
5 II 78	2	23.8 Predump 1	S	No	2	140	7,100	20	230	21
5 II 78	2	23.8 Predump 1	S	Yes	26	200	710	10	280	22
5 II 78	3	16.8 Predump 1	S	No	1	140	1,400	10	230	68
5 II 78	3	16.8 Predump 1	S	Yes	35	210	760	10	230	20
6 II 78	6 dump C	2.4 Postdump 1	S	Yes	5	120	2,700	20	275	25
6 II 78	6 dump C	2.4 Postdump 1	S	No	< 1	160	3,700	20	230	19
6 II 78	8 dump D	6.7 Postdump 1	S	No	1	140	4,100	30	320	16
6 II 78	8 dump D	6.7 Postdump 1	S	Yes	13	240	2,300	30	350	25
7 II 78	dump G	24.7 Postdump 1	S	No	5	210	2,600	20	320	24
7 II 78	dump G	24.7 Postdump 1	S	Yes	5	220	1,100	20	370	25
8 II 78	23	57.2 Postdump 1	S	No	2	200	930	20	280	69
8 II 78	23	57.2 Postdump 1	S	Yes	3	200	440	20	350	30
8 II 78	23	57.2 Postdump 1	S	Yes	3	310	750	40	440	29
9 II 78	24	0.5 Postdump 2	S	Yes	16	130	10,000	10	430	25
9 II 78	24	0.5 Postdump 2	S	No	1	170	30,000	10	480	21
9 II 78	24	0.5 Postdump 2	S	No	3	160	16,000	<10	430	19
9 II 78	24	0.5 Postdump 2	S	Yes	15	260	16,000	10	510	50
9 II 78	24	5.25 Postdump 2	10m	No	1	140	591	20	320	51
9 II 78	24	5.25 Postdump 2	30m	No	2	75	2,500	20	270	48
9 II 78	24	5.75 Postdump 2	S	Yes	7	120	1,800	10	340	28
9 II 78	24	5.75 Postdump 2	S	No	1	140	1,500	<10	220	21

*S designates surface sample.

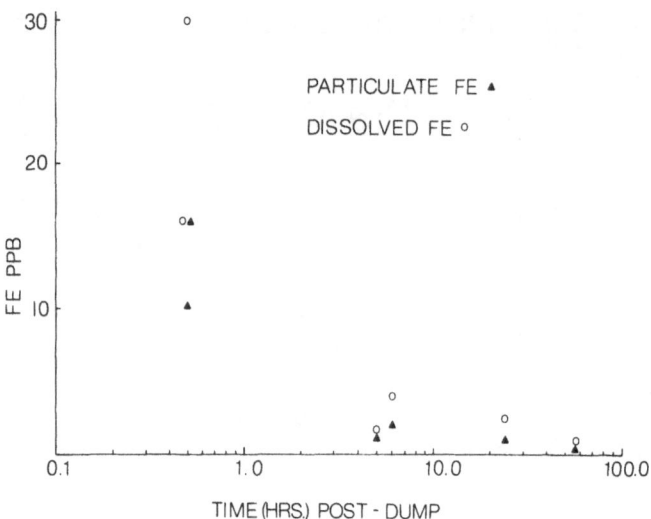

Fig. 2.   Fe concentration in waste plume surface following dumping on
          February 6 and 9, 1978.

half-life predicts that the Fe concentration should fall to the
relatively innocuous level of 100 ppb within 25 minutes. This may
explain why so little iron was found in the post dump 1 samples which
were taken 2.4 hours after dumping began.

     The suspended matter data (Table 4) shows low levels of TSM, as
would be expected in clean open-ocean water. The values are much
higher, however, after the second dump when the red colored waste was
clearly visible on the filters. Analyses of the filters shows that
the increased Fe levels recorded by the unfiltered water samples was
largely particulate and that particulate Fe alone is responsible for
most of the increase in TSM. Lead and Mn also increased in the particulate phase after the second dump, but there is so much scatter in
the other metal data that no conclusions can be drawn. Random variations due to plankton or other large particles is always a problem in
low TSM samples and this may obscure some trends in the data. The
effect of the Fe-rich waste is, however, very marked, even though it
is transitory.

     The Puerto Rico zooplankton (Table 5) are similar to zooplankton
from California and Texas in their Cu, Cd, Mn and Pb content when the
natural variability in these metal levels is considered. Zooplankton
from Texas seem to be enriched in Fe according to Table 5, but this
is a consequence of clay incorporation from the relatively muddy
Texas waters, as can be shown by Al values. On the other hand, the
one Fe enriched sample from Puerto Rico almost certainly resulted

Table 4. Suspended matter metal levels at the Puerto Rico Dumpsite.

	Depth	Time Post Dump (HRS)	TSM (µg/l)	Cd A	Cd B	Cu A	Cu B	Fe A	Fe B	Mn A	Mn B	Pb A	Pb B
Pre Dump													
4 II 78													
Station 1	10 m		87	2.5	0.03	5	0.06	680	7.8	24	0.28	4	0.05
	110 m		71	3.1	0.04	14	0.2	230	3.2	12	0.17	2	0.03
5 II 78													
Station 3	Surface		36	0.08	0.002	4	0.11	3,300	92	5	0.14	10	0.28
Post Dump													
6 II 78													
Station 8	Surface	2.4	87	0.02	0.0002	1	0.011	790	9.1	2	0.02	3	0.034
Station 8	Surface	2.4	87	1.1	0.01	17	0.20	980	11	4	0.05	5	0.06
7 II 78													
Station G	Surface	2.4	36	0.03	0.008	2	0.055	33	0.92	6	0.2	<1	<0.03
Station G REP	Surface	2.4	56	0.2	0.004	7	0.13	23	0.41	5	0.09	<1	<0.02
Post Dump 2													
9 II 78	Surface	0.5	362	1	0.003	17	0.047	14,000	39	98	0.27	33	0.09
9 II 78	Surface	0.5	194	0.74	0.004	7	0.04	8,500	44	45	0.23	8	0.04
9 II 78	Surface	0.5	202	0.2	0.001	11	0.05	16,000	79	47	0.23	20	0.10
10m Nylon	10 m	5.25	58	0.27	0.005	12	0.21	330	5.7	15	0.26	6	0.10
30m Nylon	30 m	5.25	50	1.2	0.02	16	0.32	110	2.2	14	0.28	4	0.08
9 II 78	Surface	5.80	78	0.08	0.001	13	0.17	1,300	17	16	0.21	4	0.05

A = Concentration of particulate metal in ng/1 of water.
B = Concentration of metal in dry suspended matter g/kg.

Table 5. Zooplankton trace metals data

Sample Number	Location	Description	Concentration (ppm dry weight)					
			Cd	Cu	Fe	Mn	Pb
D-1	Puerto Rico	Before Dumping	8.0	35	350	0.75	5.5
D-2	Puerto Rico	Before Dumping	0.45	45	200	1.1	< 0.01
D-3	Puerto Rico	7 hours Post Dump 1	0.70	55	2400	1.2	< 0.01
D-4	Puerto Rico	24 hours Post Dump 1	4.0	35	250	2.5	< 0.01
D-5	Puerto Rico	30 min Post Dump 2	1.8	20	600	2.3	< 0.01
	Texas	BLM STOCS 1977	3.0	45	3700	---	16
	California	Martin & Knauer (1972)	4.1	10.5	197	4.4	3.3

from incorporation of the waste material from the 6 II 78 dump. The sample (D-3 in Table 5) was collected seven hours after the dump in apparently clean water, but it was greatly contaminated with Fe. Sample D-5 was collected about an hour after the 9 II 78 dump and it does not show elevated Fe levels. Obviously, some time is required for the waste to sink and be incorporated into zooplankton.

In conclusion, the industrial wastes dumped at the Puerto Rico dumpsite appear to have negligible effects on levels of the toxic metals Cd, Cu, Pb, Hg, and Ni. On the other hand, the wastes have an easily measured short term effect on particulate and dissolved Fe levels. Zooplankton seem to take up the Fe, but within 24 hours after a dump advection has greatly changed the location of water, particles and plankton. A more extensive sampling program than the one conducted would be needed to determine the fate and effect of the Fe in such a dynamic system.

REFERENCES

Anderson, P. W. and R. T. Dewling (1981) Industrial ocean dumping in EPA Region II - Regulatory Aspects. In: Ocean Dumping of Industrial Wastes, B. H. Ketchum, D. R. Kester, and P. K. Park, editors, Plenum Press, New York. This volume, pp. 25-37

Atlas, E, G. Martinez, and G. S. Giam (1981) Chemical characterization of ocean-dumped waste materials. In: Ocean Dumping of Industrial Wastes, B. H. Ketchum, D. R. Kester, and P. K. Park, editors, Plenum Press, New York. This volume, pp. 275-293.

Bewers, J. M. and B. Sunby (1974) Trace metals in the waters of the Gulf of St. Lawrence, Can. J. Earth Science, 11, 939-950.

Bewers, J. M., B. Sunby and P. A. Yeats (1976) The distribution of trace metals in the Western North Atlantic off Nova Scotia. Geochimica et Cosmochimica Acta, 40, 687-696.

BLM STOCS (1977) Zooplankton trace metal data from the South Texas Outer Continental Shelf. B.J. Presley and P.N. Boothe, Texas A&M University, unpublished data.

Boyle, E. A. and J. M. Edmond (1977) Determination of copper, nickel, and cadmium in seawater by APDC chelate coprecipitation and flameless atomic absorption spectrometry. Analytica Chimica Acta, 91, 189-197.

Hatcher, P. G., G. A. Berberian, A. Cantillo, P. A. McGillivary, P. Hanson, and R. H. West (1981) Chemical and physical processes in a dispersing sewage sludge plume. In: Ocean Dumping of Industrial Wastes, B. H. Ketchum, D. R. Kester, and P. K. Park, editors, Plenum Press, New York. This volume, pp. 347-378.

Kester, D. R., R. C. Hittinger and P. Mukherji (1981) Effect of acid-iron waste disposal on transition and heavy metals at Deep Water Dumpsite 106. In: Ocean Dumping of Industrial Waste, B. H. Ketchum, D. R. Kester, P. K. Park, editors, Plenum Press, New York. This volume, pp. 215-232.

Martin, J. and G. Knauer (1972) A comparison of inshore vs.
    offshore levels of 21 trace and major elements in marine
    plankton. In: Baseline Studies of Pollutants in the Marine
    Environment, E. D. Goldberg, editor, NSF-IDOE, Brookhaven,
    New York, p. 35.
Strickland, J. D. H. and T. R. Parsons (1968) A Practical
    Handbook of Seawater Analysis, Fisheries Research Board of
    Canada, Ottawa, p. 57.

CHEMICAL AND BIOLOGICAL ASPECTS OF OCEAN DUMPING

AT THE PUERTO RICO DUMPSITE

Claude R. Schwab, Theodor C. Sauer, Jr., Guinn F. Hubbard,
Hussein Abdel-Reheim and James M. Brooks

Department of Oceanography
Texas A&M University
College Station, TX 77843

ABSTRACT

Chemical and biological studies at the Puerto Rico Ocean Dumpsite during February 1978 are reported. Dye and various chemical and
biological parameters were measured with time within the waste plume
after the dumps. Several parameters (e.g., ammonia, POC, DOC, volatile organics, and transmissometry traces) were elevated in the waste
plume compared to the baseline values and were, therefore, very useful in tracking the dispersion of the waste.

The volatile organics constitute the major non-aqueous fraction
of the waste. Butanol and dimethyl aniline constitute about 70% of
this fraction. Dimethyl aniline was very persistent in the dumpsite
having a concentration of $0.1 \ \mu g \cdot l^{-1}$ three days after dumping. Evaporation studies indicate that benzene and toluene concentrations decrease to background levels within 12 hrs following dumping. Many
hydrocarbons, halocarbons, alcohols, esters, ketones and nitrogen
compounds were identified in the waste and surface waters of the
dumpsite.

Phytoplankton studies indicate no significant differences existed between samples taken from similar depths inside or outside of
the dumpsite. Laboratory toxicity tests using ^{14}C uptake and ATP
indicate that 0.14% of the composite waste causes a 76% decrease in
growth of the phytoplankton Skeletonema costatum.

INTRODUCTION

In early February 1978, a study was conducted aboard the R/V KNORR at the Puerto Rico ocean dumping site. The location of the dumpsite in relation to the island of Puerto Rico is shown in Fig. 1. Several days were devoted to hydrographic, chemical and biological characterization of the site to provide background for post-dump studies. Two barges of waste material occurred at a three day interval during this study. At initiation of a dump, the research vessel moved into the plume created by the towed barge and followed the barge for approximately 2 miles. At this point, drogues were launched and an attempt was made to follow and sample this specific section of the plume. Some preliminary laboratory results are reported on the toxicity of the wastes to phytoplankton cultures as measured by ^{14}C uptake and ATP levels.

THE DUMPSITE

The Puerto Rico Ocean Dumpsite was established in 1972 as an interim disposal location for wastes resulting primarily from pharmaceutical production. The dumpsite is located over the Puerto Rico

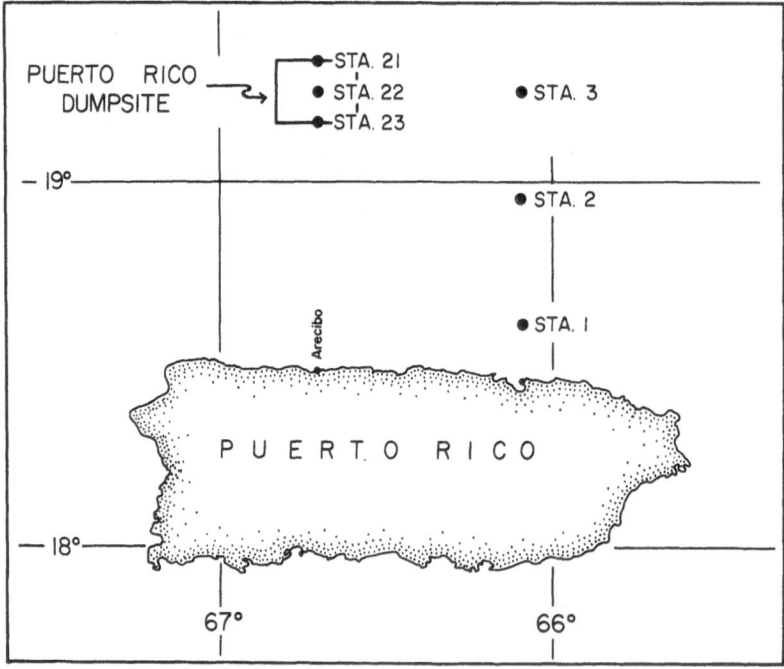

Fig. 1. Location of the Puerto Rico chemical waste ocean dumpsite.

Trench in about 6000 m of water, 42 mi north of Arecibo, Puerto Rico between 19°10' to 19° 20'N, and 66°35' to 66°50'W (140 mi^2). The Antilles Current moves through the site in a westerly direction. The material dumped is a variable combination of batch wastes from several pharmaceutical and chemical companies. The wastes are pooled in a large onshore storage tank and barged to the site every few days. Approximately 5.3 x 10^5 wet tons year^{-1} are dumped into the surface waters at a rate of 35,000 gallons nmi^{-1}.

METHODS

Sampling. Samples for hydrographic, chemical and biological measurements were collected using a 12-bottle, 12-1 Niskin Rosette Sampler fitted with a Plessey 9040 STD. Samples were drawn for analyses of dissolved oxygen, salinity, nutrients, organic carbon, light hydrocarbons, N_2O, total Kjeldahl nitrogen (TKN), and total suspended matter (TSM). Volatile organic water samples were collected with specially built, glass-lined Niskin bottles (NOAA/AOML) attached to the Rosette Sampler. The volatile organic samples were drawn under pressure (purified N_2) from the bottles through copper tubing into volatile-free, 2-1 glass bottles. Sodium azide was added to prevent biodegradation of the organic material, and the stoppered bottles were stored at ~4°C for later analysis. Dye samples were taken from both glass-lined and 12-1 Niskin bottles. Typically, water samples were taken at 1, 5, and 10 m depths in the waste plumes.

Hydrographic and Chemical. Oxygen samples were analyzed by the modified Winkler methods (Strickland and Parsons, 1972). Nutrients were analyzed onboard using a Technicon Autoanalyzer II (Strickland and Parsons, 1972). Salinities were determined on a Plessey 6210 inductive salinometer. Dissolved (DOC) and particulate organic carbon (POC) concentrations were determined after Fredericks and Sackett (1970). TSM was determined by filtration of a known water volume onto preweighed 47 mm Nuclepore filter (0.4 μm). The filters were washed with distilled water and stored along with blanks in dust-free containers. Light hydrocarbons and N_2O were determined by the procedures of Brooks et al. (1977). TKN samples taken and frozen in 500 ml plastic bottles were analyzed by the procedures of Strickland and Parsons (1972). Transmissometry was conducted with the Martek Model XMS In Situ Transmissometer system. Temperature was determined by a temperature probe on the transmissometer and by XBTs.

Volatile Organics. The analytical method used to determine volatile organics in seawater is described by Sauer et al. (1978), Sauer (1978), and Atlas et al. (1980). It involves a combination and modification of the dynamic headspace sampling techniques presented by Bertsch et al. (1975) and May et al. (1975). The volatile organics were stripped from a heated (70°C) 2-1 sample by an ultra-pure helium stream at 135 ml·min^{-1} for 100-120 min and collected on Tenax-GC. The

components trapped on the Tenax-GC were desorbed at 250°C for 15 min
and retrapped on a liquid nitrogen cooled pre-column. When the de-
sorption process was complete, the pre-column was switched into the
chromatographic stream by a six-port valve. The trapped components
on the pre-column were injected onto the chromatographic column by
replacing the coolant with 150°C mineral oil. A 3.2 mm O.D., 4.6 m
copper column packed with 10% SP-2100 on 80/100 mesh Supelcoport was
used to separate and quantify the volatiles using a Hewlett-Packard
5700A gas chromatograph with dual flame ionization detectors (FIDs).
The column was temperature programmed at 0°C for 2 min, 0-180°C at
$4°C \cdot min^{-1}$ and 180°C for 16 min.

For GC/MS identification of volatile components, the entire GC
column with the cryogenically trapped components was transferred to a
GC interfaced to a mass spectrometer. A Hewlett-Packard 5980A dodec-
apole mass spectrometer interfaced to a 5710A GC with a single stage
glass jet separator and supported by a 5933A Data System was used for
volatile organic identifications. Mass spectra were recorded at the
rate of one per 2.0 sec from 40 to 350 amu with an electron ionization source voltage and temperature of 70 EV and 170°C, respectively.
Many of the component identifications were confirmed by GC retention
times. Mass spectra identifications were assisted by Cornu and Massot (1966), Imperial Chemical Industrial Ltd. (1970), and McLafferty
(1973). Peak areas were measured by a 3933A Hewlett-Packard integrator, peak height x 1/2 peak height width, or planimetry. Concentrations were determined by ratios of peak areas to those of standard
alkanes. The component sensitivity of the entire stripping and analytical method was below $1 ng \cdot 1^{-1}$.

The volatile organic components determined range in boiling
point from near n-hexane (69°C) to L.C.-tetradecane (254°C). Organic
compounds outside this range are essentially unascertainable. However, some compounds with boiling points below n-hexane, such as acetone, can be qualitatively determined. The recovery of volatiles by
the stripping procedure depends on the type of organic compound. For
hydrocarbons and halocarbons, the recovery is greater than 85% (Sauer
et al., 1978). The more soluble ketones and alcohols have lower recoveries.

Biological. ATP was determined by the technique of Holm-Hansen
and Booth (1966) using a JRB Inc. ATP Photometer. One-1 samples were
filtered onto 0.40 μm, 47 mm Nuclepore filters. Immediately after
filtration, the filters were placed in 5 ml of boiling Tris-buffer
for 5 min, and the vials frozen at -20°C until analysis. Chlorophyll
was estimated by extractive spectrophotometry using the procedures of
Strickland and Parsons (1972) and the equations of Parsons and
Strickland (1963). Chlorophyll and phaeophytin were also determined
by fluorometry as described by Yentsch and Menzel (1963) and Holm-
Hansen et al. (1965). The nano portion of these parameters was that
portion which passed through a 20 μm mesh.

Phytoplankton Counts and Species Identification. Samples were taken from Niskin bottles and preserved with buffered formalin. Laboratory analysis proceeded, after agitation of a 500 ml sample for one min, by settling 10, 50, or 100 ml aliquots in settling chambers. The volume settled depended on the apparent relative density of cells in each sample container. Two replicates of each sample were settled, one being used for identification and species counts and the other for rough cell counts. Species were identified and counted on each chamber until at least 300 individuals were encountered (exceptions occurred whenever the chamber contained less than 300 total cells). Phytoplankton in the rough cell counts were classified as to diatoms, dinoflagellates, coccolithophorids, and others. With both replicates, the area of the chamber used for the counts was determined and calculations of cells per liter accomplished using number of cells, area counted, and volume settled.

Dye. Rhodamine WT dye concentrations were analyzed with a Fischer Spectrophotometer, with concentrations measured to a lower limit of 0.01 ppb. A standard curve was constructed with known concentrations of dye in seawater. Dye deterioration was not evident during the sampling intervals.

Laboratory Tolerance Studies. Studies were conducted on the short term acute effects of the waste materials on the productivity of a common phytoplankton species, Skeletonema costatum (SS10 from Department of Oceanography Marine Phytoplankton Culture Collection, Texas A&M University). Grund marine growth media (McLachlan, 1973) was prepared with $35^\circ/_{oo}$ seawater diluted to 90% with double glass distilled water by combining the diluted seawater and additives, and then autoclaving. The phytoplankton were cultured at room temperature (23-24°C) under cool white fluorescent lights, 2200 lux illumination, with an 8 hr dark, 16 hr light cycle. Cultures were transferred periodically to maintain log-phase growth and to reduce bacterial contamination to very low levels as determined by microscopic examination. Growth media, ^{14}C labeled $NaHCO_3$ solution, varying levels of waste materials, and phytoplankton were added to 150 ml, ground-glass stoppered, clear glass bottles. The bottles were incubated under the same conditions as the stock cultures, with illumination maintained throughout the 9 hr experiments. Bottles were periodically rotated in place and position in the light box to minimize any inconsistencies in illumination.

ATP samples were taken at intervals throughout each run. After thorough mixing by repeated inversion, 2 ml of the mixed culture solution from each bottle was drawn for each ATP replicate and injected into 3 ml of boiling Tris Buffer (pH = 7.75). ATP samples were frozen and stored at -20°C for analysis (Holm-Hansen and Booth, 1966).

After 9 hrs the bottles were placed in the dark and all filtrations were performed in dim light to minimize variable light

Table 1. Some hydrographic, chemical and biological measurements at the Puerto Rico chemical waste dumpsite (19°15'N, 66°42'W). Chlorophyll a was determined by fluorometry.

Depth (m)	Silicate (µM)	Phosphate (µM)	Nitrate (µM)	Salinity (°/oo)	Oxygen (ml/1)	DOC (mg/1)	POC (µg/1)	ATP (ng/1)	Chloro. a (mg/m³)	Methane (nl/1)
0	2.3	0.56	0.45	35.455	4.85	1.08	26.1	42.5	0.173	58
30	2.3	0.48	0.45	35.450	4.71	1.38	27.5	47.0	0.230	54
60	2.1	0.48	0.45	35.685	4.78	1.47	25.1	48.3	0.403	55
80	1.8	0.48	0.50	36.794	4.83	1.05	19.6	95.8	0.518	86
100	1.8	0.43	0.50	36.917	5.07	0.98	30.4	45.6	0.690	72
120	1.6	0.43	1.35	37.048	4.48	1.09	14.8	42.9	0.460	71
140	1.6	0.43	1.35	37.079	4.55	0.93	18.0	48.8	0.173	65
160	1.6	0.43	1.70	37.098	4.54	0.95	14.0	40.9	0.173	60
180	1.6	0.48	2.20	37.046	4.39	0.99	14.4	39.3	0.115	53
200	1.8	0.48	2.20	36.778	4.24	1.14	14.2	44.3	0.115	57
250	2.0	0.52	3.45	36.538	4.20	0.84	17.6	4.6	0.058	55
300	2.5	0.66	6.60	36.515	4.33	0.85	20.5	4.5	—	—
400	4.0	1.08	12.10	36.130	4.09	0.73	15.4	4.5	—	51
600	9.6	1.55	22.00	35.988	3.61	0.76	12.1	4.3	—	30
700	14.4	1.88	28.00	35.972	3.29	0.68	5.7	4.5	—	26
800	17.2	2.02	28.30	35.499	3.41	0.74	7.3	4.4	—	46
900	17.7	1.97	28.50	35.158	3.79	0.71	4.2	4.5	—	—
1100	13.7	1.64	21.80	35.019	5.31	0.89	5.1	4.5	—	—
1300	12.2	1.55	20.20	35.023	4.85	0.71	11.9	4.4	—	23
1500	13.0	1.50	19.80	35.014	5.98	0.79	11.3	6.1	—	20
2000	15.6	1.50	19.80	35.002	6.08	0.80	9.0	4.4	—	21
2500	21.0	1.50	19.80	34.952	6.29	0.81	4.3	4.2	—	16
3000	21.4	1.50	19.80	34.929	6.27	0.64	5.8	4.1	—	16
4000	26.4	1.50	20.20	34.902	6.21	0.78	5.8	4.3	—	—

exposure. After the final ATP sampling, the remainder of the solution was filtered through Millipore HA (0.45 µm) filters using a suction less than 30 cm Hg. Each filter was rinsed with filtered seawater. Two sets of controls were included in each experiment, one set grouped at the beginning and one at the end, in order to span the approximately 1 hr time lag in the filtration process. The filters were placed directly into scintillation vials and dried in a desiccator. Ten ml of Beckman Filter Solv Solution were added and the samples counted on a Beckman LS-100B Liquid Scintillation Counter.

These experiments were conducted with waste that had been filtered through a series of glass fiber filters (Whatman GF/D and GF/F), if necessary, and finally through Millipore HA (0.45 µm).

SITE CHARACTERIZATION

Some hydrographic and chemical data collected in February 1978 at a site characterization station in the dumpsite is presented in Table 1. A mixed layer existed to about 70 m in February with a slight temperature inversion immediately below this layer (Fig. 2). Many of the biologically associated parameters appear to be related to the strong pycnocline in the 60 to 100 m depth interval below the mixed layer. The strong halocline below the mixed layer (Fig. 2) results from Subtropical Underwater. The pycnocline creates a density gradient upon which suspended material can accumulate as it sinks from the mixed layer. The increased organic carbon levels in the upper part of the halocline may be due to the density gradient and/or to higher productivity associated with regeneration of nutrients from falling organic detritus. Sharp maxima in POC, chlorophyll, and ATP are observed within the pycnocline. The ATP maximum at 80 m (Table 1) results from a peak in the larger (> 20 µm) size fraction. The ATP level in the nano (< 20 µm) size fraction remains rather constant through approximately the upper 200 m and, except for those depths which have peaks in ATP, it comprises most or all of the ATP present. This suggests that the ATP maxima result from large phytoplankton and/or zooplankton, and that the smaller phytoplankton and bacteria are distributed rather uniformly. ATP levels drop off sharply at about 200 m and are near constant to the bottom. The methane peak at 80 m appears to be associated with the biological activities.

Preliminary results indicate that no significant difference existed in the phytoplankton populations between stations inside (Station 22) and outside (Station 3) of the dumpsite. The main difference in a series of stations (Stations 1 through 3) taken on a transect out from San Juan was an increase in the depth of the maximum phytoplankton numbers. Maxima in cell numbers tended to agree with chlorophyll a maxima. The coccolithophorids Gephyrocapsa huxleyi and Umbellosphaera irregularis as well as unidentified flagellates and monads occurred persistently as the numerically dominant groups at

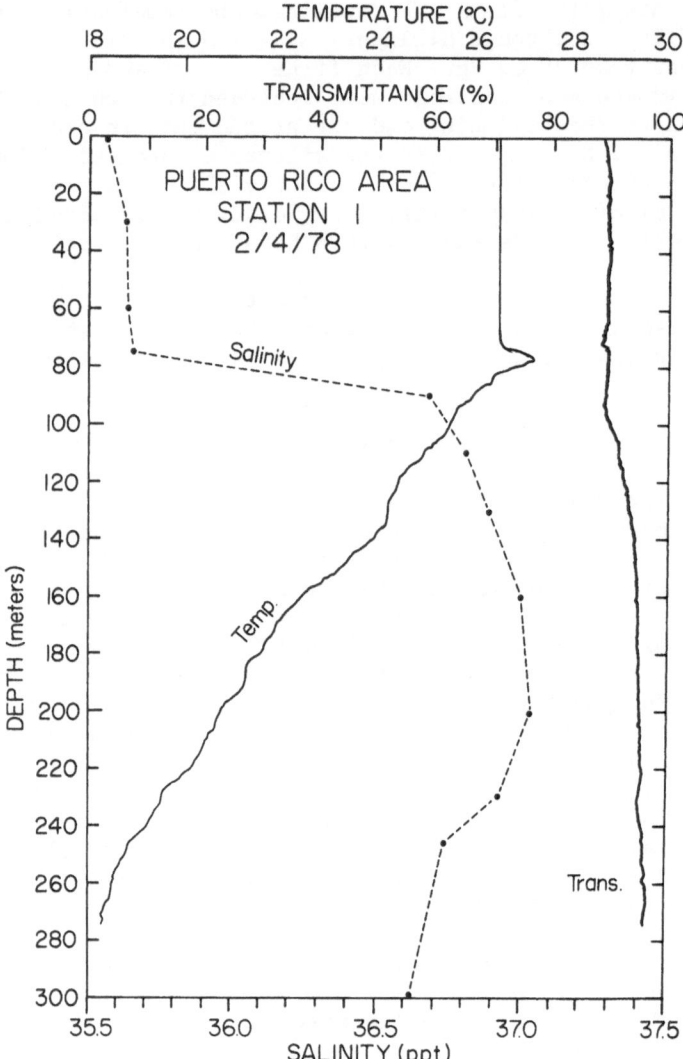

Fig. 2.   Depth profiles of temperature, salinity, and transmittance
          near the Puerto Rico chemical waste dumpsite, 4 February
          1978.

all depths and stations, each usually accounting for over 5% of the
cells in a sample. Most species were less than 1% of cell counts.
Table 2 shows species counts for Station 22. The number of empty
cells increased with depth. The Shannon-Wiener species diversity
indices (H') were examined:

$$H' = \sum_{i=1}^{n} \frac{n_i}{N} \log_2 \frac{n_i}{N}$$

where n_i = number of individuals of i and N = total number of individuals. The H' values determined on each sample were uniformly high compared with the theoretical maximum for a given number of species present (H_{max} = $\log_2$ S, with S = number of species). Equitability, the ratio of H' to H_{max}, was consistently high indicating a diverse phytoplankton assemblage. No differences in species compositions, total cells, or the empty to full cell ratio were observed in samples taken in the waste plume following dumping compared to the Station 22 distributions.

WASTE TRACKING STUDIES

The parameters determined during the site characterization phase were also monitored following two different dumps of waste materials. The objective was to determine which chemical parameters are useful in characterizing the persistence and dispersion of the waste. Some results from these studies are presented in Table 3. Stations DS/1 through DS/8 were taken within the waste plume following the first dump (6 February 1978) and Stations DS/9 through DS/12 were taken after the second dump (9 February 1978). Most parameters in the waste plume did not differ significantly from background levels. No significant differences from background levels or changes with time after a dump were seen for silicate, nitrate, salinity, oxygen, ATP, chlorophyll, plant carotenoids, phaeophytin, gaseous hydrocarbons, and N_2O. Some parameters, such as methane, which were useful in tracking the Gulf of Mexico dumps (Atlas et al., 1980) were ineffective at the Puerto Rico site because of the low bacterial activity in the waste.

The most useful real-time tracer of the waste was transmissometer measurements. Although the waste contained few particulates, the plume was easily identified by light transmission readings. Background transmission (Fig. 2) gave a constant reading of ~90% in the mixed layer. Fig. 3 shows representative traces for the 6 February dump. The dumped material was very unevenly distributed in the plume, transmission varied significantly between up and down traces as the ship drifted. The waste was always concentrated above 20 m, with subsurface maxima often present. The plume was identifiable up to 12 hrs following the first dump by transmissometry. The portion of the plume tracked during the second dump (Fig. 4) gave much larger transmission decreases. The greater levels of waste in this plume is also seen in chemical measurements (Table 3). As during the first dump, most particulates were concentrated in the upper 20 m, although

Table 2.   Phytoplankton distribution at Station 22 in center of
           dumpsite.   Species lists only provided for 0 m depth.

Depth (m)	Species	Full Cells	Cells/ Liter	% of Total	Empty Cells
0	**Diatoms**				
	Hemiaulus membranaceus	1	39	0.327	0
	Nitzschia bicapitata	6	234	1.961	3
	Nitzschia kolaczekii	2	78	0.654	0
	Nitzschia spp.	2	78	0.654	0
	Thalassionema nitzschioides	5	195	1.634	0
	Dinoflagellates				
	Blepharocysta paulseni	3	117	0.980	0
	Dinophysis parva	1	39	0.327	0
	Goniaulax spp.	1	39	0.327	0
	Gymnodinium fuscum	3	117	0.980	0
	Gymnodinium oceanicum	1	39	0.327	0
	Gymnodinium spp.	13	506	4.248	0
	Oxytoxum laticeps	1	39	0.327	0
	Oxytoxum variabile	2	78	0.654	0
	Oxytoxum viride				
	Oxytoxum spp.	4	156	1.307	0
	Prodinophysis spp.	1	39	0.327	0
	Protoperidinium tuba	2	78	0.654	0
	Protoperidinium spp.	1	39	0.327	0
	Scrippsiella trochoidea	14	545	4.575	4
	Coccolithophorids				
	Anthosphaera oryza	3	117	0.980	0
	Calyptrosphaera oblonga	3	117	0.980	0
	Corisphaera sp. A.	1	39	0.327	0
	Coronosphaera mediterranea	4	156	1.307	0
	Cyclococcolithus leptoporus	3	117	0.980	0
	Discosphaera tubifera	11	428	3.595	0
	Gephyrocapsa huxleyi	59	2297	19.281	0
	Gephyrocapsa oceanica	35	1363	11.438	0
	Helicosphaera carteri	1	39	0.327	0
	Helicosphaera hyalina	2	78	0.654	0
	Michaelsarsia elegans	0	0	0.000	1
	Ophiaster hydroideus	1	39	0.327	1
	Rhabdosphaera stylifer	4	156	1.307	0
	Syracosphaera pulchra	6	234	1.961	0
	Thoracosphaera heimii	5	195	1.634	0
	Umbellosphaera irregularis	37	1441	12.092	0
	Coccolithophorid sp. A	2	78	0.654	0

Table 2. (continued)

Depth (m)	Species	Full Cells	Cells/ Liter	% of Total	Empty Cells
0	Others				
	"Colonial Greens"	2	78	0.654	0
	Flagellates & monads	58	2259	18.954	0
	Pterosperma cf. moebii	2	78	0.654	0
	Radiolaria	0	0	0.000	1
	Totals	306	11916		10
	$H'=3.929$; $H_{max}=5.322$; $E=0.738$				
30 m	Diatoms	22	705	7	2
	Dinoflagellates	47	1508	15	2
	Coccolithophorids	172	5514	54	2
	Others	73	2340	23	1
	Totals	314	10067		7
	$H'=3.562$; $H_{max}=5.000$; $E=0.712$				
60 m	Diatoms	11	337	3	1
	Dinoflagellates	54	1657	17	11
	Coccolithophorids	168	5154	53	9
	Others	85	2608	27	3
	Totals	318	9756		24
	$H'=3.202$; $H_{max}=5.044$; $E=0.651$				
80 m	Diatoms	3	124	1	0
	Dinoflagellates	52	2137	16	5
	Coccolithophorids	173	7108	54	9
	Others	91	3739	29	2
	Totals	319	13108		16
	$H'=2.928$; $H_{max}=4.644$; $E=0.630$				
100 m	Diatoms	18	440	6	6
	Dinoflagellates	48	1175	15	4
	Coccolithophorids	150	3916	49	23
	Others	100	2448	30	5
	Totals	326	7979		38
	$H'=3.424$; $H_{max}=5.044$; $E=0.679$				

Table 2. (continued)

Depth (m)	Species	Full Cells	Cells/ Liter	% of Total	Empty Cells
120 m	Diatoms	23	115	12	0
	Dinoflagellates	11	55	6	0
	Coccolithophorids	73	365	38	3
	Others	86	430	44	2
	Totals	193	965		5
	$H'=3.165$; $H_{max}+4.524$; $E=0.700$				

a few minima in the transmission were seen as deep as 40 m.

Several chemical parameters were useful in tracking the waste plume. Ammonia was elevated by a factor of four in surface waters after the first dump and by a factor of forty in the second dump. Ammonia showed a definite concentration of the plume in the upper 10 m (Table 3). Only Station DS/5, 7 hrs after dumping, showed high ammonia levels below 10 m. In both dumps, ammonia returned to near background levels after 10 to 12 hrs. Since ammonia is rapidly utilized by phytoplankton and also can be oxidized by various biological and chemical processes, the rapid decrease may be affected by processes other than dispersion. It is interesting to note that the ammonia data indicate that the waste plume was missed at Station DS/2. However, surface phosphate values were elevated compared to background. This may indicate that the elevated phosphate was associated with a surface slick and that for Station DS/2, at least, the slick was not over the main subsurface plume.

POC and DOC concentrations were also elevated in stations within the plume, especially in the second dump, and the concentration levels confirm the shallow penetration of the waste. At Station DS/5, high POC concentrations were observed at 30 m, as were high ammonia levels.

VOLATILE ORGANICS

GC-MS ion spectra were determined on a sample of the barge waste (Fig. 5) and a 1 m sample (Station DS/9) taken 30 min after the dump (Fig. 6). Station DS/9 contained the highest concentration of volatile organics. The highest dye concentration (1 ppb) was also measured at Station DS/9. However, these concentrations are three to four orders of magnitude less than the barge concentrations. An initial dilution of the waste material occurs almost immediately following discharge into the barge wake (see Csanady, 1981).

Table 3. Selected parameters at stations taken following dumps at the Puerto Rico dumpsite.

Sta.	Depth (m)	Time (hr)	Si (µM)	PO$_4$ (µM)	NO$_2$ (µM)	NH$_3$ (µM)	DOC (mg/l)	POC (µg/l)	ATP (ng/l)	TKN (mg/l)	TSM (µg/l)	Chloro. a (mg/m^3)	CH$_4$ (nl/l)
DS/1	1	0.9	2.4	0.95	0.50	13.7	2.01	32.4	47.2	<0.07	225	0.230	51
	10	0.9	2.4	0.48	0.45	10.6	1.44	23.0	59.1	<0.07	216	0.288	49
	20	0.9	2.4	0.48	0.45	6.1	1.28	22.1	44.8	<0.07	216	0.403	67
	30	0.9	2.2	0.48	0.45	6.1	1.23	22.6	39.8	<0.07	196	0.230	50
DS/2	1	2.1	2.4	0.90	0.45	6.1	1.48	20.1	73.3	0.16	188	0.230	58
	15	2.1	2.4	0.38	0.45	6.1	1.45	20.2	72.8	<0.07	167	0.230	56
	25	2.1	2.4	0.48	0.45	6.1	1.23	22.2	45.7	0.18	188	0.230	38
	30	2.1	2.2	0.38	0.50	9.1	1.38	20.8	36.3	<0.07	214	0.230	55
DS/3	1	3.2	2.4	0.56	0.45	21.4	1.55	37.6	41.9	<0.07	232	0.173	52
	10	3.2	2.4	0.52	0.45	9.1	1.46	30.3	53.0	0.17	207	0.230	55
	20	3.2	2.2	0.38	0.50	7.6	1.66	21.6	99.2	<0.07	251	0.230	54
	30	3.2	2.2	0.38	0.50	6.1	1.32	27.3	47.5	<0.07	--	0.230	48
DS/4	1	5.2	2.4	0.66	0.50	12.2	1.28	23.2	42.0	0.13	123	0.115	51
	10	5.2	2.4	0.52	0.50	19.8	1.52	35.9	96.2	0.16	271	0.230	64
	20	5.2	2.4	0.38	0.50	7.6	1.52	29.8	68.9	0.18	112	0.230	63
	30	5.2	2.2	0.43	0.50	7.6	1.44	19.8	43.7	<0.07	187	0.230	63
DS/5	1	7.0	2.4	0.56	0.50	10.6	1.22	38.3	40.4	0.16	155	0.173	59
	10	7.0	2.4	0.56	0.50	16.8	1.69	47.5	75.6	<0.07	148	0.230	59
	20	7.0	2.4	0.43	0.50	9.1	1.72	31.5	62.8	0.19	265	0.230	51
	30	7.0	2.2	0.43	0.60	15.2	1.47	39.9	49.2	<0.07	203	0.173	51
DS/6	1	13.2	2.4	0.76	0.50	7.6	1.15	30.8	58.7	0.12	187	0.115	71
	10	13.2	2.4	0.43	0.60	7.6	1.33	42.9	50.7	0.15	124	0.115	60

Table 3 (continued)

Sta.	Depth (m)	Time (hr)	Si (μM)	PO_4 (μM)	NO_2 (μM)	NH_3 (μM)	DOC (mg/l)	POC (μg/l)	ATP (ng/l)	TKN (mg/l)	TSM (μg/l)	Chloro. a (mg/m³)	CH_4 (nl/l)
DS/7	1	15.8	—	—	—	—	1.09	35.8	36.5	0.21	95	0.115	71
	10	15.8	—	—	—	—	1.21	31.5	51.0	0.10	95	0.115	65
	20	15.8	—	—	—	—	1.58	34.4	42.4	0.12	122	0.115	65
	30	15.8	—	—	—	—	1.36	28.1	41.1	<0.07	199	0.115	66
DS/8	1	25.6	2.4	0.66	0.45	7.6	1.18	16.9	66.0	0.30	101	0.173	50
	10	25.6	2.4	0.48	0.50	7.6	1.37	16.9	38.0	0.16	46	0.115	57
	20	25.6	2.4	0.48	0.50	6.1	1.22	25.7	69.9	<0.07	99	0.173	52
	30	25.6	2.2	0.38	0.50	4.6	1.27	15.3	60.2	0.16	176	0.173	60
DS/9	1	0.5	2.3	1.46	0.45	189	2.60	92.8	33.5	0.31	453	—	64
	5	0.5	2.3	0.56	0.50	184	2.52	86.9	39.7	0.32	431	—	63
	10	0.5	2.3	0.56	0.80	12	1.13	27.2	40.5	<0.07	305	—	52
	15	0.5	2.3	0.52	0.50	4.6	1.04	22.0	53.4	0.18	274	—	55
DS/10	1	1.8	2.3	2.35	0.45	74	1.79	49.9	39.1	0.22	312	—	53
	5	1.8	2.3	0.56	0.45	35	1.37	32.7	59.3	0.19	310	—	59
	10	1.8	2.3	0.52	0.45	21	1.23	30.7	70.3	<0.07	112	—	52
	15	1.8	2.3	0.38	0.50	15	1.13	29.7	49.4	0.19	231	—	50
DS/11	1	5.0	2.3	0.52	0.45	18	1.26	27.1	37.7	0.19	202	—	61
	5	5.0	2.3	0.48	0.50	29	1.40	31.1	36.1	<0.07	214	—	61
	10	5.0	2.3	0.38	0.60	18	1.22	26.0	48.1	0.25	274	—	58
DS/12	1	10.7	2.3	0.43	0.45	4.6	1.01	35.1	41.3	<0.07	169	—	67
	10	10.7	2.3	0.66	0.45	6.1	1.04	22.9	42.0	<0.07	149	—	55

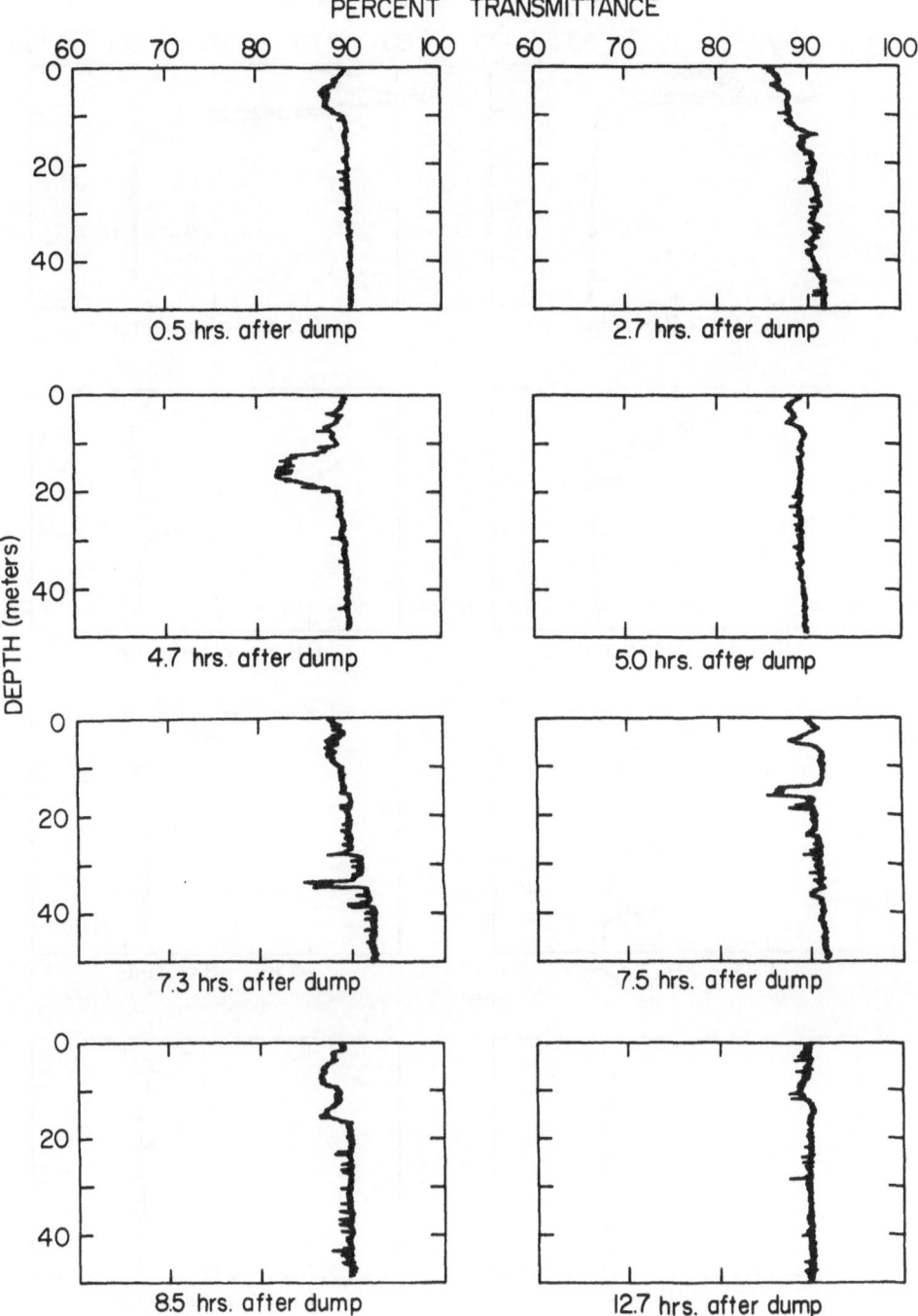

PERCENT   TRANSMITTANCE

Fig. 3.   Transmissometry traces in the waste plume after the dump of
6 February 1978.

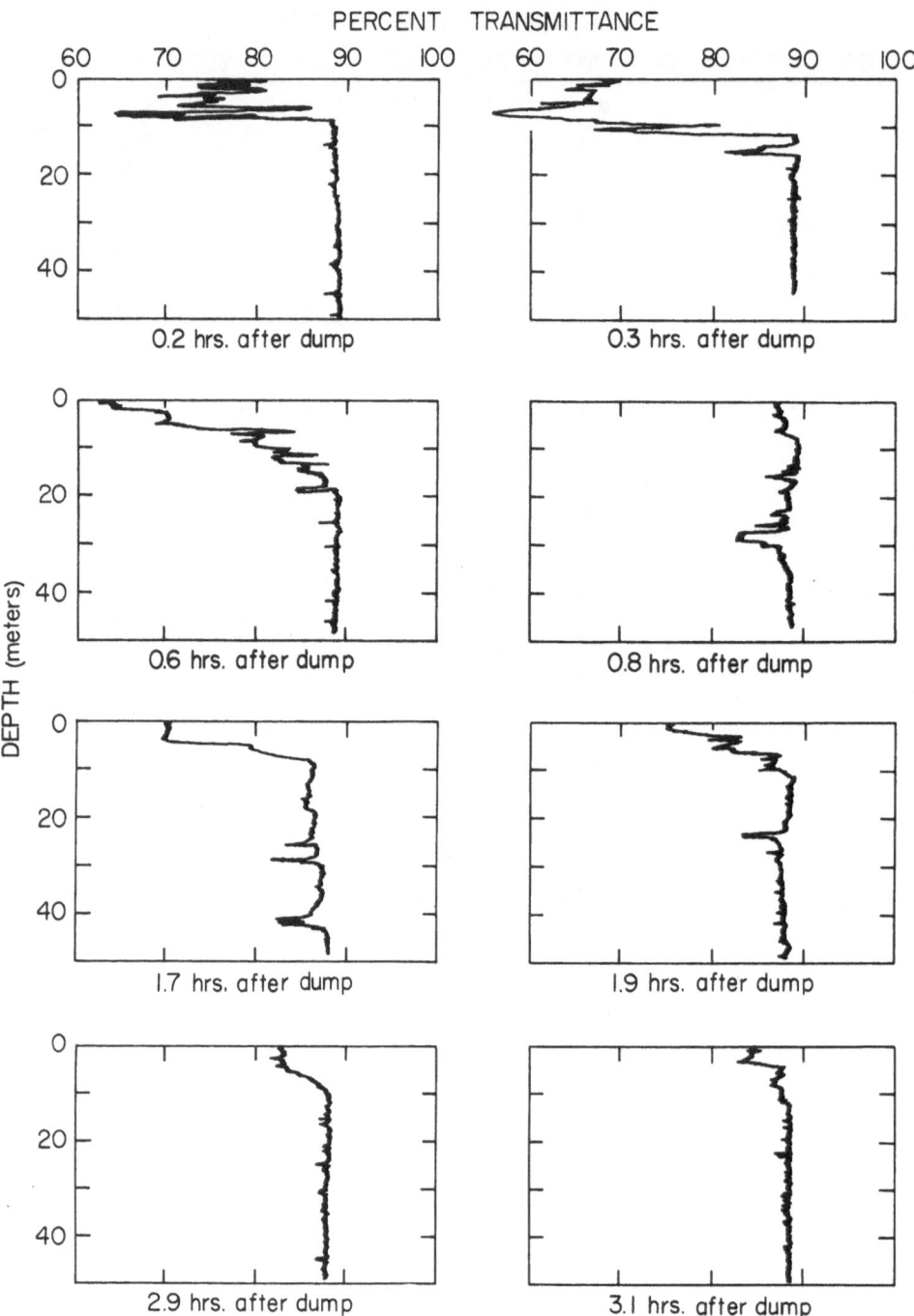

Fig. 4.   Transmissometry traces in the waste plume after the dump of
          9 February 1978.

Compounds identified by GC-MS techniques in the waste material and at Station DS/9 are presented in Table 4 (numbers correspond to numbered peaks in Figs. 5 and 6). Identified volatile organic compounds include hydrocarbons, chlorinated hydrocarbons, alcohols, esters, ketones and nitrogen compounds. A comparison of the two . samples shows that there are many compounds present in sample DS/9 that are not present in the barge sample. The absence of these compounds is probably due to unrepresentative sampling of the barge wastes, which were sampled from the fill pipe as the barge was loaded. Most of the compounds in the DS/9 water sample, but not in the barge sample, are not found in unpolluted oceanic waters, and therefore they most likely result from the dumping. Approximately 99% of the waste dumped is composed of volatile organic solvents and water; the remainder is higher-molecular-weight pharmaceutical compounds.

Quantitative determinations were made on some volatile organics following the second dump. Fig. 7 shows chromatograms at the same attenuation of two successive 100 min strips of water sample DS/9.

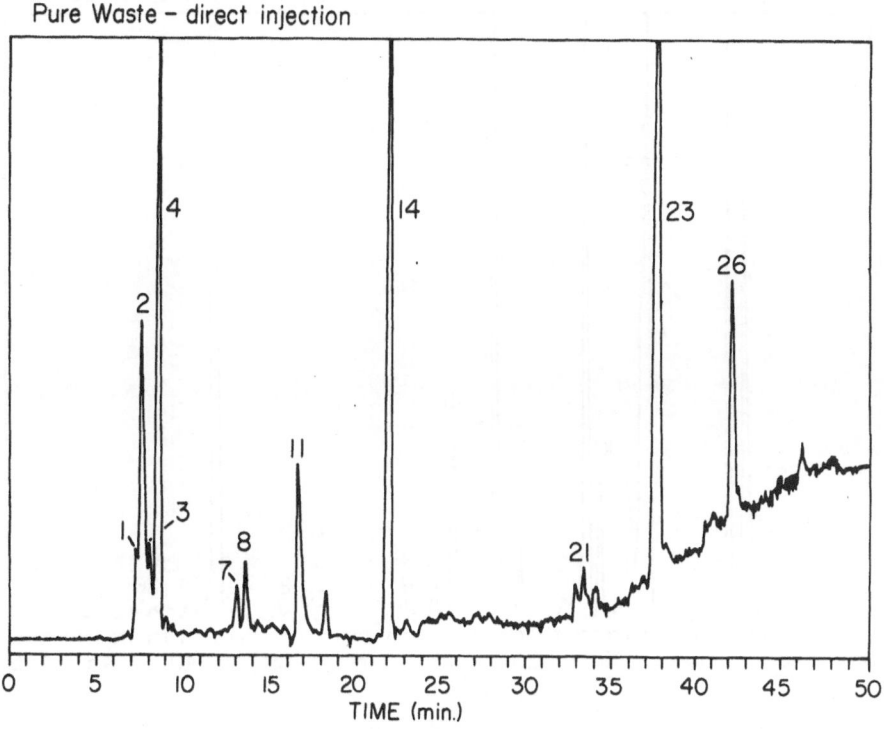

Fig. 5.   GC-MS ion spectrum of barge waste material taken 8 February 1978 (peak numbers correspond to components in Table 4).

The first strip displays all the volatile organic compounds. The 100 min stripping time removes from the sample all the volatile hydrocarbons and chlorinated hydrocarbons (Sauer et al., 1978). These compounds, such as benzene (peak 10) and toluene (peak 14), are not observed in the second chromatogram. Concentrations of benzene and toluene at Station DS/9 (1 m) were 0.56 $\mu g \cdot l^{-1}$ and 10.5 $\mu g \cdot l^{-1}$, respectively, quantities 2 to 3 orders of magnitude greater than oceanic baseline values of 10-40 $ng \cdot l^{-1}$ (Sauer et al., 1978). However, many of the more soluble compounds with low vapor pressures, such as acetone, ethyl acetate, n-butanol, methyl isopropyl ketone and dimethylaniline, are still evident in the second chromatogram. The technique used did not quantitatively determine these more soluble volatile components, since only a small percentage were removed during each successive strip. By increasing the stripping rate and using additional Tenax-GC, one would be able to quantitatively determine the more soluble components. Estimates for the amounts of butanol and dimethylaniline are around 1 to 10 $mg \cdot l^{-1}$.

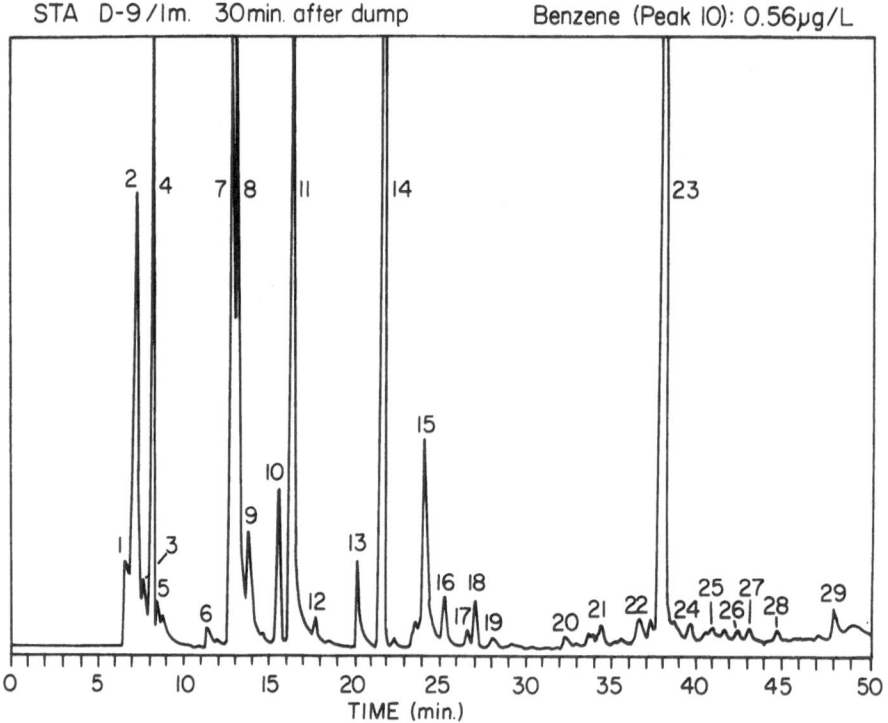

Fig. 6. GC-MS ion spectrum of a surface sample at Station DS/9 taken 30 minutes after the dump (peak numbers correspond to components in Table 4).

Table 4.   Volatile organic peak identifications from Puerto Rico ocean dumping cruise.

Peak #	Compound
1	Acetone
2	Many possible alcohols and ketones
3	Ethyl mercaptan
4	Dichloromethane
5	n-Propanol
6	Unknown
7	Chloroform
8	Ethyl Acetate
9	Tetrahydrofuron
10	Benzene
11	n-Butanol
12	Trichloroethylene
13	Methyl isobutyl ketone
14	Toluene
15	N-butyl or iso-butyl acetate
16	Column bleed
17	Ethyl benzene
18	m,p-Xylenes
19	o-Xylene
20	C_3-Benzene
21	Column bleed
22	Dimethyl pyridine
23	Dimethyl aniline
24	C_4-Benzene
25	Alkenyl (C_4)-benzene
26	Naphthalene
27	C_5-benzene
28	Ethyl cumene
29	Unknown

Approximately 70% of all the organic solvents in the waste consists of these two compounds.

Water samples for baseline comparison were taken in the center of the chemical waste dumpsite before the second dump. To insure contamination-free samples, surface seawater was taken 200 m upwind and upcurrent of the research vessel from a small rubber boat. Samples were collected directly in volatile-free, 2-1 glass bottles by lowering them below the surface over the bow of the boat and filling at arms length. Fig. 8 (Station 22) shows a chromatogram of this sample (the attenuation is almost 2 orders of magnitude less than the chromatograms in Fig. 7). The number of volatile compounds are not as numerous as samples in the waste plume, and concentrations are

considerably lower. The concentrations of benzene, toluene, and
xylenes are 8, 16 and 54 ng·l^{-1}, respectively, which are typical of
unpolluted coastal waters in the Gulf of Mexico. Not typical, however, is the presence of dimethylaniline. In fact, even in polluted
waters of the Mississippi Delta (Sauer et al., 1978) and coastal
waters of Massachusetts (Schwarzenbach et al., 1978) dimethylaniline
has not been found. This baseline sample was taken 3 days after the
first dump, and indicates that solvents such as dimethylaniline with
solubilities similar to chloroform and vapor pressure lower than the
xylenes, persist in surface waters for extended periods of time.
The baseline concentration of the dimethylaniline is approximately
0.1 µg·l^{-1}. This concentration is 4 to 5 orders of magnitude less
than at Station DS/9 taken 30 min after the dump.

    The persistence of selected volatile organics were studied fol-
lowing the second dump. Station DS/9 samples had the highest concentrations of volatiles and dye. These concentrations decreased gradually to approximately baseline values some 24 hours after the dump
(Table 5). However, as discussed previously, some solvent did linger

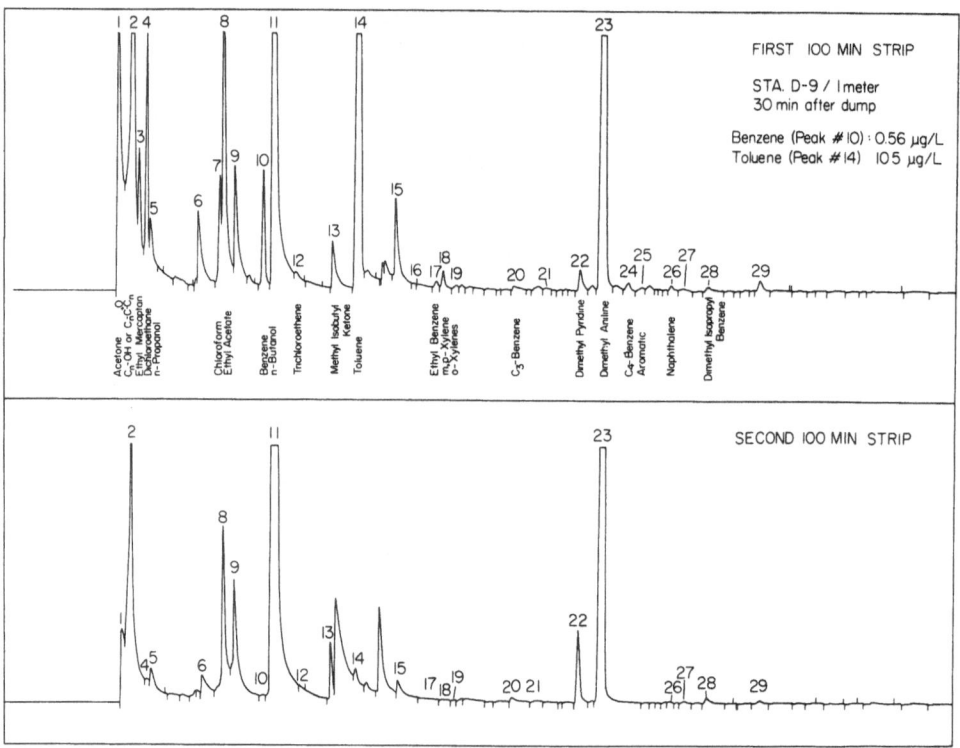

Fig. 7.   Chromatograms of surface water taken at Station DS/9, 30
          minutes after dump (peaks identified in Table 4).

in surface waters for a longer period of time (Fig. 8). The gradual
decrease (dilution) in dye concentration is due principally to the
entrainment of surface water with the waste plume. Since the dye is
a conservative tracer, it will show only the physical effects of
mixing. Volatile organics, however, undergo both dilution and
evaporation.

   In Figs. 9 and 10, benzene and toluene concentrations are plotted against dye concentration with time. Since the dye concentrations in samples DS/9 to DS/13 covered a two order of magnitude
range, the concentrations of the volatiles selected for the evaporation studies need to have at least a 2 order of magnitude difference
between maximum concentration and background. Benzene and toluene
meet this prerequisite, but the xylenes decrease to background concentrations before the effect of evaporation becomes evident. · The
dilution lines in Figs. 9 and 10 represent the decrease of dye and
volatile concentrations due to simple two-component mixing (i.e., no
effect from evaporation). The amount of volatile hydrocarbons lost

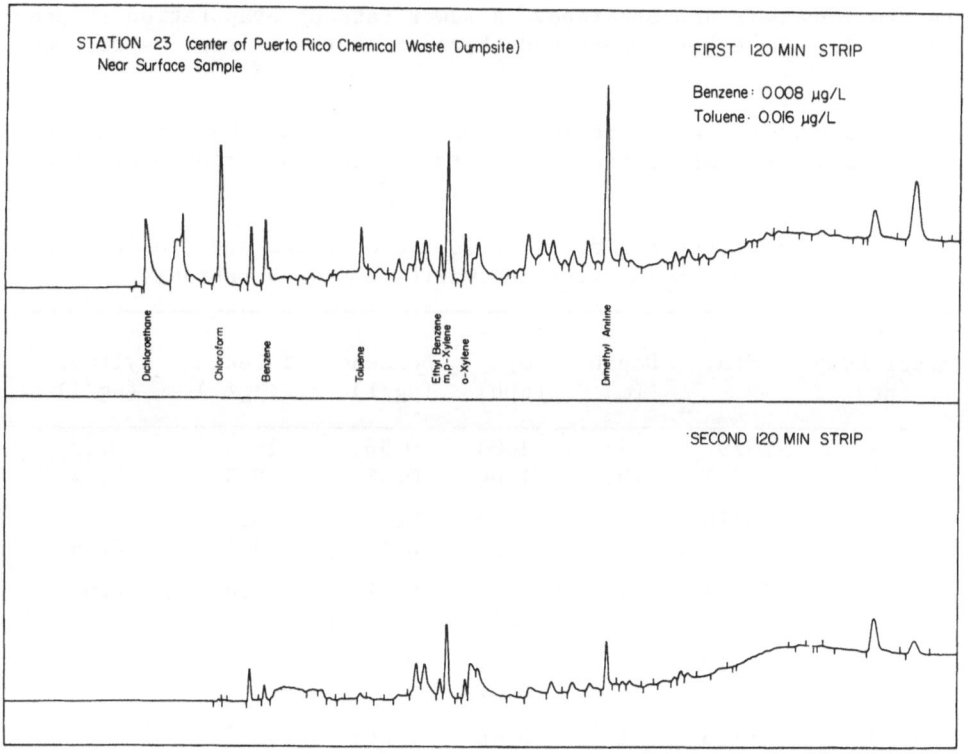

Fig. 8. Chromatograms of surface water at the center of the Puerto
         Rico dumpsite three days after a dump.

from evaporation is calculated to be the difference between the concentration predicted from the dilution line and that determined as background. This difference could be due to separation of dissimilar phases in the water column; one phase being water insoluble containing the organic waste and the other being a water soluble phase containing the dye. The phase separation of these two components could give the impression of volatile loss due to evaporation. This possibility is remote since volatile organic concentrations in the water are well below saturation.

For both benzene and toluene, loss due to evaporation is not evident in the first 2 hrs after the dump. Around the fifth hour, 27% of the benzene was lost apparently through evaporation. After 10 hrs benzene levels decreased to background. Toluene, on the other hand, does not show discernible evaporation until 10 hrs after the dump and does not reach background levels within the 12 hrs of the experiment. The different rates of evaporation between the two volatiles result from different vapor pressures. Benzene, with a higher vapor pressure, has a higher rate of evaporation from the water column. Evaporation rates for non-hydrocarbon volatiles, such as butanol and dimethylaniline, were not determined since quantitative concentrations were not obtained. A lower rate of evaporation is predicted for these heterocompounds because of their lower vapor pressure.

Evaporation rates are important in assessing the potential effects of toxic components found in the wastes on marine organisms.

Table 5. Concentrations of volatile hydrocarbons and dye in water samples taken after the waste dump of 9 February.

Time After Dump (hr)	Sta.	Depth (m)	Dye (ppb)	Benzene (μg/l)	Toluene (μg/l)	Xylenes (μg/l)
0.5	DS/9	1	1.05	0.56	10.5	0.15
		10	1.04	0.46	9.9	0.14
1.8	DS/10	1	0.43	0.22	4.7	--
		5	0.28	0.13	3.1	0.06
5.0	DS/11	1	0.27	0.11	2.8	0.05
		5	0.14			
10.7	DS/12	1	0.01	0.008	0.030	0.07
		10	0.01			
12.0	DS/13	1	0.06	0.008	0.044	--
Background		1		0.008	0.016	0.054

Since volatile hydrocarbons (e.g., benzene and toluene) evaporate faster than more soluble heterocompounds such as butanol or dimethyl-aniline, their potential toxic interactions with organisms is reduced

PHYTOPLANKTON TOXICITY

Preliminary laboratory toxicity studies were conducted using the phytoplankton <u>Skeletonema costatum</u> with Puerto Rican waste material. Waste CIA was a composite sample taken during barge onloading on 5 February 1978, and stored in a gas-tight glass bottle, cold and dark. Waste AM was from the American Cyanamid plant and was stored in a plastic bottle, cold and dark. The 100 relative percent ATP and ^{14}C uptake line (Fig. 11) represents an average of the two control

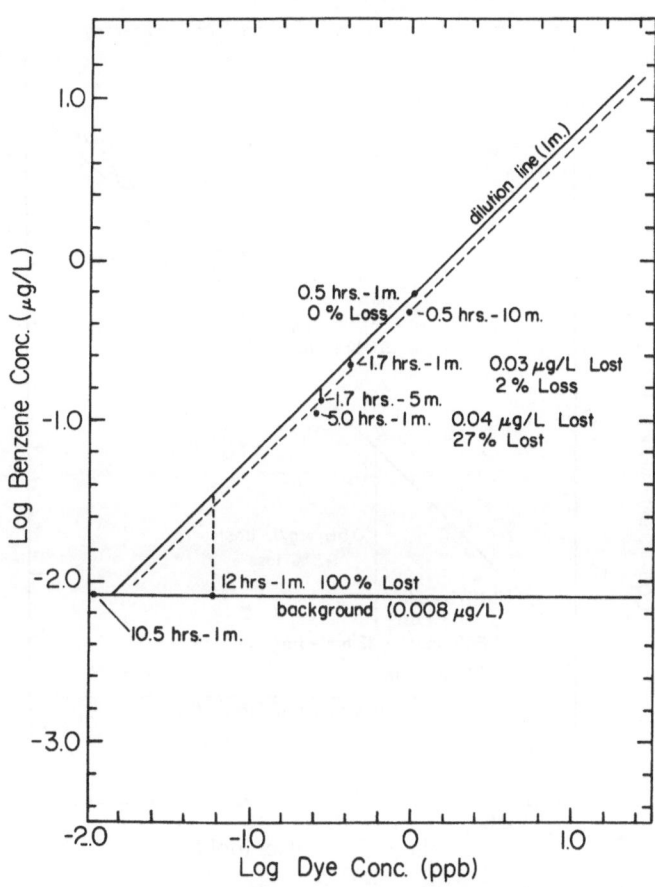

Fig. 9. Benzene versus dye concentrations in the waste plume following a dump on 9 February 1978.

groups in each experiment. By ratioing the results of each run to
the control average, any variability due to experimental procedures
or conditions is removed, and a direct comparison can be made between
different experiments. Three replicate bottles were run for each
level of waste studied.

The ^{14}C uptake results represent an integrated, overall value
for the carbon fixed, either currently living or dead, from the time
of ^{14}C addition (one hr after exposure) until the end of the experiment after 9 hrs. The ATP results represent a relatively instantaneous measurement of the living biomass. Carbon fixation by photosynthesis is a rather complex procedure which requires control and organization of a long series of metabolic activities. Because of
this, the ^{14}C uptake results are an indication of the phytoplankton's ability to function, while the ATP results are a measure of the

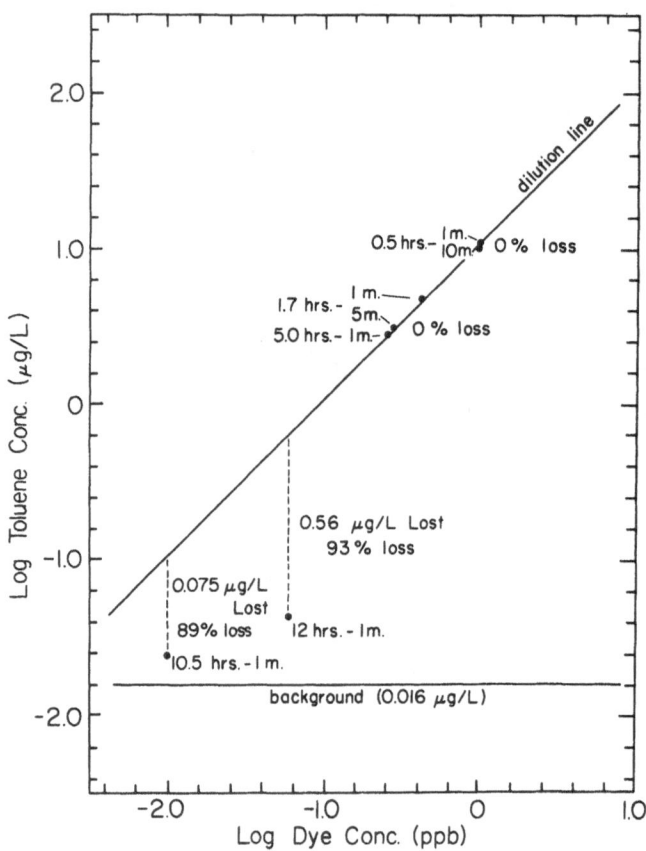

Fig. 10. Toluene versus dye concentrations in the waste plume
following a dump on 9 February 1978.

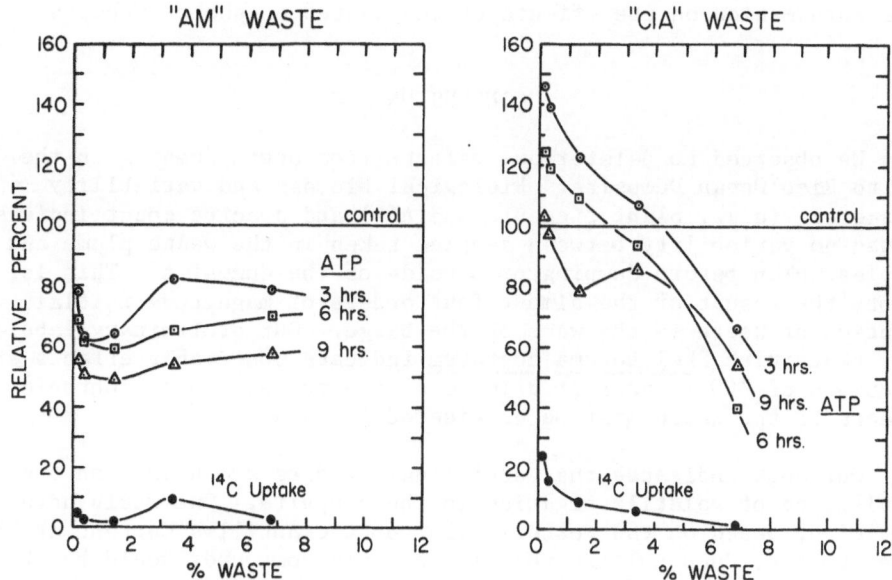

Fig. 11. Toxicity of waste to the phytoplankton <u>Skeletonema</u> <u>costatum</u> as measured by ^{14}C uptake and ATP levels.

ability to survive and the response to stress.

The results for CIA waste show how ATP and ^{14}C uptake vary in response to changes in waste concentration and exposure time (Fig. 11). The ^{14}C uptake is reduced to 24% of the control at a waste level of 0.14%, and drops rapidly at higher levels. Relative ATP, however, is significantly greater than the control at low levels of waste and short exposure times. These elevated ATP levels appear to be in response to metabolic stress. As waste levels or exposure time increase, the ATP concentrations drop off rapidly. The slight tint in the culture media from the blue-black color of the waste could not have reduced the available light enough to have significantly affected the results. The AM waste results are somewhat different, but again, ^{14}C uptake is greatly inhibited at all levels of waste addition (Fig. 11). The ATP results seem to indicate that the most toxic level of waste is around 1.4%, although inhibition is observed at all levels of waste addition.

Although both wastes are toxic to phytoplankton, it would appear from these limited results that the toxicity of the individual wastes and the composite may vary considerably. Other individual wastes and different composite waste samples are being tested to assess this variability. The toxicity of the wastes after various treatments will be studied to help identify the toxic components of the waste and the timing and duration of ^{14}C addition will be varied to gain more information on the effects of the wastes on photosynthesis.

CONCLUSION

We observed no deleterious effects from ocean dumping in the Puerto Rico Ocean Dumpsite. Biological biomass and variability parameters (e.g., plant pigments and ATP) and species count indices showed no variability between samples taken in the waste plume and samples taken before dumping or outside of the dumpsite. This is, no doubt, the result of the almost four orders of magnitude initial dilution of waste in the wake of the barge. Our preliminary laboratory tests with Skeletonema costatum indicate observable effects at dilutions of 700:1; greater dilutions were not examined. Chronic effects of the waste must be considered in future work.

Our work indicates that additional studies are needed on the persistence of volatile organics in the dumpsite. One would have predicted, based on the westerly flow of the Antilles Current at the site, that no dump related volatile organic compounds would be observed within a day after dumping. However, 0.1 μg/1 of dimethylaniline, a major constituent of the waste, was observed 3 days following a dump at the site. The 1-10 ppm concentrations of butanol and dimethylaniline in the waste plume are very significant with respect to the viability of organisms in contact with the plume. Although most hydrocarbon and halocarbon volatiles are lost within 12 hrs following dumping, the more soluble heterocompounds may be very persistent in the dumpsite.

ACKNOWLEDGEMENTS

Dr. Greta Fryxell's extensive personal library was employed for verification of cell identifications. We wish to thank Dr. W. M. Sackett for use of his laboratory for some of the studies. This work was supported by the NOAA Ocean Dumpsite Reseach and Monitoring Program (Contract 04-8-M01-55).

REFERENCES

Atlas, E., J. M. Brooks, J. Trefry, T. C. Sauer, C. R. Schwab, B. B. Bernard, J. Schofield and C. S. Giam (1980) Environmental aspects of ocean dumping in the Western Gulf of Mexico. Journal of the Water Pollution Control Federation, 52, 329-350.
Bertsch, W., E. Anderson and G. Holzer (1975) Trace analysis of organic volatiles in water by gas chromatography — mass spectrometry with glass capillary columns. Journal of Chromatography, 112, 701-718.

Brooks, J. M., B. B. Bernard and W. M. Sackett (1977) Input of low-molecular-weight hydrocarbons from petroleum operations into the Gulf of Mexico. In: Fate and Effects of Petroleum Hydrocarbon in the Marine Ecosystem and Organisms, D. A. Wolfe, editor, Pergammon Press, pp. 373-384.

Cornu, A. and R. Massot (1966) Compilation of mass spectral data. Heyden Sons Ltd., London, 500 pp.

Csanady, G.T. (1981) An analysis of dumpsite diffusion experiments. In: Ocean dumping of industrial wastes, B. H. Ketchum, D. R. Kester, and P. K. Park, editors, Plenum Press, New York, pp. 109-129. This volume.

Fredricks, A. D. and W. M. Sackett (1970) Organic carbon in the Gulf of Mexico. Journal of Geophysical Research, 75 (12), 2199-2206.

Holm-Hansen, O., C. J. Lorenzen, R. W. Holmn and J. D. H. Strickland (1965) Fluorometric determination of chlorophyll. J. Conseil, Conseil Perm. Interm. Exploration Mer., 30, 1-15.

Holm-Hansen, O. and C. R. Booth (1966) The measurement of ATP in the ocean and its ecological significance. Limnology and Oceanography, 11, 510-519.

Imperial Chemical Industries, Ltd. (1970) Eight peaks index of mass spectra, Vol. 1 and 2, Mass Spectrometry Data Center, U. K., 1500 pp.

McLafferty, F. W. (1973) Interpretation of mass spectra. W. A. Benjamin, Inc., London, 278 pp.

McLachlan, J. (1973) Growth media - marine. In: Handbook of Physiological Methods, Culture Methods and Growth Measurements, J. R. Stein, editor, Cambridge University Press.

May, W. E., S. N. Chester, S. P. Cram, B. H. Gump, H. S. Hertz, D. P. Enagonio and S. M. Dyszel (1975) Chromatographic analysis of hydrocarbons in marine sediments and seawater. Journal of Chromatographic Science, 13, 535-540.

Parsons, T. R. and J. D. H. Strickland (1963) Discussion of spectrophotometric determinations of marine - plant pigments, with revised equations for ascertaining chlorophylls and carotenoids. Journal of Marine Research, 21, 155-163.

Sauer, T. C., Jr. (1978) Volatile liquid hydrocarbons in the marine environment. Ph.D. dissertation, Texas A&M University, College Station, TX 317 pp.

Sauer, T. C., Jr., W. M. Sackett and L. M. Jeffrey (1978) Volatile liquid hydrocarbons in the surface coastal waters of the Gulf of Mexico. Marine Chemistry, 7, 1-16.

Schwarzenbach, R. P., R. H. Bromund, P. M. Gschwend and O. C. Zafiriou (1979) Volatile organic compounds in coastal seawater, preliminary results. Journal of Organic Geochemistry, 1, 93-107.

Strickland, J. D. H. and T. R. Parsons (1972) A practical handbook of seawater analysis. Bulletin Fisheries Research Board of Canada, 167, (2nd edition), 310 pp.

Yentsch, C. S. and D. W. Menzel (1963) A method for the determination of phytoplankton chlorophyll and phaeophytin by fluorescence. Deep-Sea Research, 10, 221-231.

CHEMICAL CHARACTERIZATION OF OCEAN-DUMPED WASTE MATERIALS

E. Atlas, G. Martinez, and C. S. Giam

Department of Chemistry
Texas A & M University
College Station, TX 77843

ABSTRACT

This paper describes methods for chemical characterization and organic "finger printing" techniques for complex ocean-dumped wastes. The methods employ wet chemical separations and gas chromatography combined with mass spectrometry and selective GC detectors. Application of these methods is demonstrated for the characterization of a pharmaceutical waste and for tracking two types of chemical wastes in actual ocean dumping stations.

INTRODUCTION

Industrial wastes dumped in the ocean are typically a complex mixture of inorganic and organic chemical byproducts, cell debris, or a combination of chemical and biological wastes. Chemical characterization of individual components of these wastes is a difficult task and to evaluate the effects of the waste on marine organisms in seawater is even more challenging. Concentrations of specific inorganic chemical components, such as Hg, Cd, and, Zn, can be easily monitored. Measurement of the organic components of industrial wastes, however, is usually restricted to monitoring bulk organic properties such as total organic carbon, oil and grease, or biological oxygen demand. Recent advances in organic separations and analytical instrumentation can provide a more sensitive and specific approach to the characterization of the organics in the complex wastes before or during the dumping. In particular, gas chromatography combined with mass-spectrometric and element-specific detectors has a very valuable application to the characterization of ocean-dumped waste.

Table 1. Reported contributions to ocean-dumped waste at Puerto Rico Dumpsite.*

Company	Products	Reported Wastes	Waste Volume (Wet tons/yr)
I	Lincomycin Clinomycin	Pyridine, Methanol, Palmitic Acid	150,000
II	L-Methyldopa	Vanillin, Dimethylsulfoxide, Ethylamine	83,000
III	Ethylene Propylene Butadiene	Sodium Hydroxide (pH ≅ 12)	70,000
IV	Gentamicin Sisomicin	—	62,000
V	Doxepin Chlorpropamide Meclysine	Tetrahydrofuran, Xylenes	21,000
VI	Penicillin Derivatives	β-Naphthalene, Methylisobutyl Ketone Phenylglycine, Triethylamine	12,000
VII	Penicillin Derivatives	Methanol, Acetone, Xylene, Isopropyl Acohol, other Organic Solvents	8,400
VIII	Levamisole Tetramisole	—	—

*(Personal communication, T. O'Connor, 1978)

We have investigated two types of industrial wastes – a "biological" sludge dumped in the Gulf of Mexico and a "pharmaceutical" waste dumped near Puerto Rico (Table 1). Our objectives were to characterize the organic components in the waste material and to use the organic components as chemical tracers of waste during an experimental dump. We were most interested in the high-molecular weight, solvent-extractable organic components in the waste materials, where many of the synthetic organic components will be found. Volatile compounds were only briefly examined in the laboratory. In this paper we will present a summary of the results of our laboratory characterization of the industrial wastes, limiting our discussion to the Puerto Rican pharmaceutical waste. Then we demonstrate the use of specific chromatographic detectors in tracing the different wastes after they are dumped in the ocean. The purpose of this paper is to present an overview of our approach to the problem of characterizing complex materials and demonstrate an application to ocean-dumped wastes. More detailed procedures and discussions are presented by Giam et al. (1979).

METHODS

A 50 ml aliquot of each waste sample was made basic (pH 11) with a 10% KOH solution and extracted three times with a total of 15 mls of MCB "Pesticidequality" petroleum ether. The aqueous portion was acidified to pH 2 with concentrated H_2SO_4 and extracted with petroleum ether as above. The acid extracts were esterified with BF_3/MeOH and the resultant methyl esters were examined by gas chromatography. The base/neutral extracts of the individual wastes were examined by gas chromatography without further treatment. The base/neutral extracts of the composite wastes were acidified and re-extracted with petroleum ether to separate the neutral and basic compounds. The neutral components were chromatographed on a silica-gel/alumina column (Giam et al., 1976). Forty ml volumes of the following solvents were used to elute the column: pentane; 1:1 (v/v) pentane: benzene; 1:3 (v/v) methanol: hexane; 1:1 (v/v) methanol: hexane; 3:1 (v/v) methanol: hexane. The eluates were evaporated under nitrogen to near-dryness and dissolved in iso-octane for further analysis.

The samples were analyzed primarily using glass capillary gas chromatography and several detectors. Different columns were used in the analyses: 30 m OV-101 WCOT, 30 m SP-2100 WCOT, and 30 m SP-2100 SCOT. For gas chromatography using flame ionization (FID) and nitrogen selective (NPD) detectors the following conditions were used: initial temperature, 70°C; initial time, 2 min; program rate: 3°/min; final temps 250°C. The chromatographs were a Hewlett-Packard 5830 or a Perkin-Elmer Sigma 1.

For GC/MS analyses we used a Hewlett-Packard 5992B gas

Table 2. Volatile organic compounds in the Puerto Rican Waste.
          (Identifications from Schwab et al., 1981).
          Peak numbers refer to Fig. 1.

Peak Number	Compound
1	acetone
2	unknown
3	alcohol or ketone
4	methylene chloride
5	n-propanol
6	unknown
7	chloroform
8	ethyl acetate
9	unknown
10	benzene
11	n-butanol or 1-chlorobutane
12	trichloroethylene
13	methyl iso butyl ketone
14	toluene
15	unknown
16	dimethylaniline

chromatograph/mass spectrometer. The conditions were: initial
temperature, 90°C, program rate: 4°C/min; final temperature: 270°C,
ionizing voltage, 70 ev; scan rate: 190 amu/sec.

While GC/MS is an established powerful tool in examining un-
knowns, we wish to emphasize that GC/MS evidence is not conclusive
confirmation for identifying compounds. Thus, in this paper, we use
GC/MS data as supportive evidence in our chemical characterization.
The compounds we "identify" are consistent with the data we examined.
Much further analysis of these mixtures is necessary though to confirm compound identities - for example, isolation of the compound,
elemental analysis, UV, NMR spectroscopy etc.

RESULTS

Laboratory Characterization (Puerto Rico Waste)

Sampling variation. Improper sampling of the waste for laboratory analysis can potentially change its chemical composition. For

Fig. 1. Volatile compounds in headspace gas of composite waste
         samples. Note loss of peak 14 (toluene) between samples
         stored in glass versus plastic.

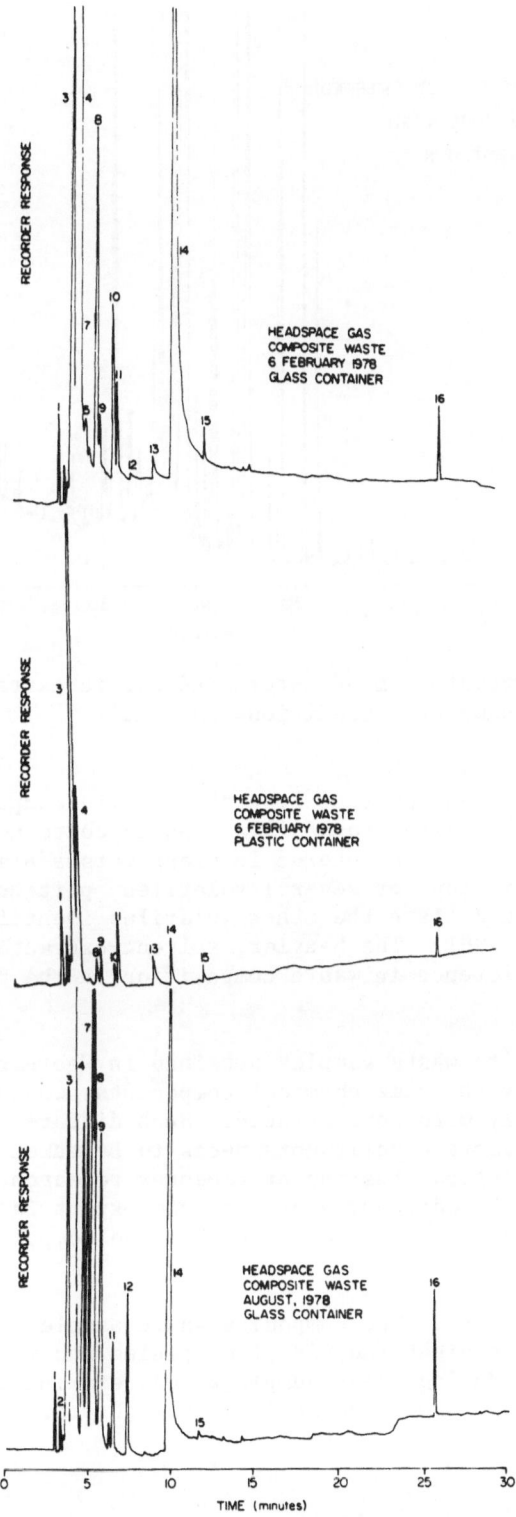

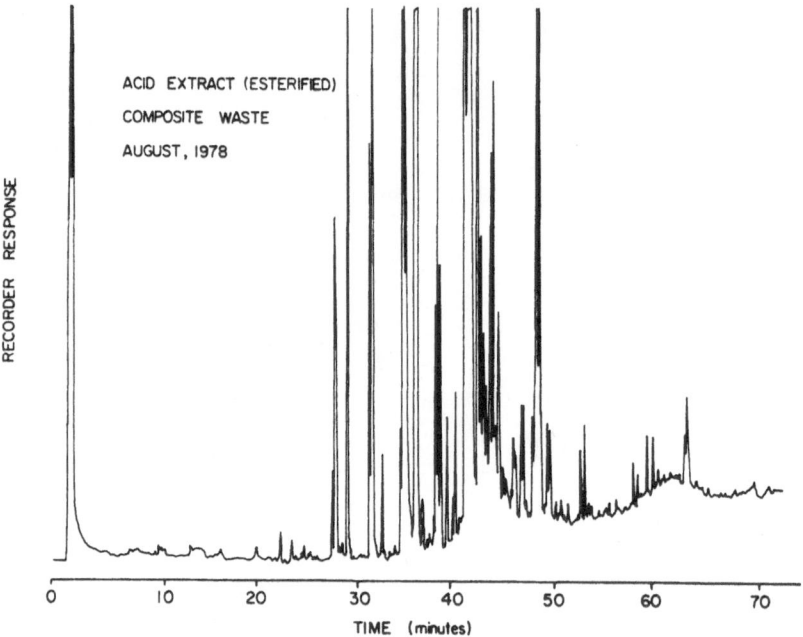

Fig. 2. FID chromatogram of esterified acidic extract of composite waste sample. (Conditions in text).

example, we found an apparent loss of volatile compounds in a sample of pharmaceutical waste stored in a plastic container. Headspace-gas analysis of waste samples stored in glass versus plastic containers revealed specific loss of several volatiles, particularly toluene (Fig. 1). Table 2 lists the other volatiles identified in the waste (Schwab et al., 1981) The heavier, solvent-extractables showed no significant difference in waste composition in the plastic containers.

The composite waste samples obtained in February and August 1978 showed basically the same chemical components, though the August sample was generally more concentrated. Such differences in the concentration of the waste constituents needs to be taken into consideration during biological testing or bioassay research. Because of the very nature of the waste product, one can expect fairly large variations in the concentration (and possibly the composition) of the waste.

Composite waste. The composite waste sample is a mixture of wastes from up to eight individual companies and is the material actually disposed during ocean-dumping. Our examination of the waste demonstrated its extreme complexity. Gas chromatography/mass

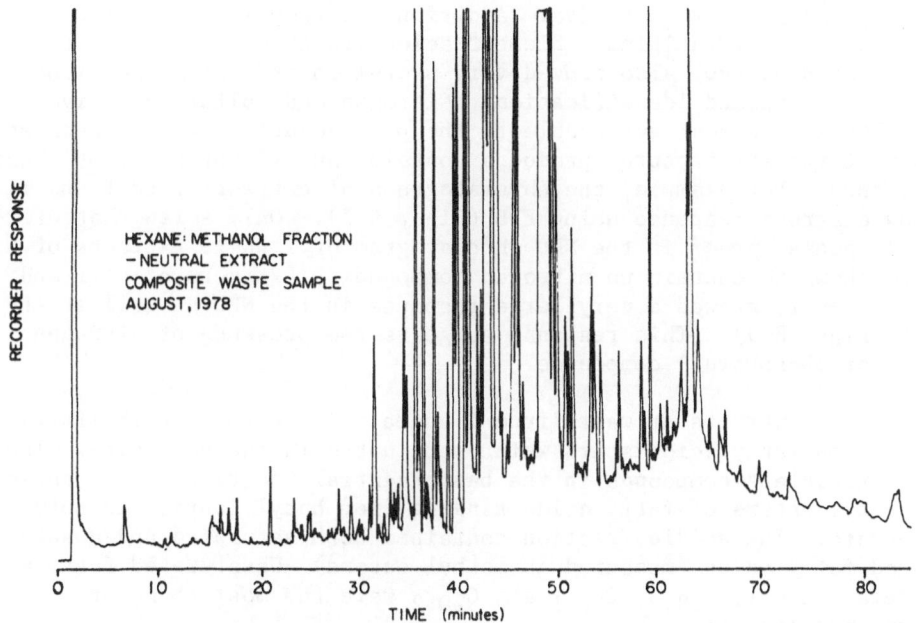

Fig. 3.   FID chromatogram of 3:1 (v/v) hexane:methanol eluate from
silica gel column chromatography of composite waste-neutral
extract. (Conditions in text).

spectrometry was used to examine these fractions, though the complexity of the waste allowed identification of only a portion of the
peaks. In addition to the composite sample, we looked in more detail
at the waste from individual companies to identify the major components of the waste. The individual companies will be discussed in
the next section. In the composite, most compounds occur in the
acidic and neutral, polar fractions of the waste (Figs. 2-3). The
major components in these fractions appear to be saturated and unsaturated fatty acids and fatty acid esters. The basic fraction contains only a few compounds - the main one is dimethylaniline (Fig.
4). Normal hydrocarbons (C_{15}-C_{32}) are also observed in the
neutral, aliphatic fraction (Fig. 5).

   Individual wastes. Because there were so few compounds in the
basic fraction of the composite waste, we examined the entire base-
neutral extracts from each company's waste without further separation
The acid extracts were derivatized as in the composite waste. We
used GC/MS as the primary means of compound identification. Chemical
methods of confirmation, such as alkaline hydrolysis, were used where
applicable. When possible, authentic standards were used to support
the identification. We reiterate that compound identification at

this point is only tentative. Additional workup is required to confirm compound identities. Element-selective chromatographic detector
(such as NPD, FPD) also proved very useful in providing additional
data for compound identification. Nitrogen and sulfur selective
detectors were most applicable to the pharmaceutical wastes because
most of the manufactured products contain one, if not both, of these
elements. For example, the chromatograms of extracts from I and II
show a strong response using FID (Figs. 6-7). Only a few characteristic peaks appear in the NPD chromatogram of I while extracts of II
were shown to contain no nitrogen compounds. Extracts of VIII and
VI, however, showed a very large response in the NPD as well as the
FID (Figs. 8-9). This response suggests the presence of nitrogen
(and/or phosphorus) compounds.

     I. Extracts of waste from I contained the bulk of the fatty
acids and fatty acid esters which were noted in the composite. The
most prevalent compounds in the basic/neutral fraction were a series
of butyl esters of fatty acids ranging from butyl laurate to butyl
stearate. The acidic fraction contained both saturated and unsaturated fatty acids (measured as methyl esters). Unsaturated fatty acid
esters - $C_{18:1}$, $C_{18:2}$, $C_{20:1}$ and $C_{20:2}$ were the most abundant
compounds identified.

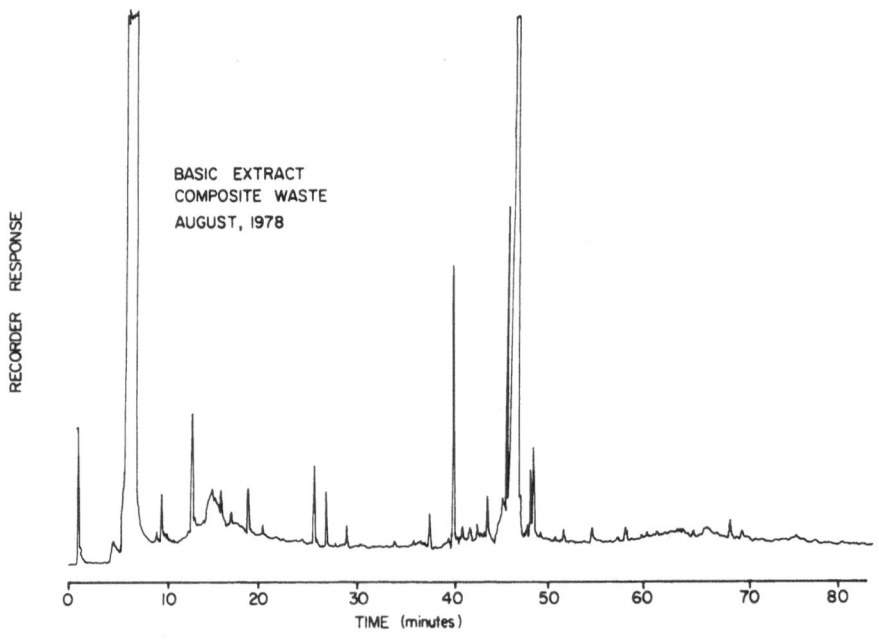

Fig. 4. FID chromatogram of basic extract of composite waste. Large
         peak at 7 minutes is dimethylaniline. (Conditions in text).

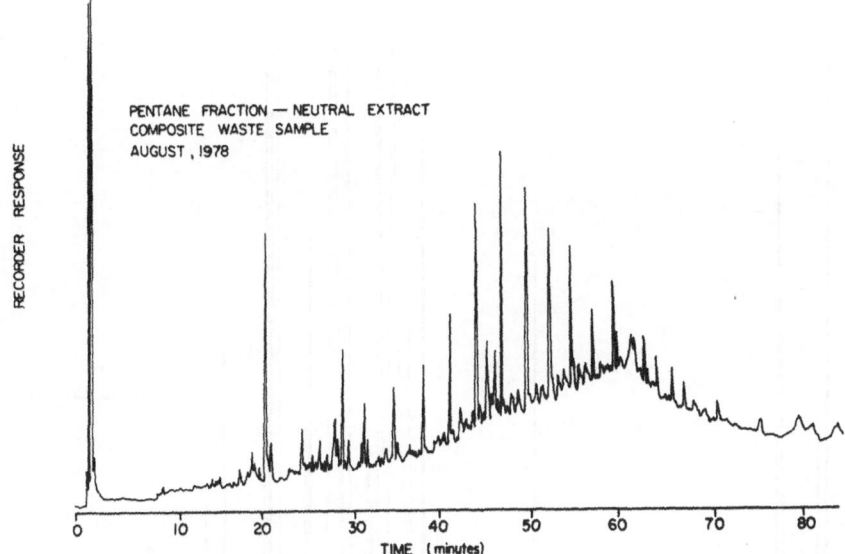

Fig. 5.   FID chromatogram of pentane eluate from silica gel column
chromatography of composite waste. (Conditions in text).

II.   There was one main component found in the extract of II,
methoxy-4-hydroxyphenylacetone. This compound is reported as a
starting product for the synthesis of L-methyl dopa. Additional
minor compounds in the extract were unidentified.

VI.   The main contribution of VI to the composite waste is
dimethylaniline. The primary isomer is N,N-dimethylaniline, though
it is likely that other isomers are present. Confirmation of N,N-di-
methylaniline was obtained by comparing an authentic standard using
GC/MS, GC/NPD, and NMR spectroscopy. Smaller quantities of methylaniline or toluamide were also identified in the waste. Bis-4,4'-
(dimethylamino)methane was identified by GC/MS using library spectra.

VIII.   Mass spectrometry, nitrogen selective detection and
flame photometric detection (FPD) (sulfur specific) were used to
determine the presence of ($\pm$)-2,3,5,6-tetrahydro-6-phenyl imidazol
[2,1-b]thiazole. This drug is the main product of the plant in Puerto
Rico. The drug is used primarily in veterinary medicine as an anti-
helminthic. This compound was not, however, detected in the composite
samples received.

IV, V.   Thus far no compounds have been identified in these
wastes. Extracts of waste from IV showed no peaks using FID, FPD,
NPD, and MS detectors. V contained a detectable peak in the GC/MS,
which has not yet been identified. Nitrogen-containing compounds

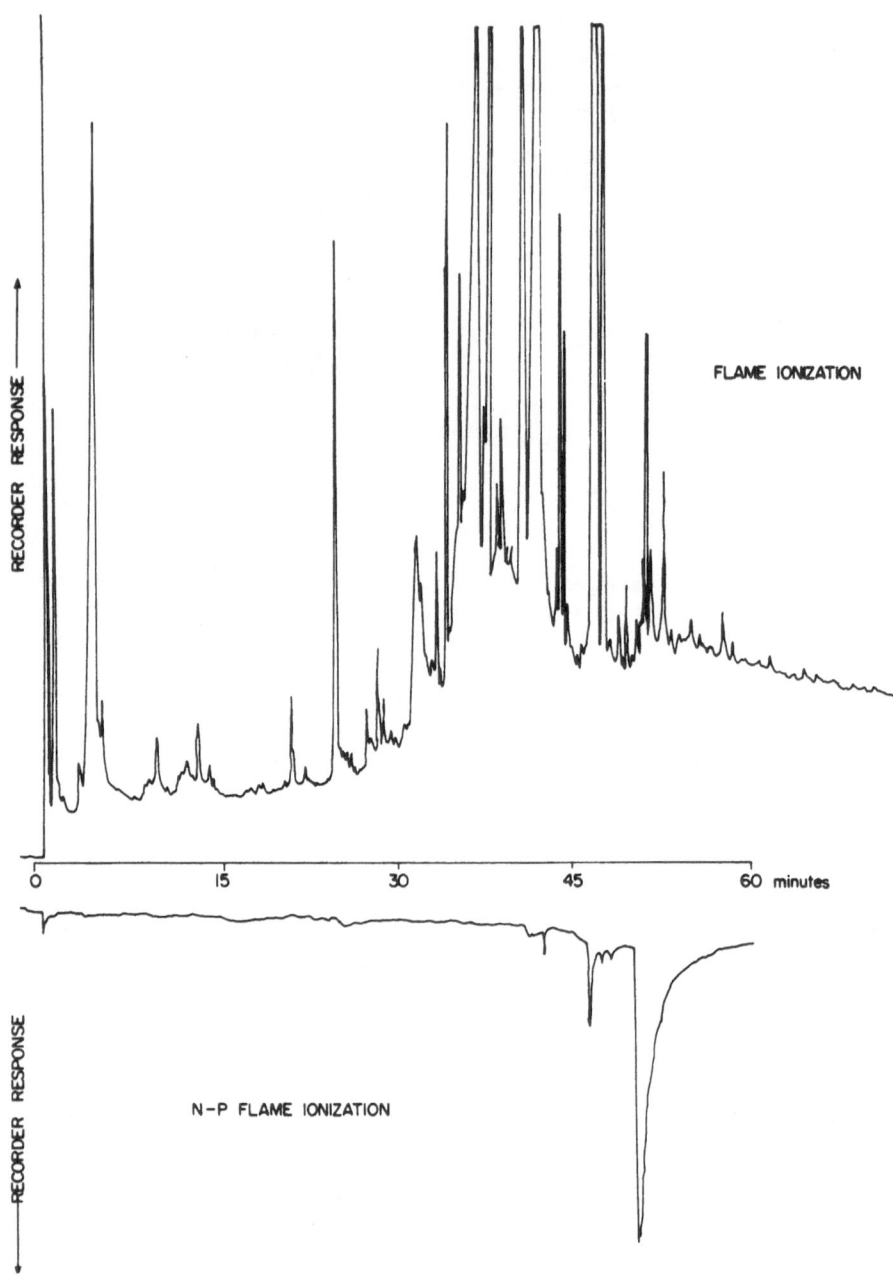

Fig. 6. FID (upper) and NPD (lower) chromatograms of base/
         neutral extract of waste from I. (Conditions in text).

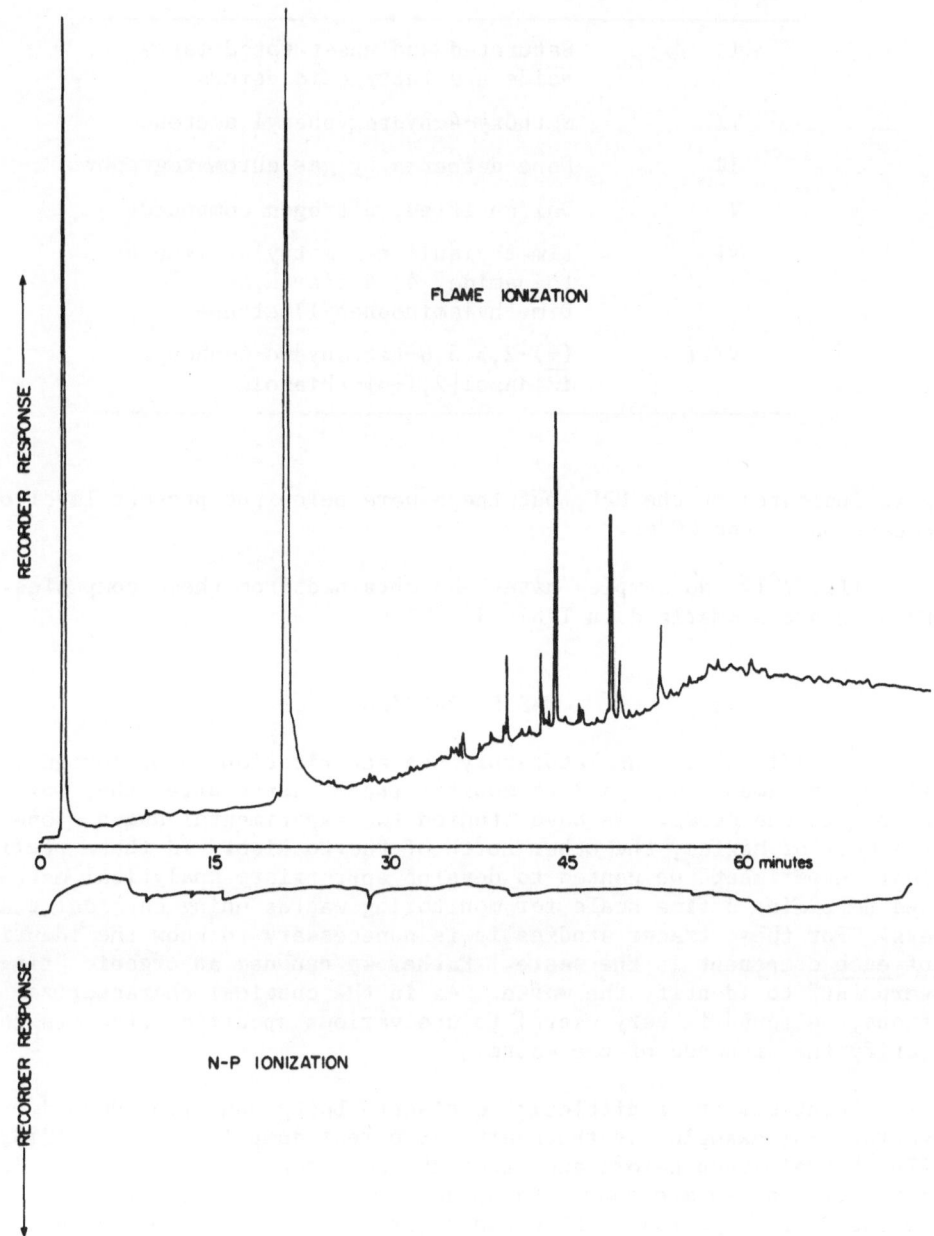

Fig. 7.   FID (upper) and NPD (lower) chromatograms of base/
          neutral extract from II waste. (Conditions in text).

Table 3. Contribution of individual companies to compounds in
          Puerto Rico waste.

I	Saturated and unsaturated fatty acids and fatty acid esters
II	Methoxy-4-hydroxyphenyl acetone
IV	None detected by gas chromatography
V	Unidentified, nitrogen compounds
VI	Dimethylaniline, methylaniline or toluamide, 4, 4-Bis-(N,N-Dimethylaminophenyl)methane
VIII	(+)-2,3,5,6-tetrahydro-6-phenyl imidazol[2,1-6]-thiazole

were indicated by the NPD, but these were below the present level of
detection of the GC/MS.

    III, VII. No samples have been obtained from these companies.
Results are summarized in Table 3.

                              WASTE TRACKING

    In addition to our laboratory characterization we performed
field experiments designed to monitor these wastes after they were
dumped in the ocean. We have studied two experimental dumps – one in
the Gulf of Mexico, the other north of Puerto Rico. In these preliminary experiments we wanted to develop appropriate analytical methods
and determine a time scale for monitoring wastes using chemical tracers. For these tracer studies it is unnecessary to know the identity
of each component in the waste. Rather we can use an organic "fin-
gerprint" to identify the waste. As in the chemical characteriza-
tions, we found it very useful to use various specific detectors to
verify the presence of the waste.

    Sometimes it is difficult to discern background from dumped
waste. For example, at the Puerto Rico test dump in February 1978,
FID chromatograms before and after the dump contain many peaks. In
fact, only a few are common to an actual sample of waste taken prior
to the dump (Fig. 10). NPD chromatograms, however, simplify the
interpretation considerably (Fig. 11).

Fig. 8. FID (upper) and NPD (lower) chromatograms of base/
         neutral extract from VI waste. (Conditions in text).

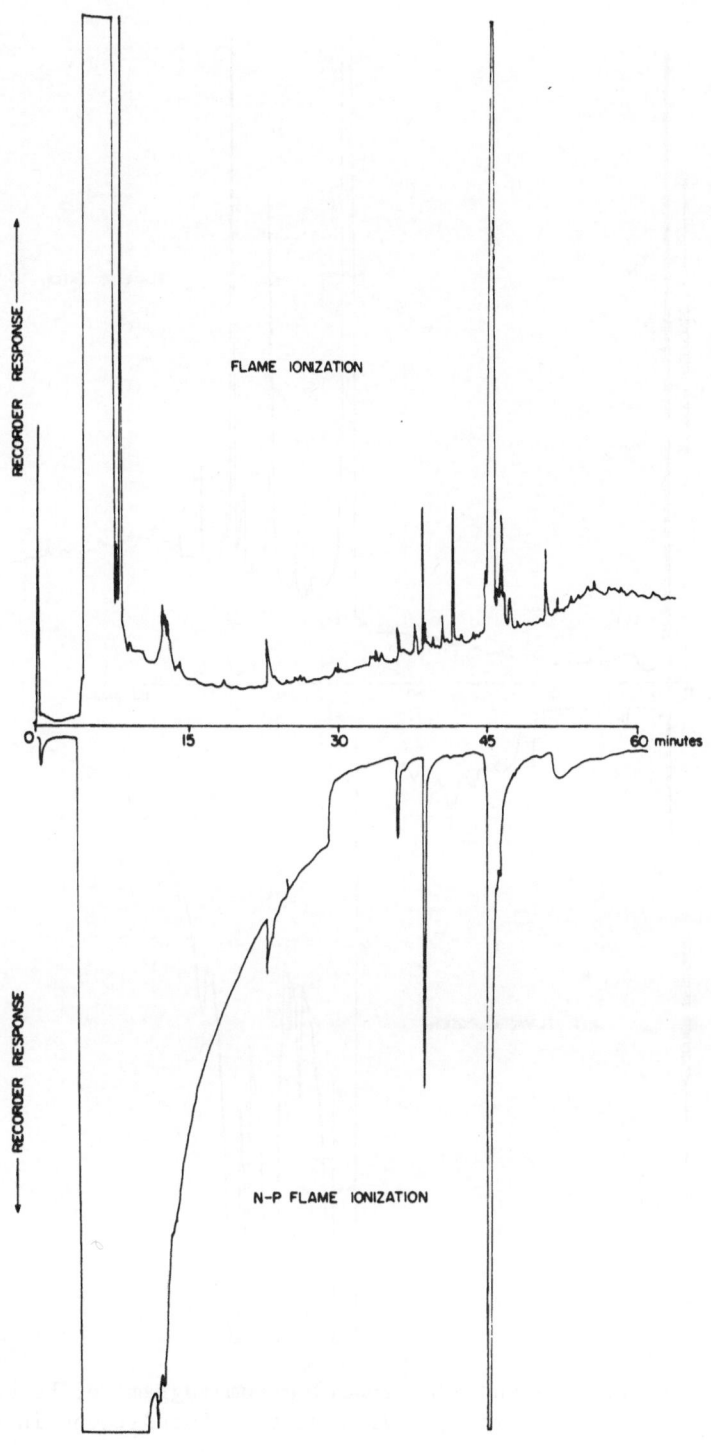

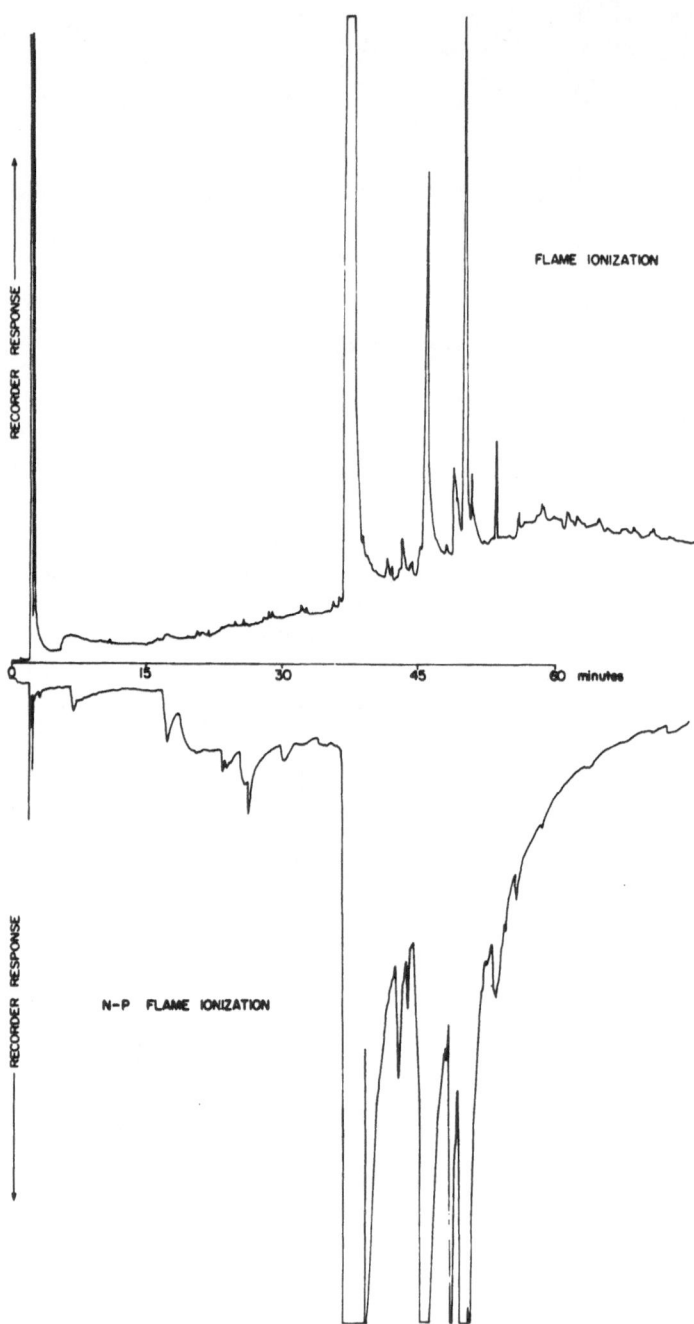

Fig. 9.   FID (upper) and NPD (lower) chromatograms of base/
          neutral extract from VIII waste. (Conditions in text).

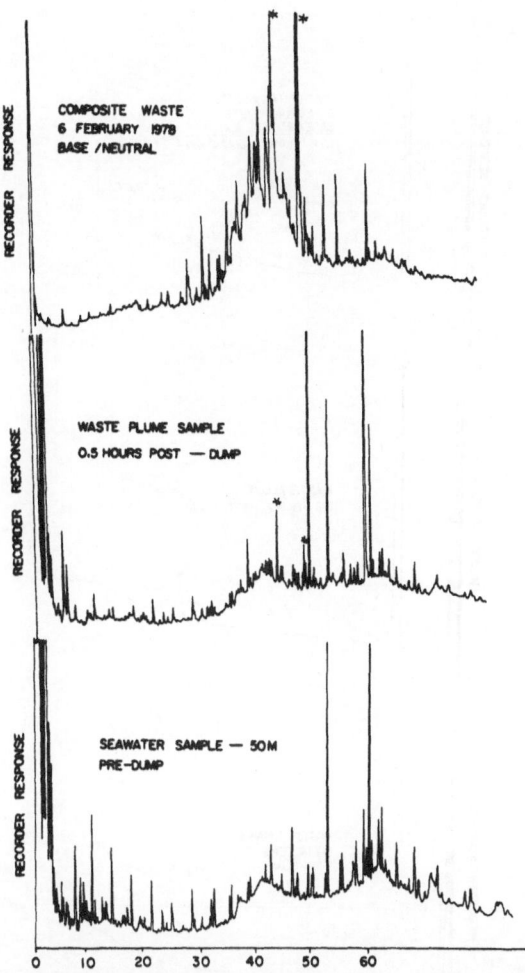

Fig. 10.   FID chromatograms of composite waste sample extract
(upper), waste plume sample (middle) and seawater sample prior to dumping (lower). (Conditions in text).
Peaks marked with * in the waste plume sample give a
point of reference to the actual waste. Note many
additional peaks.

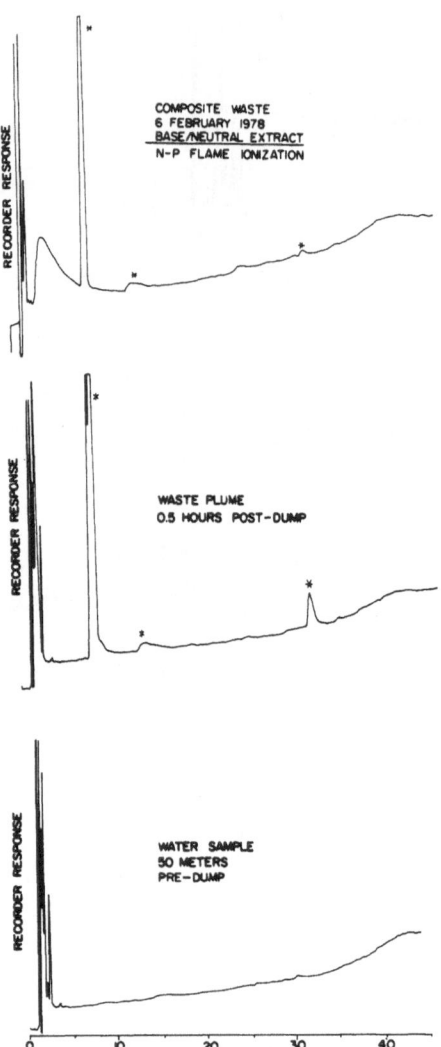

Fig. 11. NPD chromatograms of same samples as in Fig. 10.
          Note simpler identification of compounds in waste
          plume. (Conditions: 6' Apiezon–L column, 150–270°
          at 3°/min).

Fig. 12. ECD chromatograms of sludge and waste plume samples in Gulf
          of Mexico, July 1977. Left: "dissolved fraction 230°C;
          Right: "particulate fraction, 200°C. (other conditions: 3%
          OV-17, 6' glass column; 60 mls/min 95% Ar/5% CH$_2$.

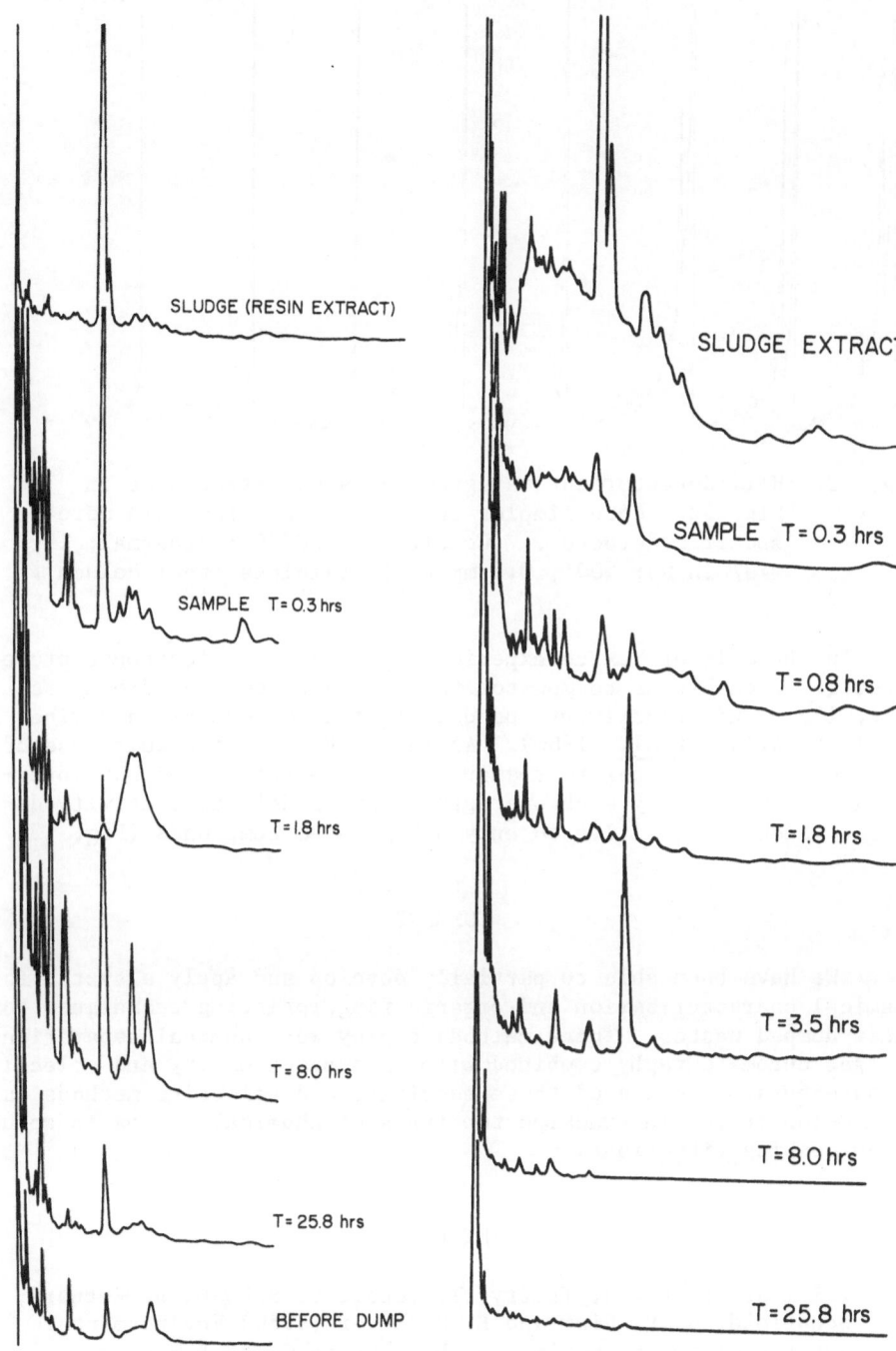

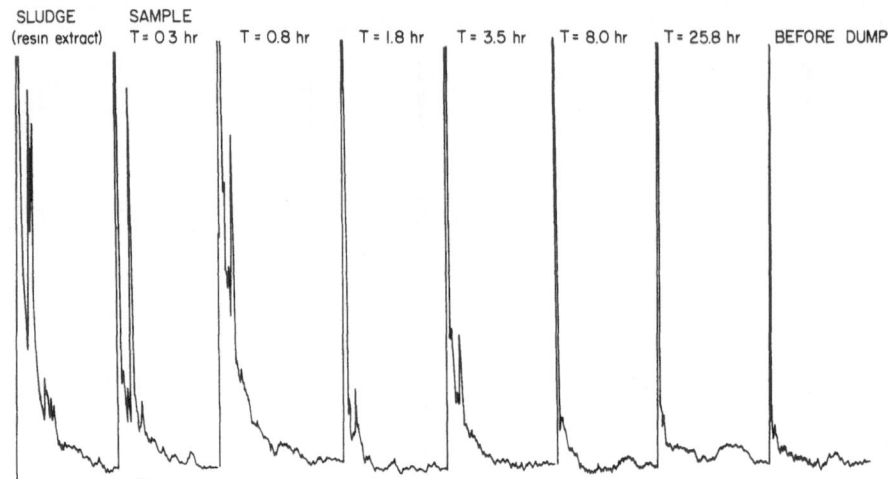

Fig. 13. Hall detector chromatograms of same extracts as in
          Fig. 12. Note simpler chromatograms using this more
          specific detector. (Conditions: 200°C isothermal; 15
          mls/min N_2; 500', 0.5 mm WCOT stainless steel column).

In the Gulf of Mexico experiment, we used an electron capture
detector to enable us to monitor the dumped waste over 3-8 hours
(Fig. 12). This experiment and descriptive methods are described in
detail by Atlas et al. (1980). As in the Puerto Rico dump, use of a
more selective detector to examine the waste simplified our interpretation as shown by a chromatogram using a Hall conductivity detector which is sensitive to only halogenated compounds (Fig. 13).

SUMMARY

We have been able to partially develop and apply a system for
chemical characterization and organic fingerprinting techniques for
ocean-dumped wastes. These methods employ wet chemical separations
and gas chromatography combined with mass spectrometry and selective
GC detectors. The use of these sensitive and selective methods has
been demonstrated in tracking two types of chemical wastes in actual
ocean dumping situations.

REFERENCES

Atlas, E., J. Brooks, J. Trefry, T. Sauer, C. Schwab, B. Bernard, J.
    Schofield, C. S. Giam and E. R. Meyer (1980) Environmental
    aspects of ocean dumping in the western Gulf of Mexico.
    Journal of the Water Pollution Control Federation, 52, 329-350.

Giam, C. S., E. Atlas and G. Martinez (1979) Final Report to NOAA
    Ocean Dumping, Contract #04-8-M01-55.
Giam, C. S., H. S. Chan and G. Neff (1976) Distribution of
    n-paraffins in selected marine benthic organisms. Bull. of
    Environ. Contam. and Toxicol., 16 (1): 37-43.
Schwab, C. R., T. C. Sauer, Jr., G. F. Hubbard, H. Abdel-Reheim, and
    J. M. Brooks (1981) Chemical and biological aspects of ocean
    dumping at the Puerto Rico Dumpsite. In: Ocean Dumping of
    Industrial Wastes, B. H. Ketchum, R. D. Kester, and P. K. Park,
    editors, Plenum Press, New York. This volume, pp. 247-273.

OCEAN DISPOSAL OF ORGANOCHLORINE WASTES

BY AT-SEA INCINERATION

Kenneth S. Kamlet

National Wildlife Foundation
1412 16th Street, N. W.
Washington, DC   20036

ABSTRACT

Among the industrial wastes with a history of ocean dumping in the U.S. and abroad, organochlorine (and other organohalogen) compounds are perhaps the most troublesome in terms of potential environmental impact. This paper describes the initial experience of the United States with at-sea incineration as a new method of ocean disposal for organochlorine wastes. It traces the development of a program for regulating such disposal under the Ocean Dumping Law, and it discusses the results of several at-sea incineration operations. The paper also briefly describes current efforts to develop international regulations and guidelines to regulate at-sea incineration throughout the world.

INTRODUCTION

As a legal term "ocean dumping" refers to the transportation of wastes (e.g., via ship, barge, or aircraft) for the purpose of disposing such wastes in ocean waters. The discharge of wastes through marine outfall pipes is not "ocean dumping" within the meaning of the Ocean Dumping Law (technically known as the Marine Protection, Research, and Sanctuaries Act), because outfalls have no transportation component. Pipe discharges are subject to regulation under another statute: the Federal Water Pollution Control Act -- also called the Clean Water Act.

Historically, most ocean dumping in the United States and elsewhere has taken the form of waste transport in self-propelled or

Table 1. Ocean dumping of industrial wastes, 1973-77 (all U. S.
          dumpsites). Source: U. S. Environmental Protection
          Agency (1978) Sixth Annual Report to Congress on Ocean
          Dumping in the United States, 53 pp.

Year	Quantity (tons)
1973	5,050,800
1974	4,579,700
1975	3,441,900
1976	2,733,500
1977	1,843,800

towed barges with the direct discharge of such wastes into the ocean.
In the U. S., the dumping of industrial wastes has been on the decline within recent years (Table 1). Indeed, with the exception of
the handful of industrial dumpers whose wastes may be able to satisfy
the fairly stringent requirements of the Environmental Protection
Agency's (EPA) Ocean Dumping Criteria, all industrial chemical dumping is required by regulation to be phased-out by the end of 1981.
If legislation presently pending in the Congress is approved, the
phase-out deadline will shortly become a statutory as well as a regulatory requirement.

    Consequently, although it may be useful to discuss the charac-
teristics and behavior of currently ocean-dumped industrial wastes
from the standpoint of expanding our basic understanding of the fate
and effects of such wastes in the marine environment, as a practical
matter such ocean dumping will shortly be of historical interest only
since, with minor exceptions, it will no longer occur off the coasts
of the United States.

    Accordingly, this paper will confine itself to the one form of
industrial chemical ocean "dumping" -- at-sea or ocean incineration
-- which can be expected to continue into the indefinite future, and
which as a consequence deserves study and evaluation for practical as
well as academic reasons.

                    AT-SEA INCINERATION AS "OCEAN DUMPING"

    At-sea incineration is now firmly established in the United
States as being subject to the requirements of the Ocean Dumping Law
and the Ocean Dumping Criteria. Internationally, protocols are in
the process of being drafted under the London Ocean Dumping Convention to cover ocean incineration. The manner by which at-sea incineration came to be regulated as ocean dumping is a matter of some interest.

The Shell Chemical Company facility at Deer Park, Texas, had a number of years prior to December 1973, barged certain organic chloride constituents (among other waste streams) to an ocean dumpsite in the Gulf of Mexico. With the advent of the Ocean Dumping Law (which became effective in April 1973), Shell was granted a brief temporary ocean dumping permit by EPA which expired in November 1973. Thereafter, EPA refused to renew the permit for Shell's organochlorine wastes, and Shell was forced to store its wastes in above-ground tanks at its manufacturing facility and its associated refinery. As of September 1974, Shell had accumulated more than 19,000 tons of this waste material and was continuing to generate the material at the rate of approximately 2,100 tons per month. The waste consisted of a mixture of chlorinated hydrocarbons resulting from the plant's production of glycerin, vinyl chloride, epichlorohydrin, and epoxy resins, with trichloropropane, trichloroethane, and dichloroethane predominating (Wastler et al., 1975).

The five individual Shell organic waste streams and the approximate proportion of each waste stream to the total waste are given in Table 2. Table 3 lists the major components of each waste stream (i.e., all components making up 10 percent or more of a given waste stream). These data are derived from unpublished analyses compiled by W. E. Roberts of the Shell Chemical Company for wastes covering the period January-March 1974. The analyses were performed in response to a request from the Regional Administrator of EPA Region VI (Dallas).

In July 1974, Shell contracted with Ocean Combustion Service BV (OCS), a subsidiary of Hansa Lines of the Netherlands, for the use of the motor vessel VULCANUS, which was designed for high-temperature at-sea incineration of organic chloride wastes from manufacturing operations in Europe. OCS complied with all applicable international

Table 2. Shell organic waste streams. (Source: Shell Submission to EPA Region VI, March 1974)

Process Stream Designation*	Approximate Proportion of Total Waste (%)
C Light Ends (volatile fraction)	15
C Heavy Ends	25
VCM Heavy Ends	25
VCM Tars	25
D-D® Flasher Bottoms	10

*C = "C" Section of the plant
VCM = Vinyl chloride monomer plant
D-D® = Soil fumigant (nematocide) section of the plant

safety regulations and modified the VULCANUS to comply with additional requirements imposed by the U. S. Coast Guard. By memorandum
dated January 23, 1974, EPA concluded that neither the Ocean Dumping
Law nor the Clean Air Act covered the proposed at-sea incineration
activity, apparently leaving Shell and OCS free to proceed without
further governmental regulation.

The EPA legal opinion was based on the lack of evidence in the

Table 3. Major constituents of Shell waste streams. (Source: Shell
          submission to EPA Region VI, March 1974)

Component	%w
C Light Ends	
2-Chloropropane	17
Ethyl Chloride	17
2-Chloropropene	17
1-Chloropropane	22
3-Chloro-1-propene	18
1,2-Dichloropropane	11
2,3-Dichloropropene	14
C Heavy Ends	
1,2,3-Trichloropropane	70
1,2-Dichloro-3-propanol	10
Tetrachloropropyl Ethers	14
VCM Heavy Ends	
1,2-Dichloroethane	15
1,1,2-Trichloroethane + 1,1,1,2-Tetrachloroethane	58
VCM Tars	
1,2-Dichloroethane	36
1,1,2-Trichloroethane + 1,1,1,2-Tetrachloroethane	15
$C_3-C_6Cl_x$	14
D-D® Flasher Bottoms	
1,2-Dichloropropane	17
cis-1,3-Dichloropropene	13
trans-1,3-Dichloropropene	15
Freon-Soluble Material	24
Freon-Insoluble Material	12

Ocean Dumping Law and its legislative history that "Congress [in-
tended] to deal with airborne pollutants in the ocean dumping act."
According to the memorandum, "if the product of high temperature incineration aboard ships constitutes ocean dumping, on the grounds
that those products eventually wind up in the marine environment,
then there would be a host of other, land-based activities which
would also necessarily constitute 'dumping'" --an argument which EPA
felt proved too much. As to the Clean Air Act, the EPA memorandum
concluded "that it has no applicability beyond the territorial jurisdiction of the United States" and that, in any event, "we do not believe that a ship can plausibly be described as a 'stationary source'
[of air pollution], even with respect to activities within the territorial sea."

     In the late summer of 1974, the National Wildlife Federation
(NWF) learned both that EPA had disavowed jurisdiction over at-sea
incineration and that the VULCANUS was already en route from Europe
to pick up and incinerate Shell's wastes. NWF responded on September 17, 1974, with a letter to EPA Administrator Russell E. Train,
urging EPA to affirm in writing its jurisdiction over at-sea incineration of wastes transported from the United States. NWF argued that
the VULCANUS operates and was designed to operate to assure maximum
mixing of incinerated wastes with sea water, and that as a legal matter, at-sea incineration is in fact "dumping" within the meaning of
the Ocean Dumping Law. NWF also alerted Representative John Dingell
(then Chairman of the House Subcommittee charged with oversight of
EPA's implementation of the Ocean Dumping Law and also a chief archi-
tect of the ocean dumping legislation) to the problem.

     Congressman Dingell likewise wrote Administrator Train urging
EPA to assert jurisdiction over "'indirect' dumpers-by-incineration
such as Shell." In support of his contention that the Ocean Dumping
Law applied, Chairman Dingell cited Section 101(a) of the Act which
bars the unpermitted "transportation from the United States" of "any
... material for the purpose of dumping it into ocean waters," and
Section 3(f) which defines "dumping" to mean "a disposition of material." He concluded that the Shell waste, "when carried out on a vessel for incineration at sea is certainly being transported 'for the
purpose of dumping it into ocean waters.' The fact that the particulates are first projected into the atmosphere before falling or being
rained down into the sea does not make the operation any less of a
disposition in ocean waters." He added that, "if EPA lets its position on ship-board incineration stand, the use of catapults could,
with equal justification, become the standard technique for waste
dumping at sea."

     On September 24, following a meeting by representatives of EPA,
NWF, and Shell, EPA acceded to the position taken by NWF and Chairman
Dingell. It was agreed that EPA would withdraw its January 23, 1974
Legal Memorandum and substitute a new one asserting jurisdiction over

ocean incineration of the VULCANUS sort. EPA further agreed to pre-
cede any permit to Shell with public notice, hearing and a new ocean
disposal site designation -- all to be carried out on an expedited
basis. It was also agreed that the disposal operation would be
closely monitored. Finally, NWF, in view of the responsible attitude
of EPA and Shell, agreed to support a carefully monitored ocean
incineration operation.

An unprecedented joint press release was issued on September 27,
1974, by EPA, Shell, and NWF announcing "a joint proposal to evaluate
a new alternative to the ocean dumping of organic chloride wastes --
incineration at sea." Public hearings on the Shell proposal were
scheduled for October 4th in Houston.

On October 3, 1974, EPA issued a revised legal memorandum on the
applicability of the Ocean Dumping Law to ocean incineration of
wastes. The memorandum concluded as follows: "In any case where it
can reasonably be anticipated that the incineration of wastes at sea
will result in any of such material, or the emissions of the incineration of such material, entering ocean waters, such incineration will
constitute a disposition of material in ocean waters subject to the
provisions of the Act and the Convention and, accordingly, is prohibited in the absence of an appropriate permit issued under the Act and
the Convention."

The memorandum stated that the previous EPA position on this
issue was being modified "because of [EPA's] concern that the failure
to regulate recently developed waste disposal techniques involving
ocean incineration would frustrate the purposes of the Act and Convention." The earlier memorandum was said to have been based on the
concern that regulation of ocean waste incineration would require
regulation of land-based incineration activities as well. The October opinion disagreed, stating that "[s]ince the Act regulates transportation of materials for the purpose of disposal such a conclusion
is not required" given that "[t]he ocean fall-out of pollutants from
the atmosphere attributable to land-based sources does not constitute
'transportation' within the meaning of the Act."

Such were the tortuous beginnings of governmental regulation of
a waste disposal technique which gives every indication of increasingly being used by industrial facilities plagued with waste disposal
problems of highly toxic substances such as organochlorine wastes.

THE FIRST SHELL OPERATION

Following the October 4 hearing in Houston, the EPA hearing
panel, consisting of 5 scientists and chaired by an EPA pesticides
lawyer, issued (on October 9) its report to the Administrator. The
report recommended that EPA issue Shell a "research permit" for the

initial incineration of 4,200 metric tons of organochlorine wastes
(one-quarter of the 16,800 tons proposed to be incinerated by Shell
and representing one shipload) at a new dumpsite in the Gulf of Mexico approximately 130 miles from land, with permission to incinerate
the remaining 12,600 metric tons of material contingent upon the successful completion of the initial test "burn." The major constituents of the mixed waste proposed for incineration are tabulated in
Table 4. The initial burn would require "detailed monitoring" and
would have to display a combustion efficiency of greater than 99.9
percent. In addition, further at-sea incineration would be condi-
tional on a showing that the "pH level in the waters in the immediate vicinity of the VULCANUS does not drop by more than 0.5 pH unit"
(a principal combustion product of organochlorines is hydrogen chloride gas -- a strong acid when mixed with water), and upon the failure to detect any "significant effects upon the marine environment."
A maximum feed rate was later set of first 20 and then 25 metric tons
per hour and the ship was required to be operated to ensure an effective wind velocity over the incinerator stacks of 10 knots.

On October 10 the EPA Administrator announced his decision on
the Shell proposal, noting the novelty and importance of the issue as
raising "for the first time in the United States the question of the
environmental acceptability of incinerating complex and otherwise
dangerous chemicals on the high seas." In adopting the findings and
recommendations of the hearing panel, the Administrator emphasized
four facts: "First, extremely efficient incineration at sea of organochlorine wastes may well be the least environmentally offensive
means of disposing of these troublesome materials. Second, there was

Table 4. Major constituents of Shell mixed waste used in the
          October 20-28, 1974 test burn aboard the VULCANUS.
          (Source: EPA permit)

Chemical Characteristics	Concentrations (wt/wt)
Constituents	Not to Exceed
Trichloropropane	700,000 ppm
Tetrachloropropyl Ether	120,000 ppm
Dichloroethane	290,000 ppm
Trichloroethane	280,000 ppm
Dichlorobutene & Heavier	350,000 ppm
Dichloropropene & Lighter	380,000 ppm
Alkyl Chloride	50,000 ppm
Dichlorohydrin	120,000 ppm
Metals	
Cadmium	0.04 ppm
Mercury	0.01 ppm

little substantive opposition voiced at the public hearing to a properly supervised research incineration.... Third, the European experiences with ocean incineration [for two years, the vessel had incinerated similar wastes in the North Sea for companies in the Netherlands, Great Britain, and Scandinavia] provide us with significant
information about this method of disposal, but more information
should be obtained and a research 'burn' is the most desirable means
of obtaining this data. And finally, I am continually mindful of the
responsibility Congress has bestowed upon this agency to use all
caution in allowing an ocean discharge of wastes such as Shell pro-
poses to incinerate, however small or in whatever form."

The Administrator added that "the monitoring by the Permittees,
EPA and other interested governmental agencies should be exhaustive
and subject to review by government experts and interested parties
before the additional 12,600 metric tons of waste are incinerated."
He directed EPA's Office of Research and Development "to use its extensive scientific expertise in a far-reaching monitoring of the research incineration."

In a separate action, a new site designation for the Shell burn
was docketed at the Federal Register office.

The incineration began on October 20, and ended on October 28.
Several difficulties arose. A dispute developed between EPA and
Shell over feed rates and burn temperatures, and at one point EPA
considered calling the VULCANUS back to port. The permit specified
that no incineration was to take place at less than 1200°C, and that
the permittee was to "maintain a minimum average combustion temperature of 1400°C (a running four-hour average) ...." Apparently one
problem involved a difference of opinion as to whether the 1200°C
minimum was to be based on a "wall" temperature (in the incinerator
stack) or on a "flame" temperature. This problem was ultimately resolved with Shell agreeing to measure "wall" temperatures. The State
of Louisiana had developed concern about the impact of the HCl plume
on migratory waterfowl crossing the Gulf of Mexico and was threaten-
ing to go to court for an injunction to halt the burn.

Technical problems were experienced in monitoring the burn, including problems with visualizing the HCl plume, a broken crystal in
an HCl detector to be used on a spotter plane, the breakdown of a
condensation nuclei detector, and the blow-out of a piston on the EPA
surveillance plane as it was taking off to do some infrared scans.
There were also leaks in a storage tank valve, a shut-off of water to
a water-cooled probe resulting in a burn-up of the probe and its
associated thermocouple, the burn-up of the standby thermocouple 18
hours later, a plug-up of the waste feed line for a day and then
again for 4-1/2 days, destruction of the backup probe by corrosion,
and the development of leaks in the sample line near the gas analyzer.

The NOAA research vessel, OREGON II also had its problems. The cruise was hampered by inexperience on the part of its scientific crew, by delays occasioned by feedline plugging on the VULCANUS, and by inadequate ship-to-ship communications which resulted in many wasted hours of chasing the wrong radar blip. The second of the two OREGON II cruises during the first burn was hampered by last-minute planning, difficulties in assembling equipment and reagents, by the need to make measurements in the dark and under rough sea conditions, and by a dispute between EPA and NASA personnel over the use of the adenosine triphosphate (ATP) data.

In addition, the VULCANUS operation was impaired in some cases by communication difficulties between the ship's crew and EPA and Shell scientific personnel.

Last but not least were the delays, mixups, and oversights that resulted from non-productive conflicts within EPA --involving the Water and Hazardous Materials Office, the Office of General Counsel, the Office of Research and Development, and the regional staff in Dallas.

Technical meetings for evaluating the results of the Shell test burn were subsequently convened on November 7, in Washington, D. C., and on November 14-15, in Houston, Texas.

Although some dissatisfaction was expressed at these meetings regarding the adequacy of the data gathered in monitoring, EPA nevertheless concluded "that the conditions and criteria of the initial research permit had been met, and that no information gathered in Research Burn I in any way changed or called into question the findings and conclusions of the original hearing panel" (Wastler et al., 1975). Because of acknowledged shortcomings in the monitoring efforts, EPA decided to authorize Shell to conduct another test burn under a second "research" permit. Further monitoring and evaluation would determine whether Shell would be permitted to incinerate the remaining two shiploads of organochlorine wastes.

A "Supplementary Decision" of the EPA Administrator, issued November 27 (and relying on a hearing officer's report of the same date on the November 14 technical meeting), concluded, in view of the criticisms levelled at the monitoring of the first burn, that "the wisest course is to limit the additional authorized incineration to one shipload of 4200 metric tons pursuant to a research permit and that thorough monitoring should be carried out again, with measures taken to avoid the problems which may have hampered the monitoring during the first incineration."

Commenting on the propriety of the research permit approach for what in fact were waste disposal operations, the Administrator stated the Agency's perspective as follows: "We are looking at the general

problem of the disposal of chemical wastes. Ocean dumping is only
one element in a larger interrelated problem. This is shown by the
fact that the availability of alternative means of disposal is one
element in determining whether any particular ocean dumping should be
allowed. We envision ocean incineration as an alternative to other
disposal methods such as direct ocean dumping, land disposal or land
incineration. Yet, while having obvious possibilities, such incineration is a method about which we need more knowledge regarding its
applications and effects. The research aspect of the permit is not
limited to detecting the effects or determining the efficacy of the
monitoring efforts. It also encompasses the expansion of our knowledge regarding the ability of such ships to effectively dispose of
the wastes in the manner represented by the ship owners. Only then
will it be possible to determine if it is in fact a viable alternative. Of course, our desire for information on the performance of
the vessel cannot be satisfied at the expense of the environment and
the incineration will not be allowed if it does cause harm."

A second research permit was granted on November 27 and the second research burn took place on December 2-9, 1974. On December 10,
representatives of EPA and the Gulf Coast states met to consider the
results. "Their unanimous conclusion was that incineration by the
VULCANUS of Shell's remaining 8,400 MT of organochlorine wastes, under the conditions imposed by EPA in the two research permits, was an
environmentally compatible means of disposing of the wastes" (Wastler
et al., 1975).

On December 12, EPA issued a permit to Shell for incineration of
the remaining wastes. These wastes were then incinerated in two
loads, on December 19-26 and December 31, 1974 – January 4, 1975. In
July 1975 EPA's Office of Water and Hazardous Materials issued a report (Wastler et al., 1975) on the monitoring results derived from
the two Shell test burns. The major conclusions of this report are
briefly summarized below.

In the first research burn, the efficiency of incineration was
calculated in two ways: as the overall efficiency of combustion and
as the degree of oxidation of organochlorides. In the second research burn, combustion was considered complete if stack emissions
contained less than 1,000 ppm of carbon monoxide, 3 to 10 percent
oxygen, and less than 10 ppm of organochlorine compounds. For the
first burn, combustion efficiencies ranged from 99.92 to 99.98%
(99.95% average); for the second burn, destruction efficiencies
(based on the ratio of organochloride atoms in the stack gas to those
in the feed) ranged from 99.987 to greater than 99.998%. For the
second burn, carbon monoxide levels were 25 to 75 ppm; oxygen levels
were 9.0 to 12.5%.

During the first burn, the waste plume was monitored 6 meters
above the sea surface by the OREGON II using a Geomet hydrogen

chloride monitor supplied and operated by NASA. Sampling was confined mainly to a 90-degree arc downwind of the VULCANUS beginning a few
hundred meters behind the ship and extending about 3 nautical miles.
The HCl detector uses a chemiluminescent reaction to monitor ambient
air concentrations of HCl. It has a minimum detection limit of 10
ppb. The Geomet instrument provides a continuous read-out on a strip
chart recorder which was marked at 5-minute intervals simultaneously
with navigational readings from the bridge. During the first cruise,
winds were from the east generally at a speed of 8 to 10 knots, while
during the second cruise of Burn I they were from the east-southeast
at speeds of 17 to 21 knots. In both cases, the plume was found directly downwind of the VULCANUS at distances apparently related to
wind speed when the VULCANUS was drifting; with the VULCANUS underway, the plume was found downwind at the resultant of the vectors of
wind velocity and vessel movement. The maximum HCl concentrations
measured were found during the second cruise and ranged from 0.01 to
7 ppm -- the highest at 0.4 n. mi. from the ship.

During the second burn, a twin Turbo-Beach EPA aircraft made
crosswind and axial passes through the plume during the first three
days of incineration. The plane was equipped with a coulometer and a
chemiluminescent analyzer (detection limit about 0.01 ppm) to measure
HCl. An Environment One Corporation condensation nuclei monitor was
used to track the plume. The data collected showed that the top of
the airborne plume trailed back from the VULCANUS stack at an angle
of about 20 degrees above the horizontal, reached a maximum altitude
of 850 meters, and fanned out horizontally to a width of about 1,200
meters at a distance of 2,400 meters downwind from the stack. The
maximum HCl concentration measured in the VULCANUS plume was 3 ppm
the first and third day and 1.8 ppm the second day. All three maxima
were encountered at about the same relative position each day -- 100
to 240 meters in altitude, and between zero and 400 meters downwind.
Grab samples collected in the plume for later analysis indicated that
the samples were low in pollutants.

Measurements of pH and chlorinity of seawater by the OREGON II
where the plume touched down, revealed a maximum pH depression of
0.15 unit and an increase in chlorinity of about 500 ppm (compared to
the chlorinity of seawater which is about 20,000 ppm). Additional
measurements by the research vessel ORCA showed no differences between fallout and control areas.

Water samples collected by the OREGON II and the ORCA were analyzed for organochlorides. Results in all cases were below detection
limits (i.e., below 0.5 ppb for the former and 25 ppb for the latter).

Eight heavy metals were also tested for; again no significant
variations were found.

A sampling grid of 16 stations was also laid out to include the area covered by the plume during the last 24 hours of the first burn. This area was also downwind and downcurrent of the "dump" site. No significant changes in pH, chlorinity, organochlorides, trace metals, phytoplankton, chlorophyll-a or ATP could be detected, suggesting the absence of cumulative impacts from the burn. Wastler et al. (1975) noted, however, that there was very little life in the dump site and that, it was therefore possible that effects might occur in more populated areas.

The first U. S. use of at-sea incineration for the disposal of organochlorine wastes was, therefore, rated a success and an environmentally acceptable practice when closely monitored and regulated.

AT-SEA INCINERATION OF AGENT ORANGE HERBICIDE

The second U. S. use of at-sea incineration involved the disposition of stockpiled "Agent Orange" herbicide, or "orange herbicide". Agent Orange is a herbicide which consists of approximately equal amounts of the phenoxy herbicides 2,4,5-T (the n-butyl ester of 2,4,5-trichlorophenoxyacetic acid) and 2,4-D (the n-butyl ester of 2,4-dichlorophenoxyacetic acid), along with trace amounts (i.e., about 2 mg/kg) of the highly toxic contaminant TCDD or "dioxin" (2,3,7,8-tetrachlorodibenzo-p-dioxin) -- all of which are organochlorines. In April 1970, the Secretaries of Agriculture, HEW, and the Interior jointly announced the suspension of certain uses of 2,4,5-T. The Department of Defense subsequently suspended the use of Agent Orange which had been used as a defoliant in Vietnam. At the time of this suspension, the Air Force had an inventory of 1.4 million gallons of Orange in South Vietnam and an additional 0.86 million gallons in Gulfport, Mississippi. The Vietnam stores were moved to Johnston Island in the Central Pacific Ocean for storage in April 1972.

Initially, the Air Force proposed to dispose of the Orange by incineration at a commercial facility in the United States. Because of intense public concern, the Air Force embarked on a more extensive study of incineration as well as alternative disposal methods (Air Force, 1972a, 1974b). The Environmental Impact Statement (EIS) concluded that "Orange destruction efficiencies of 99.9 percent or better appear[ed] feasible for a large scale incineration project," and proposed incineration of the Orange "upon the open tropical sea west of Johnston Island on a specially designed vessel" -- an approach which, according to the EIS, could be shown by a "dispersion zone model" to produce "no significant environmental impact upon either the air or ocean environment" (Air Force, 1974b). Alternatively, should the EPA Administrator decide "not to issue a permit for incineration at sea, the Air Force [would] pursue the principal alternative of incineration in [a] facility that would be constructed on Johnston Island" (Air Force, 1974b).

The fact that Agent Orange was a chemical warfare agent complicated matters since the Ocean Dumping Law expressly prohibits the
ocean dumping (and the transportation for the purpose of dumping) of
chemical, biological, and radiological warfare agents. Ultimately
the ocean incineration approach had to be justified on the basis that
it was not the warfare agent itself, but its combustion products that
were being transported for "dumping" and would in fact be ocean-
"dumped."

The Air Force applied to EPA on January 9, 1975, for a permit to
ocean-incinerate 2.3 million gallons of Herbicide Orange. Following
a public meeting in February 1975, and public hearings on April 25,
and April 28, 1975, the Air Force was requested to explore the feasibility of reprocessing Herbicide Orange before final action would be
taken on the application. On October 12, 1976, the Air Force filed
an Amendment to the Final Environmental Statement on the Disposition
of Orange Herbicide by Incineration (Air Force, 1976), which proposed
to reprocess the herbicide via charcoal adsorption to remove the dioxin contaminant so that the herbicide would meet registration and
use requirements of EPA and could be returned to the marketplace. A
reclamation technique using coconut charcoal adsorption seemed promising and was found to be technically feasible; however, the Air
Force was unable to find a technique for destroying the 1000 car-
tridges containing 1.3 million pounds of carbon to which were adsorbed up to 50 pounds of highly toxic TCDD. Thus, this approach was rejected and the Air Force requested EPA to reconvene the hearing on
ocean incineration.

A research permit was issued by the EPA Administrator on April
25, 1977, authorizing the incineration of up to one shipload (4,300
metric tons) of Herbicide Orange stocks located at Gulfport, Mississippi. Some 3,520 metric tons were thereupon incinerated in the Pacific Ocean during the period of July 14-24, 1977, at a designated
site 120 miles west of Johnston Atoll.

After completion of the heavily monitored test burn, a Preliminary Report on the monitoring results was issued by EPA in August
1977 (Wyer, 1977), which recommended that the Air Force be authorized
to burn the remaining Orange material. A permit authorizing the at-
sea incineration of the 8,700 metric tons of remaining material (2
shiploads) was issued on August 4. The second burn was carried out
from August 6-16, 1977; followed by a third burn on August 23 - September 3, 1977. This completed the at-sea destruction of the stockpiled Orange.

A final report on the Orange incineration operation (prepared by
TRW, Inc., under contract to the Air Force) was published by EPA in
April 1978 (Ackerman et al., 1978). The principal conclusions of
this report can be summarized as follows:
    1. The average incineration rate was 14.5 metric tons per hour.

For all three burns, the average flame temperature was 1500°C as determined by daily optical pyrometer measurements. The average incinerator wall temperature was 1273°C for all three burns (as measured by controller thermocouple). The average calculated incinerator
residence time was 1.0 second.

2. Average overall combustion efficiency for the three burns was
99.990% (based on CO and CO_2 levels). Destruction efficiencies averaged greater than 99.999% based upon both Herbicide Orange and
chlorinated hydrocarbon destruction, greater than 99.93% based on
TCDD destruction (chemical interferences during the analyses may account for the somewhat lower number), and 99.985% based on total hydrocarbon destruction.

3. Biological monitoring was limited to the collection of plankton samples in the burn site before and after the first burn. No
consistent differences between pre-burn and post-burn tows could be
found.

4. The VULCANUS was equipped with a sealed "black box" display-
ing indicator thermocouple readings for each incinerator; day, month,
and time; status (on/off) of the waste pumps; and the vessel's position by the Decca Navigator (at least in Europe, where the Decca Navigator can receive suitable land-based signals). An 8 mm movie camera within the black box photographed the panel every 15 minutes during incinerator operations. Films were turned over to and retained
by EPA.

5. Miscellaneous (mostly minor) problems were experienced, especially during the first burn. These included loose fittings, radio
interference with onboard gas chromatograph, electrical interference
by solenoid valves on the gas conditioners with on-line instrumentation recorders, frequent partial plugs in the heat traced lines due
to condensate and particulate material shed by the incinerators, corrosion of stainless steel fittings due to corrosiveness of combustion
effluent, and condensate damage to one of the CO analyzers. Several
flameouts occurred during the first burn due to water floating on the
top of the herbicide in certain tanks.

The operation was rated an overall success, the permit conditions were deemed to have been completely satisfied, and at-sea incineration was regarded as having again been demonstrated to be an
environmentally safe disposal method for organochlorines.

THE SECOND SHELL OPERATION AND OTHER DOMESTIC DEVELOPMENTS

During the period from March - April 1977 Shell conducted a second at-sea incineration operation in the Gulf of Mexico at the same
site as the first series of burns. Four shiploads of organochlorine
wastes (similar in composition to those burned the first time),
amounting to some 16,000 metric tons were burned during this operation, with heavy monitoring taking place during the first burn of
some 4,100 tons. The results of this monitoring have been reported

by EPA (Clausen et al., 1977). In general, they were quite similar to those associated with the first operation. In addition, further biological studies were carried out, as yet unpublished, described in a preliminary report to EPA by the TerEco Corporation, College Station, Texas. Some of the tentative results of these studies are of interest since they provide the first evidence of measurable biological impacts of at-sea incineration on marine organisms.

Substantial numbers of Fundulus grandis were transported to the Gulf incineration site and exposed within "biotal ocean monitors" to the VULCANUS plume while the Shell waste was being burned. The frozen livers were then assayed for the activities of three enzymes: Catalase, ATP-ase, and Cytochrome P-450. Of these, only the P-450 showed a significant response (showing a nearly 3-fold increase in activity relative to controls), although catalase did show some depression. The report noted, however, that exposed fish that were returned live and acclimated for a few days in the laboratory before being tested had depurated and showed control levels of all three enzymes. The report emphasized the importance of selected metabolic enzymes as early warning signals of "untoward responses of animals to chemical pollutants in the water column," and concluded that the Shell PVC-derived organochlorine waste "generates a definite stress within the organism [i.e., Fundulus] either at high concentrations for short periods or low concentrations for long periods." The authors nevertheless concluded that "incineration of organochlorine wastes has only temporary effects on the marine encironment...," and that one would expect a potentially serious problem to exist only "with benthic animals that are exposed to toxicants in the sediments for prolonged periods or with pelagic animals that are exposed to repeated injections of toxicants into the water column or with pelagic animals that must drift with a polluted water mass that maintains its integrity for prolonged periods" -- conditions not likely to be associated with sporadic, short-lived, at-sea incineration operations.

These results, nevertheless, highlight the need for caution and for further research into possible impacts of ocean incineration on aquatic organisms -- particularly, if at-sea incineration becomes a more widespread and frequent practice in the future.

A few other developments of domestic interest should be noted.

First, a Final Environmental Impact Statement on "Designation of a Site in the Gulf of Mexico for Incineration of Chemical Wastes" was published by EPA in July 1976 (EPA, 1976). This document should be read by those interested in the considerations which led to the selection of the incineration site utilized in the two Shell operations.

Second, the Maritime Administration within the U.S. Department of Commerce, in July 1976 published a Final Environmental Impact

Statement on "Maritime Administration Chemical Waste Incinerator
Ship Project" (MarAd, 1976). This EIS discusses proposals being considered by MarAd to support "the development of a U.S. capability to
incinerate toxic chemical wastes at sea" by one of three mechanisms:
(1) the sale of one or more small dry cargo vessels from the National
Defense Reserve Fleet (NDRF) for conversion to chemical waste incinerator ships; (2) the granting of government Title XI mortgate loan
guarantees for construction of one or more chemical waste incinerator
ships; and (3) the sale of one or more NDRF small dry cargo vessels
with the granting of government Title XI mortgage loan guarantees for
conversion to chemical waste incinerator ships. The EIS indicates
that "[a]s many as four incinerator vessels may eventually be needed
nation-wide." This illustrates the potential for future expansion of
the use of at-sea incineration by U.S. producers of hazardous wastes.

Third, the current Ocean Dumping Criteria, issued in January
1977, specify (Section 220.3(f)) that "[p]ermits for incineration of
wastes at sea will be issued only as research permits or as interim
permits until specific criteria to regulate this type of disposal are
promulgated, except in those cases where studies on the waste, the
incineration method and vessel, and the site have been conducted and
site has been designated for incineration at sea in accordance with
the procedures of § 228.4(b). In all other cases the requirements of
Parts 220-228 apply." Efforts to promulgate "specific criteria" for
ocean incineration as part of EPA's domestic regulations have been
placed in abeyance pending the completion of ongoing efforts to establish international ocean incinerations guidelines under the London
Dumping Convention. (The first Shell operation proceeded under two
research permits and one interim permit; the Agent Orange operation
proceeded under one research and one special permit; and the second
Shell operation proceeded under a special permit.)

Finally, EPA and the Department of State jointly published Draft
and Final Environmental Impact Statements "For the Incineration of
Wastes at Sea Under the 1972 Ocean Dumping Convention," in October
1978, and February 1979, respectively. These impact statements discussed the proposed acceptance by the U.S. of international regulations (under the Ocean Dumping Convention) governing at-sea incineration.

INTERNATIONAL DEVELOPMENTS

As previously noted, at the time the first Shell operation began
in the United States, at-sea incineration was already an accepted (if
not widespread) practice in Europe. In addition to the VULCANUS, a
series of vessels in the "Matthias" series --built and operated by a
competitor firm -- has been active in European ocean incineration activities. Clearly, the technique is one of potential world-wide
applicability.

This fact has been recognized by the parties to the London Ocean Dumping Convention. At the First Consultative Meeting of the Contracting Parties to the London Convention (held in September 1976), it was agreed that international actions to develop procedures to control incineration should be undertaken without delay. A resolution was adopted requesting that the Secretary General of IMCO, in concert with experts from individual nations, should undertake to examine the provisions of the Convention which are applicable to at-sea incineration; consider and draft any special provisions necessary to prevent marine and atmospheric pollution from at-sea incineration; and submit the special provisions to the next consultative meeting (Sept. 1977).

The U. S. agreed to participate in this work and convened a Working Group of the Advisory Committee on Ocean Dumping to prepare appropriate documents on specific regulatory and technical provisions relating to ocean incineration. This was done.

An intersessional "Consultation on Incineration at Sea" took place in London from March 21-25, 1977. This meeting generated draft technical guidelines for the parties to review in preparation for the second consultative meeting.

At the second Consultative meeting an ad hoc working group was convened to refine the technical guidelines prepared during the intersessional meeting. At the urging of the U. S. Delegation it was decided to pursue the subject both intersessionally and at the Third Consultative Meeting, with a view to the development of an amendment or protocol to the Convention on this subject which would make the present guidelines an integral part of the legal obligations accepted by a nation in ratifying or acceding to the Convention.

A meeting of the Joint Ad Hoc Group on Incineration at Sea was held at IMCO Headquarters, London from June 21-23, 1978. This Group adopted several amendments to the Technical Guidelines and recommended to the Third Consultative Meeting the adoption of the Technical Guidelines as amended. The U. S. delegation, in cooperation with the Canadian delegation, also prepared draft regulations and guidelines for consideration at the Third Consultative Meeting which was held October 9-13, 1978.

The Third Consultative Meeting subsequently adopted amendments to the Convention concerning incineration at sea. These amendments took the form of: (1) a new Paragraph 10 to Annex I (which exempts from the prohibition against direct dumping as other than "trace contaminants," organohalogens, mercury and cadmium compounds, oils, and persistent floatable synthetic materials, where they are incinerated at sea under a "special permit" in accordance with specified regulations and technical guidelines); (2) a new Paragraph E to Annex II (which allows at-sea incineration of Annex II substances and

materials under a "special permit," again in accordance with specified regulations and technical guidelines); and (3) an Addendum to
Annex I (specifying mandatory regulations for the control of at-sea
incineration).

Because of lack of time, the technical guidelines were not re-
drafted or voted upon by the parties to the Convention. These were
to be taken up at a February 1979 intersessional meeting --presuma-
bly, to be voted on at the Fourth Consultative Meeting in the fall of
1979.

The regulations which were adopted impose detailed technical and
procedural requirements upon at-sea incineration permitted or conducted by parties to the Convention. Among other things, these requirements include mininum combustion temperatures, combustion efficiencies, and monitoring requirements.

Regrettably, the regulations do not include biological monitoring requirements, but rely instead upon operational controls to limit
the escape of toxic constituents into the air or ocean. Even more
unfortunate is the fact that the regulations fail to restrict at-sea
incineration to organohalogens and other organic compounds which can
be destroyed thermally and with which there is operational experience. Indeed, they can be read to allow at-sea incineration of Annex
I and II materials other than organohalogens and pesticides to be
carried out subject only to controls satisfactory to the Country issuing the special permit. (Wastes not referred to in Annex I and II
may be incinerated under a far less restrictive "general permit").
This means that wastes containing high levels of mercury and cadmium
compounds (Annex I materials) or significant amounts of arsenic,
lead, copper or zinc (Annex II materials) can arguably be incinerated
at-sea without any international controls -- despite the obvious inability of at-sea incineration to destroy these elemental materials.

U. S. officials involved in developing the international regulations insist that the intent was that wastes containing Annex I and
Annex II constituents -- other than organohalogens, oils, and pesticides -- would still be subject to the prohibition against disposal
of these constituents as other than "trace contaminants." This is
far from clear on the face of the regulations nor is it clear, under
the U. S. interpretation, whether the constituents may exceed "trace
contaminant" levels going into the incinerator. It is to be hoped
that these questions are clarified either in the Technical Guidelines
or in a separate Memorandum of Agreement, rather than being left to
the divergent interpretations of the parties.

                                CONCLUSIONS

The U. S. (and European) experience to date with at-sea

incineration gives cause for optimism that this technique will emerge
as an environmentally sound alternative to other, potentially more
dangerous, methods of toxic waste disposal (including direct ocean
dumping). In the case of organochlorine wastes, an obvious advantage
of ocean incineration over land-based methods of thermal destruction
is that the former do not require cumbersome stack scrubber devices
to prevent the spread of noxious and corrosive hydrogen chloride gas
into populated areas. Where incineration takes place at sea far from
land, the combined dispersive and dilutional capabilities of the atmosphere and ocean are available to dissipate HCl emissions. In addition, scrubber-equipped land-based incinerators involved in the
high-temperature combustion of organochlorine wastes, would be expected to be subject to frequent operational problems due to scrubber
failure associated with the emission of highly corrosive HCl vapors,
and to the generation of large quantities of scrubber residues requiring disposal in scarce and often inadequate landfill facilities.
A final consideration is that, freed of the need to attach scrubbers,
shipboard incinerators can be designed so as to maximize combustion
efficiency -- without being limited to configurations able to accommodate scrubbers.

To be sure, at-sea incineration is not an unmixed blessing.
Even if vessel operations proceed flawlessly, there is always the
risk of accidents at sea. The ecological effects of a spill of 4,000
tons of Agent Orange Herbicide on phytoplankton productivity could be
substantial. Stringent controls on the loading of incinerator ships
(e.g., to avoid spills in the course of transfer) are essential, as
is the need to ensure that improper tank washing, surreptitious dumping, and the like do not occur. Equally important is the need to ensure that this disposal technique, which may work fine for one type
of waste (e.g., organochlorines) will not be unjustifiably applied to
other waste types (e.g., heavy metals).

Finally, it will become increasingly important to monitor subtle
biological effects, such as impacts on enzyme function and bioaccumulation of persistent contaminants. Such studies could conceivably
require a reassessment of current thinking about the environmental
acceptability of incineration at sea.

In the meantime, at-sea incineration of industrial wastes is
something we can expect to hear a great deal more about in the future.

REFERENCES

Ackerman, D. G., H. J. Fisher, R. J. Johnson, R. F. Maddalone, B. J.
    Matthews, E. L. Moon, K. H. Scheyer, C. C. Shih, and R. F.
    Tobias (1978) At-sea incineration of herbicide orange onboard
    the M/T VULCANUS. U. S. EPA, Research Triangle Park, NC,

(EPA-600/2-78-086), 273 pp.

Clausen, J. F., H. J. Fisher, R. J. Johnson, E. L. Moon, C. C. Shih,
R. F. Tobias, and C. A. Zee (1977) At-sea incineration of
organochlorine wastes onboard the M/T VULCANUS. U. S. EPA,
Research Triangle Park, NC, (EPA-600/2-77-196), 95 pp.

U. S. Air Force (1976) Amendment to the final environmental state-
ment on the disposition of orange herbicide by incineration.
Office of the Special Assistant for Environmental Quality,
SAF/ILE, Washington, DC.

U. S. Air Force (1974a) Revised draft environmental statement on
disposition of orange herbicide by incineration. Office of
the Special Assistant for Environmental Quality, SAF/ILE,
Washington DC.

U. S. Air Force (1974b) Final environmental statement on disposi-
tion of orange herbicide by incineration. Office of the
Special Assistant for Environmental Quality, SAF/ILE,
Washington, DC.

U. S. Department of Commerce, Maritime Administration (1976) Final
environmental impact statement on maritime administration
chemical waste incinerator ship project. (#MA-EIS-7302-76-
041F), 2 vol.

U. S. Environmental Protection Agency (1976) Final environmental
impact statement of designation of a site in the Gulf of
Mexico for incineration of chemical wastes. EPA Division of
Oil and Special Materials Control, Office of Water and Hazardous
Materials, Washington, DC.

Wastler, T. A., C. A. Offutt, C. K. Fitzsimmons, and P. E. Des
Rosiers (1975) Disposal of organochlorine wastes by incineration
at sea. U. S. Environmental Protection Agency, Washington, DC,
(EPA-430/9-75-014), 238 pp.

Wyer, R. H. (1977) Preliminary report on at sea incineration of
herbicide orange onboard the M/T VULCANUS. 31 pp.

STABILIZED POWER PLANT SCRUBBER SLUDGE AND

FLY ASH IN THE MARINE ENVIRONMENT

Iver W. Duedall, Frank J. Roethel, James D. Seligman,
Harold B. O'Connors, Jeffrey H. Parker, Peter M. J.
Woodhead, Ramesh Dayal, Bart Chezar[1], Beverly K.
Roberts[2], and Hugh Mullen[2]

Marine Sciences Research Center
State University of New York
Stony Brook, New York 11794

[1]Power Authority of the State of New York
10 Columbus Circle
New York, New York 10019

[2]IU Conversion Systems, Inc.
115 Gibraltar Road
Horsham, Pennsylvania 19044

ABSTRACT

Solid (stabilized), brick-like forms of scrubber sludge and fly
ash from coal combustion were tested for their physical properties,
leaching behavior in seawater, effect on organisms, and long term
stability in a marine environment. Laboratory results showed that
trace elements did not leach significantly from test samples. In the
field, colonization and overgrowth of test blocks progressed very
rapidly, and block integrity, as determined by compressive strength,
was maintained for over 500 days.

These preliminary results and other results from ongoing studies
suggest that marine disposal of scrubber sludge and fly ash, in the
stabilized form, may be environmentally acceptable.

INTRODUCTION

The National Energy Plan calls for a very rapid increase in coal production and utilization. A requirement in the plan is that new coal burning power plants be fitted with flue gas desulfurization scrubbers which would remove harmful sulfur oxides.

According to the U. S. Environmental Protection Agency, the use of scrubbers is a reliable technology (Herlihy, 1977). The combustion of coal in boilers fitted with scrubbers, however, results in the production of huge amounts of scrubber sludge, namely calcium sulfate and calcium sulfite. Fly ash, another waste product of coal combustion, is also produced in large volumes.

A major unsolved problem in the widespread use of scrubbers is the disposal of the calcium sulfate–sulfite sludge and fly ash in areas with minimal land available for disposal. One prime area is coastal areas. This problem must be overcome in order to develop the full potential for coal combustion using scrubbers.

Several strategies for the disposal of these coal wastes have been suggested and discussed by Lunt et al. (1977). They include disposal in mine shafts, land-fill applications and ocean disposal. The present paper addresses ocean disposal of stabilized scrubber sludge and fly ash.

Lunt et al. (1977) suggest that the disposal of untreated sludges (liquid slurrys or dewatered sludges) on the continental shelf is probably environmentally unacceptable. A principal problem is the potential depletion of dissolved oxygen in the water column due to the rapid oxidation of sulfite to sulfate. Other potentially serious problems include increased turbidity, toxicity of sulfite, smothering of benthic communities, and the release of heavy metals that could be harmful to organisms. The disposal of treated (stabilized) brick-like scrubber sludge in the ocean, however appears promising (Lunt et al., 1977).

For the past two years, we have been investigating the composition, physical and chemical behavior and biological acceptability of stabilized scrubber sludge and fly ash (SSFA) supplied by IU Conversion Systems, Inc. Here 'stabilized' is a term used to describe a solid admixture of scrubber sludge and fly ash. Fig. 1 is a photograph of some of the test samples we have investigated. We have utilized laboratory measurements as well as experiments in the marine environment at a site in Conscience Bay, Long Island Sound.

Specifically we sought to measure:
1. the composition of SSFA, with respect to selected major and minor components, and its physical properties (density, porosity, permeability);

2.   the compressive strength of SSFA exposed to seawater in the laboratory for short periods of time and of samples placed in Conscience Bay for periods up to 17 months;

3.   the leaching of metals, sulfite, and sulfate from SSFA in laboratory experiments;

4.   the toxicity of SSFA elutriates to a laboratory cultured marine diatom; and

5.   the biological colonization of 0.028 m^3 (1 ft^3) SSFA and concrete blocks placed _in situ_ in an estuarine environment.

The goals of the work were to obtain preliminary measurements of some aspects of the physical, chemical and biological behavior of

Fig. 1.   Cut sections of stabilized scrubber sludge and fly ash of varying sulfate/sulfite ratios. (See Tables 2 and 3 for composition and physical properties.)

SSFA placed in the marine environment. The primary use of these preliminary findings was in the design of a comprehensive experimental
approach to determine environmental acceptability of SSFA for marine
disposal and as a material for constructing a large, demonstration
artificial reef.

LABORATORY METHODS

The test samples (Fig. 1) of stabilized scrubber sludge and fly
ash used in our work had a fly ash to sludge ratio of 5:1 (wt/wt).
Part of the ash contained in mix 6 was bottom ash. The scrubber
sludge and ash were obtained from the Elrama Power Plant in Pitts-
burgh, Pennsylvania. For purposes of comparison of analytical techniques and interpretation, samples of construction grade concrete
were prepared using methods given by American Society of Testing and
Materials (ASTM). Details of the methods used in the laboratory
characterization studies were given by Seligman (1978).

Cured samples of SSFA and concrete were ground, digested, and
analyzed for Ca, Cd, Cu, Cr, Fe, Mn, Ni, Pb, and Zn by flame or
flameless atomic absorption spectrometry (AAS). Hg was determined
using a small volume mercury apparatus adapted from Hawley and Ingle
(1975) for a cold vapor determination by AAS (Marine Sciences Research Center, 1978). Sulfite was converted to sulfate (ASTM, 1974)
and total sulfate was determined gravimetrically. Total carbon was
determined on powdered samples using a CHN analyzer standardized with
acetanalide. $CaCO_3$ was determined using the gas buret method of
Hulsemann (1966).

In the seawater analyses Ca was determined by adding 1 mg of
$LaCl_3$ to the filtered seawater samples and then aspirating the solution into the flame of the AAS. Cu, Fe, and Ni were determined
directly (without extraction) in seawater using NH_4NO_3 as a com-
plexing agent for the NaCl in seawater and then selectively volatil-
izing the NH_4Cl and $NaNO_3$ using a variable temperature program on
the graphite furnace. Sulfite was determined colorimetrically by
end-point titration; and sulfate was determined turbidimetrically
(ASTM, 1974).

When required, pH was determined with a combination glass electrode. Eh was determined with a combination platinum electrode
standardized against Zobell's solution (Zobell, 1946).

Density and porosity of the SSFA were determined using a water
absorption procedure and by measuring the change in weight of the
samples (ASTM, 1974). The coefficient of permeability (K) was determined using the standard falling head permeability technique and
calculated using Darcy's equation (Verbeck, 1956):

$$\frac{dg}{dt} \frac{1}{A} = K \frac{\Delta h}{L}$$

where dg/dt is the rate of flow of water in $cm^3 sec^{-1}$, A is the cross-sectional area of the sample in cm^2, Δh is the drop in hydraulic head through the sample, in centimeters, L is the thickness of the sample in cm, and K is expressed in cm per second.

Compressive strength of SSFA was determined by measuring the weight needed to achieve total failure of the sample when the load was applied to the sample's vertical axis (ASTM, 1974). A Riehle Universal testing apparatus was used. SSFA samples were selected to undergo compressive strength testing for the following conditions: 30 days and 150 days curing in a desiccator containing water to provide a humid atmosphere, 120 days curing with 30 days exposure to seawater and 120 days curing with 30 days exposure to seawater. The salinity of the seawater was 34°/₀₀.

Percolation leaching experiments were performed at laboratory temperatures (22 ± 2°C) on SSFA mixes 5 and 7 (Fig. 1) in order to maximize the dissolution of the components present throughout the internal structure of each test sample. Leaching of the more soluble components can be greatly enhanced by percolation since leaching is not limited to the dissolution of the outside surfaces of the test samples. The experiment was based on similar studies (Mahloch and Averett, 1975; Helm et al., 1975; and Beers et al., 1974) in which seawater was forced, using pressurized N_2, through the samples of SSFA sealed into the bottom of PVC columns, 1 m in height. Samples for analysis of the leached components were collected over a 15-30 day period, depending on the mixes. In these experiments, pH, Eh and the concentrations of Ca, SO_3^{2-}, SO_4^{2-}, Ni, Cu, and Fe were measured in the percolated seawater using the methods previously described.

Samples of mixes 5 and 7 were placed in unstirred tanks containing 3 liters of seawater (at 22 ± 2°C) for one month in order to ascertain the short term leaching rates of major and minor soluble components from SSFA's. Filtered aliquots of seawater were withdrawn at 1, 3, 5, 10, 20, and 30 day intervals and analyzed for SO_3^{2-} + SO_4^{2-}, Ca, Cu, Fe, and Ni; before taking an aliquot, the tank was well stirred.

An experiment was conducted to determine toxic effects, if any, of soluble SSFA components upon the growth of the diatom Thalassiosira pseudonana, which is cultured and utilized routinely at the Marine Sciences Research Center for similar tests on several toxic chemicals. The experiment represented a "worst case" situation in terms of rapid exposure of a marine photosynthetic organism to any potentially toxic substances incorporated in SSFA. Any observed toxicity might imply that the SSFA surfaces would be less acceptable than surfaces with less toxic or non-toxic characteristics of

Table 1. SSFA elutriate toxicity experiments.

Experiment number	SSFA mix tested	% Elutriate concentration in treated cultures	Replicate 400 ml cultures for each treatment or control
1	5	10	3
2	7	5, 10, 20	2

colonization by other photosynthetic algae.

The method involved using different concentrations of a sea-water-SSFA elutriate. The elutriate leaching (elutriate test) procedure reported by Keeley and Engler (1974) is supposed to provide a rapid means of measuring the pollution potential of soil-like materials introduced into seawater. SSFA samples were ground to a fine powder. For each SSFA mix tested, 150 cm^3 of the powdered sample were added to 600 ml of filtered seawater whose salinity was approximately 34‰ ; the seawater-SSFA mix was stirred for 24 hours before testing its effect on organisms. Table 1 presents the specific details of each experiment which consisted of replicate 400 ml volumes each of treated (measured volume of elutriate added) and control (equal volume of filtered Long Island Sound water added) suspensions of Thalassiosira pseudonana cells (clone 3H) grown in nutrient en-riched Long Island Sound water. The growth medium was equivalent to Guillard and Ryther's (1962) f/2 formulation. Control and treated volumes were placed in a 23°C, fluorescent tube-illuminated incuba-tor. The suspended cell volume concentration (particle concentration, ppm by volume), chlorophyll a concentration (μg l^{-1}), and quantity of ^{14}C incorporated [counts minute^{-1} ml^{-1} or cpm ml^{-1}] were determined daily in each control and treated cell suspension using methods outlined in Strickland and Parsons (1972). The amount of carbon-14 incorporated is an index of the rate of photosynthesis in the cultured algal cells.

LABORATORY RESULTS

The concentrations of selected acid leachable major and minor components are presented in Table 2. Average values are reported based on replicate samples and sub-samples of each mix type. There was no significant variation (one-way analysis of variance (ANOVA) α = .05; Sokal and Rohlf, 1969) between samples of the same mix type for any of the components measured, indicating that a fairly uniform

Table 2. Concentrations of selected major and minor components[a].

Sample Mix Type	Bulk Content (mg/g)				Acid Leached Components[b] (μg/g)											
	CaCO3	Carbon Content			Ca	SO3	SO4[d]	Fe	Mn	Zn	Cu	Cr	Pb	Ni	Cd	Hg
		Total	CO3	Organic[c]												
4	118	15	14	1	52200	23600	2600	3660	58.6	23.2	19.1	16	11	7	0.6	.048
5	118	17	14	3	66600	26700	3000	4750	53.0	30.5	22.7	18	19	7	0.7	.059
6	15	20	2	18	33200	16300	24500	4930	35.1	13.7	12.3	7	12	6	0.4	.272
7	19	25	2	23	37400	16300	24500	3340	33.1	14.1	10.9	10	10	3	0.2	.299
Concrete[e]	–	–	–	–	111300	–	6400	5780	312.5	18.4	5.2	24	6	5	<.2	.010
Detection Limit	5	1	0.5	1	400	100	100	30	.3	.1	.4	2	3	2	.2	.002
C.V.[f] (%)	9.5	15.7	9.5	15.7	10.5	10.7	10.7	4.9	4.1	21.4	7.2	9.4	20.8	18.9	17.3	12.1

[a] Concentrations are presented on a dry weight basis and are average values of sample and subsample replicates.

[b] Nitric acid was used for heavy metal leaching and hydrochloric acid was used for calcium and sulfur oxide leaching.

[c] Organic carbon is determined by subtracting carbonate carbon from total carbon.

[d] Total SO_x was measured analytically; reported values of SO_3^{2-} and SO_4^{2-} are based on $SO_3:SO_4$ ratios supplied by I. U. Conversions (S. Taub, personal communication).

[e] Concrete component concentrations are based upon total weight including aggregates. The component concentrations presented would be approximately two times higher if the aggregate materials were not included.

[f] C.V. is the weighted average coefficient of variation ($\frac{s.d.}{\bar{y}}$ x 100) for the replicate samples and subsamples of each mix type for each component measured.

< Denotes values below the detection limit.

product was achieved in the production of the samples.

The high sulfite mixes (mixes 4 and 5) had relatively high $CaCO_3$ concentrations. In contrast, the high sulfate mixes (mixes 6 and 7) had low $CaCO_3$ concentrations. The difference between these mixes is probably due to different operating stoichiometry in the scrubber. The high sulfate mixes (mixes 6 and 7) had a high organic carbon residue due to incomplete coal combustion and a high sulfate: sulfite ratio due to inefficient boiler operation (S. Taub, personal communication). The mole ratios of Ca: (sulfite + sulfate) were highest in the high sulfite mixes, 5:1 as opposed to 2:1 in the high sulfate mixes. These ratios are a function of the different scrubber reaction efficiencies and the amount of added stabilizers which are high in Ca. This excess Ca is very important in determining the degree of stabilization and the resulting physical properties of the blocks.

The high sulfate mixes were lower in heavy metals except for Hg, the concentrations of which were about five times higher than those observed in the high sulfite mixes. Cd, Cu and Hg are the heavy metals present in SSFAs in the highest concentrations compared to concrete. The remaining heavy metals which we measured had concentrations similar to or lower than those of the concrete test samples.

The bulk densities (Table 3) of SSFAs were lower than concrete due to the added fly ash in the SSFAs and the absence of high density aggregate materials. Mix 7 was found to have greater porosities and permeabilities than mixes 4 and 5 (Table 3). Additionally the SSFAs were found to be 3-4 times more porous and 10-650 times more permeable than the concrete test samples.

Table 3. Selected physical properties of test samples.

Sample[a] mix type	Bulk density (g cm^{-3})	Porosity vol. (%) of water (permeable voids)	Coefficient of permeability K(cm sec^{-1})x10^{-7}
4	2.17	34	3.9
5	2.18	37	1.0
6	1.99	58	-
7	2.03	48	65.0
Concrete	2.70	13	0.1

[a]All samples were cured 90 days.

Compressive strength values for the SSFA mixes and concrete are presented in Table 4. The compressive strength of all of the samples increased significantly with curing time. After 30 days curing in air the SSFA had a compressive strength of 25-75% that of concrete. After five months of curing, the compressive strengths of mixes 4, 5, and 7 were 1.2-2 times greater than those observed at 30 days. Mix 5, however, showed a threefold increase in compressive strength while concrete increased fivefold. Compressive strength of concrete and similar cementing compounds varies as a function of curing time, humidity, temperature, water content, and additive content.

The observed differences between mix types and concrete is due to the variations in the composition of the samples. These variations lead to different hardening processes which result in the variations in compressive strength. For example, concrete continues to harden for many months due to the slow bonding kinetics associated with the Ca-Si-Al-Fe system and the hydration processes involved. The increased abundance of available Ca for bonding in Portland cement accounts for its greater hardness. Excess Ca in SSFA forms Ca-Si bonds, as in concrete, through pozzolanic reaction (Neville, 1973).

Mix 7 and concrete were also subjected to 120 days immersion in seawater. Here it was found that mix 7 lost no additional strength but concrete lost an additional 22% of its optimum strength. The loss in strength of seawater-exposed concrete is well documented (Swenson, 1968). The process involves sulfate ions in seawater which disrupt the bonds in cement. The loss in strength occurs as a result of lattice expansion due to the formation of calcium sulfoaluminate precipitate. This precipitate is formed by sulfate ions reacting with the tricalcium-aluminate in cement, which is one of the bonding compounds, to form solid ettringite ($3CaO \cdot Al_2O_3 \cdot 3CaSO \cdot 31H_2O$) in situ, which occupies a 14% greater volume than the tricalcium-aluminate originally present. As this and other solution reactions occur, the concrete expands and loses its strength due to internal pressures. Consequently cementing bonds and lattice structures are broken. The reason that the SSFA test samples do not continue to lose strength upon continued exposure to seawater is that they already contain minimal lime available to react with sulfate ions in the seawater. They also have high porosities which permit lattice expansion with less internal pressures. Except for an initial loss of strength there was no evidence in the laboratory study that continued exposure of SSFA to seawater will result in the continued rapid loss of strength which was observed for concrete. Similarly, results for the in situ study in Conscience Bay, to be discussed later in this paper, show that after an initial decrease there has been a steady increase in compressive strength of the SSFA blocks which have been submerged in the sea for 1.5 years.

The percolation leaching results were examined as a funtion of

Table 4. Compressive strength of test samples.

Days Cured	Days in Seawater	Age of Sample	Compressive Strength (psi)[a]					Concrete
			4	5	6	7		
30	0	30	110	320	160	200	425
150	0	150	215	920	205	355	2160
120	30[b]	150	–	–	–	–	–
120	30[c]	150	–	–	–	–	–
120	30[d]	152	180	840	–	320	1780
30	120[d]	152	–	–	–	310	1520
120	0[e]	150	–	–	–	245	1650

[a]The coefficient of variation for compressive strength values, based on three replicates of one mix type, was 7%.

[b]Samples were tested immediately upon removal from the seawater.

[c]Samples were allowed to dry for 3 hours after removal from seawater before testing.

[d]Samples were allowed to dry for 2 days after removal from seawater before testing.

[e]These samples underwent 20 cycles of rapid freezing and thawing in air requiring 30 days.

the volume of seawater percolated rather than percolation time because the leaching rate of each component is a function of the experimental conditions and will vary with the height of the seawater hydraulic head and the permeabilities of the samples. In this investigation, the 0.84 liters of seawater allowed to percolate through SSFA mix 5 required 30 days, and the 1.68 liters of seawater that were allowed to percolate through mix 7 required 15 days. The dimensions of the SSFA fixed in the PVC columns were 7.6 cm (dia) x 7.6 cm (height) (Seligman, 1978). The results of the percolation experiment are given in Figs. 2 and 4 in which $\Sigma a_n/A_o$ x 100 is plotted as a function of V_n (Mahloch, 1976) where: A_o = the initial mass of component a_n present in the test sample and Σa_n = the cumulative mass of a_n that is leached from the test sample; thus $\Sigma a_n /A_o$ x 100 is the cumulative percentage of a_n that is leached from the sample. V_n is the volume of seawater percolated through the test sample.

The results of the percolation experiment show that the heavy metals Fe, Cu and Ni leached very little or not at all from the SSFA test samples during the period of this investigation. Fe, for instance, did not leach from the samples at all, but instead was removed from the percolating seawater by the SSFA (Fig. 2a). Cu, on the other hand, leached from the samples initally but then for mix 7 (high SO_4) reached a steady state concentration after 0.5 liters of seawater percolation (Fig. 2b). As a result, less than 0.5% of the total Cu initially present in the SSFA sample was released before a steady state concentration was achieved. Ni leaching followed a similar pattern (Fig. 2c) but the total release of Ni from the SSFA samples was considerably less than that for Cu.

The leaching behaviors of Fe, Cu and Ni appear to be interrelated and dependent upon the Eh of the percolating seawater. In the Eh-pH ranges encountered in this experiment (Fig. 3), the speciation of these heavy metals in seawater (Stumm and Morgan, 1970) can account for the observed leaching behaviors. The maximum release of Cu and Ni from the samples occurred at the lowest Eh conditions, when Fe loss from the seawater was a minimum. As the Eh increased, the increased Fe loss from seawater was due probably to the formation of insoluble ferric oxides in the seawater which adsorbed on the internal surfaces of the SSFAs. Additionally, Cu and Ni release from the SSFAs, which was facilitated by low Eh conditions, decreased as the Eh of the percolating seawater increased. This may be due to a coprecipitation or adsorption of Cu and Ni with the Fe oxides, which are known to be scavengers in seawater (Goldberg, 1954).

Leaching of sulfite + sulfate (Fig. 4a) and Ca (Fig. 4b) were found to be relatively constant and rapid compared to the minor components during the percolation experiment. Seligman (1978) has shown that the concentrations of these major components is controlled by their solubility.

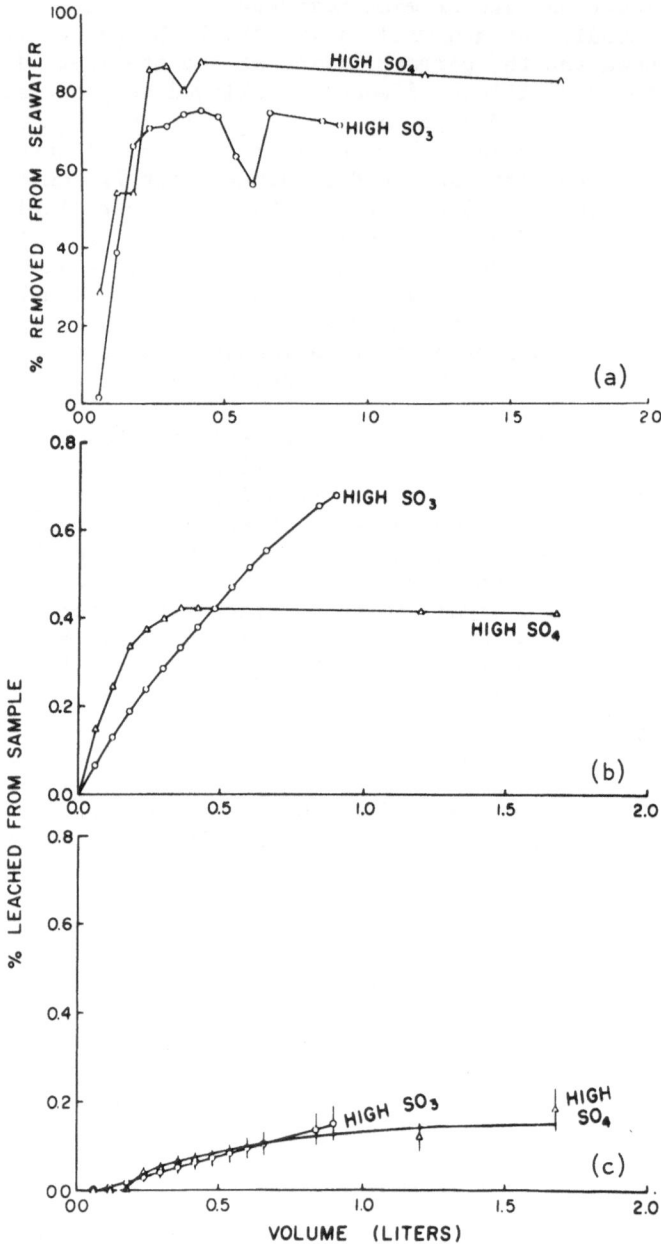

Fig. 2. Leaching behavior of trace elements during percolation
         experiments: (a) Iron absorbed from seawater; (b) Copper
         released from SSFAs; (c) Nickel released from SSFAs. High
         SO_3 and high SO_4 refer to SSFA mixes 5 and 7, respectively.

Immersion of the SSFA blocks in a tank of seawater provided further information on chemical leaching. The dimensions of the SSFA blocks in the tank were 7.62 cm in length by 7.62 in diameter. There

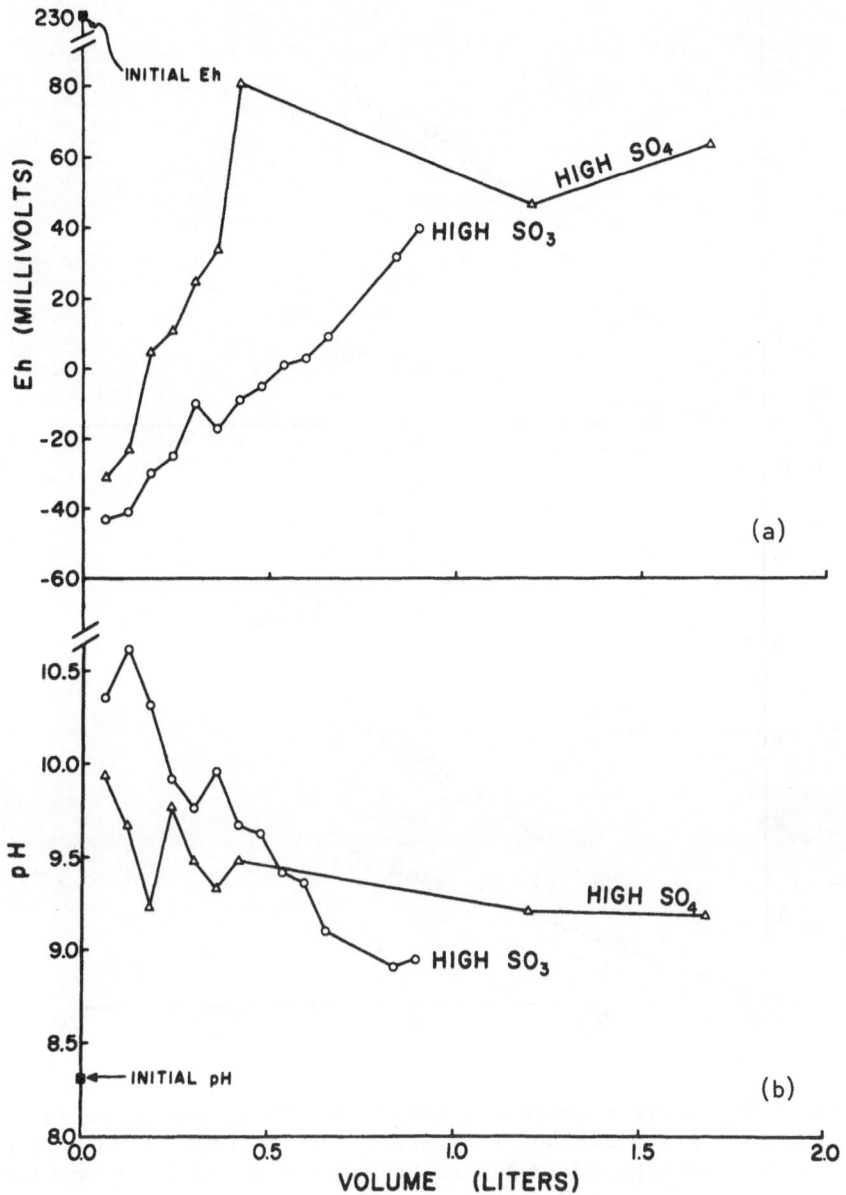

Fig. 3.   Eh (a) and pH (b) of percolated seawater.

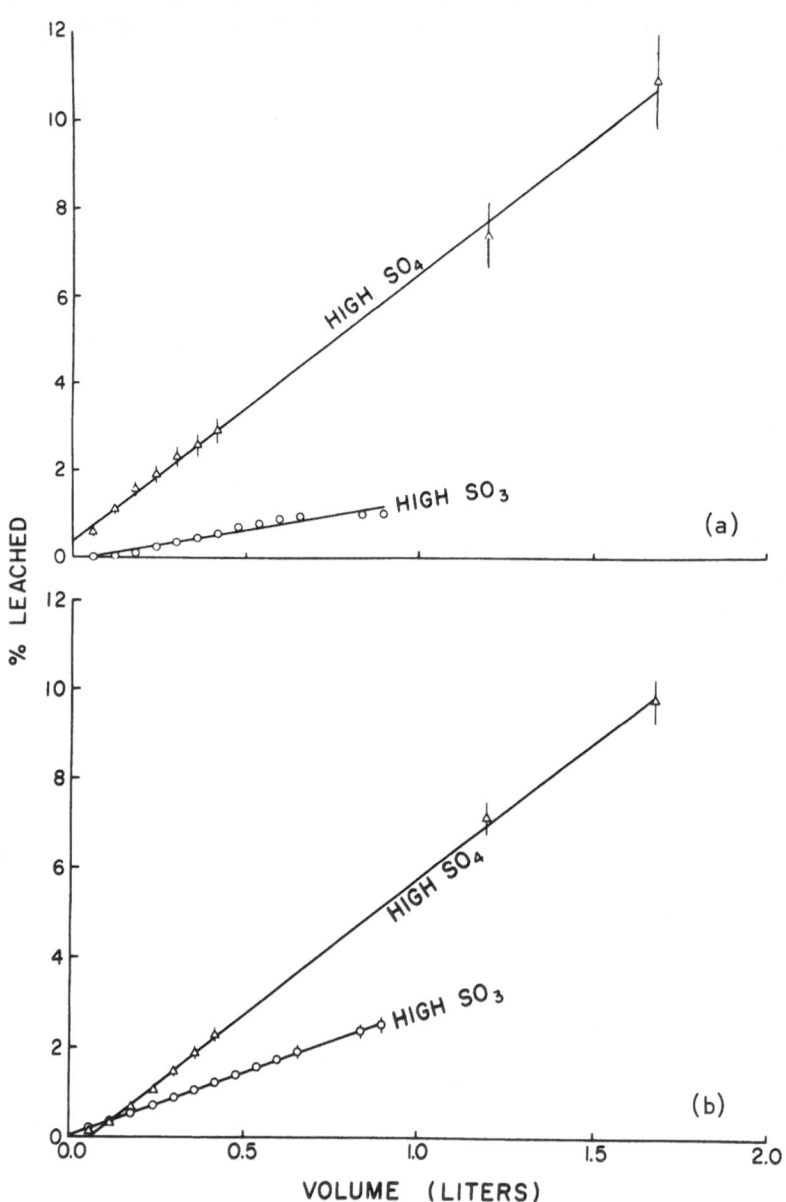

Fig. 4.   (a) Sulfite + sulfate leached from SSFAs during percolation;
          (b) calcium leached from SSFAs during percolation. High
          SO_3 and high SO_4 refer to SSFA mixes 5 and 7, respectively.

was an initial increase of dissolved Fe and Ni within the first few days after exposure to seawater, but after approximately 10 days the concentrations of these components decreased to levels near the

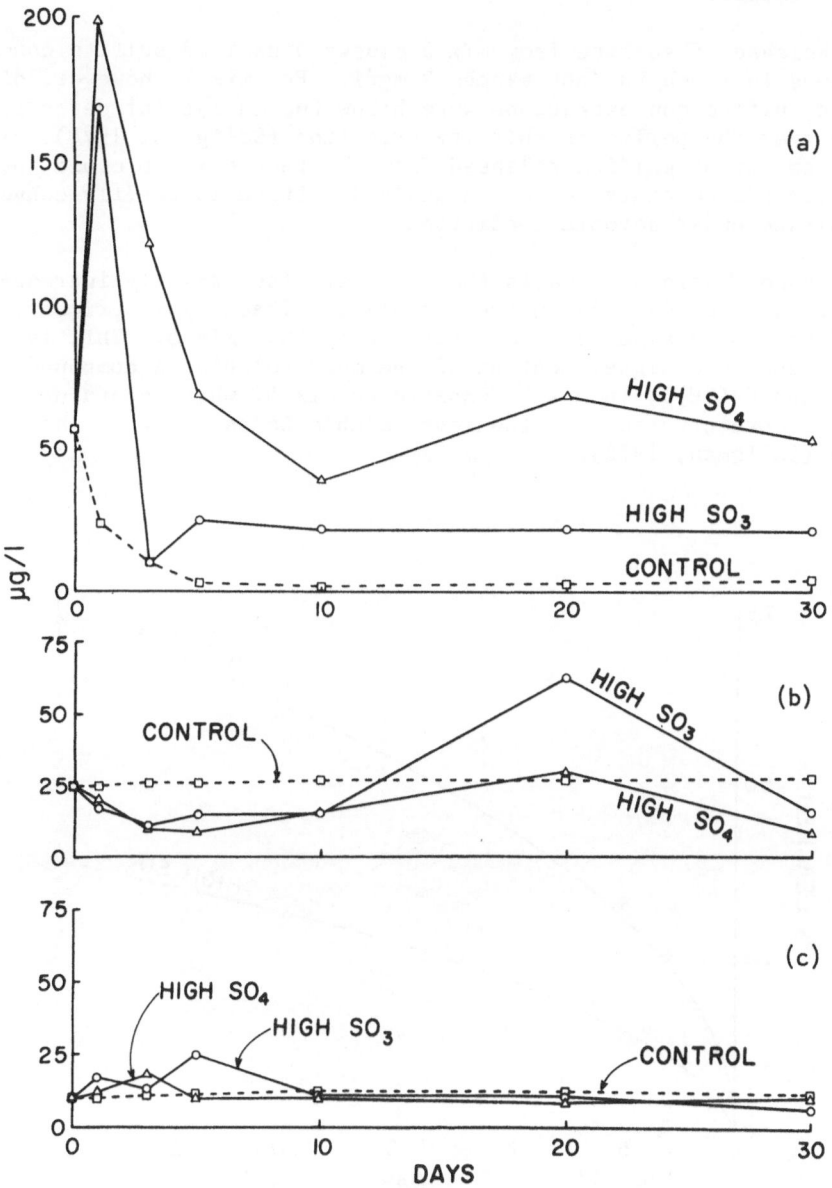

Fig. 5.   Concentrations of dissolved trace elements in tank experiments: (a) iron; (b) copper; and (c) nickel.

original seawater concentrations (Fig. 5). The concentration of Cu, however, decreased during the first 10 days of the experiment. The behavior of dissolved Fe, Cu, and Ni concentrations is probably due to desorption-adsorption processes or precipitation reactions occurring on the surfaces of the SSFAs and the suspended particulates that were released.

Release of sulfite from mix 5 caused dissolved sulfite concentrations to reach but not exceed 2 mg/l. For mix 7, however, dissolved sulfite concentrations were below the analytical detection limit over the period of this investigation (Seligman, 1978). The total amount of sulfite released from the test samples could not be determined accurately because dissolved sulfite is readily converted to sulfate under aerobic conditions.

Concentrations of Ca in the seawater also steadily increased with time (Fig. 6). As in the percolation leaching experiment, mix 7 was found to release Ca at a greater rate than mix 5. This is primarily due to a higher content of the more soluble Ca compounds i.e., $CaSO_4$ and $Ca(OH)_2$, in mix 7 compared to mix 5, which contains greater concentrations of the less soluble Ca salts, $CaSO_3$ and $CaCO_3$ (Seligman, 1978).

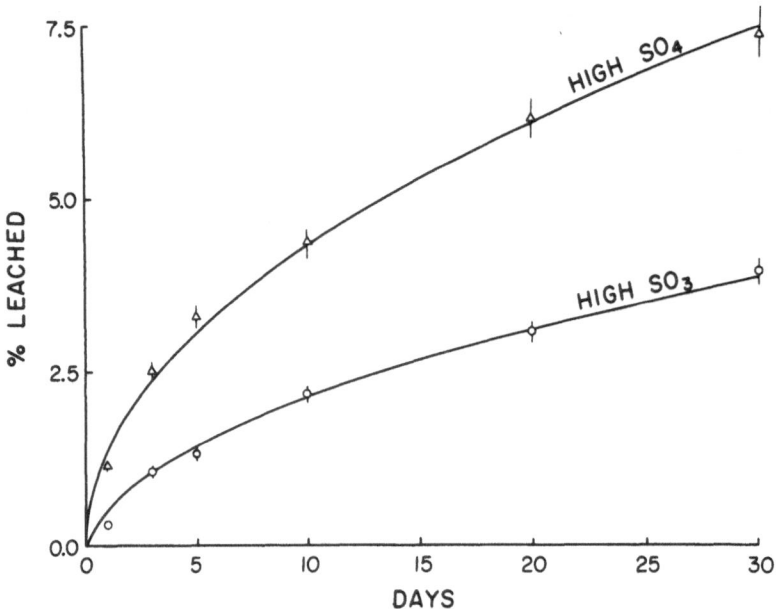

Fig. 6.   Calcium, expressed as percent of total calcium in test
          sample, leached from SSFAs in tank.

The marine diatom Thalassiosira pseudonana was tested with an elutriate mix to determine concentrations of cell volume, chlorophyll a and ^{14}C uptake rates (Fig. 7). The time series measurements of biomass (suspended cell volume and chlorophyll a concentrations) were used to calculate the diatom's growth rate μ, as doublings per day using

$$\mu = 35 \; \frac{\ln C_T - \ln C_o}{\Delta t}$$

where C_T = cell volume or chlorophyll a concentration at the end time interval Δt, C_o = cell volume or chlorophyll a concentration at the beginning of time interval Δt, and Δt = time interval between measurement, in hours.

It was found following the initiation of experiment 1 that the growth rate (doublings day^{-1}) was suppressed by a small, but statistically significant, amount from day 1 to day 2 and day 0 to day 1, respectively, for cell volume and chlorophyll a concentrations (Fig. 7). However, by day 4 the growth rate for the treated sample was greater than the control.

In experiment 2 the effects of three concentrations of mix 7 elutriate were measured. Treated cell suspensions containing 5%, 10% and 20% elutriate were prepared, along with appropriate control. Statistically significant redutions in cell volume and chlorophyll a concentration, as well as ^{14}C incorporation, were measured in the 20% elutriate treated suspensions during the interval day 0 to day 1. Cell volume and chlorophyll a concentrations were reduced, but ^{14}C incorporation was not diminished during the interval day 1 to day 2 for that dilution (Fig. 7 and Table 5). The suspensions treated with 10% elutriate showed reduced ^{14}C incorporation and chorophyll a concentration for the day 0 to day 2 intervals. The suspensions receiving 5% elutriate showed increased levels of cell volume and chlorophyll a concentration, as well as ^{14}C incorporation early in the experiment.

Reduced cell volume growth rates (Fig. 7d) were measured for 20% elutriate treated suspensions for day 0 to day 2. The reduced cell volume growth rate measured in the 5% elutriate treated suspension for day 1 to day 2 probably resulted from reduced nutrient levels in the medium rather than from elutriate effects. The elevated growth rate for day 0 to day 1 could have easily exhausted the nutrients supplied.

The 5%, 10% and 20% elutriate concentrations utilized in these experiments far exceed any concentration found in the natural environment (Seligman, 1978). The slow release of soluble components together with diffusive and advective processes occurring in situ

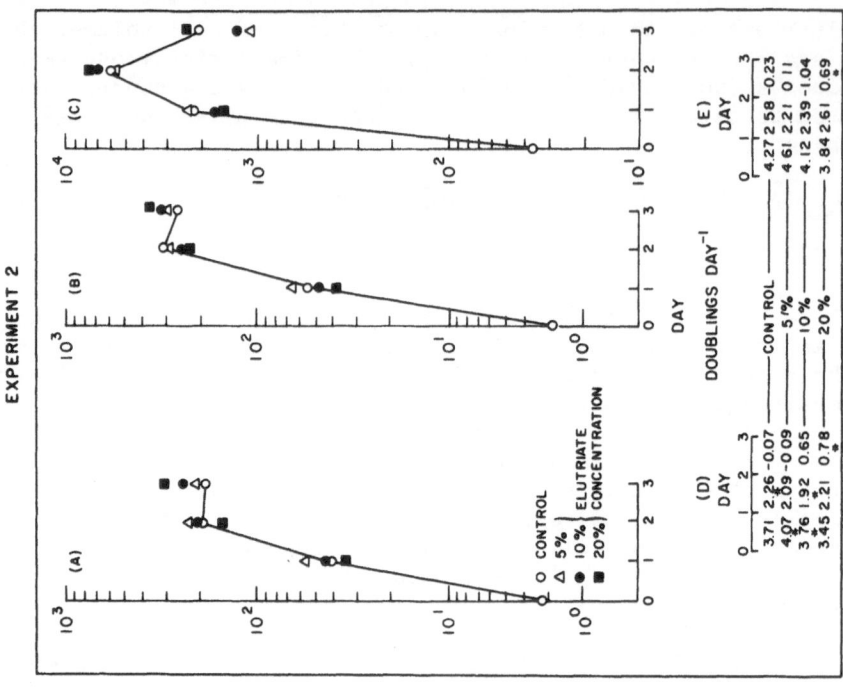

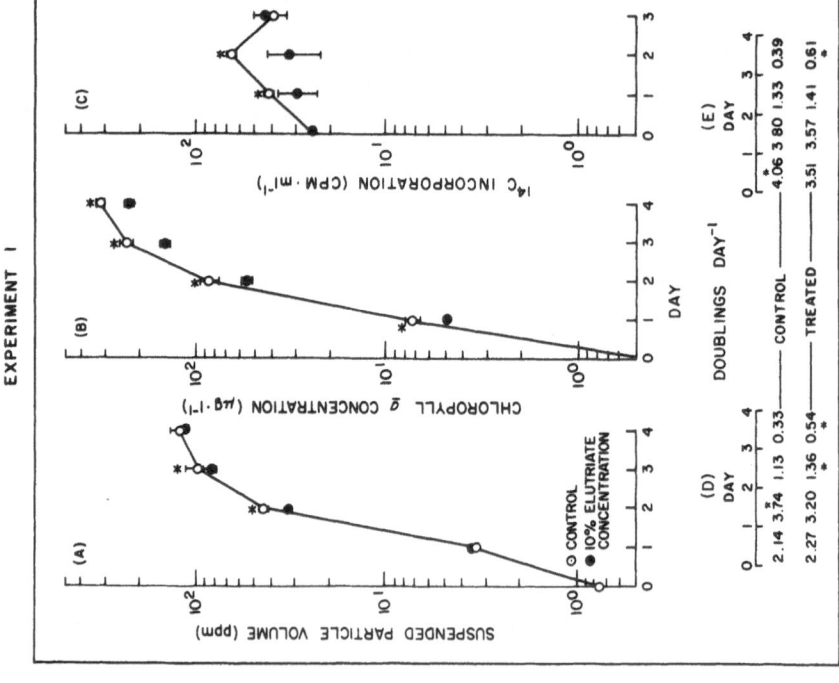

Fig. 7. Growth of Thalassiosira pseudonana in SSFA elutriates: (a) Experiment 1 using mix 5; (b) Experiment 2 using mix 7. Time series measurements of (A) suspended particle (cell) volume, ppm; (B) chlorophyll a concentration, μg l^{-1}; and (C) ^{14}C incorporation, CPM ml^{-1}. Growth rates were calculated as doubling day^{-1} based on (D) cell volume and (E) chlorophyll a concentration. In experiment 1, an * above values in (A), (B), and (C) or above a treated growth rate in (D) and (E) indicates that a value for the treatment mean, $\bar{X}_T$ was significantly less than that of the control mean $\bar{X}_C$. An * below the treated values indicates that the treated value was greater than the control. The absence of an asterisk indicates no significant difference between treated and control values. Statistical analysis for experiment 2 is given in Table 5.

Table 5. Results of statistical analysis of data presented in Fig. 7 [(b)A,B,C].

Treatments, Elutriate Concentration	DAY								
	0-1			1-2			2-3		
	P^a	Chl $\underline{a}^b$	$^{14}C^c$	P	Chl $\underline{a}$	^{14}C	P	Chl $\underline{a}$	^{14}C
20%	$*^d$	*	*	*	*	0	0	0	N.S.
10%	N.S.^{e}	N.S.	*	N.S.	*	0	0	N.S.	*
5%	0^f	0	0	0	N.S.		N.S.	N.S.	*

[a] P is particle (cell) volume concentration (ppm).

[b] Chl $\underline{a}$ is Chlorophyll $\underline{a}$ concentration ($\mu g\ l^{-1}$).

[c] ^{14}C is uptake (CPM ml^{-1}).

[d] * indicates that the value for the treated mean $\bar{X}_T$ was significantly less than that of the control mean $\bar{X}_C$.

[e] N.S. indicates no significant difference between the means tested. The hypotheses $H_o:\bar{X}_C = \bar{X}_T$, $H_a:\bar{X}_C > \bar{X}_T$ or $\bar{X}_C < \bar{X}_T$ was tested using "Student's" t distribution, $\alpha = 0.10$.

[f] 0 indicates that $\bar{X}_T > \bar{X}_C$.

would immediately reduce the exposure of any marine organisms to any potentially toxic components of the SSFA.

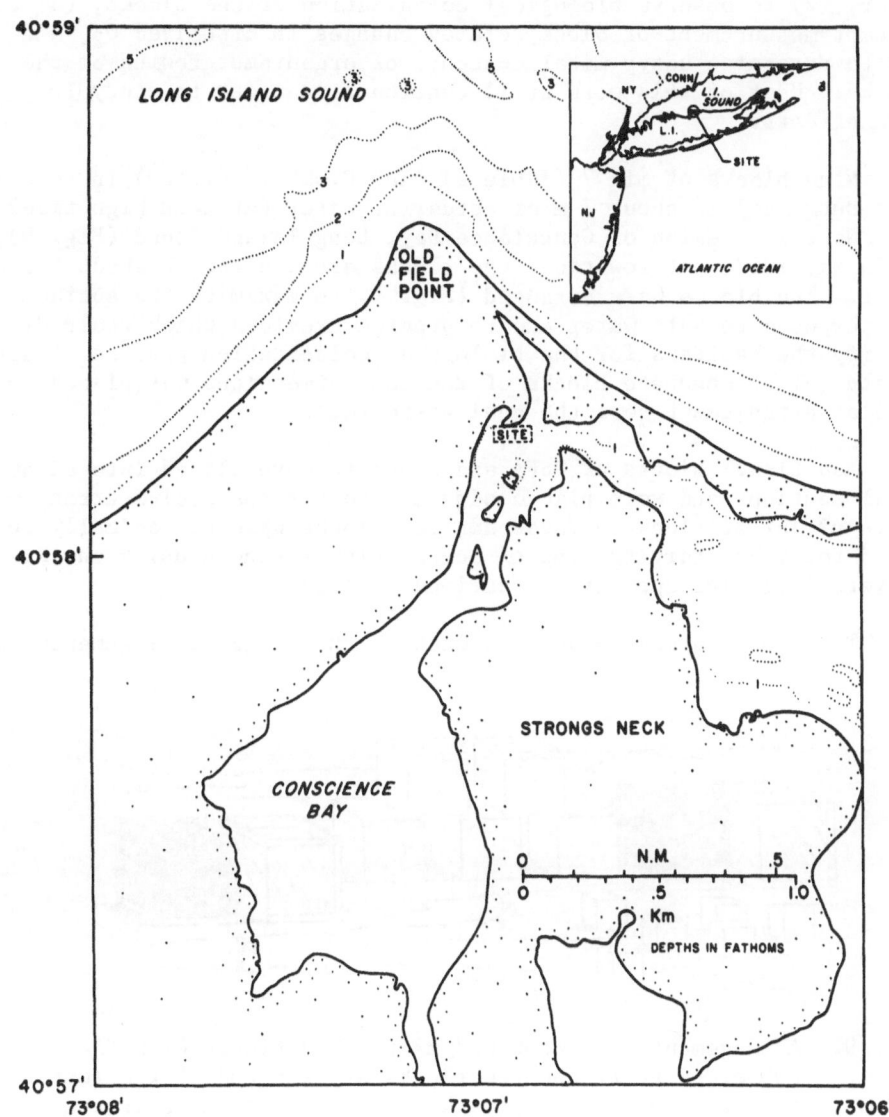

Fig. 8.   Conscience Bay test site: adjacent to Port Jefferson Harbor on Long Island Sound.   Depths in fathoms at mean low water (MLW).

FIELD STUDY METHODS

     We placed the SSFA and concrete control blocks in an estuarine
environment (1) to monitor changes with time in the physical integrity and compressive strength of SSFA during in situ exposure to salt
water, (2) to observe biological colonization of the blocks, (3) to
attempt measurement of block-related changes in dissolved O_2, and
(4) to determine heavy metal contents of organisms growing on the
blocks. Most of this work still continues in order to study long
term effects.

     Nine blocks of mix 7 (Table 2) each 0.028 m^3 (1 ft^3) in volume,
were submerged in about 7 m of estuarine water (at mean high tide) in
the "Narrows" region of Conscience Bay, Long Island Sound (Fig. 8),
on 25 May 1977. At low tide, the blocks are located in about 5 m of
water. The blocks were arranged (Fig. 9) to maximize the surface
area exposed to salt water and to provide crevices which would di-
versify the habitats for the biological colonization process. A duplicate set of concrete blocks of the same dimensions was placed in a
similar arrangement near the coal waste reef.

     Additional blocks of SSFA and concrete were sliced into eight
equal sections and were placed adjacent to the respective arrange-
ments of larger blocks. These smaller blocks were periodically removed for laboratory testing of compressive strength using the Riehle
universal testing apparatus described earlier.

     The two uppermost blocks in each of the larger arrangements had

Fig. 9.   Arrangement of nine 0.028 m^3 (1 ft^3) blocks of SSFA as a
          submerged reef; the two blocks at each end of the reef had
          respiration chambers (dark and light) placed over the SSFA
          blocks. The quartered slabs on the top of the center blocks
          were used for colonization studies. The smaller cubic
          blocks in front of the larger blocks were used for compres-
          sive strength measurements. Nearby a duplicate reef (not
          shown) made of concrete blocks served as a control for the
          experiment.

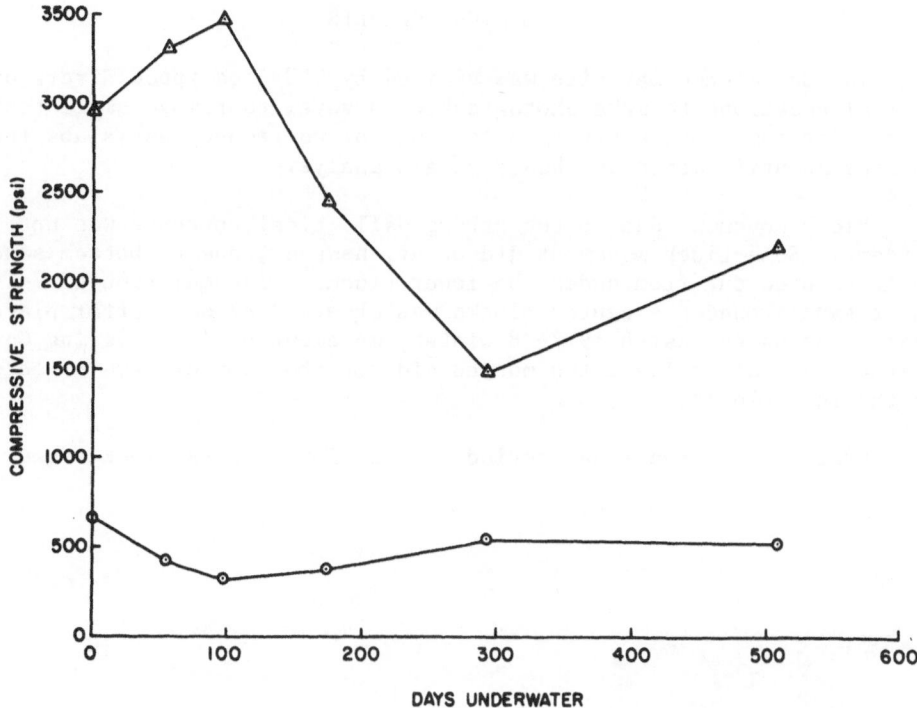

Fig. 10. Compressive strength of in situ test blocks in Conscience
Bay: triangles- concrete; circles- SSFA.

a 5 cm (2 inch) slab sliced from the top of the block. The slice was
quartered and fastened to the parent block with plexiglass corners
and a large rubber band. Quarters were periodically removed for examination and photography of the colonizing organisms.

Samples of encrusting organisms were removed from SSFA and concrete blocks at the center of each arrangement during the August and
September visits. The organisms sampled encompassed a sufficient
area of the blocks to be representative of the invertebrates and
algae present. In the laboratory, samples were rinsed in triple-
distilled water to remove surface adhering substrate particles and
salt. Organisms were freeze-dried for 48 hours; the dehydrated biomass was finely ground with a quartz mortar and pestle. Three 1 g
portions of each sample were weighed, digested with concentrated
HNO_3 and analyzed for Cd, Pb, Cu, Cr, Zn, Hg and Ag using AAS
(Marine Science Research Center, 1978).

FIELD STUDY RESULTS

The Conscience Bay site was visited by SCUBA equipped divers on several occasions to make photographic surveys, to remove small test blocks for compressive strength testing, or to remove test slabs for species identification and heavy metals analysis.

Block movement due to the strong daily tidal currents was not evident. Some block movement did occur, however, due to bottom sand being scoured out from under the lower blocks. Accumulation of 7-10 cm of sand around the center blocks was observed 82 days after placement. During the harsh 1977-78 winter, movement of ice covering Conscience Bay for at least two months did not show any extreme effects on the reef blocks.

Over the observational period, no erosion of SSFA block edges

Stabilized Scrubber Study
and Fly Ash

Concrete

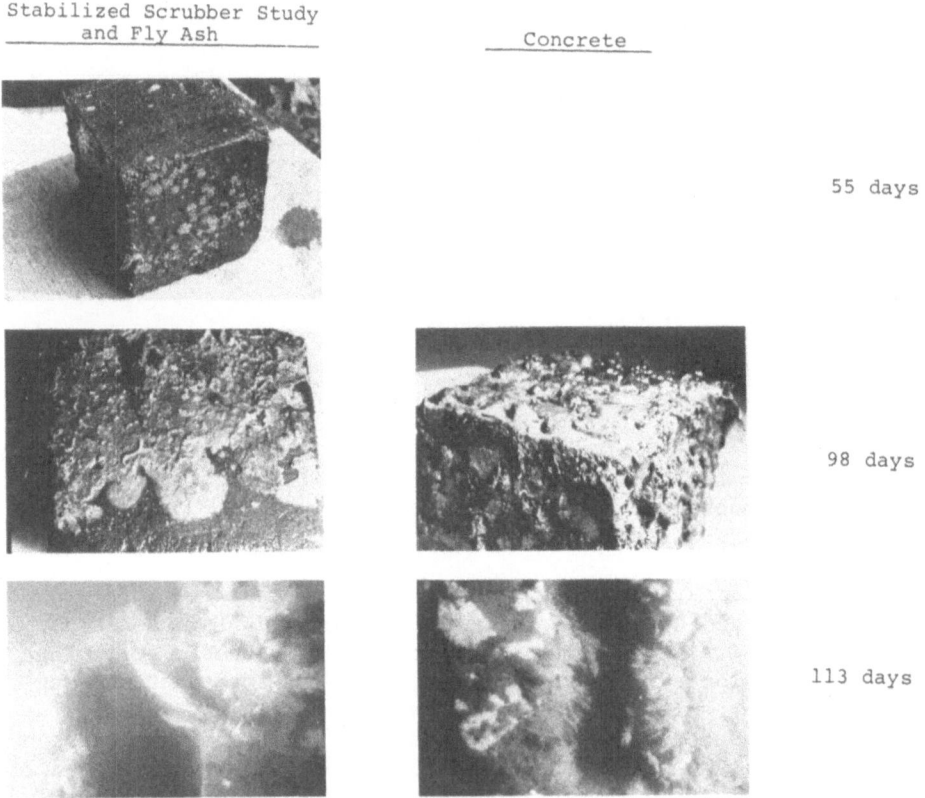

55 days

98 days

113 days

Fig. 11. Time history of the colonization on SSFA [mix 7 (Fig. 1)] and on concrete.

was observed. However, some surface softening was detected during
the dives on days 96 and 111. Compressive strength tests were
performed and results are given in Fig. 10. The results show that
during the first 100 days, compressive strength of the coal waste
blocks steadily decreased while that of the concrete increased.
After 100 days, however, the compressive strength of coal waste
stabilized and later began to increase, while the strength of the
concrete demonstrated a dramatic decrease. Approximately 300 days
after placement, duplicate samples were brought back to the laboratory for testing. The means of the duplicates were 380 and 2,452 psi
for the SSFA and concrete blocks, respectively. The differences
between the duplicate samples were 15 psi for coal waste and concrete
blocks, demonstrating that the trend we observed is significant. The
loss in strength of concrete exposed to seawater is well documented
and was discussed earlier in this paper and by Seligman (1978).

Of the invertebrates observed with the unaided eye, hydroids
were early colonizers on both SSFA and concrete blocks. These colonies were first observed at 20 days following placement and they had
an average length of about 1 cm. By day 82 the hydroid colonies had
reached a length of 5-10 cm which did not appear to increase on subsequent observations (Fig. 11). At 55 days following placement,
small patches of both red and green algae were observed on both substrates, while many small encrusting colonies of bryozoans were encountered only on the SSFA blocks.

On the 15 August visit, 82 days after placement, the algal colonies had increased in abundance and size, as had the bryozoan colonies, which occurred mainly on the SSFA blocks. By this time, the
green algae covered about 30% of the SSFA surface. Numerous slipper
limpets were attached to the concrete surfaces, but not on the SSFA
blocks. Calcareous polychaete tube worm casings occurred on both materials.

By day 111 following placement, bryozoan colonies were seen on
concrete surfaces but they were not as numerous or extensive as on
the SSFA blocks. Many limpets were encountered on concrete; none,
however, were seen on SSFA.

By day 174 there was a very heavy growth of colonizers dominated
by algae on the top surfaces of both the SSFA and the concrete
blocks. Snails and their grazed tracks could be seen on the top surfaces of these blocks. At the beginning of the second year the limpets were found on the SSFA. More than 460 days after placement, hydroid and algal growths were very extensive. The differences in colonization between SSFA and concrete which were seen at the earlier
stages of the study were no longer evident. The bryozoan colonies
and tube worms dominated the sides of the blocks, with algae and hydroids attached to the tops, sides, corners and edges. The colonizers seemed to form a stable community. Fish, crabs and gastropods

Table 6.   Organisms associated with the SSFA and concrete reef which
           have been tentatively identified.

Invertebrates

      Phylum Porifera (sponges)
      Cliona celata
      Microciona prolifera

      Phylum Coelenterata (hydroid polyps)
      Tubularia sp.
      Bougainvillea superciliaris
      Obelia dichotoma

      Phylum Annelida (segmented worms)
      Hydroides dianthus
      Serpula vermicularis
      Diopatra cupraea

      Phylum Mollusca (snails and bivalves)
      Acmaea testudinalis
      Busycon canaliculatum
      Venus mercenaria
      Ostrea virginica
      Polinices heros

      Phylum Bryozoa (Bryozoans)
      Schizoporella unicornis
      Lepralia pallasiana
      Membranipora pilosa

      Phylum Echinoderma (star fish)
      Arbacia punctulata
      Asterias forbesi

      Phylum Arthropoda (crabs)
      Pagurus longicarpus

Fish (common name)

      Blackfish/Tautog   Tautoga onitis
      Silversides   Menidia menidia
      Scrup/Porgy   Stenotomus chrysops
      Toadfish   Opsanus tau
      Mudskippers   Periophthalmus
      Bergall/Cunner   Tautogolobrus adspersus
      Winter flounder   Pseudopleuronectes americanus
      Killifish/mummichog   Fundulus heteroclitus

were commonly seen grazing and browsing on the dense growths. Table 6 gives a listing of the larger organisms, (tentatively identified) observed on, or associated with, the in situ SSFA and concrete blocks.

Results of metals analyses of colonizer biomass removed in August and September are presented in Table 7. Only two sets of analyses showed significant differences between metal concentrations in the biomass obtained from SSFA and concrete. The biomass obtained from the concrete block contained more Pb and Zn than did the biomass obtained from the SSFA during the September sampling, though the reverse occurred during the August sampling.

DISCUSSION

The major soluble compounds present in the SSFA test samples were calcium and sulfate. In the present work, the leaching of these compounds is controlled primarily by solubility. $CaSO_3$ also a major component in SSFA and important in the toxicity to marine organisms, was relatively insoluble in seawater. Under aerobic conditions, the sulfite released in the leaching process would be rapidly oxidized to sulfate, thus minimizing the impact of sulfite. However, in an anaerobic environment, concentrations of sulfite may increase due to a lack of oxygen for oxidation to sulfate.

Except for Hg, the concentrations of the heavy metals present in solid SSFA blocks were comparable to the concrete control. Hg concentrations in the SSFA ranged between 0.05 and 0.033 $\mu g\ g^{-1}$. The Hg in the concrete was 0.01 $\mu g\ g^{-1}$. However, the concentration of Hg in SSFA was less than the current permissible level of 0.75 $\mu g\ g^{-1}$ mandated by EPA (EPA, 1976) for solid waste disposal in the ocean. An elutriate test demonstrated that heavy metals were not leached appreciably from the solid phases in SSFA. Therefore, the presence of SSFA in the ocean may not pose a serious problem with respect to toxic heavy metals (Seligman, 1978).

The percolation and tank leaching experiments showed that release of trace heavy metals was higher from the sulfite mix than from the high sulfate mix. On the other hand, release of Ca and sulfur oxides was more rapid in the high sulfate mix because of the higher concentrations of the more soluble $CaSO_4$ compounds found in the high sulfate mix compared to the high sulfite mix. Dissolution of soluble components occurred primarily on the surfaces of the blocks due to the very low porosity and permeability of the blocks. Therefore the use of larger blocks with lower surface area: volume ratios would substantially increase the lifetime of the blocks in the ocean.

Although physical testing showed that concrete samples were harder than SSFA, both before and after seawater exposure, the

Table 7. Metal concentrations ($\mu g \ g^{-1}$) in encrusting biomass.
Data are means of replicate (n = 3) analyses of samples
collected on 16 August and 13 September 1977.

Metal[a]	Collection date	Concentration in SSFA biomass sample ($\bar{X}_s$)	Concentration in concrete biomass sample ($\bar{X}_{con}$)	"t" test results[b]
Cd	8/16	0.29	0.60	N.S.
	9/13	0.43	0.55	N.S.
Pb	8/16	40.2	37.7	N.S.
	9/13	33.4	40.4	*
Cu	8/16	46.2	49.7	N.S.
	9/13	59.8	63.5	N.S.
Cr	8/16	22.8	26.8	N.S.
	9/13	22.4	23.1	N.S.
Zn	8/16	287	232	N.S.
	9/13	201	239	*

[a]Hg and Ag concentrations were below instrumental detection limits.

[b]The hypotheses $H_o: \bar{X}_s = \bar{X}_{con}$, $H_a: \bar{X}_s > \bar{X}_{con}$ or $\bar{X}_c < \bar{X}_{con}$, were tested each date, using Student's "t" distribution $\alpha = 0.10$ (Sokal and Rholf, 1969), where $\bar{X}_{con}$ and $\bar{X}_s$ are mean metals concentrations measured on replicate analyses of samples obtained from concrete and SSFA surfaces, respectively. N.S. indicates no significant difference observed, i.e. H_o could not be rejected. An asterisk indicates the rejection of $H_o: \bar{X}_s = \bar{X}_{con}$; in both cases, $\bar{X}_{con}$ was significantly larger than $\bar{X}_s$.

compressive strength of SSFA exposed to seawater did not decrease as it does for exposed concrete. Based on the laboratory work it would appear that SSFA can maintain structural integrity in a seawater environment over extended periods of time. Long term in situ measurements have shown an increase in compressive strength of SSFA blocks after the initial 100 days of exposure to the marine environment (Fig. 10).

Some superficial surface softening in seawater, especially in test samples containing high concentrations of calcium sulfate, may effect the colonization of the SSFA by attached organisms in terms of species presence and in the rate of colonization. Early results obtained from the SSFA blocks in Conscience Bay indicate that surface softening is indeed occurring, and early differences were observed in the occurrence of colonizing invertebrates and algae when SSFA and concrete blocks were compared. But, as the blocks became encrusted and overgrown with attached organisms, especially those belonging to the phylum Bryozoa, the surfaces of the SSFA's blocks in Conscience Bay became similar to those of concrete.

Exposure of T. pseudonana cells to elutriates prepared from mix 5, a SSFA rich in sulfite, and mix 7, a SSFA rich in sulfate, produced modest reductions in cell volume, chlorophyll a concentrations, ^{14}C uptake and growth rates (Fig. 7). These results obtained in a laboratory situation where maximum surface area reacted with a restricted volume of seawater, indicate the SSFA may possess a potential to reduce productivity of some photosynthetic marine organisms. While an exposure to 10% elutriate concentration of mix 5 and an exposure to 20% (and 10%, to a lesser extent) elutriate concentration of mix 7 produced reduced biomass, ^{14}C incorporation, and growth rates early in the experiments, no effect was measured when cells in a later experiment were exposed to a 10% elutriate concentration of mix 7.

Our studies of the colonization process were confined to direct observations and to photographs of the algae and macro-invertebrates. These observations showed that the early stages of colonization of both SSFA and concrete materials began soon after block placement. About 55 days into the observation period, numerous bryozoan colonies were observed to be attached on the SSFA surfaces and limpets were attached only to the concrete. The different distribution of bryozoans and limpets may be related to the different surface hardness of SSFA and concrete. The sessile bryozoan colonies may prefer a softer, more easily penetrated surface, while the mobile and prehensile limpets may not be able to cope with softer surfaces that cannot provide a firm base for suction attachment. Later observations showed that limpets colonized the SSFA blocks after the block surfaces were encrusted by the bryozoans.

Because Conscience Bay was heavily iced-over during the severe

1977-78 winter, it was not possible to make underwater photographic surveys during the period January-March 1978. The dives that took place in April and May demonstrated that the colonizers survived very well the effects of the exceptionally severe winter season, and 431 days after placement. We observed a dense coverage of hydroids, algae, and other organisms on both the SSFA and concrete blocks.

CONCLUSIONS

1.   Stabilization of scrubber sludges and fly ash prevents the rapid mobilization and solubilization of the major and minor components present in SSFA.

2.   Leaching of the major calcium-containing compounds is primarily regulated by the concentration of each compound in the block, its solubility, and the surface area/volume ratio of the stabilized test blocks (Seligman, 1978). Dissolution of calcium sulfite is the only major component of environmental concern. In aerobic environments, sulfite is rapidly oxidized to sulfate.

3.   Trace metal concentrations in blocks of SSFA and leaching experiments (Seligman, 1978) suggest that heavy metal contamination by SSFA is not likely to be environmentally significant in aerobic environments.

4.   Under laboratory conditions, elutriates made from the SSFA mixes available may contain components which reduce productivity for at least one species of marine algae.

5.   SSFA maintains its structural integrity in situ for extended periods of time. An initial surface softening was observed which may have affected organism colonization; later the surfaces of the SSFA blocks were "biologically stabilized" due to the overgrowth of encrusting organisms.

6.   Early colonization of in situ SSFA and concrete surfaces began soon after placement. Differences in the occurrence of species of colonizing macro-invertebrates on SSFA and concrete surfaces were notable after about 50 days. These initial differences may be related to different surface hardness and differential selection by the settling colonizers. By day 111 after placement, clear differences could not be seen. After one and one-half years, the SSFA blocks have been completely covered by heavy growths of organisms.

7.   The concentrations of selected heavy metals in colonizer biomass removed from SSFA surfaces were not greater than concentrations measured in biomass removed from the concrete controls.

ACKNOWLEDGEMENTS

    This work was supported by grants from New York State Energy Re-
search and Development Authority (NYSERDA), New York Sea Grant Institute, the Department of Energy, the Environmental Protection Agency,
the Electric Power Research Institute, the Power Authority of the
State of New York and the Link Foundation. We are indebted to Sue
Oakley, Jackie Restivo, Mary Ann Lau and Ruth Toyama who were very
helpful during manuscript preparation. Steven Taub and John Minnick
provided valuable discussion during the initiation of this work. A
portion of this work will appear in F. Roethel's Ph.D. thesis. Marine Sciences Research Center Contribution No. 275.

REFERENCES

American Society of Testing and Materials (1974) Annual Book of ASTM
    Standards, Philadelphia, Pa.
Beers, W. F. et al. (1974) (In: Helm et al., 1975). Soil as a medium
    for the renovation of acid mine drainage water. Presented at
    Fifth Symposium on Coal Mine Drainage Research. National Coal
    Association, Washington, D.C.
Environmental Protection Agency (1976) Proposed revision of ocean
    dumping regulation and criteria. Federal Register, June 28.
Goldberg, E. D. (1954) Marine geochemistry of chemical scavengers of
    the sea. Journal of Geology, 62, 249-269.
Guillard, R. R. J. and J. H. Ryther (1962) Studies in marine plank-
    tonic diatom, I. Cyclotella nana Hustedt and Detoniela
    confervacae (Cleve) Gran. Canadian Journal of Microbiology, 8,
    229-239.
Hawley, J. E. and J. D. Ingle, Jr. (1975) Improvements in cold vapor
    atomic absorption determinations of mercury. Analytical
    Chemistry, 47, 719-723.
Helm, R. B., G. B. Keefer and W. A. Sack (1975) Environmental aspects
    of compacted mixtures of fly ash and wastewater sludge.
    Presented at 48th Annual Conference of Water Pollution Control
    Federation, Washington, D.C.
Herlihy, J. (1977) Flue gas desulfurization in power plants. Status
    Report. (April 1977) Division of Stationary Source Enforcement,
    Office of Enforcement, U.S. Environmental Protection Agency,
    Washington, D.C.
Hulsemann, J. (1966) On the routine analysis of carbonates in
    unconsolidated sediments. Journal of Sedimentary Petrology, 36,
    622-625.
Keeley, J. W. and R. M. Engler (1974) Discussion of regulatory
    criteria for ocean disposal of dredged materials: elutriate
    test rationale and implementation, March 1974. Misc. paper
    D-74-14. U.S. Army Corps of Engineers, Waterways Experiment
    Station, Vicksburg, Mississippi.
Lunt, R. R., C. B. Cooper, S. L. Johnson, J. E. Oberholtzer, G. R.

Schimke, W. I. Watson (1977) An evaluation of the disposal of
flue gas desulfurization wastes in mines and the ocean, initial
assessment. EPA Report 600/7-77-051, Research Triangle Park,
N.C.

Mahloch, J. L. (1976) Chemical fixation of FDG sludges - physical and
chemical properties. Presented at EPA symposium on Flue Gas
Desulfurization, New Orleans, La.

Mahloch, J. L. and D. E. Averett (1975) Pollution potential of raw
and chemically fixed hazardous industrial wastes and flue gas
desulfurization sludges: Interim Report. EPA-IAG Report
D4-0569. Waterways Experiment Station, Army Corps of Engineers,
Vicksburg, Ms.

Marine Sciences Research Center (1978) Aquatic disposal field
investigations: Eatons Neck disposal site, Long Island Sound,
Appendix B: Final Report (Contract No. DACW51-75-C0016, Work
Unit 1A06B), Waterways Experiment Station, Army Corps of
Engineers, Vicksburg, Ms.

Neville, A. M. (1973) Properties of Concrete. Halsted Press, New
York, N.Y.

Seligman, J. (1978) Chemical and physical behavior of stabilized
scrubber wastes and fly ash in seawater. M.S. Thesis. Marine
Science Research Center, State University of New York, Stony
Brook.

Sokal, R. R. and F. J. Rohlf (1969) Biometry: The Principles and
Practice of Statistics in Biological Research. Freeman and
Company, San Francisco.

Strickland, J. D. H. and T. R. Parsons (1972) A Practical Handbook of
Seawater Analysis. Bulletin 162 (2nd edition). Fisheries
Research Board of Canada, Ottawa.

Stumm, W. and J. J. Morgan (1970) Aquatic Chemistry, An Introduction
Emphasizing Chemical Equilibria in Natural Waters.
Wiley-Interscience, New York, N.Y.

Swenson, E. G. (1968) Performance of Concrete: Resistance of
Concrete to Sulfate and Other Environmental Conditions,
University of Toronto Press, Toronto.

Verbeck, G. J. (1956) Pore structure. ASTM special technical
publication, 1969, 136-142.

Zobell, C. E. (1946) Studies on redox potential of marine sediments.
Bulletin American Association Petrology and Geology, 30,
477-513.

CHEMICAL AND PHYSICAL PROCESSES IN A DISPERSING

SEWAGE SLUDGE PLUME

Patrick G. Hatcher[1], George A. Berberian, Adriana Y.
Cantillo[2], Philip A. McGillivary[3], Philip Hanson[4],
and Richard H. West[5]

NOAA, Atlantic Oceanographic and Meteorological
Laboratories
Ocean Chemistry Laboratory
15 Rickenbacker Causeway, Miami, Florida 33149

Present Address: 1. U.S. Geological Survey, Reston, VA 22092
                 2. NOAA, NOS, Engineering Development Lab,
                     Rockville, MD 20582
                 3. Institute of Ecology, University of Georgia,
                     Athens, GA 30602
                 4. Department of Chemistry, Arizona State
                     University, Tempe, AZ 85281
                 5. 1463 Exposition Blvd., Sacramento, CA 95815

ABSTRACT

In July 1976, an experiment was conducted to determine the
physical and chemical changes occurring during a sewage sludge dump
in a thermally stratified water column in the New York Bight with a
line dump, a modified line dump, and one spot dump. The results
indicate that a rapid dilution occurs in the line dump such that
dissolved and particulate trace metals, nutrients, and bulk organic
compounds are diluted to near background levels. Only the trace
organic compounds, ammonia, and dissolved Cu provide any indication
for presence of the plume for a period of 5 hours after dumping. The
fecal steroid coprostanol is used as a quantitative sludge tracer to
determine the initial 10^4 dilution of the plume, to measure the
amount of mixing as the sludge disperses in the stratified water column, and to determine if any other chemical compounds deviate from
conservative mixing. There is physical and chemical fractionation as
the sludge settles. Initially, a large portion rapidly settles

within a narrow zone below the thermocline. Above the thermocline in
the mixed layer, the sludge mixes and disperses more rapidly. Dissolved sludge components and the sludge plume, thereby, undergo dis-
solved/particulate fractionation. In the spot dump and modified line
dump, the particulate and dissolved sludge are diluted a factor of 10
less than in the line dump and the plume concentrations of measured
chemical parameters are a factor of 10 more elevated than background
values.

INTRODUCTION

Sewage sludge, industrial wastes, and dredge spoils have been
ocean dumped for decades along the east coast of the United States
(Pararas-Carayannis, 1973). The dispersal and chemical effects of
such wastes on the biota have been identified recently as potential
environmental problems. In the past several years, many studies have

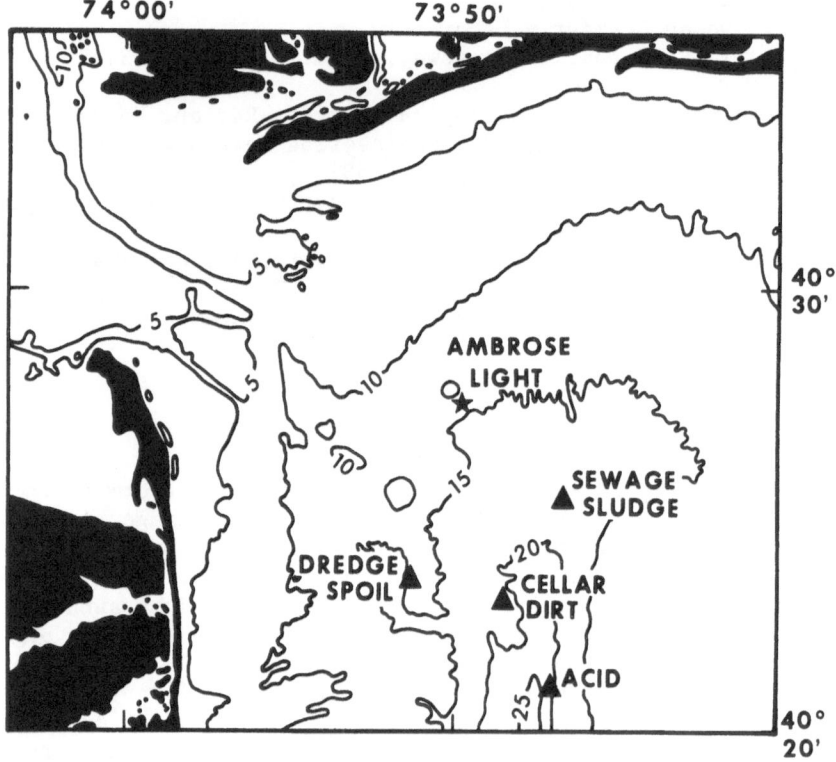

Fig. 1.   Index map of the New York Bight Apex showing the various
          waste dumping areas. Contour intervals are measured in
          meters.

been initiated to define the dispersal of these wastes and their associated "toxic" substances (Proni et al., 1976; Duedall et al., 1975; Mackay et al., 1972). In every case, the chemical data obtained was less than satisfactory due to logistical problems involved in adequately sampling the plumes. Locating the plumes and positioning sampling equipment in them is a formidable task; however, the recent use of acoustic sounding devices to guide sampling equipment has proved valuable (Proni et al., 1976).

In July 1976, an experiment was conducted to examine acoustical and chemical changes in particulate and dissolved waste immediately after sludge dumping in the New York Bight. The experiment was called STAX II for Sludge Tracking Acoustical Experiment II and was conducted in the EPA designated sludge disposal area shown in Fig. 1. The physical aspects of the dumping process have been described by Proni et al., (1980). Dissolved and particulate trace metals, nutrients, bulk organic compounds, and trace organic compounds were measured and we report the observed chemical changes in the dispersing plume of the dump.

EXPERIMENTAL DESIGN

The experiment was initally designed to examine four sewage sludge dumpings; two spot dumps and two line dumps. Due to sampling and analytical limitations, only three of the dumps could be examined (two spot dumps and one line dump). For 36 hours prior to the initial dump, no ocean dumping of sludge was permitted in the study area (Fig. 1) which has been defined as the EPA dump zone (south and east of 40°23'30"N, 73°43'45"W). On July 11, 1976, water samples were collected from this area to provide "background" information. The one station occupied was labelled Station 1 (Table 1).

On July 12, 1976, the New York City dumping vessel NORTH RIVER began a line dump, discharging sewage sludge collected from Newtown Creek sewage treatment plant at 10:19 A.M. Within 20 minutes, the NOAA Ship GEORGE B. KELEZ traversed the plume, and the water samples were collected in the plume using 10-liter, top-drop Niskin bottles and a 30-liter "GO-FLO" Niskin bottle. The bottles were mounted on a General Oceanic rosette sampler and were positioned in the plume with the aid of a 200 kHz acoustic water column profiler described by Proni et al. (1980). Subsamples of the water collected from various depths were processed according to methods outlined below. Few samples were collected on this first dump due to shipboard logistical constraints, but the samples proved to be isolated in the plume of the dump and comprised the best set of samples collected during the experiment. In all, four stations were occupied (Table 1).

On July 14, 1976, a modified line dump was sampled, whereby the dumping vessel discharged its wastes while slowly drifting. The

Table 1. Station, time, and sampling depths for STAX-II

Date	Local Time	Station	Depth (m)	Sample
7-11-76	1322-1430	1	1,12	Background
7-12-76	1044-1050	2	1,14,18	Dump 1
7-12-76	1245-1257	3	12,16,18,24	Dump 1
7-12-76	1527	5	4	Dump 1
7-14-76	1013-1023	6	1,5,13,18,24	Dump 2
7-14-76	1115-1127	7	1,4,10,14,22	Dump 2
7-14-76	1303-1311	8	1,5,8,16,23	Dump 2
7-14-76	1426-1442	9	1,6,9,16,24	Dump 2
7-14-76	1608-1620	10	1,4,6,12,24	Dump 2
7-16-76	0914-0947	15	1,4,22	Background
7-16-76	1105-1107	16	1,12,25	Dump 3
7-16-76	1208-1218	17	1,4,9,10,22	Dump 3
7-16-76	1336-1344	18	1,4,6,14,22	Dump 3
7-16-76	1507-1519	19	1,5,6,14,24	Dump 3

KELEZ began sampling the plume within ten minutes of dumping and five stations were occupied (Table 1). Acoustic tracking of this plume proved to be less successful due to electronic noise problems and other logistical constraints.

On July 16, a spot dump was tracked and a background station (Station 15) was occupied prior to release of the wastes. The sludge vessel dumped its load all at once and departed. Sampling problems similar to those of the July 14 dump were encountered, and the interpretation of data from the four stations occupied is limited.

In addition to water samples, sewage sludge samples were collected from the dumping vessel prior to discharge of the wastes. These were analyzed in the same fashion as the water samples, with slight modifications as described below.

METHODS

Sampling and Analysis of Seawater. As part of the STAX II experiment, samples of seawater were collected from 10-liter and 30-liter Niskin bottles for multiple chemical analyses. A flow diagram of the aliquots taken and the chemical analyses performed on these aliquots is shown in Fig. 2. Sampling and analysis are briefly described below for the various types of chemical compounds analyzed.

Salinity samples were collected from the 10 liter top-drop Niskin bottles and stored in aged glass bottles. They were analyzed using a Beckman Model RS-7B induction salinometer whose accuracy and precision are $0.003°/_{oo}$ and $0.004°/_{oo}$, respectively.

The temperature of the water column was recorded using XBT probes launched from the KELEZ. Dissolved oxygen samples were drawn and immediately preserved using manganous sulfate and alkaline iodide. The analysis of these samples was conducted on board by Dr. James Thomas of the National Marine Fisheries Service, Sandy Hook

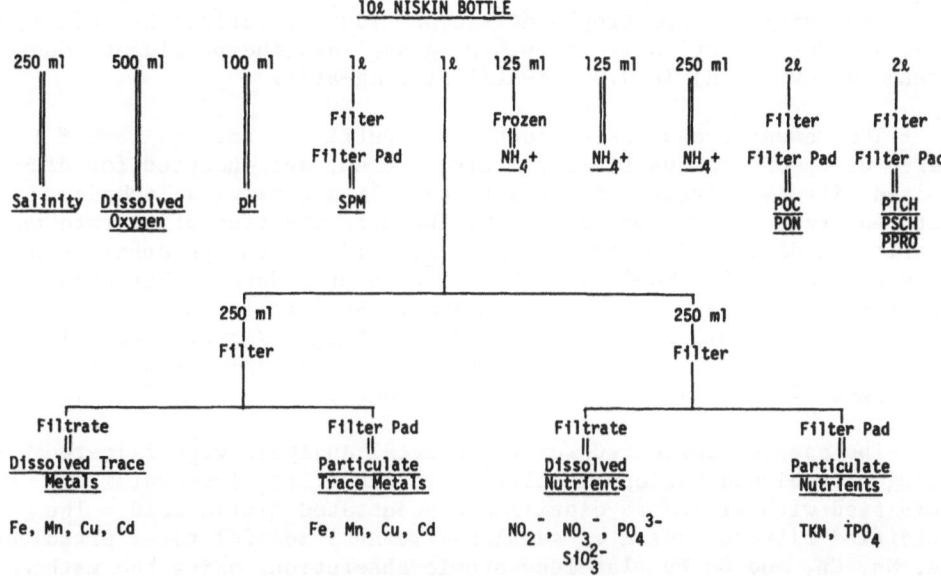

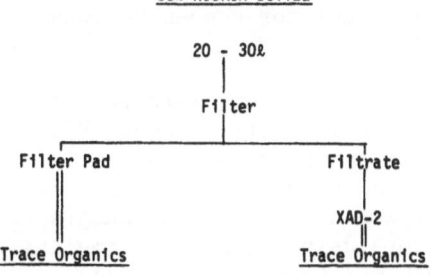

Fig. 2.   Schematic of aliquots taken from the Niskin bottles.

Laboratory, using a modified Winkler method of Strickland and Parsons (1968).

The pH samples were drawn and determined immediately with a digital pH meter and a glass electrode standardized with pH 7 and pH 10 buffers. Measurements of pH under those conditions should be viewed with caution as many problems have been encountered with pH measurements in seawater.

Three aliquots were drawn for ammonia; one was preserved with phenol/alcohol and analyzed on board 12 hours later; another was preserved with phenol/alcohol, frozen, and analyzed at our laboratory in Miami, and the third was preserved with sulfuric acid and analyzed by Mr. John VanLandingham of Connell, Metcalf and Eddy Associates of Coral Gables, Florida. The ammonia analyses performed on board are most reliable, and only these will be reported. These samples were analyzed using the spectrophotmetric procedure described by Solorzano (1967). The detection limit is 0.5 µg-at/l and the precision (one standard deviation) is 0.07 µg-at/l at 3 µg-at/l.

The seawater samples collected for nutrient analyses were filtered using GF-C glass fiber filters, frozen, and analyzed for dissolved nitrate, nitrite, phosphate and silicate using a Technicon Auto-Analyzer I. The procedure for the determination of nitrate and nitrite is described by Armstrong et al. (1967), the phosphate procedure by Grasshoff (1965), and the silicate procedure by Strickland and Parsons (1968). The detection limits and precisions are as follows: nitrate, 0.5 µg-at/l and 0.17% at 15 µg-at/l; nitrite 0.1 µg-at/l and 1.2% at 3.2 µg-at/l; phosphate, 0.05 µg-at/l and 1.6% at 0.64 µg-at/l; and silicate, 0.5 µg-at/l and 1.0% at 25 µg-at/l.

The samples collected for trace metal analyses were filtered using pre-weighed Nuclepore filters and the filtered seawater was acidified with silica re-distilled concentrated nitric acid. The acidified filtered seawater samples were analyzed for total dissolved Fe, Mn, Cu, and Cd by flameless atomic absorption, using the method of Segar and Cantillo (1976a). The approximate detection limits of the analyses are as follows: Fe, 0.4 µg/l; Mn, 0.3 µg/l; Cu, 0.5 µg/l; and Cd, 0.01 µg/l. The precision of the analyses varied, but it was always better than ± 10% for concentrations in excess of ten times the detection limit.

The pre-weighed Nuclepore filters were freeze-dried and sent to Dr. Peter R. Betzer at the University of South Florida for dissolution. A weak-acid soluble leach and a strong-acid digestion were performed on each filter pad as described by Betzer (1978), and the resulting solutions analyzed at Atlantic Oceanographic and Meteorological Laboratories (AOML) using standard flameless atomic absorption techniques (Perkin-Elmer Corporation, 1973).

One liter of seawater was filtered through a pre-weighed Nuclepore filter which was subsequently washed with distilled water, frozen, freeze-dried, and re-weighed to determine the suspended particulate load.

Particulate matter from two liters of seawater filtered through
a GF-C glass fiber filter, which was then frozen freeze-dried, was
analyzed for POC and PON by Galbraith Laboratories, Knoxville, Ten-
nessee, using the dry combustion procedure outlined by Cantillo et
al. (1980a). A precision ± 10% was determined for POC and PON
analyses.

Particulate total carbohydrates (PTCH) and Particulate Proteins
(PPRO) were obtained from two liters of seawater filtered through a
GF-C glass fiber filter which was subsequently frozen and freeze-
dried. The filters were analyzed for PTCH by the modified method of
Gerchakov and Hatcher (1972) and for PPRO by the modified method of
Greenfield et al. (1971). The precisions of the carbohydrate and
protein analyses are better than ± 20%.

Dissolved organic carbon (DOC) and dissolved inorganic carbon
(DIC) were measured in 125 ml amber glass bottles containing a
$HgCl_2$ solution, which was filled with seawater from the 10-liter
Niskin bottle. These samples were analyzed for DOC and DIC by a
Beckman Model 915 carbon analyzer. The detection limit for both DIC
and DOC was 1 mg/l and the relative standard deviation at 20 mg/l,
was ± 1%.

Usually 20 to 30 liters of seawater were collected from the 30-
liter Niskin bottle and filtered by pressure through 142 mm GF-C
glass fiber filters. The filters were frozen immediately, later
freeze-dried, and extracted with a 1:1 v/v methanol/benzene solution.
The filtrate was acidified to pH 2 and passed through XAD-2 resin
columns (2 cm x 20 cm) by nitrogen pressure at a rate of approximately 400 ml/min. The XAD columns were eluted, first with a 10% methanol in ethyl ether solution, and then with a methanol solution. The
methanol/ether eluate was subsampled for hydrocarbon, fatty acid,
steroid, and PCB analysis. The methanol eluate was subsampled for
PCB analysis.

The extracts of both the particulate matter on the filters and
the dissolved compounds adsorbed to the XAD-2 resins, were processed
in a similar fashion. The extracts were evaporated and saponified.
The solutions were then acidified, extracted with CCl_4, and evaporated to dryness. The residue was then esterified with 14% BCl_3 in
methanol, and recovered in hexane. The hexane solution was placed on
a silica gel column and fractionated into three eluates: hexane,
containing the saturated hydrocarbons; benzene, containing mostly the
fatty acid methyl esters; and methanol, containing the steroids. The
hydrocarbon fraction was concentrated and injected onto a

30 m x 0.2 mm WCOT SE-30 column and programmed from 100°C to 270°C at 6°/min with a four-minute initial hold. The fatty acid methyl esters were similarly chromatographed, but at an 8°/min program rate. The methanol eluates were evaporated to dryness and treated with BSA (N,O bis trimethylsilylacetamide) to form the trimethylsilyl ether derivatives of steroids. The BSA solution was then injected onto an SE-30 WCOT column (18 m x 0.2 mm, programmed at 200°C to 270°C at 4°/min). Individual hydrocarbons, fatty acids, and steroids were identified by retention time, and assignments were confirmed by gas chromatography-mass spectrometry. Comparisons were made with authentic standards. Quantitative measurements were made from electronically calculated peak areas using an external standard technique. Precision for measurements of relative peak concentrations (% coprostanol, % fatty acid of total fatty acids) is $\pm$ 10%. Absolute accuracy for steroid concentrations (μg/gm) is $\pm$ 20%.

Particulate and dissolved PCB analysis was conducted in a similar fashion to that reported by West et al. (1980). Both the methanol/ether and methanol extracts of the XAD resin were analyzed along with the benzene/methanol extract of the filtered particulates.

Samples of sewage sludge were collected from the barges and frozen within six hours. These samples were thawed, weighed, and centrifuged to isolate dissolved and particulate fractions for chemical analyses. The supernatant was decanted, and the precipitate was weighed and freeze-dried. A subsample of the solids was extracted with 1:1 v/v benzene-methanol and analyzed for trace organics by gas chromatography. An aliquot of the supernatant was extracted with CCl_4, and the extract analyzed for dissolved trace organics. Another aliquot of the supernatant was diluted (1:30), acidified with re-distilled concentrated nitric acid, and analyzed for dissolved trace metals. The remainder was also diluted (1:20) with distilled water, filtered, and the filtrate analyzed for dissolved nutrients. The trace metal analysis was performed by flameless atomic absorption techniques as described by the Perkin-Elmer Corporation (1973). The dissolved and particulate trace organic analyses (hydrocarbons, fatty acids, and steroids) were the same as previously described.

RESULTS AND DISCUSSION

The data obtained on sewage sludge and samples of all three dumps are presented in tabular form by Cantillo et al. (1980b). Virtually no previous chemical data other than the study of Duedall et al. (1977) existed describing rapid changes in water column chemistry due to sludge dumping. Therefore, measurements were made of a large number of organic and inorganic compounds indentified as important contributors to sewage sludge in order to determine the impact these might have on the background levels of chemical compounds in waters of the New York Bight.

Background. Due to failure of the rosette sampler, an adequate number of background samples was not collected prior to dumping. Therefore, we identified, acoustically all samples which were collected outside the plume of dumped material and assumed that they were representative of background conditions. The apparent high degree of water column variability at this site during this time of year precludes establishing an accurate measure of background concentrations. Therefore, background values are reported as a range of concentrations observed throughout the experiment (Table 2). Because the thermocline strongly stratifies the water column, the range of background values established for the surface includes all samples above the thermocline, while background values below the thermocline incorporate all samples below 15 m depth. Thermocline depths were established by CSTD profiles and XBT profiles before and during the experiments and are described in detail by Proni et al. (1980).

Data for bulk organic compounds, trace metals, and nutrients in background samples (Cantillo et al., 1980b) were generally typical for summer water conditions in the Bight where a large plankton bloom appeared to dominate the water column particulates (Segar and Berberian, 1976; Segar and Cantillo, 1976b). Photomicrographs of this particulate matter indicated the presence of large amounts of Ceratium sp. organisms thought to be responsible for the development of near anoxic conditions in the lower water column in the summer of 1976 (Malone, 1978). Low dissolved oxygen concentrations were observed below the thermocline during this experiment.

Although the bulk organic compounds and total suspended loads gave some clue of the quantity of organic and inorganic matter in the water column, the nature or source of this material was ill-defined by these parameters. The trace organic compounds were more indicative of composition. The hydrocarbon distributions shown in Figure 3 are characterized by homologous series of n-alkane peaks superimposed on a bimodal unresolved complex mixture. The distribution of n-alkanes strongly suggests oil contamination of the particulate matter, possibly by motor oil which contains high molecular weight alkanes similar to those found in the sample. The unresolved complex mixture's mode in the low molecular weight region (shorter elution time off the gas chromatograph) has been observed in sandy sediments of the Bight (Hatcher, unpublished results) and attributed to plankton or bacterial sources.

The fatty acid chromatograms are shown in Fig. 4. The distributions are dominated by C_{14}, C_{16}, and C_{18} fatty acids, typical of phytoplankton. One major difference between the 1 and 12 m samples is the relative concentration of the $C_{16:1}$ fatty acid (singly unsaturated). Loss of unsaturation in fatty acids is a well-documented phenomenon in the decomposition of organic matter (Rhead et al., 1971). It appears, then, that the organic matter at 12 m depth has undergone some decomposition since most of the $C_{16:1}$ fatty

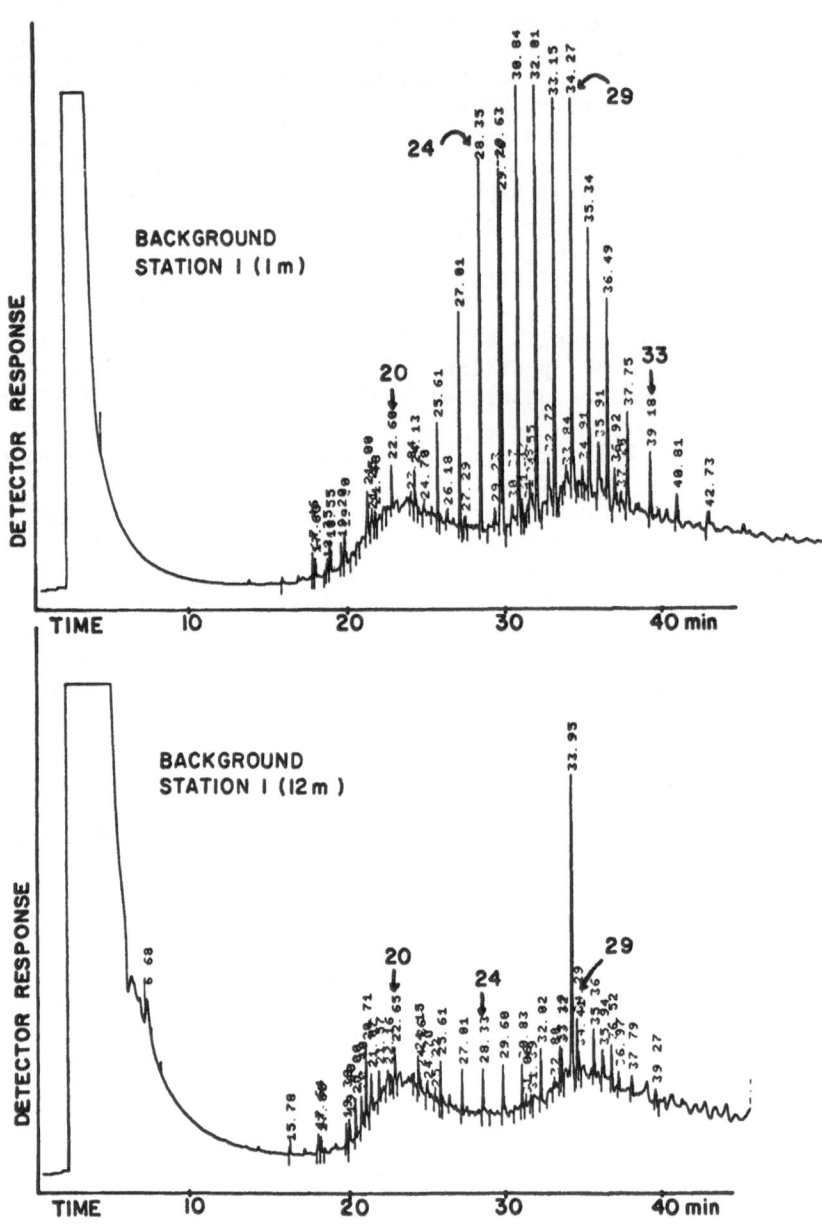

Fig. 3. The saturated hydrocarbon chromatograms for background
         particulate matter at 1 m and 12 m depths. The SE-30, 30
         m x 0.2 mm, WCOT column was programmed from 100°C to 270°C
         at 6°/min with a 4 min initial hold. The bold numbers
         identify the n-alkane carbon numbers.

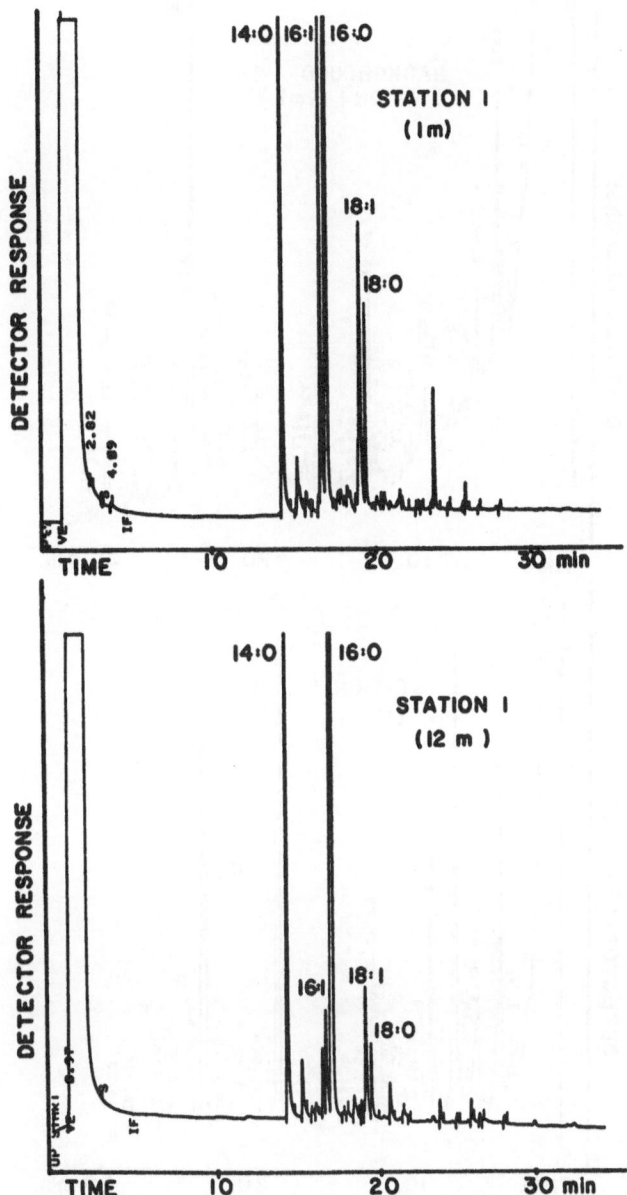

Fig. 4. The fatty acid (methyl esters) chromatograms for
background particulate matter. The SE-30, 30 m x 0.2 mm,
WCOT column was programmed from 100°C to 270°C at 8°/min
with a 4-min initial hold. The first number refers to
the fatty acid carbon number and the number following the
colon denotes the number of double bonds present.

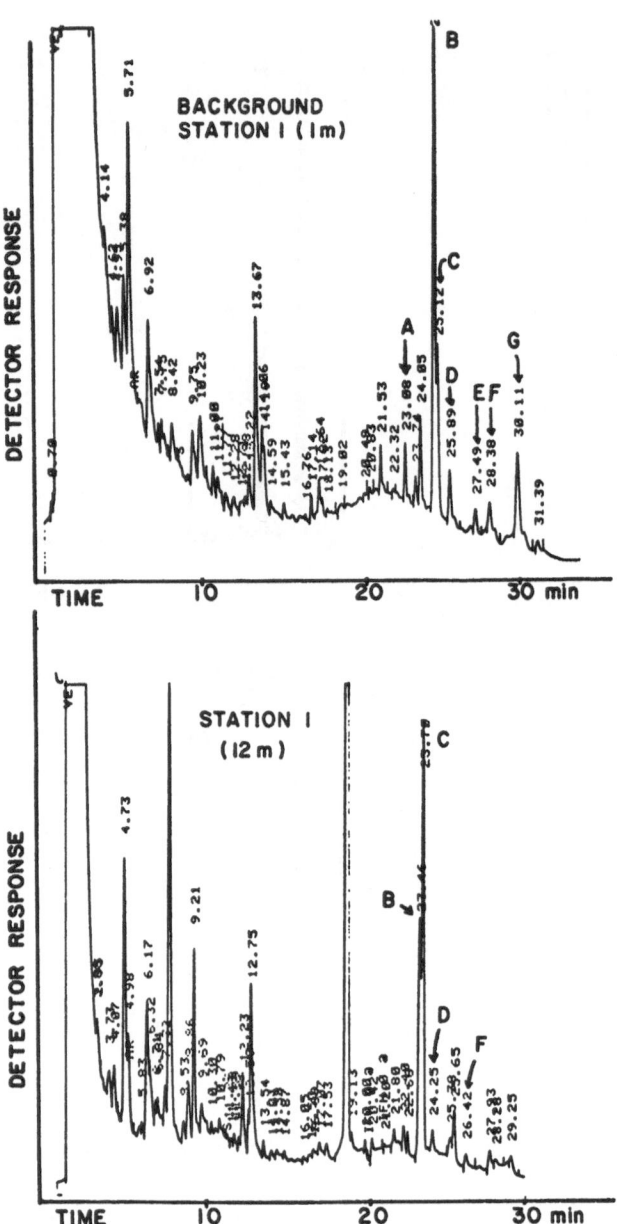

Fig. 5. The steroid (TMS derivatives) distributions for background particulate matter. The SE-30, 18 m x 0.2 mm, WCOT
column was programmed from 200°C to 270°C at 4°/min with a
4-min initial hold. The steroids are (A) coprostanol,
(B) cholesterol, (C) cholestanol, (D) brassicasterol, (E)
campesterol + 24β-ethylcoprostanol, (F) stigmasterol,
and (G) sitosterol.

Table 2. The sludge concentration, background, and measured concentrations of chemical constituents at Station 2, July 12, 1976.

Parameter	Units	Sludge Concentration	Station 2 (1 m) Dilution 1/2.4 x 10^4*				Station 2 (18 m) Dilution 1/1.0 x 10^4*		
			Sludge Contribution	Background	Measured		Sludge Contribution	Background	Measured
Particulate:									
SPM	mg/l	**3.7 - 4.2 x 10^4	1.3 - 2.1	1.7 - 3.4	2.8		3.1 - 5.2	0.73 - 1.4	9.1
POC	mg/l	**1.0 - 1.1 x 10^4	0.36 - 0.55	0.90 - 1.9	0.94		0.82 - 1.4	0.43 - 1.0	2.5
PON	mg/l	**1.3 - 1.4 x 10^3	0.05 - 0.07	0.16 - 0.31	0.18		0.11 - 0.18	0.07 - 0.12	0.40
Fe	µg/l	4.7 - 12 x 10^5	17 - 60	20 - 50	39		39 - 150	29 - 56	230.
Cu	µg/l	3.5 - 6.1 x 10^4	1.3 - 3.0	0.83 - 3.0	4.4		2.9 - 7.6	0.68 - 4.5	30.
Cd	µg/l	2.7 - 5.9 x 10^3	0.10 - 0.30	BDL - 1.0	0.49		0.23 - 0.73	BDL - 0.69	2.2
Mn	µg/l	1.3 - 1.8 x 10^4	0.46 - 0.90	3.5 - 36	2.8		1.1 - 2.2	5.2 - 16	11.
Dissolved:									
Fe	µg/l	2.2 - 2.8 x 10^3	<0.4	3.5 - 6.9	6.9		<0.4	5.2 - 8.7	12.
Cu	µg/l	5.9 - 7.1 x 10^2	<0.5	2.5 - 8.2	44.		<0.5	3.0 - 8.9	6.5
Cd	µg/l	32 - 38	<0.01	0.70 - 2.6	1.2		<0.01	0.19 - 2.4	0.53
Mn	µg/l	33 - 39	<0.3	7.8 - 36	12.		<0.3	10 - 160	33.
NH_4^+	µg-at/l	2.1 - 2.3 x 10^3	<0.5	<0.5	8.1		<0.5	<0.5	<0.5
NO_3^-	µg-at/l	40 - 46	<0.1	<0.1	<0.1		<0.1	<0.2	<0.12
NO_2^-	µg-at/l	15 - 17	<0.1	<0.1	<0.1		<0.1	<0.1	<0.1
PO_4^{3-}	µg-at/l	1.8 - 2.2 x 10^3	0.96 - 0.11	0.19 - 0.56	0.57		0.15 - 0.28	0.23 - 0.71	0.78

BDL = Below Detection Limits
*These factors calculated from coprostanol data carry on error of 20%
**From I. Duedall
‡The reported ranges of values in the Table are high and low values obtained from duplicate analyses.

acid, predominately derived from phytoplankton, has been depleted. This observation is supported, also, by the particulate carbon and nitrogen data. The C/N ratio at 1m is 4.7, typical of living phytoplankton cells. At 12 m, the C/N ratio increased to 7.2, a ratio more nearly typical of detrital material. As mentioned earlier, photomicrographs indicate that Ceratium tests are the most important contributors to the suspended matter. It is likely that changes described above are due to decomposition of Ceratium cells which have died and settled to the pycnocline near 12 m depth.

The steroid chromatograms for the background particulate material are presented in Fig. 5. At 1 m depth, cholesterol, cholestanol, sitosterol, and stigmasterol are the major steroids (cholesterol predominant), which is typical of plankton. Interestingly, a small amount of coprostanol, a fecal steroid, is present, indicating residual sewage-derived materials in the water column. At 12 m, the steroid distribution changes dramatically, with cholestanol being the predominant steroid. The presence of such large amounts of cholestanol is unreported in plankton. Since cholestanol is thought to be derived from cholesterol via decomposition reactions (Nishimura and Koyama, 1977), it is likely that the Ceratium at 12 m depth have been subjected to such decomposition and their cholesterol converted to cholestanol, an observation which tends to support the fatty acids and C/N ratio analyses discussed above.

Dump 1: 12 July 1976. This particular dump, a line dump, dispersed over a wide area in a relatively short period of time. Acoustic traces of a 90° transect of the plume (Fig. 6) shows that within 8 minutes a large fraction of dumped material penetrated the strong thermocline at about 13 m and reached the bottom. The lateral dispersion in the mixed layer was much greater than below the thermocline. Calculations made from the acoustic intensity indicated that the sludge below the thermocline was more concentrated than that in the mixed layer (J. Proni, personal communication), an indication that the greater mixing above the thermocline has effectively diluted the sludge by dispersing it over a wider area.

Initial Chemical Changes. Collecting samples of the plume above and below the thermocline within 20 minutes of the dump, we had hoped to determine the impact of dumping on ambient concentrations of a wide range of chemical constituents. We had also hoped to determine whether any of these concentrations deviated from conservative mixing. Unfortunately in this short interval, the concentrations of nearly all bulk chemical parameters such as suspended particulate matter (SPM), particulate organic carbon (POC), salinity (S°/$_{oo}$), temperature (T), dissolved oxygen (DO), dissolved inorganic carbon (DIC) dissolved organic carbon (DOC), particulate total carbohydrates (PTCH), particulate proteins (PPRO), and particulate organic nitrogen (PON) were not significantly different from background concentrations in plume samples above and below the thermocline (Table 2). The only

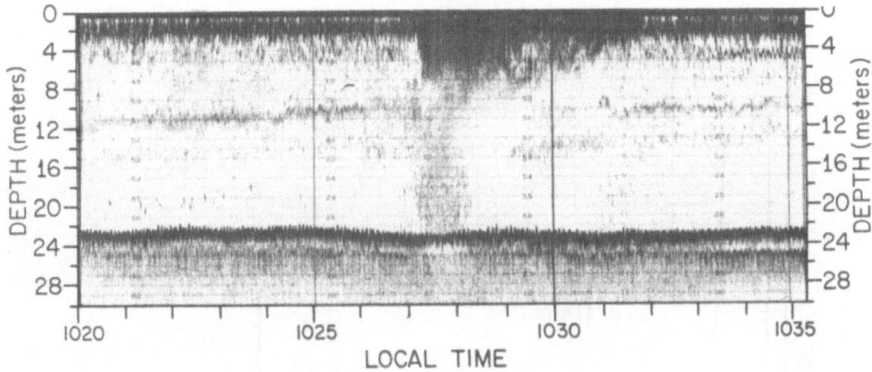

Fig. 6.   The 200 kHz acoustic return from the first perpendicular traverse of dump 1 (from J. Proni).

trace metals which showed an increase in concentration over background were dissolved Cu in the surface layers and particulate Fe at depth.   Furthermore, all nutrients, except ammonia (in the surface sample), were diluted to background levels and no elevated concentrations were detected in the plume.   Apparently the dilution was so great in this line dump and the background levels of these constituents were so high in New York Bight waters that no major changes were detectable.

Trace organic compounds such as hydrocarbons, fatty acids, steroids and PCB's are extremely sensitive to source changes as the concentrations in sludge are so much more elevated than background that dilutions of 10^6 or more can be detected.   Comparisons of hydrocarbons and fatty acid chromatograms of particulate material from the plume samples with the hydrocarbons and fatty acid chromatograms of background materials are shown in Fig. 7 and 8.   The saturated hydrocarbon distribution (Fig. 7) of the diluted material at station 2 (18 m) in the plume was identical to that of sewage sludge (Hatcher, unpublished results).   Obviously, the saturated hydrocarbon content of sludge was high enough to completely dominate the hydrocarbon character of the mixture with seawater in the plume.

The particulate fatty acid chromatograms (Fig. 8) also showed large differences before and after dumping.   Iso and anteiso C_{15} fatty acids, as well as $C_{18:1}$ and C_{18} acids, were more abundant in the sludge-seawater mixture than in the background samples.   Furthermore, the C_{20} - C_{26} series of fatty acids, even numbered acids predominating, were also significantly more apparent in the post dump particulate matter at station 2 (18 m).   These fatty acids are typical of terrestrial material or of vascular plant material found in sewage sludge.

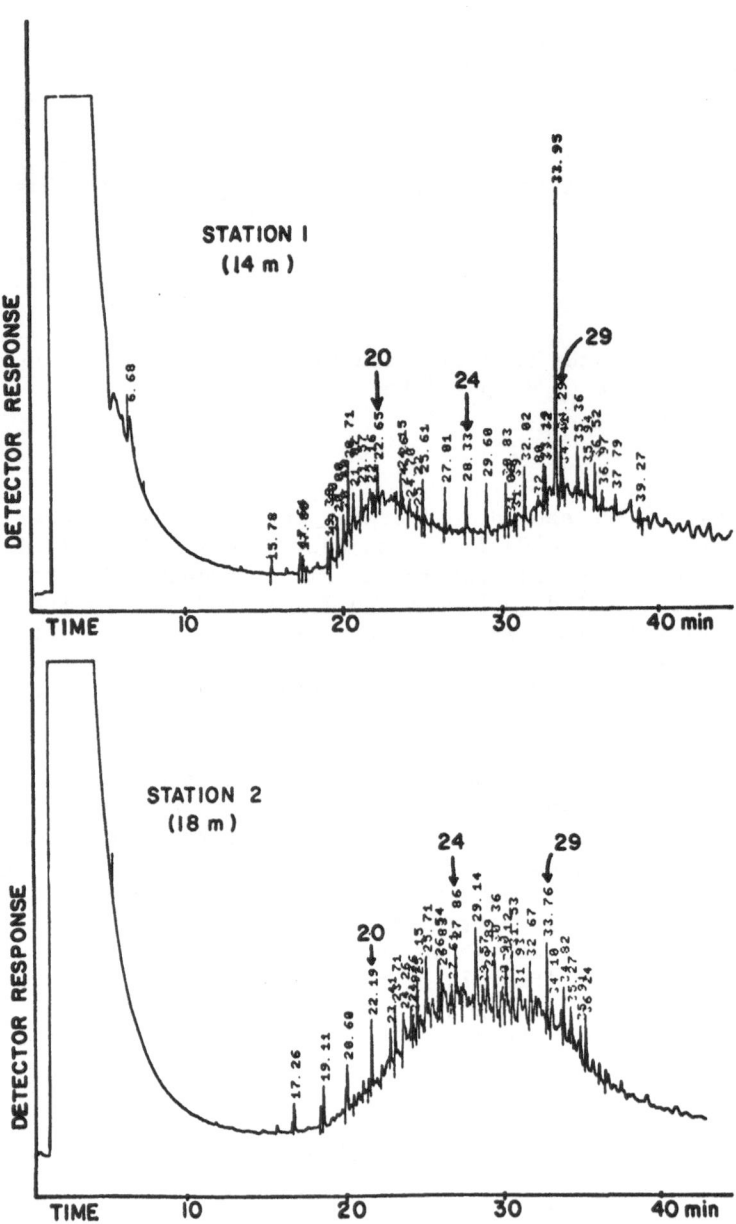

Fig. 7. The saturated hydrocarbon distributions in background and sludge plume particulate. Chromatographic conditions are the same as in Fig. 1.

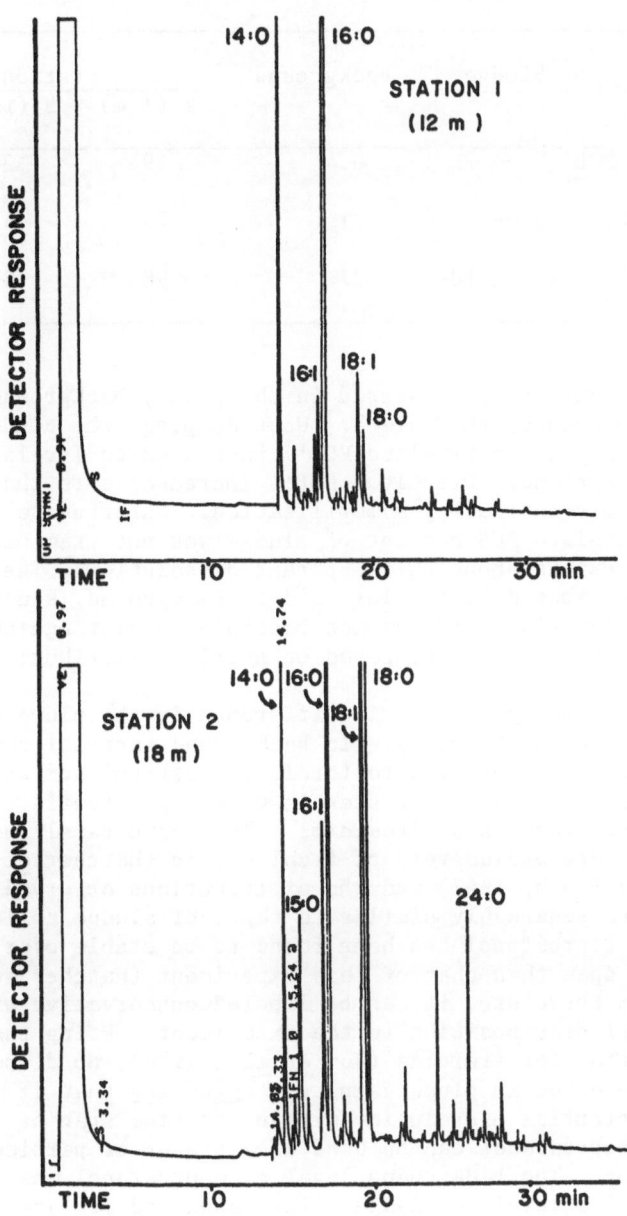

Fig. 8.   The fatty acid distributions in background and sludge
          plume particulates.   Chromatographic conditions are the
          same as in Fig. 2.

Table 3.   Total PCB concentrations (µg/l) in sludge and dump 1
           samples using Aroclor 1248 as a standard.

Fraction	Sludge	Background	Station (depth)		
			2 (1 m)	3 (12 m)	5 (4 m)
Particulate PCB	--	6.4	19	38	36
Dissolved PCB	--	21	79	47	58
Total PCB	2.6×10^6	27	98	85	94

    PCB concentrations measured in the plume, background, and sewage
sludge are presented in Table 3.   Upon dumping, the concentrations of
both dissolved and particulate PCB's increased to levels significantly above background.   Dissolved PCB's increased more than particulate
PCB's, possibly indicating solubilization.   Unfortunately, the dis-
solved/particulate PCB content of sludge was not examined, so some
uncertainty exists about this apparent dissolution.   The PCB distribution was dominated by Aroclor 1248 in background, sludge, and in
the plume.   Therefore, it was not possible to distinguish the sludge
plume from background PCBs, based on Aroclor distributions.

    Steroids showed a dramatic difference in the plume compared to
background material (Fig. 9).   In background material the major
sterols were cholesterol, sitosterol, cholestanol and stigmasterol.
In plume material, the major sterols were coprostanol, cholesterol,
24β-ethylcoprostanol and sitosterol.   The coprostanol and 24β-ethyl-
coprostanol were exclusively of fecal origin (Hatcher et al.,1977;
Murtaugh and Bunch, 1967) and the distributions observed for plume
material were remarkably similar to those of sludge collected from
the barge.   Coprostanol has been found to be stable over a much
longer time span than that of this experiment (Hatcher and McGillivary, 1979); therefore, it can be labeled conservative with respect
to biological decomposition in the experiment.   Using the XAD-2 adsorption method for steroids (Wun et al., 1976), no dissolved coprostanol was detected in plume samples, suggesting that it did not
undergo substantial dissolution within the time span of the sampling.
Therefore, coprostanol can be used as a tracer of particulate sludge
during a dump.   The background level of coprostanol was quite small
compared to its level in sludge.   This enhanced its use as a tracer.

    Because of this conservative nature, coprostanol concentrations
(Table 4) were used to calculate dilutions.   Then, applying these
dilutions to other measured parameters, deviations from conservative
mixing were noted and ascribed to possible chemical changes occurring
in the plume.   Using the coprostanol concentrations in the sludge

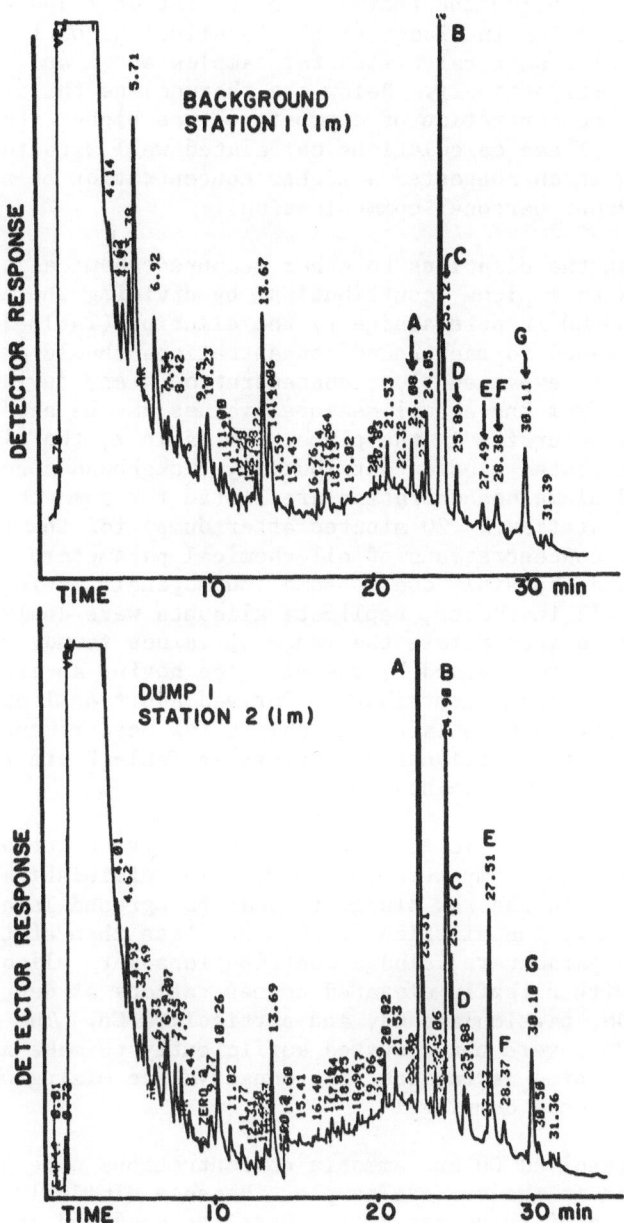

Fig. 9.   The steroid distributions in background and sludge plume
          particulates.   Chromatographic conditions are the same as
          in Fig. 3.   The steroids are (A) coprostanol, (B) choles-
          terol, (C) cholestanol, (D) brassicasterol, (E) campesterol
          + 24β-ethylcoprostanol, (F) stigmasterol, and (G)sitos-
          terol.

plume above and below the thermocline at station 2 and the coprostanol concentration in sludge (Table 4) dilutions of 1: 2.4 x 10^4
and 1: 1.0 x 10^4 were calculated for samples above and below the
thermocline, respectively. Below the thermocline the dilution was
less and the concentration of coprostanol was higher than above the
thermocline. These calculations correlated well with the acoustic
observations which suggested a higher concentration of sludge at
depths (J. Proni personal communication).

    Applying the dilutions to other measured chemical compounds we
have calculated a plume "contribution" by dividing the concentration
of each compound in pure sludge by the dilution (Table 2). These
values when added to background concentrations should provide an
estimate of the expected plume concentration. Any deviations of
these values from the actual measured values can be ascribed to chemical changes occurring in the plume. In Table 2, the sludge concentration, calculated sludge contribution, background concentration,
and measured plume concentration are listed for some chemical constituents at station 2 (20 minutes after dump) for the 1 m and 18 m
depths. The concentrations of all chemical parameters in sludge were
quite variable, possibly due to some inhomogeneities in the sludge;
however, in all instances, replicate aliquots were analyzed and the
reported errors incorporate the range of values found. When these
sludge values were divided by the dilution having an error of $\pm$ 20%,
the calculated sludge contribution derived additional errors.
Because the errors were large and only a few measurements were made,
concentrations of individual parameters in Table 2 are reported as a
range of high and low values.

    From Table 2, it is apparent that the large dilutions encoun-
tered in this line dump were responsible for diminishing the high
concentrations in the raw sludge to near background levels for almost
all parameters. The dilution at 18 m was less than at the surface,
and for some parameters, sludge contributions were slightly greater.
Parameters with slightly elevated concentrations at depth included
SPM, POC, PON, particulate Cd, and particulate Cu. The concentrations, however, were not elevated sufficiently to make any firm conclusions regarding deviations from conservative mixing which might
have occurred upon dumping.

    Only dissolved Cu and ammonia concentrations were much greater
than expected in the surface sample. Because similarly elevated
concentrations of these parameters were not observed at depth where
less dilution or dispersion had taken place, physical fractionation
between dissolved and particulate phases occurred. These dissolved
components of sludge may have remained in a relatively concentrated
form above the thermocline, while sinking particles penetrated to the
bottom layers in substantial concentrations.

    If we assume that ammonia is a tracer for the dissolved plume, and that it is conservative over the short duration of the experiment as used by Duedall et al. (1977), the dilution for dissolved substances can be calculated by comparing the observed plume values of ammonia with those of the dissolved fraction of sludge. The calculated dilution for the 1 m depth at station 2 is 1:270. Applying this dilution to concentrations of dissolved chemical species in sludge, only Fe and phosphate had sufficiently elevated concentrations in sludge to undergo such dilution and contribute to levels higher than background. However, these compounds were not found in concentrations greater than background in the plume (Table 2), suggesting that either our dilution factor is too small and ammonia was not conservative or that phosphate and Fe was scavenged or precipitated from solution. Particulate Fe concentrations at 18 m depth were slightly higher than expected (Table 2); thus, this scavenging process may have been active.

Dissolved Cu was not concentrated enough in the sludge to undergo the dilution of 1:270 and be detected above background. Nevertheless, the dissolved Cu concentration was greater than expected (Table 2). Contamination is not a satisfactory explanation, as the 14 m sample (above the thermocline) at this station contained a similar concentration of Cu (Cantillo et al., 1980b). Dissolution from the particulate phase upon contact with seawater is possible, although a concomitant decrease in particulate Cu was not observed.

Mixing Calculations. Upon dumping, sludge undergoes rapid mixing and initial dilutions on the order of 10^4 can be calculated from coprostanol concentrations. Coprostanol can also be used to determine the amount of mixing that sludge undergoes with natural background material. This mixing calculation can be made by computing the coprostanol content as a percentage of the total steroid content. Thus, a new parameter, % coprostanol, is calculated. The % coprostanol of sludge from the barge was 40%, and that of background material averaged 3% (Table 4). The resultant particulate mixture will exhibit a value of % coprostanol which is intermediate between 40 and 3%, depending on the amount of mixing. A mixing scale can thereby be established based on the two extremes; however the scale is not linear as sewage sludge particulates contain an average of 4.2 times more steroid per gram of material (SPM) than background particulates and would heavily influence the steroid distribution in the samples. A weighing factor must, therefore, be applied to the 3 to 40% scale in order to calculate the percentage of sludge particulate matter relative to background particulate matter.

Using the following mathematical logic, a relationship between particulate sludge percent in the plume and the % coprostanol content is calculated:

$$40A + 3B = X (A + B) \tag{1}$$

A = the amount of sludge particulate steriods in a sample

B = the amount of natural particulate steroids in a sample

X = the % coprostanol of the sample

From this equation, the A/B ratio is calculated and, considering the fact that steroids are 4.2 times more concentrated in sludge than in natural particulates, the A/B ratio is divided by 4.2 to calculate the ratio of sewage to natural particulate matter in the sample. Thus, the ratio of percent sludge (A') and percent natural material (B') in the particulate plume equals A/4.2 B, and we can write the following equation:

$$\frac{A'}{B'} = \frac{A}{4.2B} \tag{2}$$

Since $A' + B' = 100\%,$ $\tag{3}$

then $\dfrac{A}{4.2B} = \dfrac{A'}{100-A'}$ $\tag{4}$

From equation (1), $\dfrac{A}{B} = \dfrac{X-3}{40-X}$ $\tag{5}$

and we can substitute A/B into equation (4) to yield the following equation:

$$\frac{X-3}{4.2 \ (40 - X)} = \frac{A'}{100 - A'} \tag{6}$$

Solving for A', the percent sludge in each sample can be calculated from the % coprostanol (X) by the following equation:

$$A' = \frac{100X - 300}{165 - 3.2X} \tag{7}$$

We have calculated the % sludge, on a mass basis, in each of the samples analyzed for coprostanol (Table 4) listing the range of each value calculated from error estimates. The calculations strongly suggest that a rapid initial mixing occurs upon dumping, and that further mixing after initial dilution is relatively slow. Shortly after dumping, the sludge contributes only 30-50% of the total particulate matter in the water column. At the surface (above the thermocline), the mixing is much greater than below the thermocline, where a mixture of roughly 1:1 of sludge with background is observed. These calculations are consistent with acoustic observations, as it appears that the penetrating plume undergoes much less dispersion

Table 4. The coprostanol data and calculated sludge % in plume samples of dump 1.

Parameters	Sludge Concentration	Background (1 m)	Station (depth)			
			2 (1 m)	2 (18 m)	3 (12 m)	5 (4 m)
Local Time	–	–	1044	1050	1245	1527
Coprostanol (µg/l)	130,000	0.2	5.6	13	5.3	6.6
% Coprostanol[a]	40	3	23	32	27	26
% Sludge[b]	100	0	22^{29c}_{18}	45^{65}_{35}	30^{40}_{24}	28^{37}_{22}

a – Percent of total steroids
b – Percent of total particulate matter
c – The super- and sub-script numbers represent the high and low estimates based on errors in the analyses

and, therefore, less mixing as it settles to the lower layers, while the surface portion of the plume disperses laterally at a much more rapid rate, mixing with surface materials to a greater extent. These observations on the physical dispersion processes are discussed in detail by Proni et al. (1978).

   Temporal Changes in the Dumped Material.   After the initial mixing occurred, the concentration of sludge in the samples collected above the thermocline two hours and five hours after the dump did not change significantly. Also, the coprostanol concentration and % coprostanol did not change significantly over this time frame (Table 4), further indicating that little mixing was occuring in the plume above the thermocline, and that the sewage concentration in each of the samples was the same.

   Because the initial dilution of sludge particulates was so large, and because most of the particulate matter recovered was of non-sewage origin, the concentrations of most particulate chemical constituents (e.g., bulk parameters) was not noticeable above background. Therefore, if any major temporal changes occurred in the sludge, they would not be detectable above background. This indeed, was the case: no major temporal changes were observed in the bulk chemical character of particulate matter above the thermocline over a five-hour period (Cantillo et al., 1980b).

   Although the bulk organic characteristics of the plume's particulate matter showed little change, the concentrations of trace organics such as particulate and dissolved PCBs exhibited some interesting temporal changes (Table 3). Over a five-hour period (stations 2-5), the particulate PCB concentration increased, while the dissolved PCB concentration decreased a corresponding amount suggesting that the dissolved PCBs were scavenged from the dissolved phase by the particulate matter. This is not unusual if one considers that PCBs are hydrophobic and tend to sorb to particulate matter. The distributions of other trace organic compounds (hydrocarbons and fatty acids) remained nearly unchanged with time after initial dumping, indicating that some effects of the dumped sludge were persistent.

   The ammonia concentration in the sludge plume displayed an interesting temporal variation by nearly doubling over a five-hour span (Cantillo et al., 1980b). Although an extremely small decline in particulate nitrogen was observed, this change was not sufficient to account for the increased levels of dissolved ammonia. Two possibilities can account for these observations. First, although the acoustic record and coprostanol data confirm that the particulate matter samples were in the plume, we have no way of determining the location of our sample with respect to the highest concentration of dissolved sludge constituents. The changes in dissolved ammonia may be a result of our inability to sample the center of a dissolved plume which separates from the particulate plume. Second, an increase in dissolved ammonia may be occurring in the plume due to some chemical alteration of some dissolved organic nitrogen compounds which, unfortunately, were not measured. The chemical decomposition of urea is a possible source for the ammonia.

Dump 2: 14 July 1976. This particular dump of sewage sludge was conducted in a significantly different manner than dump 1, since it was a modified line dump. The dumping vessel emptied its tanks rapidly while drifting. High concentrations of sludge collected from the Newtown Creek treatment plant were rapidly released into the water. As no measurements of trace organic compounds were made, we have no means of normalization as in dump 1. Nevertheless, analyses were made of all other bulk parameters from five stations collected in the fringes of the plume. Due to acoustic interference with the 200 kHz system and ship-to-ship communication problems, we were not as successful in this experiment in tracking the maximum plume concentrations. Concentrations of selected parameters are plotted and contoured in Fig. 10 and 11.

High concentrations of ammonia and phosphate of 120 µg-at/l and 11 µg-at/l, respectively, were observed at the surface immediately following the dump (Fig. 10). Both of these concentrations diminished fairly rapidly. Assuming ammonia to be a conservative tracer for the dissolved plume in the initial stages of the dump, a dilution of 1:18 was calculated using the ammonia data of station 6 (1m). This relatively small dilution, compared to that of dump 1, indicates that very little dispersion of the dissolved plume has taken place.

Suspended Matter. The suspended particulate matter (SPM and POC) data for this dump are shown in Fig. 10. It appears that spot dumping has an enormous impact on the suspended load immediately following the dump. However, the concentrations drop rapidly to background levels after 3 hours. Either the sharp drop is due to rapid dilution, or to the fact that we missed the highest plume concentration. The latter appears to be the case as determined from the acoustic records. Unfortunately, realtime identification of the plume by acousic devices and subsequent sampling was hampered by communications problems. Positioning the sampling gear in the plume was virtually impossible. Nevertheless, an initial dilution for surface layers of the plume can be obtained from either the POC or SPM data. By comparison to POC and SPM concentrations in the sludge (Cantillo et al., 1980b), an initial dilution of 1:6.3 x 10^2 was obtained using both POC and SPM data at station 6 (1 m). This dilution obtained for particulate matter was substantially greater than that of dissolved components. This suggests that either a fractionation of dissolved and particulate matter occurred or that ammonia cannot be used for dilution calculations. The former explanation appears most likely, as the same dilution for dissolved components can be obtained from the phosphate and nitrate data.

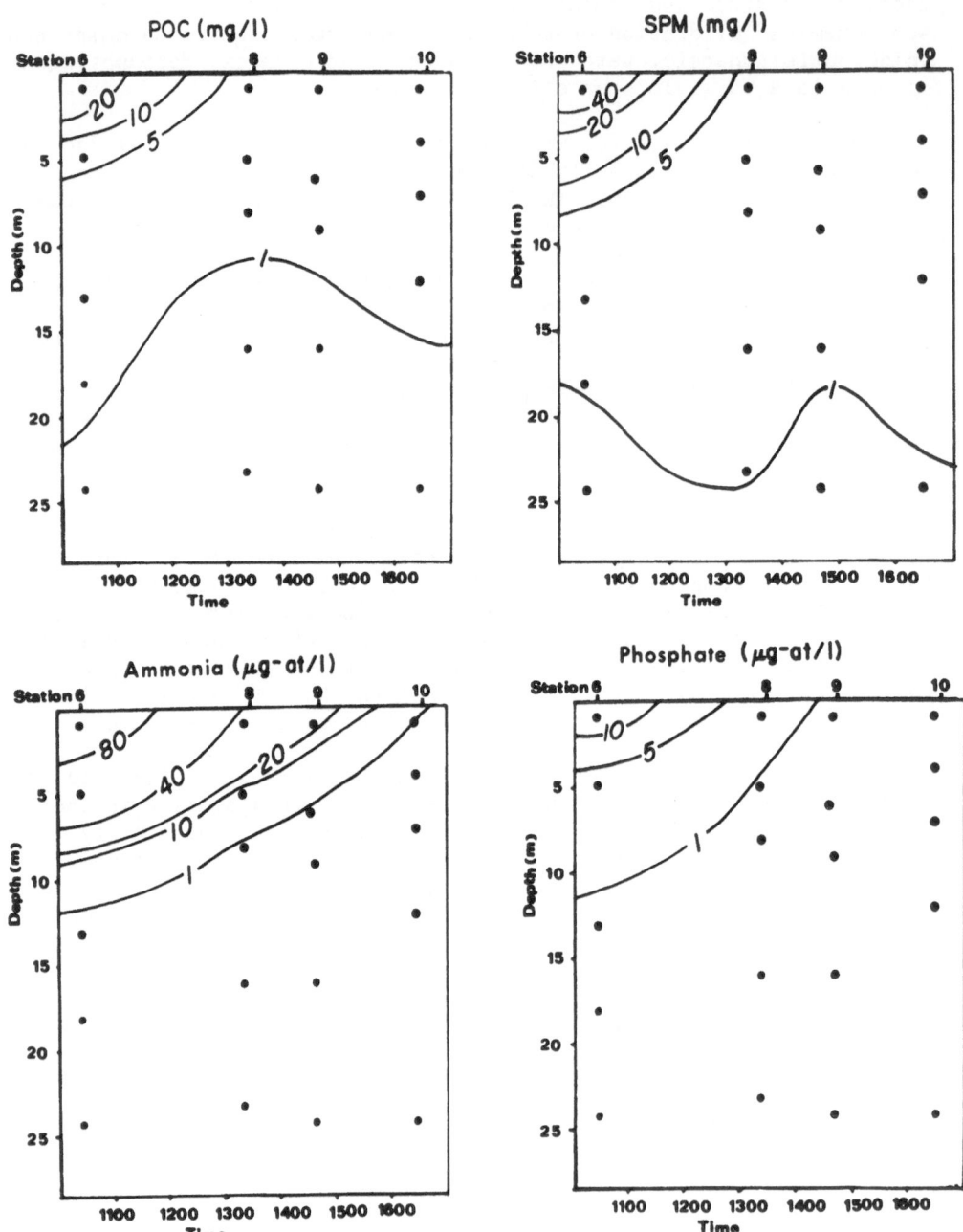

Fig. 10. The dissolved nutrient data, POC, and SPM for dump 2 on
July 14, 1976 presented as a function of depth and station
number.

Trace Metals. The dissolved and particulate trace metal data for dump 2 are shown in Fig. 11. The distribution of particulate metals plotted as a function of time after the dump was quite similar to the POC and SPM distributions (Fig. 11), and a rapid decline was observed after the initial high. As previously stated, this drop in concentration is most likely related to inadequate sampling of the plume. The dissolved metal distributions were slightly different than those of particulate metals, since the highest value at station 6 were at 5 m depth. In fact, at all stations in the plume (Fig. 11), the highest dissolved metal concentrations did not correspond to the highest particulate metal concentrations, suggesting that some fractionation may have occurred. Intestingly, the high dissolved Cu concentration was less than that observed for dump 1, even though not so much dilution had occurred in this spot dump. No explanation can be offered, other than the fact that the dissolved Cu concentration was less in this particular sludge. Unfortunately, no measurements were made of the dissolved trace metal content of this sludge sample.

Dump 3: 16 July 1976. This particular dump was a spot dump of sludge collected from Ward's Island sewage treatment plant. The physical characteristics of the dispersing plume were similar to those of dump 2 as determined from the acoustic data. Chemical concentrations were, however, slightly different. If the sludge concentrations of POC and SPM are compared, as was done for previous dumps, to their respective plume concentrations immediately after release of the sludge, a particulate dilution of 1:1.2 x 10^3 is calculated.

As in the case of dump 2, the high concentrations decreased rapidly after the first station, most likely due to plume sampling inadequacies. This is confirmed again by acoustic data, and further interpretation of the data is limited. In this particular dump, the high concentrations of dissolved compounds coincide with high concentrations of particulate compounds in the initial stages of the dump. However, this does not necessarily indicate that no particulate/dissolved fractionation occurs. This spot dump was similar to 2 except that the dilution was greater during the initial dispersion.

Using ammonia as a dissolved plume tracer, a dissolved dilution of 1:11 was calculated. This dilution is similar to that of the other spot dump (2). Because this dilution is so much less than that of particulates (1:11 as compared to 1:10^3), it appears that dissolved/particulate fractionation is an important physical process occurring in the dump.

CONCLUSIONS AND RECOMMENDATIONS

Preliminary studies of three sludge dumps in the New York Bight have provided information on dispersal characteristics of dumped sludge in the water column. Examining the concentrations of a

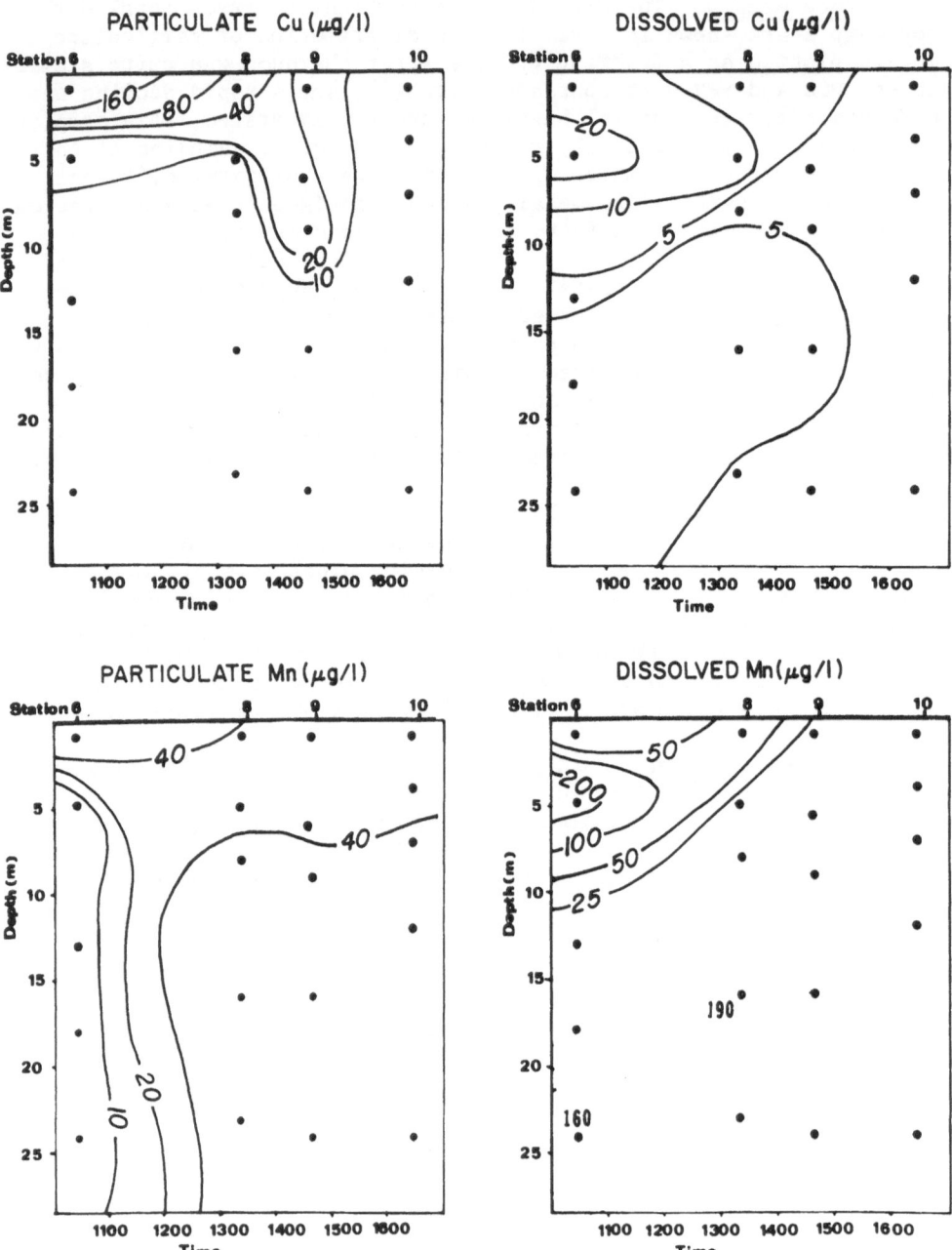

Fig. 11a. The particulate and dissolved trace metal distributions
for dump 2 as a function of depth and station number:
copper and manganese.

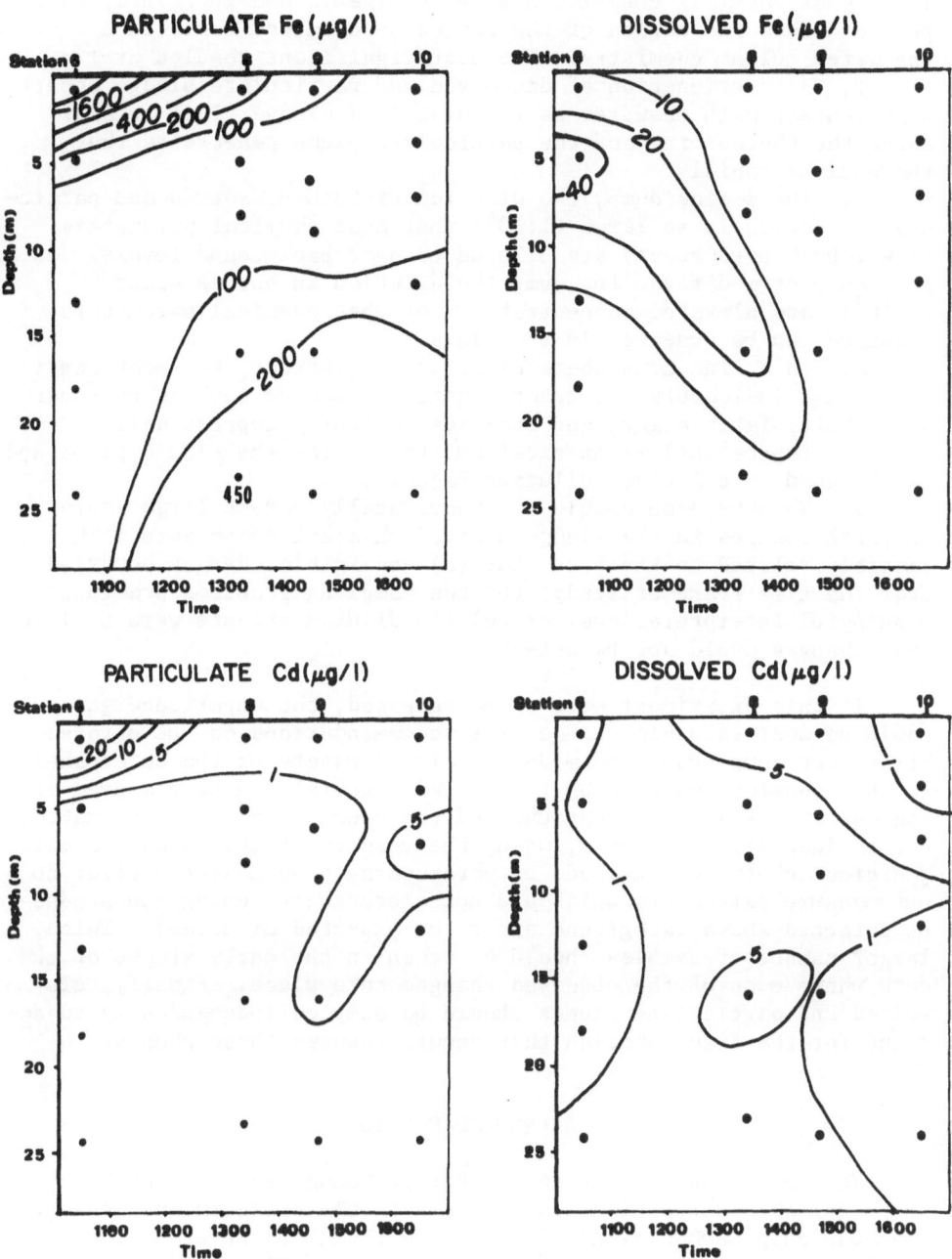

Fig. 11b. The particulate and dissolved trace metal distributions
for dump 2 as a function of depth and station number: iron
and cadmium.

variety of chemical constituents, both organic and inorganic, has
provided some indication of the impact of chemical constituents on
the water column chemistry. The most significant results are:

    1. A fractionation of dissolved and particulate sludge occurs
upon contact with seawater as the dissolved sludge plume remains
above the thermocline and the particulate plume penetrates the
thermocline rapidly.

    2. For a line dump, the dilution of both dissolved and partic-
ulate fraction is so large $(1:10^4)$ that most chemical parameters
(i.e., bulk parameters) are diluted to near background levels. In a
spot dump or modified line dump the dilution is not as great
$(1:10^3)$, and elevated concentrations of most chemical parameters
examined can be observed in the plume.

    3. In a line dump where dilution is greatest, the most sensi-
tive plume indicators are trace organic compounds such as hydrocarbons, PCBs, fatty acids, and steroids including coprostanol.

    4. Coprostanol is an excellent tracer for the sludge plume and
can be used to calculate dilution factors.

    5. We have been unable to unequivocally detect large scale
temporal changes in the sludge-derived chemical components. This is
possibly related to the fact that (a) negligible changes occurred
over the time frame of study; (b) the sampling problems precluded
meaningful interpretations; or (c) the diluton effects were so large
that changes could not be detected.

    If this experiment were to be repeated, the experience gained
would be most valuable and several recommendations can be offered.
First, the dump should be made in "clean" waters of the outer shelf
so that chemical changes of the dumped material can be measured at
concentrations above background and the plume of particulate matter
can be identified. Second, using the results of this study, a more
judicious choice can be made of parameters to be measured (i.e. do
not measure parameters which are not concentrated enough in sludge to
be detected above background after the expected dilution). Third, a
larger number of samples should be taken in the early stages of the
dump where most of the observed changes take place. Finally, dissolved and particulate plumes should be sampled independently to account for the fractionation that occurs between these phases.

ACKNOWLEDGEMENTS

    We express our thanks to Dr. Robert Young and Dr. John Proni for
their guidance as chief scientists and to the officers and crew of
the NOAA Ship GEORGE B. KELEZ. We also thank Dr. Peter Betzer and
Mr. Robert Woodwell of the University of South Florida for their
trace metal digestions and analyses for trace metals in particulate
matter. Further, we acknowledge Mr. John VanLandingham and Mr. Kevin
O'Donnell for their analytical assistance. This project was sponsored
by NOAA's Marine Ecosystem Analysis Program, New York Bight Project.

REFERENCES

Armstrong, F. A. J., C. R. Stearns and J. D. H. Strickland (1967) The measurement of upwelling and subsequent biological processes by means of the Technicon Auto Analyzer and associated equipment. Deep-Sea Res., 14, (3), 381-389.

Betzer, P. R. (1978) A study of the sources, transport and reactions of the suspended particles in waters of the New York Bight. NOAA Tech. Memo ERL-MESA 23, 37 pp.

Cantillo, A. Y., G. A. Berberian, P. G. Hatcher, L. E. Keister and D. A. Segar (1980a) MESA New York Bight Project water column chemistry data, cruise #6-12 of the NOAA Ship FERREL, April-November, 1974, NOAA Data Report (in press).

Cantillo, A.Y., Hatcher, P.G., and Berberian, G.A. (1980b) Chemical and physical processes in a dispersing sewage sludge plume: Results of the STAX II experiment. NOAA-ERL Data Report (in preparation).

Duedall, I. W., M. J. Bowman and H. B. O'Connors, Jr. (1975) Sewage sludge and ammonium concentration in the New York Bight Apex, Estuarine Coast. Mar. Sci., 3, 457-463.

Duedall, I. W., H. B. O'Connors, S. A. Oakley and H. M. Stanford (1977) Short-term water column perturbations due to sewage sludge dumping in the New York Bight Apex. Water Pollution Control Fed., 1977, 2074-2080.

Environmental Protection Agency (1976) Manual of methods for Chemical Analysis of Water and Waste. Environmental Monitoring and Support Laboratory, EPA Publ. 625/6-74-003A, Cincinati, Ohio.

Gerchakov, S. M. and P. G. Hatcher (1972) Improved technique for analysis of carbohydrates in sediments. Limnol. Oceanogr., 17, 938-943.

Grasshoff, K. (1965) Automatic determination of fluoride, phosphate, and silicate in seawater. Technicon Fifth International Symposium, London, pp. 04-307.

Greenfield, L. J., R. D. Hamilton, and C. Weiner (1970) Nondestructive determination of protein, total amino acids and ammonia in marine sediments. Bull. Mar. Sci., 20, 289-304.

Hatcher, P. G., L. E. Keister and P. A. McGillivary (1977) Steroids as sewage specific indications in New York Bight sediments. Bull. Envir. Contam. and Toxicol., 17, 491-498.

Hatcher, P. G. and P. A. McGillivary (1979) Sewage contamination in the New York Bight: Coprostanol as an indicator, Envir. Sci. Tech., 13, 1225-1229.

Mackay, D. W., W. Halcrow and I. Thornton (1972) Sludge dumping in the Firth of Clyde. Mar. Pollut. Bull., 3, 7.

Malone, T.C. (1978) The 1976 Ceratium tripos bloom causes and consequences. In NOAA Tech. Report NMFS Circ. 410, 14 pp.

Murtaugh, J. J. and R. L. Bunch (1967) Sterols as a measure of fecal pollution. J. Water Pollut. Control Fed., 39, 404-409.

Nishimura, M. and T. Koyama (1977) The occurrence of stanols in various living organisms and the behavior of sterols in

contemporary sediments. Geochim. Cosmochim. Acta, 41,
379-386.

Pararas-Carayannis, G. (1973) Ocean dumping in the New York Bight:
An assessment of environmental studies, U.S. Army, Corps of
Engineers. Tech. Memo. No. 39, 159 pp.

Perkin-Elmer Corporation (1973) Analytical methods for atomic
absorption spectroscopy using the HGA furnace. PE Publ.
No.990-9972, Perkin-Elmer Corporation, Norwalk, Conn., 43 pp.

Proni, J. R., F. C. Newman, R. L. Sellers and C. Parker (1976)
Acoustic tracking of ocean dumped sewage sludge, Science, 193,
1005-1007.

Proni, J. R., F. C. Newman, R. A. Young, D. Walter, R. Sellers, P. A.
McGillivary, P. G. Hatcher, I. Duedall, H. Stanford, C. Parker
(1980). Observations of the intrusion into a stratified ocean
of an artificial tracer and the concomitant generation of
internal oscillations. J. Geophys. Res. (in press).

Rhead, M. M., G. Eglinton and G. H. Draffan (1971) Hydrocarbons
produced by the thermal alteration of cholesterol under
conditions simulating the maturation of sediments. Chem. Geol.,
8, 277-297.

Segar, D. A. and G. A. Berberian (1976) Oxygen depletion in the New
York Bight Apex. Causes and consequences. In: Middle Atlantic
Continental Shelf and the New York Bight, H. G. Gross, editor,
The American Society of Limnology and Oceanography, Spec. Symp.
Vol. 2, pp. 220-239.

Segar, D. A. and A. Y. Cantillo (1976a) Direct determination of trace
metals in seawater by flameless atomic absorption spectrophotometry. In: Analytical Methods in Oceanography, T. R. P.
Gibb, editor, American Chemical Society, Advan. Chem. Series No.
147: 56-81.

Segar, D. A. and A. Y. Cantillo (1976b) Trace metals in the New York
Bight. In: Middle Atlantic Continental Shelf and the New York
Bight, M.G. Gross, editor, The American Society of Limnology and
Oceanography Special Symposium Vol. 2, pp. 171-198.

Solorzano, L. (1967) Determination of ammonia in natural waters by
the phenolhypochlorite method, Liminol. Oceanogr., 14, 799-801.

Strickland, J. D. H. and T. R. Parsons (1968) A Practical Handbook of
Seawater Analysis, Bulletin 167, Fisheries Research Board of
Canada, Ontario, Canada, 311 pp.

West, R. H., P. G. Hatcher and D. K. Atwood (1980) Polycholorinated
biphenyls and DDT's in sediments and sewage sludge of the New
York Bight, NOAA Data Report (in press).

Wun, C. K., R. W. Walker and W. Litsky (1976) The use of XAD-2 resin
for the analysis of coprostanol in water, Water Res., 10,
955-959.

IV

BIOLOGICAL ASPECTS OF OCEAN DUMPING

THE BACTERIAL BIOASSAY AND LABORATORY ASSESSMENTS OF

WASTE DISPOSAL ACTIVITIES AT DWD-106[*]

Ralph F. Vaccaro and Mark R. Dennett

Woods Hole Oceanographic Institution
Woods Hole, MA   02543

ABSTRACT

Changes in the bacterial uptake of ^{14}C labeled glucose in sea-water are used to quantify some sublethal consequences of Edgemoor and Grasselli waste disposal at Deep Water Dumpsite 106.

The fractional amounts of waste in seawater which led to a 50 percent reduction in ^{14}C uptake ranged from 0.01-0.02 percent for Edgemoor waste and from 0.10-0.20 percent for Grasselli waste.

Both Edgemoor and Grasselli wastes cause an inhibitory bacterial response which exceeds that associated with their respective acid and caustic chemical compositions. Heavy metals appear to be the princi-ple toxic components of Edgemoor waste whereas organic species appear to produce the toxicity of Grasselli waste.

Resistance to chemical alteration as demonstrated by ultraviolet radiation and persulfate oxidation may imply an environmental per-sistence for Grasselli waste.

Edgemoor waste reacts with seawater with a precipitation of its heavy metal content. Such behavior is certain to influence its dis-tribution kinetics and its impact on life processes of the ocean.

Mixtures of Edgemoor and Grasselli wastes impart an inhibitory response which is measurably less than that anticipated from the sum

[*]Contribution No. 4284 from the Woods Hole Oceanographic
   Institution, Woods Hole, MA 02543

of their individual effects. This suggests the possibility of positive benefits from their coordinated release within the Dumpsite
area.

An unambiguous approach to improve bioassay interpretations in
potentially reactive, multi-waste situations was developed. The
method relies upon a graphical solution to differentiate between the
net response induced by mixed wastes and a hypothetical response corresponding to the sum of the individual response patterns observed
independently.

INTRODUCTION

The ultimate source of chemical energy which sustains the oceanic food chain is the primary production of unicellular photosynthetic
algae while zooplankton consumption accounts for the major pathway
which energizes higher marine populations.

An alternate pathway, often overlooked, originates with the ex-
tracellular release of soluble organic materials at each level of the
food chain, which also are converted into cell matter by heterotrophic marine bacteria. Consumption of these microbial entities by mi-
crozooplankton grazers also serves to energize higher trophic levels
(Webb and Johannes, 1967; Lampert, 1978).

Typically the scope of the alternative energy shunt ranges from
10 percent (Parsons and Seki, 1970) to about 50 percent (Andrews and
Williams, 1971) of the total biological energy produced in the sea.
However, in times of environmental stress, when local populations are
adversely affected, organic leakage increases and the alternative
pathway takes on greater significance. This shift in bacterial activity is accomplished by species which show an enhanced tolerance
toward the prevailing stress characteristic (Vaccaro et al., 1976;
Aaronson, 1978).

Besides facilitating the utilization of soluble organics in
seawater, marine heterotrophic bacteria also offer an ideal source of
bioassay agents for measuring sublethal levels of environmental
stress. Typical of unicellular plants, marine bacteria present high
surface-to-volume ratios and enzymatic systems which operate at extreme low half-saturation constants. Such characteristics are ideally suited for bioassay analyses and promise high resolution within
relatively short contact periods.

During the past two years we have conducted microbial bioassay
observations to help evaluate potential environmental effects associated with the release of industrial wastes at DWD-106. These studies
have been supported by the National Oceanic and Atmospheric Administration (NOAA) as part of a comprehensive program which deals with

the physical, chemical and biological consequences of ocean waste
disposal. This paper will summarize the available microbiological
information which relates specifically to the Edgemoor and Grasselli
wastes of the Dupont Corporation. Together these two sources constitute a significant fraction of the total wastes discharged at the
Dumpsite. A further objective of this paper is to communicate certain bioassay design concepts insofar as they may be of assistance to
other environmental assessment programs.

Of necessity, laboratory rather than field observations are emphasized because of insufficient information on the physical and
chemical parameters which determine waste dispersal and alteration
with time within the Dumpsite area. We are cognizant, therefore,
that some of our interpretations may require future modification once
more complete information on Dumpsite kinetics becomes available. We
also anticipate the possibility of added uncertainty arising from the
extremely broad spectrum of wastes involved, the potential for chemical interaction, and the seasonal nature of their composition and
abundance.

METHODS

Heterotrophic uptake rates were measured by a radioactive tracer
technique (Parsons and Strickland, 1961; Hobbie and Wright, 1965,
Vaccaro and Jannasch, 1966) wherein the ^{14}C assimilated from uniformly labeled glucose was determined by liquid scintillation counting. Test cells for bioassay purposes were allowed to develop during
an 18-24 hour incubation period at room temperature in unenriched
seawater samples collected from the Institution dock at Woods Hole.
Overnight storage accomplished the desired increase in cellular density and placed the bacterial population in a logarithmic phase of
growth.

Acid-cleaned glass reaction tubes containing preselected amounts
of industrial waste were provided in triplicate and made up to a final volume of 10 ml with membrane filtered (0.45 µm porosity) seawater. To help establish chemical equilibria, waste-seawater mixtures were shaken over a two hour period before measured volumes of
the test cell suspension were added. The actual amount of cell suspension added was predetermined so that about 10,000 counts per minute resulted from a 30 minute test exposure to 0.05 µcuries of
uniformity labeled glucose. The total cellular exposure was 2.5
hours, with ^{14}C glucose being present during the final half hour.
The specific activity of the added glucose was such that 100 µl of
solution were equivalent to 0.05 µcuries = 0.01 µg glucose carbon.
After the 30 minute ^{14}C exposure, formalin was added to terminate
glucose uptake and the cells were collected on membrane filters (Millipore, 0.45 µm porosity) and prepared for standard scintillation
counting. Results were recorded by plotting, as a percentage, the

relation between the average ^{14}C measurement for each waste dilution vs that of a control prepared without added industrial waste.

Photoxidation of 50% Grasselli waste in distilled water for 72
hours under an Ace Hanovia U.V. lamp providing 450 watts was used in
an attempt to reduce the organic carbon content of the waste (Armstrong et al., 1966). After irradiation the sample was aerated with
washed and filtered air for 48 hours and the pH adjusted to the original value of 13.1 with NaOH.

Persulfate wet combustion was also used in an attempt to oxidize
the organic species in Grasselli waste (Menzel and Vaccaro, 1964). A
5 ml volume of waste was sealed in a 10 ml glass ampoule containing

Table 1. Metal content of Edgemoor waste barged to DWD-106, June 27,
          1977 vs. analyses recorded in Dupont Discharge Permit
          Application

Constituent	Dicharge Permit mg/1	WHOI Analyses, Barge Sample		
		mg/1	μ m/1	μ m/40 μ1*
Iron	50,400	45,000	804,000	32.160
Chromium	208	110	2,115	0.08460
Vanadium	200	450	8,826	0.35304
Lead	41	20	96	0.00384
Nickel	--	8	136	0.00544
Copper	--	4.6	57	0.00228
Cadmium	--	0.5	4.4	0.00018
Total, including iron			815,000	32.60
Total, excluding iron			11,200	0.45
EDTA neutralization requirement			8,500	0.34

*Metallic equivalents experimentally exposed to 0.34 moles EDTA in
seawater (Fig. 2).

1 gm of potassium persulfate. The ampoule was heated in an autoclave
for 30 mins at 120°C. When cool the pH was adjusted with NaOH to
correspond with the original pH of raw Grasselli waste.

RESULTS

    <u>Edgemoor Acid-Iron Waste Studies</u>. Acid-iron waste released at
Dumpsite 106 originates from the Dupont de Nemour plant at Edgemoor,
New Jersey. This waste represents residues from the titanium extraction process and has a hydrochloric acid normality of about 1.41.
Significant concentrations of heavy metal species are also included
in the following order of abundance: iron > vanadium > chromium >
lead > nickel > copper > cadmium. Table 1 shows the concentrations
of these metals, as measured in our laboratory, for a barge sample
acquired by us for test purposes from NOAA. The table also provides
a basis for comparison with related values documented in the Dupont
Discharge Permit Application.

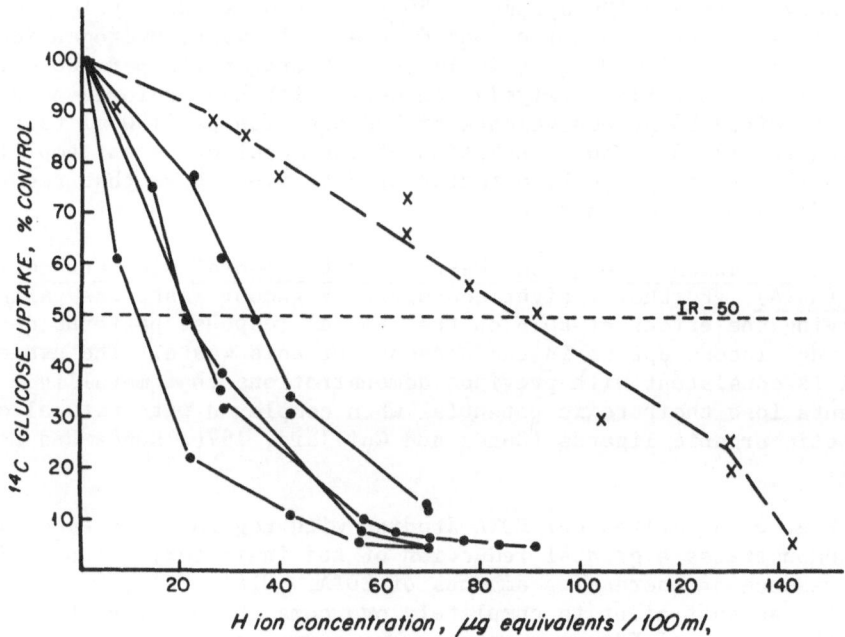

Fig. 1. Impact of known dilutions of whole Edgemoor waste (solid
         lines) and of hydrogen ions (dashed line) on a test
         bacterial population in seawater. The IR-50 level
         corresponds to an uptake reduction of 50 percent vs the
         control population.

The heavy metal matrix of Edgemoor waste is more than 99% iron yet the toxicity source is more apt to be associated with the nonferrous heavy metals rather than with iron. As shown in Table 1, the nonferrous metal components of Edgemoor waste total 11,200 mole 1^{-1} whereas iron comprises over 800,000 mole 1^{-1}. Relevant aspects of the unbalanced heavy metal distribution are discussed below.

Hydrogen Ion vs Whole Waste Effects. An important consideration concerning Edgemoor waste is its potential for both an acidic and heavy metal impact on biological systems. For this reason, a description of the relative influence of each of these characteristics was assigned an early priority.

To differentiate between the influence of acid as opposed to heavy metal plus acid effects, we examined the effect of hydrogen ion concentration on ^{14}C glucose uptake both in the presence and absence of the heavy metal component. The resulting data, shown in Fig. 1, indicate that the whole waste had a greater adverse effect on ^{14}C glucose assimilation than equal concentrations of hydrochloric acid. More specifically, the average concentration of whole waste necessary to reduce ^{14}C uptake to 50 percent of an appropriate control situation (IR-50) corresponded to a whole waste hydrogen ion concentration of 18-40 µg equivalents of hydrogen ion per 100 ml. Conversely, a parallel analysis conducted with hydrochloric acid alone required 88 µg equivalents of hydrogen ion per 100 ml to establish the IR-50. Thus, inclusion of the metal component amplified the inhibitory response by a factor of 2-4 times above that recorded for hydrochloric acid per se.

Attentuation of Edgemoor Waste with Ethylenediaminetetracetic Acid (EDTA). Further insight concerning Edgemoor waste was gained by observing the effect of EDTA on the typical response patterns describing glucose uptake in the presence of this waste. The use of -EDTA is consistent with previous demonstrations that metallic elements lose their toxic potential when complexed with natural or synthetic organic ligands (Sunda and Guillard, 1976; Sunda and Lewis; 1978).

Fig. 2 summarizes our EDTA studies with regard to Egdemoor waste and demonstrates a gradual reduction of the inhibitory response in the presence of increasing amounts of EDTA. Ultimately, 0.34 µmole of EDTA was sufficient to completely overcome the adverse effects from 40 µliters of Edgemoor waste. The above volume, according to Table 1 should have contained about 32.6 µmole of the total mixed heavy metals or 0.45 µmole of nonferrous heavy metals. Assuming a 1:1 molar relationship between EDTA and bivalent heavy metal (Gillespie and Vaccaro, 1978) the above EDTA requirement would account for the complexation of about 1 percent of the total heavy metal load or 76 percent of the nonferrous heavy metal waste content.

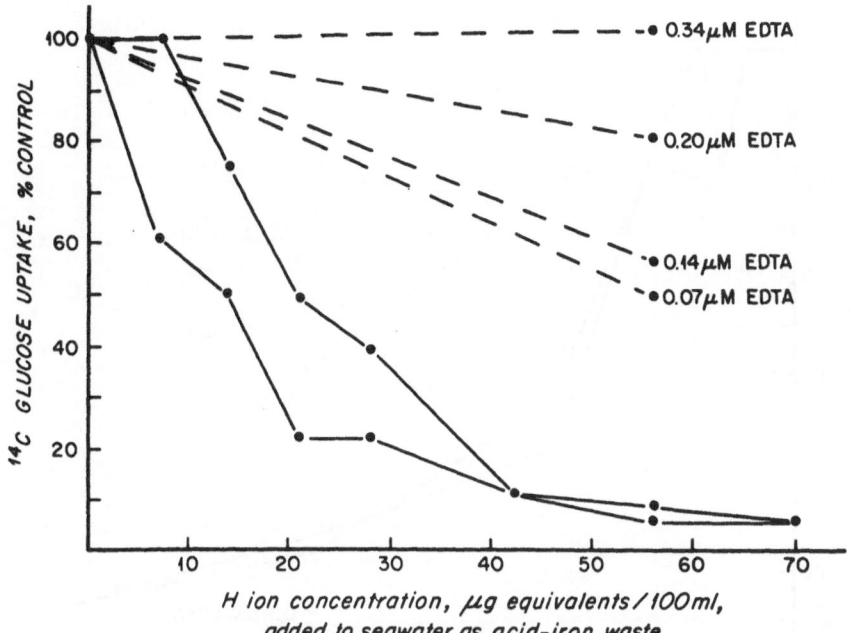

Fig. 2.   Effects of EDTA (dashed lines) on test bacterial populations
          exposed to different dilutions of Edgemoor waste (solid
          lines) in seawater.

        The buffering capacity of EDTA and its neutralizing effect on
waste acidity is an important consideration which could affect the
above interpretation.   In this case, the range of waste concentrations provided was so designed that pH differences within each of the
parallel seawater series, with and without EDTA, were minimized.   Actually, the pH resulting from the addiion of 56 μg equivalents of
hydrogen ion, the highest waste concentration used, in 10 ml of seawater was 6.50 while the addition of 0.34 μmole of EDTA led to only a
moderate pH increase to 7.60.   It appears unlikely, therefore, that
these studies were unduly influenced by pH differences even though
the presence of EDTA could have affected heavy metal solubilities.

        Studies with Dupont Grasselli Waste.   Grasselli waste, which
also originated from the Dupont Corporation, contains end-products
associated with the manufacture of the organic intermediates anisole
(phenyl methyl ester) and DMHA (N, O-dimethylhydroxylamine).   It contains high concentrations of sodium sulfate along with sodium hydroxide, water soluble organics and μmolar amounts of the heavy metals
Ni, Cu, Zn, Cr, Cd and Hg.   Typically the pH of this waste exceeds 13
and its dissolved organic carbon content measures about 2,500 ppm.

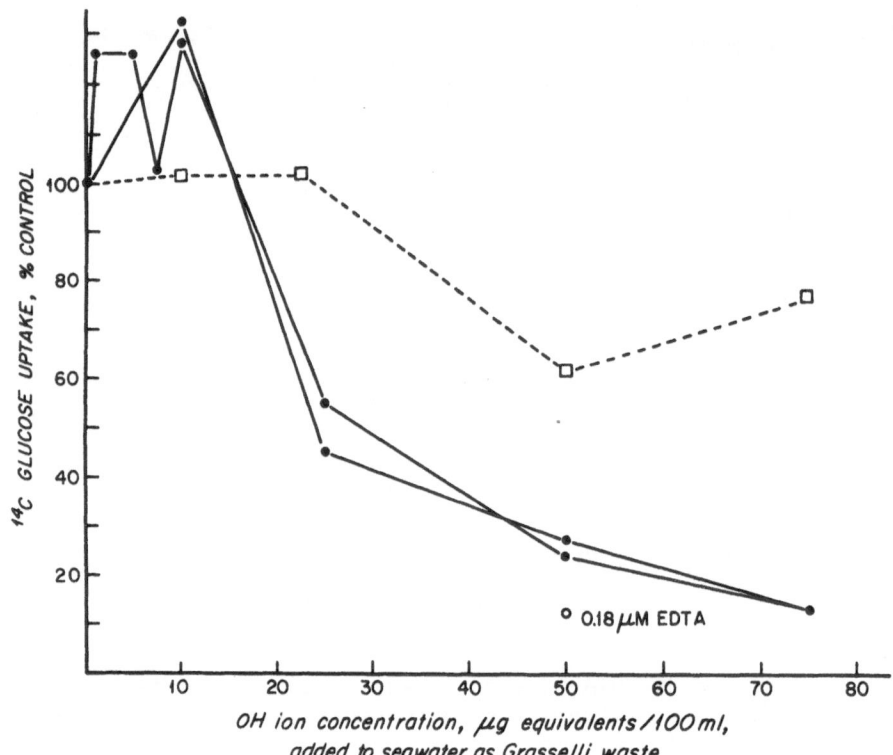

Fig. 3. Impact of experimentally added Grasselli waste (solid
         lines), hydroxyl ions (dashed lines) and EDTA-treated
         Grasselli waste (open circle) on a test bacterial population
         in seawater.

Hydroxyl Ion vs Whole Waste Effects. An effort was made to distinguish between the inhibitory impact of whole Grasselli waste as
opposed to that exerted by its hydroxyl ions. Results from this
study, summarized in Fig. 3, show a measurably greater inhibition
from whole waste as compared with that observed for hydroxyl ions
presented independently. Of added interest is that small concentrations of whole Grasselli waste, within the range 0-15 μg equivalents
of hydroxyl ion per 100 ml, stimulated ^{14}C uptake above that recorded for the waste-free controls. Apparently, in trace amounts,
the organic complement of Grasselli waste can accelerate carbon assimilation by heterotrophic marine bacteria.

Grasselli Waste Attenuation With EDTA. The average molar concentrations of heavy metals in the Grasselli barge samples analyzed
by the Surveillance and Analysis Division of EPA Region II totaled 4
μmole/100 ml. Our data indicate that a .05-.20% (by volume)

Grasselli waste concentration lowers the ^{14}C bioassay response to about 25% of a waste-free control. On this basis, the presence of 0.18 µmole EDTA in our bioassay should have been adequate to complex most if not all of the excess heavy metal activity. However, as shown in Fig. 3, additions of EDTA in this case, showed no effect on the inhibition associated with Grasselli waste. Apparently organic components, rather than heavy metal activity are the more important consideration with regard to this waste.

Attenuation of Grasselli Waste by Ultraviolet Radiation and Potassium Persulfate Oxidation. In an attempt to disrupt the organic carbon linkages of Grasselli waste a 50 percent dilution of this waste in a quartz vessel was exposed to 72 hours of ultraviolet irradiation frm an Ace Hanovia lamp (450 watts). However, a follow-up bioassay observation, shown in Fig. 4a, failed to describe significantly different response patterns between the ultraviolet and non-ultraviolet treated waste solutions. This observation indicates a highly refractory nature for Grasselli waste despite remaining uncertainty concerning the possible effects of longer and more powerful radiation exposures.

Wet combustion in the presence of potassium persulfate was also attempted as a means of oxidative removal of the toxic organic ingredients of Grasselli waste. Combustion was conducted in a sealed glass ampoule at an autoclave temperature of 120°C. After cooling, NaOH was added to restore the original waste pH of 13.

Potassium persulfate treatment according to results shown in Fig. 4b, only partially reduced the amount of inhibition recorded thereby suggesting that the extent of oxidation achieved was incomplete. Both the ultraviolet and persulfate oxidation data, therefore, suggest significant durability and a prolonged environmental longevity for Grasselli waste.

Inhibitory Responses from Edgemoor-Grasselli Waste Mixtures. Bioassay response patterns were also determined for Edgemoor-Grasselli waste mixtures in seawater. Each subsample contained a different amount of a waste mixture presented at a ratio of 1:10. The ratio selected reflects an earlier observation that 10 times more (by volume) Grasselli waste than Edgemoor waste is required to produce comparable amounts of bacterial inhibition. In this case, the experimental pH range for the resulting seawater dilutions was 8.04-7.53.

Fig. 5 describes the bioassay response obtained with known concentrations of the acid-iron and Grasselli wastes presented independently and as a. mixture of the two waste types. Mixing, in this case, did not appreciably alter the original effect from either of these wastes. However, the response to other than a 1:10 mixture has yet to be determined.

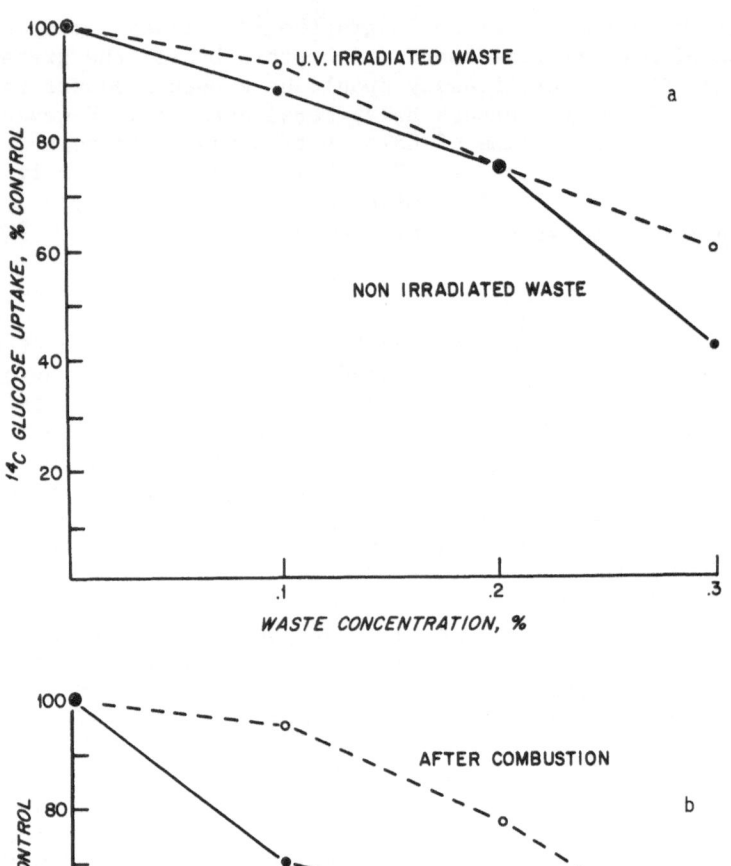

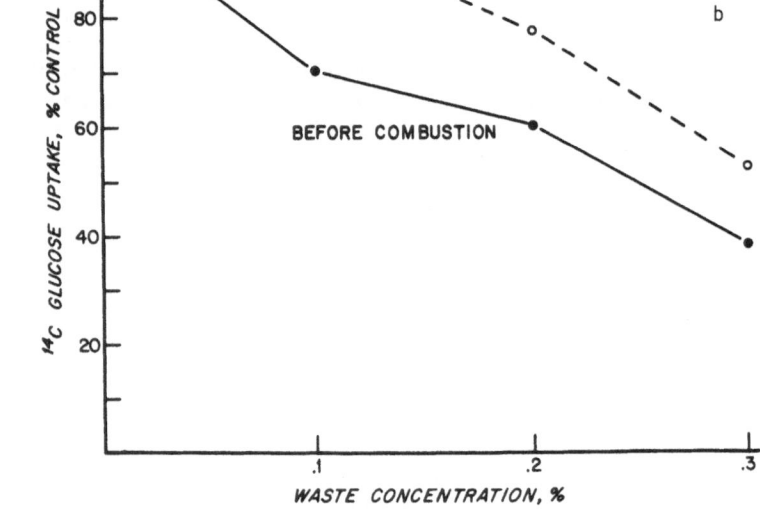

Fig. 4a.   Effect of Grasselli waste before (solid line) and after
           (dashed line) exposure to ultraviolet radiation on a test
           bacterial population.   4b. Effect of Grasselli waste before
           (solid line) and after (dashed line) wet combustion with
           potassium persulfate on a test bacterial population.

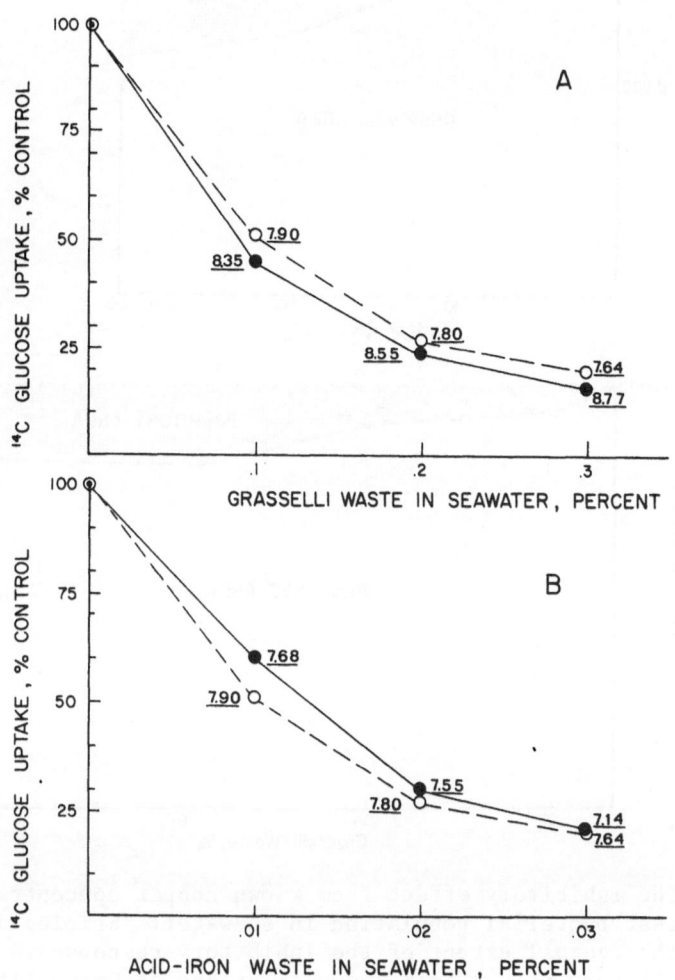

Fig. 5.   Effects on a test bacterial population of A: Grasselli waste
          (solid dots) and a 1:10 mixture of acid-iron plus Grasselli
          waste (open circles) in seawater; and B: acid-iron waste
          (solid dots) and the 1:10 mixture of the wastes. The under-
          lined numbers give pH values.

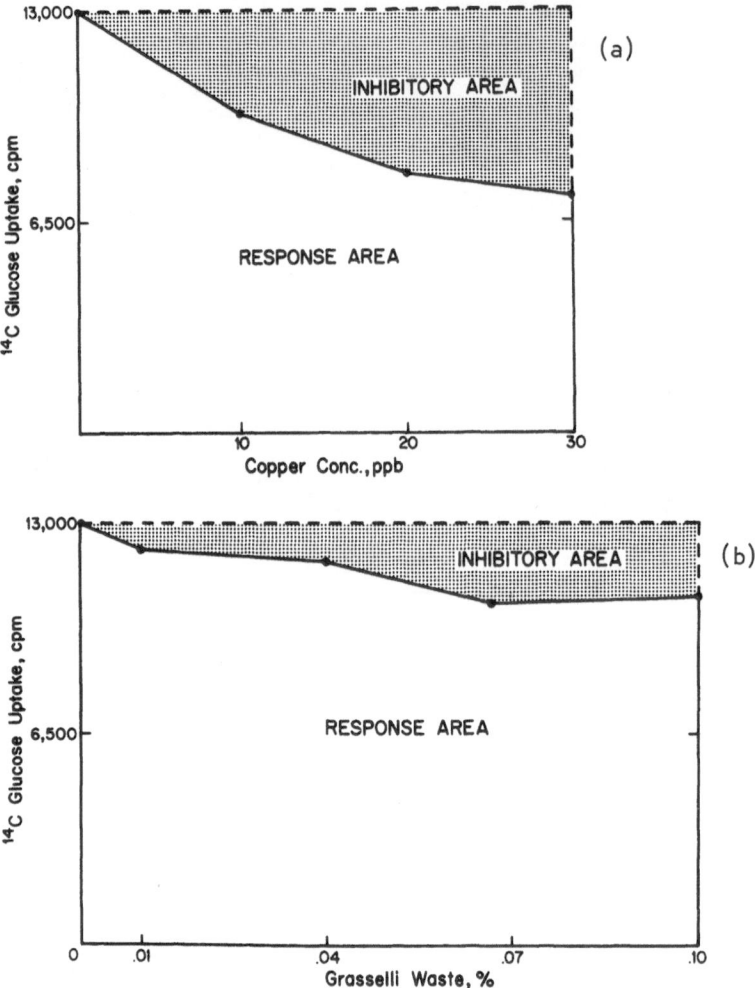

Fig. 6a. The inhibitory effect from known copper concentrations on a
         test bacterial population in seawater. Stipled area defines
         the overall extent of the inhibitory responses. 6b. The in-
         hibitory effect from known dilutions of Grasselli waste on a
         test bacterial population in seawater. Stipled area defines
         the overall extent of the inhibitory response.

    Bioassay Interpretations of Inhibitory Patterns in Multi-stress
Situations. A complicating factor affecting bioassay interpretations
of Dumpsite activities is the extent of uncertainty arising from multiple stress sources and the need to discriminate between net and additive stress effects. In responding to this problem we have developed the analytical technique described below.

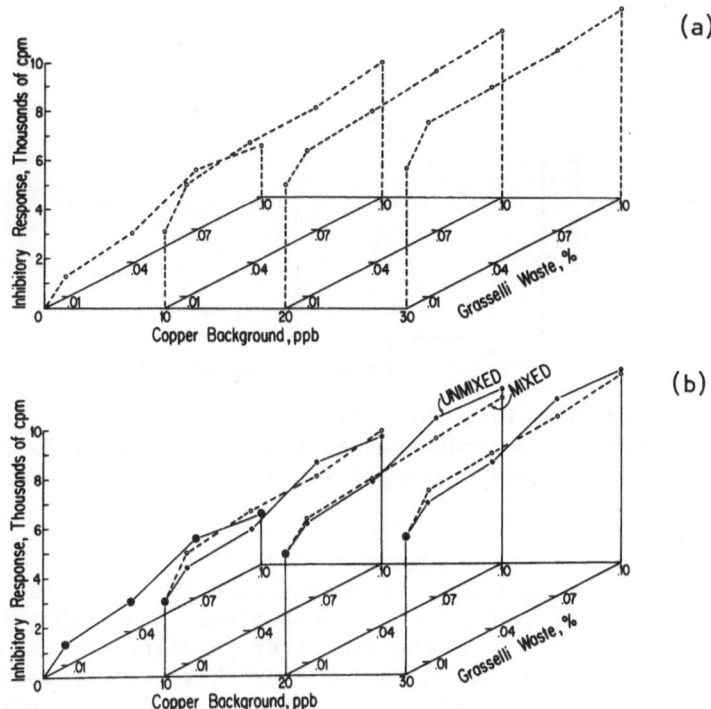

Fig. 7a. Additive inhibitory effects from known amounts of Grasselli
         waste added to four different copper concentrations in sea-
         water. 7b. Additive and net inhibitory effects resulting
         from mixed and unmixed quantities of copper and Grasselli
         waste in seawater.

When biological systems are exposed to more than one reactive
substrate, the net response can be additive, enhanced or compensatory
in terms of the sum of the individual response patterns. For exploratory purposes we envisioned and examined a hypothetical situation
wherein both copper and Grasselli waste were studied separately and
collectively. Analytically, the reacting bioassay information is
first assessed as the sum of the individual response patterns. In
turn, these hypothetical measurements are summed and used as a basis
for comparison with the net response pattern obtained by assaying the
combined wastes.

The impact pattern for copper at the ppb level on ^{14}C uptake
from labeled glucose by a test population of bacterial heterotrophs
is shown in Fig. 6a. The stippled area corresponds to the overall
extent of inhibition encountered since without any inhibition all responses would have been constant at the 13,000 counts per minute

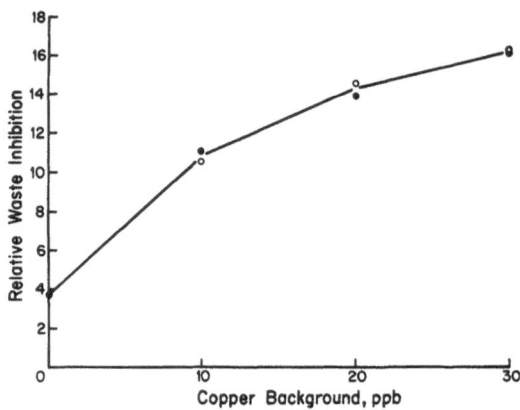

Fig. 8.   Relative values for additive and net inhibitory responses
          experimentally determined from mixed and unmixed quantities
          of copper and Grasselli waste.

(cpm) level. The actual extent of inhibition exerted at a given copper concentration is therefore represented by the difference between
13,000 and the ^{14}C count actually observed. Comparable information
compiled from exposure of the same test population to known dilutions
of Grasselli waste is shown in Fig. 6b. With interpolation the data
of Figs. 6a and b can provide the additive values for all conceivable
combinations within the ranges provided for the independently presented stress sources.

     Additive stress effects summed from Fig. 6a and b are graphed in
the three dimensional perspective shown in Fig. 7a. The amounts of
Grasselli waste used experimentally are scaled obliquely and adjacent
to each of four different abscissa-oriented copper concentrations.
The ordinate records the sums of the individual inhibitory response
patterns in cpm as extracted from Fig. 6a and b.

     Fig. 7b is identical to Fig. 7a except that both the additive
and combined bioassay response patterns are plotted along the ordinate. Differentiation between the additive and combined responses,
in terms of area, is obtained by planimetering each of the four panel
areas.

     Fig. 7a shows that a small but discernable enhancement of the
inhibitory effect occurred at Grasselli waste concentrations below
0.04 percent. Above this 0.04 percent level, however, the summed response from the individually present wastes exerts a somewhat greater
inhibitory effect. The above information plotted in Fig. 8 emphasizes the relatively small differences encountered between the additive
and combined responses in this instance.

SUMMARY

At DWD-106 waste released from the barge is diluted 1:10,000 with seawater within four hours and the concentration remains stable for several days. Laboratory bioassay observations conducted with Edgemoor waste demonstrate a measurable inhibitory effect (20 percent above that of a waste-free control) from seawater dilutions containing 0.01% (volume basis) of waste. In terms of acidity, the above dilution corresponds to a hydrogen ion concentration of about 14 μg equivalents per 100 ml seawater. Comparisons between whole waste responses and the independent effects of hydrochloric acid indicate that hydrogen ion content is not the major inhibitory agent of this waste. Other studies involving the use of EDTA to overcome biological toxicity due to heavy metals, have helped confirm that the heavy metal content accounts for most of the inhibition encountered in our studies.

Inclusion of an EDTA titration within our bioassay design enabled us to estimate the extent of heavy metal inhibition associated with Edgemoor waste. We observed that about 8.50 μmoles of EDTA was required to completely remove heavy metal stress from 1 liter of waste. Assuming that the responsible heavy metals react with EDTA at a 1:1 molar ratio, the equivalent heavy metal inhibitory source associated with this waste would also approximate 8.50 μmoles. Actually, the total heavy metal content including iron, in Edgemoor waste approaches 815 m mole 1^{-1}. Hence the heavy metal correspondence to EDTA accounts for only about 1 percent of the total complexable heavy metal loading. A more realistic interpretation may result if it is assumed that all of the iron present in Edgemoor waste undergoes precipitation in seawater thereby becoming unavailable for complexation with EDTA. Under these circumstances the EDTA equivalent takes on greater importance and corresponds to about 76 percent of the non-ferrous heavy metal content.

Of further interest is the nearly 1:1 correspondence encountered between the EDTA: heavy metal equivalence and the vanadium content of Edgemoor waste shown in Table 1. However, without substantiation, this equality is probably more fortuitous than real. It seems unlikely that all of the added EDTA would react exclusively with vanadium or that vanadium supplied the only source of bacterial inhibition.

Grasselli waste analyses indicate that caustic alkalinity is not the dominant factor leading to impared bacterial heterotrophic activity and systematic efforts to overcome the inhibitory impact of this waste by ultraviolet radiation and by persulfate oxidation have not proved fully successful. Such resistance of organic decomposition suggests highly refractory organic properties for this waste and predicts an extended half life in the oceanic environment.

    To help elucidate future bioassay interpretations regarding
mixed stress situations this paper also records a graphical solution
which permits differentiation between the net bioassay response and
that registered by isolated stress components. In the future, we
hope to incorporate this technique into our bioassay procedures to
determine whether additive, enhanced or compensatory responses obtain
when more than a single stress source is operative.

                                  REFERENCES

Aaronson, S. (1978) Excretion of organic matter by phytoplankton in
    vitro. Limnol. Oceanogr., 23, 838.
Andrews, P. and P. J. LeB. Williams (1971) Heterotrophic utilization
    of dissolved organic carbon in the sea. II. Measurements of
    the oxidation rates and concentrations of glucose and amino
    acids in seawater. J. Mar. Biol. Assoc. U.K., 51, 121-126.
Armstrong, F. A. J., P. M. Williams and J. D. H. Strickland (1966)
    Photooxidation of organic matter in seawater by ultraviolet
    radiation, analytical and other applications. Nature, 211,
    481-483.
Gillespie, P. A. and R. F. Vaccaro (1978) A bacterial bioassay for
    measuring the copper-chelation capacity of seawater. Limnol.
    Oceanogr., 23, 543-548.
Hobbie, J. E. and R. T. Wright (1965) Bioassay with bacterial uptake
    kinetics: glucose in freshwater. Limnol. Oceanogr., 10,
    471-474.
Lampert, W. (1978) Release of dissolved organic carbon by grazing
    zooplankton. Limnol. Oceanogr., 23, 831-834.
Menzel, D. W. and R. F. Vaccaro (1964) The measurement of dissolved
    organic and particulate carbon in seawater. Limnol. Oceanogr.
    9, 138-142.
Parsons, T. R. and J. D. H. Strickland (1961) On the production of
    particulate organic oceanic carbon by heterotrophic processes in
    seawater. Deep-Sea Res., 8, 211-222.
Parsons, T. R. and H. Seki (1970) Importance and general implications
    of organic matter in aquatic environments. In: Organic Matter
    in Natural Waters, D. W. Hood, editor, Univ. Alaska, 1-27.
Sunda, W. G. and R. R. L. Guillard (1976) The relationship between
    cupric ion activity and the toxicity of copper to phytoplankton.
    J. Mar. Res., 34, 511-529.
Sunda, W. G. and J. A. M. Lewis (1978) Effect of complexation by
    natural organic ligands on the toxicity of copper to a
    unicellular algal Monochrysis lutheri. Limnol. Oceanogr., 23,
    870-876.
Vaccaro, R. F. (1966) Studies on heterotrophic activity in seawater
    based on glucose assimilation. Limnol. Oceanogr., 11, 596-607.
Vaccaro, R. F., F. Azam and R. E. Hodson (1977) Response of natural
    marine bacterial populations to copper: Controlled ecosystem
    pollution experiment. Bull. Mar. Sci., 27, 17-22.

Webb, K. L. and R. E. Johannes (1967) Studies of the release of dissolved free amino acids by marine zooplankton. Limnol. Oceanogr., 12, 376-382.

THE EFFECTS OF INDUSTRIAL WASTES

ON MARINE PHYTOPLANKTON

Lynda S. Murphy[2], Peter R. Hoar[1], and Rebecca A. Belastock

Woods Hole Oceanographic Institution
Woods Hole, MA. 02543
[1]Amherst College, Amherst, MA 01002
[2]Bigelow Laboratory for Ocean Sciences, West Boothbay
 Harbor, Maine 04575

ABSTRACT

     Nineteen clones in three marine phytoplankton species were used
in bioassays of DuPont Grasselli waste. We determined that low doses
of waste stimulate growth; higher concentrations were inhibitory.
The transition from stimulation to inhibition in culture occurs at
approximately the highest concentration that is sustained in the environment. Clones from polluted estuaries are less sensitive to the
waste than are clones of the same species from other environments.

INTRODUCTION

     Deep Water Dumpsite (DWD) 106 lies along the boundary between
the neritic and the oceanic domains, each of which contains its own
characteristic flora (and fauna). The ecological boundary then coin-
cides with numerous species boundaries. Phytoplankters indigenous to
slope water frequently are mixed with and must compete with phytoplankters from oceanic water brought in by warm core rings, and from
coastal water entrained from the shelf (Murphy et al., 1978). Even
those phytoplankton species that appear to cross the boundary consist
of distinct ecological races, each with its own genetically based
physiological characteristics (Murphy and Guillard, 1976). An evaluation of toxic effects in such a region should take into account the
environmental origins of the organisms being tested.

This study reports the development of a bioassay system that is internally consistent and that allows the monitoring of growth under controlled conditions with excellent replication. The bioassay system is used to compare the responses of a number of closely related clones isolated from a wide range of environments to an industrial waste that is discharged at DWD 106.

METHODS

We describe first our methods of maintaining clones for testing, and secondly, the bioassay conditions. An explanation of the data analysis, and a description of the clones follows.

Clone Maintenance. The growth medium is prepared from surface Sargasso Sea (SS) water, filtered through a 0.8 µm filter, autoclaved in teflon, and then allowed to equilibrate overnight. The water is then enriched to F/5 of Guillard and Ryther (1962), but with silicate enriched to F/2, by sterile addition of nutrients from sterile stocks. In this medium, phosphate is the limiting nutrient. The medium is then dispensed into 25 x 120 mm pyrex test tubes. Growth is monitored with a Turner Design Model 10 Fluorometer. When chlorophyll is excited by light transmitted at 430 nm, it emits light at 670 nm. The intensity of this fluorescence is approximately proportional to the amount of chlorophyll present under standardized conditions. Thus, an increase in fluorescence can be used as an index of growth of chlorophyll. We have determined that the amount of fluorescence present is proportional to the cell count, both under normal (control) conditions, and in the presence of DuPont Grasselli waste at 0.1%.

Stock clones are monitored daily and transferred to fresh medium while still in log phase. Thus, the cells are never allowed to reach nutrient limitation. The clones are maintained in this semi-continuous culture in enriched Sargasso Sea water and grown in a constant-temperature room at 21° ± 0.5°C and 12,000 lux provided by cool-white fluorescent tubes on a 14:10 light:dark cycle.

The Bioassay. For each experiment, carried out in triplicate, clean, sterile tubes are filled with SS medium, and 1% v/v of the appropriate dilution of DuPont Grasselli waste is added to the "treated" tubes. An equal volume of sterile, distilled water is added to the "control" tubes. The tubes are agitated and then inoculated with cells from log phase cultures that have been acclimatized to the appropriate experimental conditions. The volume of the inoculum is calculated to give an initial reading of 1.0 ± 0.3 fluorometric units. After reading, the tubes are placed back in the culture room. At 24 hour intervals, they are removed, allowed to equilibrate five minutes, read in the fluorometer, and then returned to the culture room. This procedure is repeated for five days, or until the end of log

Table 1. Effect of Grasselli waste on pH of growth medium.

Final Concentration (%)	pH
0	8.4
0.01	8.4
0.05	8.6
0.1	8.8
0.5	9.3
1.0	9.9

phase growth.

The Grasselli waste is a clear liquid with a density of 1.15 g/ml. The pH is approximately 12.5. The pH of the final dilution in enriched SS medium is shown in Table 1. A pH of 8.4 is typical of surface waters, but it can be higher in the presence of high photosynthetic activity. However, pH's greater than 9.0 are not encountered in the open ocean, and therefore, concentrations greater than 0.1% were not used in this study.

Data Analysis. For each run, the daily fluorescence readings were normalized to an initial (t_0) value of 1.0 units. The growth rate, μ, was determined using the logarithm of the fluorescence in a linear regression program and is expressed as μ_{72}, the average growth rate in divisions per day from 0 to 72 hours. The depression in growth rate caused by the waste is expressed as μ_T/μ_C (the ratio of growth in the treated tubes to growth in the control tubes). The ratio of the fluorescence at 72 hours of the treated to the control experiment is expressed as FL_T/FL_C. This ratio approximates the reduction in chlorophyll due to the waste.

Description of Clones. We report here experiments with 19 clones of three species of marine phytoplankton from the culture collection of the Woods Hole Oceanographic Institution. The species and the origins of the clones are shown in Table 2.

RESULTS

Effect of Solution on Sensitivity. Undiluted Grasselli waste, when added to seawater, produces a precipitate. In order to achieve uniform solution of the waste in seawater medium for greater pipetting accuracy, the waste was first diluted with distilled water to a concentration 100 times that desired in seawater. This distilled water stock was then added to the enriched seawater in a 1:100 dilution, thus allowing dissolution without causing an appreciable

Table 2. Origin of clones used in this investigation.

Species	Clone	Origin	Year	Isolator
Thalassiosira pseudonana	3H	Moriches Bay Long Island, NY	1958	R. Guillard
	W	Wümme River Estuary Bremmen, Germany	1973	E. Paasche
	C5	Chincoteague Island Lagoon, Virginia	1964	D. Wilson
	STX-97	St. Croix Reef Virgin Islands	1969	K. Haines
	Swan-1	Swan River Estuary S. W. Australia	1965	R. Davis
	13-1	33°11'N, 65°15'W Sargasso Sea	1958	R. Guillard
	58-102	39°05'N, 71°56'W Continental Slope	1976	L. Murphy
	35-10b	39°13'N, 69°20'W Continental Slope	1977	P. Hoar/ L. Murphy
	35-11a	39°13'N, 69°20'W Continental Slope	1977	P. Hoar/ L. Murphy
	35-81	36°11'N, 69°35'W Sargasso Sea	1977	P. Hoar/ L. Murphy
	35-128b	39°20'N, 71°00'W Continental Slope	1977	P. Hoar/ L. Murphy
Skeletonema costatum	Skel	Long Island Sound New York	1956	R. Guillard
	FHS21	Friday Harbor Washington	before 1964	C. McAllister/ R. Guillard
	Fry-2	Freeport, Texas	1972	G. Fryxell
	FH-1	Falmouth Harbor Massachusetts	1974	L. Murphy
	G44S	Sargasso Sea	1968	P. Hargraves
	35-24a	39°12'N, 69°20'W Continental Slope	1977	P. Hoar/ L. Murphy
Emiliania huxleyi	451B	Oslofjord, Norway	1973	E. Paasche
	BT-6	32°10'N, 64°30'W	1960	R. Guillard

Table 3.   Comparison of effect of Grasselli waste at a final con-
           centration of 0.1%, when first diluted (1:10) in distil-
           led water (DW) and in seawater (SW).

Clone	μ_T/μ_C		FL_T/FL_C	
	DW	SW	DW	SW
Thalassiosira pseudonana				
3H	0.91	0.90	0.57	0.61
13-1	0.81	0.77	0.35	0.31
58-102	0.70	0.75	0.19	0.20
Emiliania huxleyi				
451B	0.81	0.91	0.83	0.67
BT-6	died	0.12	0.02	0.05

decrease in salinity. We have determined that toxicity of the Grasselli waste was not significantly different when it is first diluted
with distilled water before mixing with seawater than when it was
mixed directly in seawater (Table 3).

Effect of Dosage on Toxicity. Two clones of the coccolithophorid Emiliania huxleyi were tested against the Grasselli waste at concentrations from 0.001% to 0.1%, which ranged one order of magnitude
above and below the concentration (0.01%) found to persist in the
wake of the barge 17 hours after discharge (Kohn and Rowe, 1978). As
shown in Figs. 1-3, both clones were stimulated by the waste at concentrations between 0.001% and 0.01%. Between 0.01% and 0.1%, the
waste is toxic -- more so to the oceanic clone BT-6 than to the estuarine clone 451B.

Effect of Clonal Origin on Sensitivity. We tested seven clones
each of Thalassiosira pseudonana and Skeletonema costatum with three
concentrations of the Grasselli waste. Tables 4 and 5 show considerable variability among the oceanic and neritic clones in the ranges
of growth rate depression and fluorescence reduction. No significant
reduction in either fluorescence or growth rate occured in any clones
exposed to the lowest concentration, 0.01%. However, at 0.05%, fluorescence at 72 hours was less than 60% of the controls in two neritic
and two oceanic clones of T. pseudonana., and in one neritic and one
oceanic clone of S. costatum. At 0.1%, all of the clones except 3H,
W and Skel showed fluorescence less than 60% of the controls.

DISCUSSION

None of the clones were inhibited at 0.01%, the highest concentration that Kohn and Rowe (1978) indicated is environmentally significant. In fact, many clones were stimulated at this and lower concentrations. Pollution damage is generally expressed as some function of inhibition, although initial stimulation followed by inhibition at higher doses has been reported before (Dunstan et al., 1975). Any alteration in the growth rate of a clone may be considered potentially damaging to the community structure from which that clone was isolated. It has been demonstrated that a sensitive species in a mixed culture can be inhibited (relative to other species in the mix) at concentrations of a pollutant that do not inhibit it in pure culture (Mosser et al., 1972; Fisher et al., 1974). In the next phase of our work, we will look at competitive reactions, at lower dosage, but over a longer time span. We are interested in the possibility of altered community structure in the absence of change in total chlorophyll and in the effect that such altered structure may have on the food web.

Slobodkin and Sanders (1969) hypothesized that organisms adapted to stable, predictable environments would be more "fragile" or vulnerable to environmental stress than organisms adapted to unstable, unpredictable environments. Fisher et al. (1973) and Fisher (1977) found that diatom clones established from stable, oceanic environments were more sensitive to chemical stress than were closely related clones established from more variable estuarine environments. This

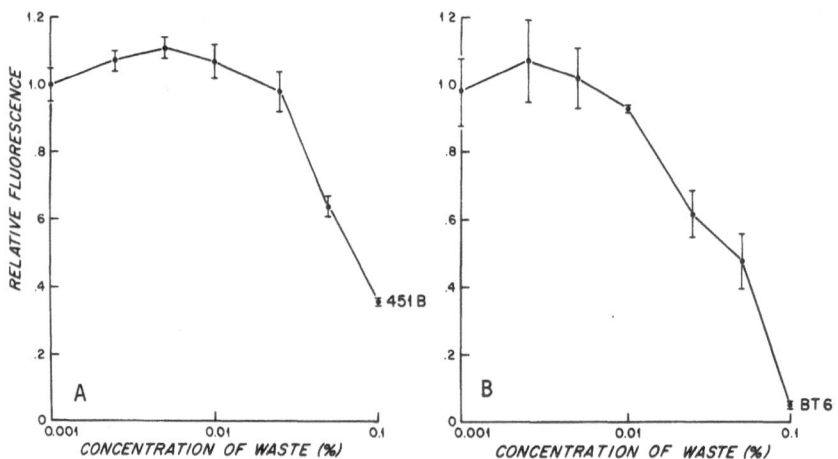

Fig. 1.   Relative fluorescence of Emiliania huxleyi clones at 72 hours: the ratio of the treated to the control fluorescence (FL_T/FL_C) as a function of dose of Grasselli waste. A. Clone 451B (neritic);   B.   Clone BT-6 (oceanic).

relationship held whether the chemical stress was one to which the diatoms may have had previous exposure (PCB's), or whether the stress was an exotic chemical, to which no previous exposure could be expected.

Jensen et al. (1974) showed that a clone of the marine diatom Skeletonema costatum isolated from a zinc-polluted fjord was several orders of magnitude less sensitive to zinc than was another clone of the same species isolated from a nearby, but unpolluted, fjord. The two Norwegian fjords were similar in overall varibility and thus, this resistance appears to be a specific adaptation to a chemical stress.

We do not know the extent to which exposure to one chemical stress predisposes resistance to other chemical stresses. Possibly, resistance involves changes in membrane structure and permeability so that the effect is generalized and exposure to one chemical stress confers non-specific resistance to other chemical stresses independent of other variables in the environment. Fisher's estuarine clones were clearly less sensitive, but his results do not distinguish between two separate potential effects: 1) Estuarine clones may be genetically adapted to change per se; they are from unstable environments and may be capable of withstanding other changes as well and, 2) The estuarine clones chosen happened to have come from polluted estuaries and may have become resistant to related chemicals.

Using our bioassay system, we have compared two of the clones

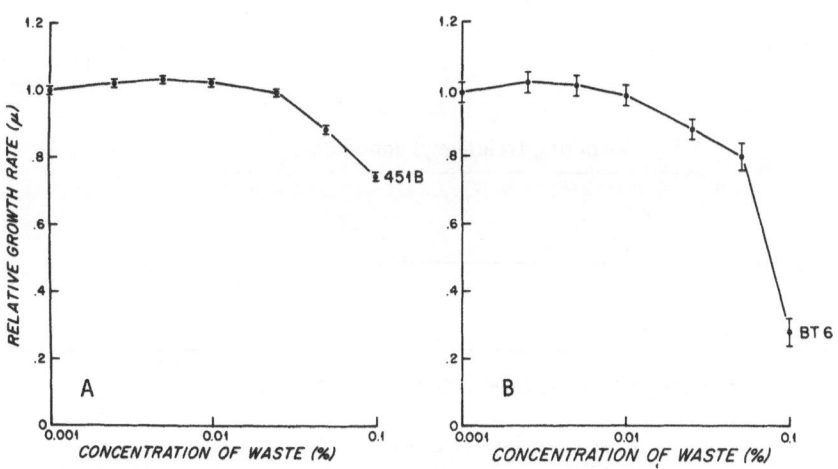

Fig. 2.   Relative growth rate of Emiliania huxleyi clones during initial 72 hours: the ratio of the treated to the control growth rate (μ_T/μ_C) as a function of dose of Grasselli waste. A. Clone 451B (neritic); B. Clone BT-6 (oceanic).

used by Fisher and have found the same trend: the estuarine clone of
the diatom Thalassiosira pseudonana (3H) is less sensitive than the
oceanic clone (13-1). The same relationship holds for the estuarine
clone (Skel) and oceanic clone (G44S) of the diatom Skeletonema
costatum, and for the estuarine clone (451B) and the oceanic clone
(BT-6) of the marine coccolithophorid Emiliania huxleyi (see Figs.
1-3). However, when other neritic and oceanic clones of T. pseudonana and S. costatum are compared, a more complex picture emerges.

    The most resistant clones originate from polluted estuaries in
temperate climates. They are: 3H from Moriches Bay, Long Island,
Skel from Long Island Sound, and W from the Wümme River Estuary,
Germany. Other neritic clones are significantly more sensitive.
They show a range of responses and clones from Falmouth Harbor, Mass-
achussetts and Freeport, Texas are less sensitive than clones from
the Swan River Estuary, Southwest Australia and Friday Harbor, Washington. There is no correlation between sensitivity and either latitude or salinity. The oceanic clones also show a range of responses.
Sensitivities overlap those of the neritic clones (Fig. 3) and four
of the eight oceanic clones are less sensitive than two of the eight
neritic clones. Sensitivity does not increase with distance from
land nor decrease with latitude. In fact, the most sensitive clones
were isolated from the upper slope (clone 35-128a from typical slope
water and clones 58-102 and 35-24a from warm core rings).

    We have attempted to correlate sensitivity with the length of
time a clone has been in culture, since it may be that oceanic clones
gradually lose sensitivity to certain toxicants after long exposure

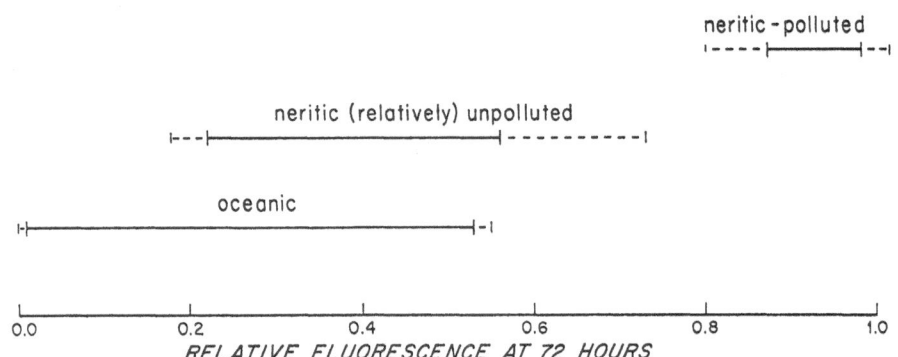

Fig. 3. Relative fluorescence at 72 hours of Thalassiosira pseudo-
         nana and Skeletonema costatum clones: the ratio of the
         treated to the control fluorescence (FL_T/FL_C) after
         exposure to 0.1% Grasselli waste in clones from polluted
         neritic, (relatively) unpolluted neritic, and oceanic
         environments.

Table 4. Relative fluorescence (FL_T/FL_C) at 72 hours as affected by Grasselli waste.

Clone	Concentration		
	0.01%	0.05%	0.1%
Thalassiosira pseudonana			
polluted neritic			
3H	1.05 ± 0.09	0.88 ± 0.08	0.87 ± 0.07
W	0.98 ± 0.07	1.03 ± 0.10	0.98 ± 0.10
other neritic			
C5	0.94 ± 0.21	0.67 ± 0.27	0.41 ± 0.17
STX-97	0.99 ± 0.03	0.45 ± 0.04	0.28 ± 0.05
Swan-1	0.88 ± 0.08	0.40 ± 0.04	0.22 ± 0.02
oceanic			
13-1	0.93 ± 0.13	0.58 ± 0.07	0.25 ± 0.03
58-102	0.73 ± 0.11	0.34 ± 0.04	0.09 ± 0.01
35-10b	0.84 ± 0.14	0.53 ± 0.07	0.25 ± 0.04
35-11a	0.79 ± 0.09	0.58 ± 0.08	0.35 ± 0.06
35-81	0.96 ± 0.13	0.61 ± 0.09	0.19 ± 0.03
35-128b	0.84 ± 0.11	0.17 ± 0.01	0.01 ± 0.003
Skeletonema costatum			
polluted neritic			
Skel	1.00 ± 0.04	0.98 ± 0.06	0.91 ± 0.09
other neritic			
FHS21	0.93 ± 0.10	0.45 ± 0.05	0.22 ± 0.04
Fry-2	1.05 ± 0.13	0.79 ± 0.12	0.44 ± 0.17
FH-1	1.01 ± 0.16	0.84 ± 0.13	0.56 ± 0.17
oceanic			
G44S	0.98 ± 0.06	0.80 ± 0.08	0.53 ± 0.02
35-24a	0.94 ± 0.12	0.09 ± 0.01	0.01 ± 0.001

Table 5. Relative growth rates (μ_T/μ_C) during initial 72 hours
as affected by Grasselli waste.

Clone	Concentration		
	0.01%	0.05%	0.01%
Thalassiosira pseudonana			
polluted neritic			
3H	1.01 ± 0.02	0.98 ± 0.02	0.93 ± 0.07
W	1.00 ± 0.01	1.01 ± 0.02	1.00 ± 0.02
other neritic			
C5	0.98 ± 0.04	0.89 ± 0.08	0.80 ± 0.08
STX-97	1.00 ± 0.01	0.83 ± 0.02	0.74 ± 0.04
Swan-1	0.97 ± 0.02	0.80 ± 0.02	0.69 ± 0.02
oceanic			
13-1	0.98 ± 0.03	0.94 ± 0.08	0.72 ± 0.02
58-102	0.95 ± 0.02	0.82 ± 0.02	0.60 ± 0.03
35-10b	0.97 ± 0.03	0.88 ± 0.02	0.73 ± 0.03
35-11a	0.95 ± 0.02	0.89 ± 0.03	0.80 ± 0.03
35-81	0.99 ± 0.03	0.86 ± 0.04	0.56 ± 0.04
35-128b	0.97 ± 0.02	0.72 ± 0.01	0.24 ± 0.07
Skeletonema costatum			
polluted neritic			
Skel	1.00 ± 0.01	0.98 ± 0.01	0.96 ± 0.01
other neritic			
FHS21	1.00 ± 0.04	0.75 ± 0.04	0.51 ± 0.06
Fry-2	1.00 ± 0.02	0.96 ± 0.02	0.86 ± 0.06
FH-1	0.99 ± 0.03	0.96 ± 0.03	0.93 ± 0.05
oceanic			
G44S	0.98 ± 0.01	0.96 ± 0.01	0.82 ± 0.01
35-24a	0.98 ± 0.03	0.55 ± 0.02	0.16 ± 0.004

to coastal water. We know that a loss of heterozygosity does occur
in culture (Murphy, 1978); other genetic changes may occur as well.
The physiological results of these genetic changes have not been determined. In this study, we found no evidence of such loss of sensitivity. Skeletonema costatum clone G443 was maintained in enriched
Narragansett Bay water in the culture colletion of the University of
Rhode Island from 1968 to 1975, and in enriched Vineyard Sound water
at the Woods Hole Oceanographic Institution since then, and is the
least sensitive of the oceanic clones. In comparison, the conspe-
cific clone 35-24a is highly sensitive and has been in culture only
one year. However, Thalassiosira pseudonana clone 13-1 has been
maintained in the culture collection of the Woods Hole Oceanographic
Institution since 1958 and is more sensitive than two recent conspe-
cific isolates (35-10b and 35-11a).

CONCLUSIONS

1) Low doses of DuPont Grasselli waste stimulate growth of marine phytoplankton; higher concentrations are inhibitory. In pure
culture, the crossover point occurs at approximately 0.01% waste.

2) Clones from polluted estuaries are less sensitive to the
Grasselli waste than are clones of the same species from other origins. The short-term pollution history of the original environment is
at least as important in determining sensitivity of a clone as is the
long-term stability and predictability of that environment.

3) In determining bioassay conditions, care should be taken to
choose appropriate test organisms, that is, organisms from appropriate environments since intra-specific variability in response does
occur.

ACKNOWLEDGEMENTS

We thank K. Mlodzinksa for her assistance. This work was supported by NOAA Grant #04-8-M01-41.

REFERENCES

Dunstan, W. M., L. P. Atkinson, and J. Natoli (1975) Stimulation and
    inhibition of phytoplankton growth by low molecular weight hy-
    drocarbons. Mar. Biol., 31, 305-310.
Fisher, N. S. (1977) On the differential sensitivity of estuarine
    and open-ocean diatoms to exotic chemical stress. Amer. Natur.,
    111, 871-895.
Fisher, N. S., E. J. Carpenter, C. C. Remsen, and C. F. Wurster
    (1974) Effects of PCB on interspecific competition in natural

and gnotobiotic phytoplankton communities in continuous and
batch cultures. Microb. Ecol., 1, 39-50.

Fisher, N. S., L. B. Graham, E. J. Carpenter, and C. F. Wurster
(1973) Geographic differences in phytoplankton sensitivity
to PCBs. Nature, 241, 548-549.

Guillard, R. R. L. and J. H. Ryther (1962) Studies of marine planktonic diatoms. I. Cyclotella nana Hustedt and Detonula
confervacea (Cleve) Gran. Can. J. Microbiol., 8, 229-239.

Jensen, A., B. Rystad, and S. Melsom (1974) Heavy metal tolerance
of marine phytoplankton. I. The tolerance of three algal
species to zinc in coastal seawater. J. Exp. Marine Biol.
Ecol., 15, 145-157.

Kohn, B. and G. T. Rowe (1978) Dispersion of two liquid industrial
wastes dumped at Deep Water Dumpsite 106, off the coast of New
Jersey, U.S.A. Final report submitted to Ocean Dumping Program., Rockville, Md.

Mosser, J. L., N. S. Fisher, and C. F. Wurster (1972) Polychlorinated
biphenyls and DDT alter species composition in mixed cultures of
algae. Science, 176, 533-535.

Murphy, L. S. (1978) Biochemical taxonomy of marine phytoplankton by
electrophoresis of enzymes. II. Loss of heterozygosity in
clonal cultures of Thalassiosira pseudonana and Skeletonema
costatum. J. Phycol., 14, 247-250.

Murphy, L. S. and R. R. L. Guillard (1976) Biochemical taxonomy of
marine phytoplankton by electrophoresis of enzymes. I. The
centric diatoms Thalassiosira pseudonana and T. fluviatilis. J.
Phycol., 12, 9-13.

Murphy, L. S., R. R. L. Guillard, H.-t. Lee, and L. E. Brand (1978)
Distribution of electromorphs and growth rate characteristics in
isolates of the diatom Thalassiosira pseudonana from the
neritic-oceanic boundary. J. Phycol., 14 (Suppl.), 26.

Slobodkin, L. B. and H. L. Sanders (1969) On the contribution of
environmental predictability to species diversity. Brookhaven
Symp. Biol., 22, 82-95.

THE EFFECTS OF POLLUTANTS ON MARINE ZOOPLANKTON

AT DEEP WATER DUMPSITE 106: PRELIMINARY FINDINGS

Judith M. Capuzzo and Bruce A. Lancaster

Woods Hole Oceanographic Institution
Woods Hole, MA. 02543

ABSTRACT

The toxicity of duPont's Grasselli waste to the marine copepods
Pseudocalanus sp. and Centropages typicus has been evaluated in 96
hour bioassays. Test organisms were exposed to concentrations of
waste ranging from 1 to 1000 ppm and the resulting effects on survival and feeding rates were monitored. Complete mortality of Pseudocalanus was observed at an exposure concentration of 1000 ppm; no
mortality was detected among control organisms or those exposed to
lesser concentrations. Significant reductions in filtration activity, however, were detected with exposure to 10 and 100 ppm. Significant mortality of Centropages was detected at 100 ppm and exposure to
higher concentrations resulted in 100% mortality. Feeding rates of
Centropages as measured by both consumption of Artemia nauplii and
fecal pellet production rates were significantly lower than control
values with exposure to 50 and 100 ppm. The findings of other investigators dealing with the effects of dumped wastes on marine zooplankton are discussed.

INTRODUCTION

Ocean dumping of industrial wastes provides an alternative to
land based disposal of such toxic materials, but the impact of this
waste disposal on the marine ecosystem is not well understood. Because of their fragile nature and ecological importance in marine
food chains, the sensitivity of plankton populations to industrial
wastes is a problem of great concern. Exposure to industrial wastes
may result in either reduced plankton biomass at several trophic
levels or accumulation and transfer of toxic materials from one

411

trophic level to another; either response may lead to significant
imbalances or alterations in food chain dynamics.

A research program was begun in March 1978 to identify the responses of zooplankton populations to wastes commonly discharged at
Deep Water Dumpsite 106 (DWD-106). The objectives of the research
were to determine changes in the rates of survival, feeding, growth
and egg production of marine copepods as a result of exposure to either duPont's Grasselli or Edgemoor waste. Our preliminary results
with marine copepods exposed to duPont's Grasselli waste are presented in this paper.

METHODS

Two species of copepods were selected as the test organisms:
Pseudocalanus sp., a herbivore, and Centropages typicus, an omnivore.
Both species of copepods are abundant in continental slope waters and
have been collected at DWD 106; they may also be collected in nearshore Cape Cod waters for part of the year —Pseudocalanus may be collected in large numbers during March and April and Centropages becomes abundant from July through October. Both species were collected using a 316 μm plankton net, towed in Vineyard Sound, MA at a
depth of 5 meters. Pseudocalanus was collected at 8°C, acclimated to
laboratory conditions, and adult males and gravid females were isolated in pairs and placed in 12 liter glass carboys with 1 μm filtered seawater. The copepods were then maintained at 10°C and fed mixtures of Monochrysis lutheri and Phaeodactylum tricornutum (1×10^5
cells ml^{-1}). Egg production and naupliar development were recorded
every 48 hours. Centropages was collected and maintained at 20°C
under conditions similar to those described for Pseudocalanus with
the exception that 100 Artemia nauplii per liter were also added to
the culture carboys daily.

According to T. P. O'Connor of the NOAA Ocean Dumping Division,
field studies at DWD 106 have indicated that the Grasselli and Edgemoor wastes are diluted by a factor of 10^4 and 10^5, respectively,
within 4 hours after dumping and level off at these concentrations
for one to several days. These dilutions are confirmed by the findings of Csanady, Hatcher et al. and Kester et al. reported in this
volume. Therefore 96 hour bioassays at concentrations ranging from 1
to 1000 ppm were chosen as being representative of concentrations and
exposures experienced under field conditions.

A sample of Grasselli waste was obtained during April 1978 and
used in bioassays from April to October 1978. All waste dilutions
were made up in seawater (salinity = 30-31°/₀₀, pH = 7.9-8.1) and
measurements of pH and dissolved oxygen were made on all dilutions
prior to the addition of test organisms. The bioassay chambers were
500 ml Fleakers ® (Corning Glass), immersed in a temperature

controlled water bath. For Pseudocalanus 50 copepods were added to each Fleaker and bioassays were conducted at 10°C; for Centropages 10 copepods were added to each Fleaker and maintained at 20°C. At the end of the 96 hour exposure period survival of copepods at each of the waste dilutions was determined using a neutral red vital staining technique (Capuzzo, 1979) and compared with control animals.

To test the sublethal effects of waste exposure on feeding rates, the filtration rates of Pseudocalanus and the feeding rates and fecal pellet production rates of Centropages exposed to concentrations of waste ranging from 1 to 100 ppm were determined and compared to control animals. For Pseudocalanus filtration rates were determined daily during a 96 hour exposure period using an indirect method, monitoring the reduction in cell concentration per unit volume during a set time interval. Copepods in the assay chambers were fed a suspension of Phaeodactylum tricornutum (1×10^5 cells ml^{-1}) and the filtration rate was calculated from the following equation:

filtration rate = $(\log C_o - \log C_t)$ V/log e·t, where,

C_o = initial concentration of cells in suspension;

C_t = concentration of cells in suspension after time t;

V = volume of media (milliliters);

t = time (hours).

For Centropages, copepods exposed to the various waste dilutions were maintained on a diet of 100 Artemia nauplii per assay chamber and 10^5 cells ml^{-1} Phaeodactylum tricornutum. Fecal pellet production and feeding rate on Artemia nauplii were determined at the end of the 96

Table 1. Measurements at 25°C of pH and dissolved oxygen concentration of mixtures of duPont Grasselli waste and seawater.

Waste Concentration (ppm)	pH	Dissolved Oxygen (ppm)
control	7.90	6.7
1	7.80	6.6
5	7.85	6.8
10	7.80	6.7
50	7.80	6.7
100	7.85	6.6
500	7.90	6.6
1000	8.05	6.8
5000	8.45	6.7

hour exposure period. Significant differences in the various parameters measured from both control and exposed organisms were determined
by analysis of variance (Dixon and Massey, 1969).

RESULTS AND DISCUSSION

Measurements of pH and dissolved oxygen content of the seawater
dilutions of duPont Grasselli waste are presented in Table 1. Significant increases in pH were detected only at concentrations of 1000
and 5000 ppm. No significant difference in oxygen content was detected in any of the waste dilutions. Therefore toxic responses can
not be attributed simply to pH stress or oxygen deficiency.

The mortality of Pseudocalanus and Centropages exposed to dilutions of duPont Grasselli waste is presented in Fig. 1. Mortality of
Pseudocalanus was only detected at an exposure concentration of 1000
ppm; no organisms survived the 96 hour exposure to this concentration. No mortality was detected among control animals or those exposed to lesser concentrations. Significant mortality of Centropages
was detected at 100 ppm (46%) and exposure to higher concentrations

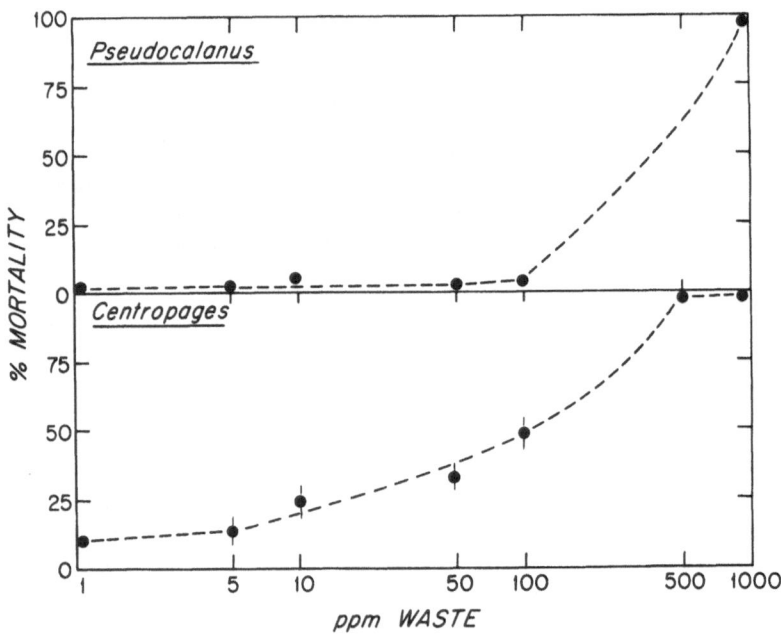

Fig. 1. Per cent mortality of Pseudocalanus and Centropages exposed to Grasselli waste; each point is the mean of 6 determinations; standard error bars are presented with the
Centropages data.

resulted in 100% mortality. Control mortality and that at lesser
concentrations was 20-30%, possibly due to some degree of cannibalism
among the test organisms.

     Filtration rates of Pseudocalanus and feeding rates and fecal
pellet production rates of Centropages are presented in Fig. 2. Significant reductions (P < 0.01) in filtration rates of Pseudocalanus
were detected with exposure to 10 and 100 ppm. Feeding rates of Centropages, as measured by removal of Artemia nauplii, were significantly (P < 0.01) lower than control values among copepods exposed to
concentrations of waste exceeding 5 ppm, the greatest reduction being
measured among copepods exposed to 100 ppm. Fecal pellet production
was variable among the test organisms from the control group and the
lower concentration exposure groups; significant decreases (P <0.01),
however, were detected among copepods exposed to 50 and 100 ppm.

     The greater sensitivity of Centropages to Grasselli waste might
be explained as a function of exposure temperature. The exposure

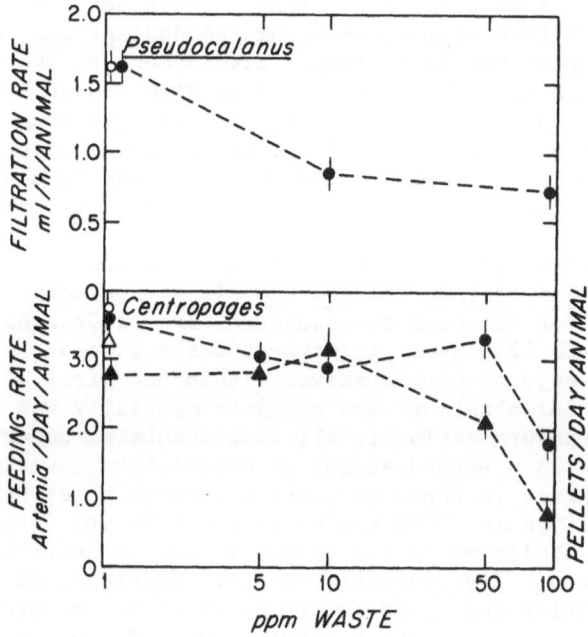

Fig. 2. Filtration rates of Pseudocalanus and feeding rates and
         fecal pellet production rates of Centropages exposed to
         Grasselli waste; for Pseudocalanus, each point is the
         mean of 24 determinations; for Centropages each point
         is the mean of 6 determinations; ● -- feeding rates,
         ▲ -- fecal pellet production rates; bars represent
         standard error of each set of data.

temperature for Centropages was 20°C, whereas that for Pseudocalanus
was 10°C. Increased metabolic rates and possible uptake rates of
Grasselli waste at higher exposure temperatures will be further ex-
plored.

It is apparent from these findings that although acute toxic effects of Grasselli waste to marine copepods only occur with exposure
to concentrations > 100 ppm, sublethal stress as indicated by reduced
feeding rates can be detected within the concentration range of 5-100
ppm. Reduced feeding rates result in less energy available for metabolism, growth and reproduction and thus long-term changes in zooplankton populations may occur.

Previous studies on the effects of dumped wastes on marine zooplankton have dealt with the effects of short-term exposure on survival and subsequent effects on growth and reproduction. Grice et
al. (1973) investigated the toxic effects of acid waste on the marine
copepods Calanus finmarchicus, Temora longicornis and Pseudocalanus
sp. Their results showed that significant mortality of the test organisms only occurred at concentrations of waste higher than those
normally experienced in field conditions. In an earlier study (Vaccaro et al., 1972) long term exposure of Pseudodiaptomus coronatus to
acid iron wastes resulted in prolonged development of embryonic and
juvenile forms; however, this could only be detected at higher than
normal concentrations. In a field study of the same problem Wiebe et
al., (1973) reported no significant difference in zooplankton biomass
or community structure between acid dumping grounds in the New York
Bight and control areas.

Lawson (1976) investigated the effects of Grasselli waste on the
copepods Centropages typicus and Acartia clausi. Specimens of Acartia were exposed for 96 hours to concentrations of Grasselli waste
ranging from 500 to 1250 ppm. Effects of waste exposure on swimming
behavior of this copepod were observed within the first 24 hours of
exposure to all concentrations and complete mortality was observed at
the end of the exposure period in all test chambers; control mortality equalled 15%. In a second series of experiments, adult Centropages were transferred in rapid succussion through three concentrations of Grasselli waste: 2000 ppm (3 min.), 1000 ppm (5 min.) and
300 ppm (5 min.), followed by two seawater rinses; control organisms
were pipetted through five seawater rinses. Observations on survival
and swimming behavior were made for 10 hours after transfer to uncontaminated seawater. No effect on survival or swimming behavior of
copepods could be detected. Lawson concluded that only prolonged exposure to high concentrations (> 500 ppm) of Grasselli waste was detrimental to copepod populations. A summary of the findings of these
studies is presented in Table 2.

It is apparent that the acute toxic effects of exposure of marine zooplankton to dump wastes are minimal. Little information is

Table 2. Summary of zooplankton responses to dumped wastes in laboratory bioassays.

Species (Source)	Waste	Time (hr)	Concentration (ppm)	Response
Acartia clausi (Lawson, 1976)	Grasselli	96	> 500	100% Mortality
Centropages typicus (Lawson, 1976)	Grasselli	0.25	300–2000	No Effect
Centropages typicus (this study)	Grasselli	96	> 100 < 100	100% Mortality Reduced Feeding
Calanus finmarchicus (Grice et al. 1973)	Acid-Iron	0.25 24 24	300–2000 > 500 < 250	No Effect 100% Mortality No Mortality
Pseudocalanus sp. (Grice et al. 1973)	Acid-Iron	24	> 200	100% Mortality
Pseudocalanus sp. (this study)	Grasselli	96	> 100 > 10	100% Mortality Reduced Feeding
Pseudodiaptomus coronatus (Vaccaro et al., 1972)	Acid-Iron	504	100	Prolonged Development
Temora longicornis (Grice et al. 1973)	Acid-Iron	48	> 500 < 250	100% Mortlity No Mortality

available, however, on the sublethal responses of zooplankton to such
waste exposures. The response of zooplankton populations to environmental stress may be manifested in reduced egg production rates, lower hatching survival, delayed development, extended generation times
and an increase in the number of abnormal offspring leading to unstable reproductive rates for this trophic level. Reeve et al.
(1977) suggested that the most sensitive short-term index of sublethal stress to adult copepod populations was egg production and reduced rates were measured among several copepod species (Acartia,
Paracalanus and Temora) exposed to sublethal concentrations of copper
and mercury. D'Agostino and Finney (1974) found that exposure of the
harpacticoid copepod Tigriopus japonicus to sublethal concentrations
of either copper or cadmium resulted in an inhibition of growth and
reduced fecundity. In a recent study completed in our laboratory,
exposure to chlorinated seawater resulted in reduced egg production
rates and increased doubling times of the amictic rotifer Brachionus
plicatilis (Capuzzo, 1979). These findings support the hypothesis
that pollution stress may result in serious imbalances in the rates
of secondary production in marine ecosystems. The impact of dumped
wastes at DWD 106 on fecundity and reproductive potential of copepod
populations warrant further consideration.

REFERENCES

Capuzzo, J. M. (1979) The effects of halogen toxicants on survival,
    feeding and egg production of the rotifer Brachionus
    plicatilis. Estuarine and Coastal Marine Science, 8, 307-316.
D'Agostino, A. and C. Finney (1974) The effect of copper and cadmium
    on the development of Tigriopus japonicus. In: Pollution and
    Physiology of Marine Organisms. Academic Press, New York,
    pp. 445-461.
Dixon, W. J. and F. J. Massey (1969) Introduction to Statistical
    Analysis, 3rd edition. McGraw-Hill, New York.
Grice, G. D., P. H. Wiebe and E. Hoagland (1973) Acid-iron waste as
    a factor affecting the distribution and abundance of zooplankton
    in the New York Bight. I. Laboratory studies on the effects of
    acid waste on copepods. Estuarine and Coastal Marine Science,
    1, 45-50.
Lawson, T. J. (1976) Pollution at deep water dumpsite 106. Report
    on laboratory studies of the effects of duPont waste on
    plankton. Unpublished manuscript, Woods Hole Oceanographic
    Institution.
Reeve, M. R., M. A. Walter, K. Darcy and T. Ikeda (1977) Evaluation
    of potential indicators of sub-lethal toxic stress on marine
    zooplankton (feeding, fecundity, respiration, and excretion):
    Controlled Ecosystem Pollution Experiment. Bulletin of Marine
    Science, 27, 105-113.
Wiebe, P. H., G. D. Grice and E. Hoagland (1973) Acid-iron waste as
    a factor affecting the distribution and abundance of zooplankton

in the New York Bight. II. Spatial variations in the field and implications for monitoring studies. Estuarine and Coastal Marine Science, 1, 51–64.

Vaccaro, R. F., G. D. Grice, G. T. Rowe and P. H. Wiebe (1972) Acid–iron waste and the summer distribution of standing crops in the New York Bight. Water Research, 6, 231–256.

GROSS AND MICROSCOPIC OBSERVATIONS ON SOME BIOTA

FROM DEEP WATER DUMPSITE 106

S.A. MacLean, C.A. Farley, M.W. Newman, and A. Rosenfield

National Oceanic and Atmospheric Administration
Northeast Fisheries Center, Oxford Laboratory
Oxford, Maryland 21654

ABSTRACT

Euphausiids were infested on the appendages with phoronts of an apostome ciliate and also had foci of melanization in the gills. The isopod, Idotea, showed patchy lack of pigmentation, particularly in the telson, and edematous gills. Heteropod mollusks had larval cestode infestions, neoplasialike lesions, and other cell abnormalities. Examination of leptocephali revealed possible traumatic lesions and developmental anomalies. Microbial testing for the presence of mutagens promises to be a very useful method for monitoring pollution in the ocean.

INTRODUCTION

The National Ocean Pollution Research and Development and Monitoring Planning Act of 1978, signed into law on May 8, 1978, as Public Law 95-273, begins with the following statement: "Many activities in the marine environment can have a profound short-term and long-term impact on such environment and greatly affect ocean and coastal resources therein." A great deal of information has now been accumulated that relates to the physical, chemical, and even biological effects of some types of man's activities, including dumping into inshore coastal areas. Little information, however, exists regarding effects of man's activities at offshore locations such as DWD 106. As part of a broader study to determine the biological effects of ocean dumping on biota from DWD 106, it was decided that gross and microscopic observations on selected species of biota collected before and after a test dumping not only would provide valuable

information on possible acute or immediate effects, but help develop
new strategies that might be useful for future biological effect
studies at other ocean dumping locales. In addition to the above observational approaches, attempts were made to develop rapid screening
methods to detect the presence of mutagens in or adsorbed to some of
the biota sampled from DWD 106 and to correlate, if possible, the
presence of these compounds with histopathologic findings.

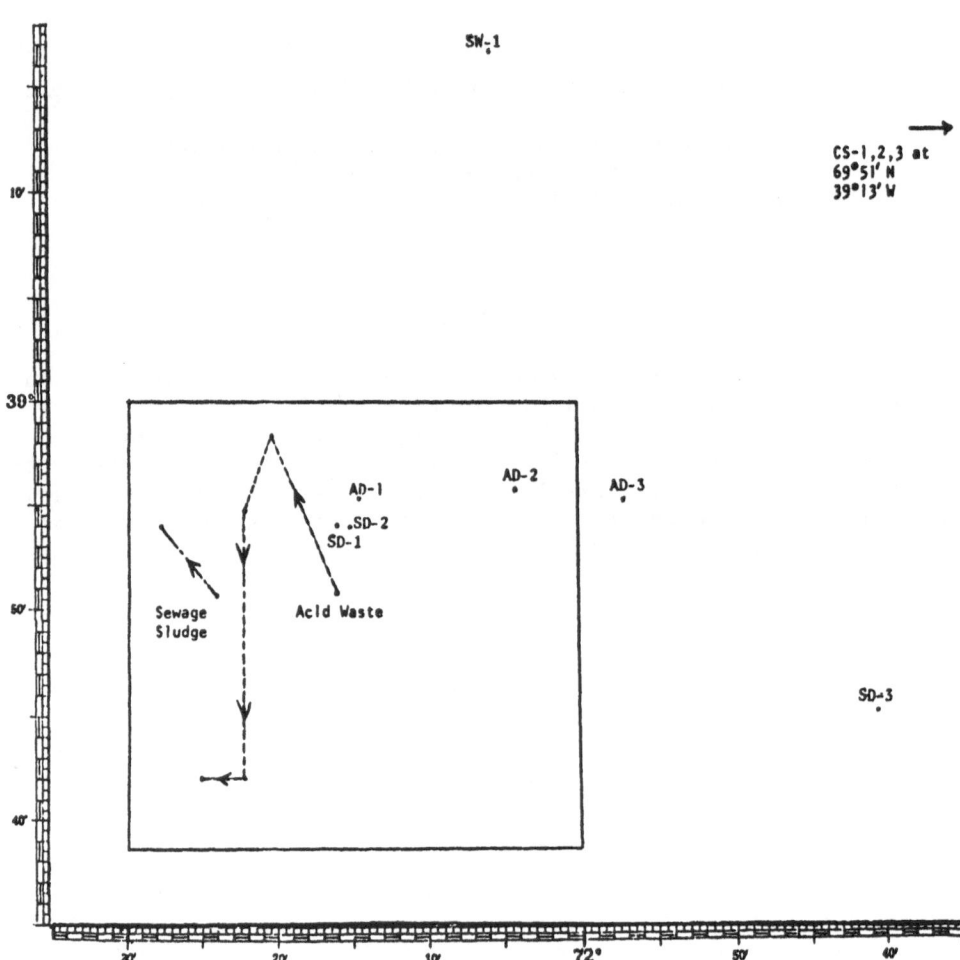

Fig. 1. Deepwater Dumpsite 106. Barge courses of acid waste and
         sewage sludge dumps and control stations (CS-1-3; SW-1) and
         tracked acid (AD-1-3) and sewage (SD-1-3) stations.

MATERIALS AND METHODS

Sampling of planktonic biota at DWD 106 began in July 1977 when two test dumps, one of sewage sludge and the other of acid-iron wastes, were studied (Fig. 1). Dispersal patterns of the dump material were determined by acoustically tracking the sludge solids or the hydrous ferric oxide floc of the acid-iron waste reaction with seawater. Dump stations (SD 1-3; AD 1-3) extended beyond the boundaries of DWD 106 due to the transport of wastes by anticyclonic movement of a Gulf Stream eddy in the site. Control stations were located in another eddy (CS 1-3) in attempts to obtain similar species and in a slope water area (SW-1).

Planktonic organisms were collected using standard mesh neuston or bongo nets (10 or 60 m depth), or the Isaacs-Kidd trawl (50 or 100 m depth). Specimens were fixed on shipboard in 4% seawater-formalin and later stored in 70% ethanol. Various Crustacea, mollusks, and ichthyofauna were selected for examination on the basis of abundance, occurrence at test stations, or relative importance to the ecosystem.

Specimens were examined in the laboratory for gross abnormalities using a dissecting microscope or were processed by normal histological procedures for examination of tissues. Sections were cut at 6 μm and stained by hematoxylin-eosin or the feulgen picromethyl blue reaction (Farley, 1969).

Tissues and whole organisms taken for mutagenic testing were frozen on shipboard. In the laboratory, tissues were minced and extracted with absolute ethanol (1 part by weight tissue to 2 parts by volume abs. EtOH). Supernatants were sent to England and tested by J. M. Parry, University of Swansea, U.K., for mutagenic activity. The test system utilized histidine-deficient bacteria cultured on medium lacking histidine. A predictable natural mutation rate allows some bacteria to grow and form colonies. Mutagens present in the test extracts, incorporated in the medium, increase the mutation rate thereby forming a greater number of colonies. The samples were further tested on yeast cultures (Parry et al., 1976).

RESULTS

Crustacea. Table 1 lists the species examined and type station of collection. Euphausia krohnii, Idotea metallica, the Sergestidae, various Euphausiacea, and copepods were the most abundant crustaceans collected.

Gross observations of euphausiids revealed numerous white, oval bodies on the setae of the thoracic and abdominal appendages. Hematein-stained whole mount preparations (Fig. 2) showed these to be encysted apostome ciliates. The ciliates were common on E. krohnii

Table 1. DWD 106 — July 1977. Gross examinations completed —
          Crustacea.

Species (total)	Total by station*			
	CS	SD	AD	SW
Euphausia krohnii (94)	0	61	33	0
Oplophorus spinosus (2)	2	0	0	0
Idotea metallica (26)	4	8	14	0
Thysanopoda acutifrons (4)	3	0	1	0
T. raschii (5)	0	0	5	0
Sergestidae (26)	26	0	0	0
Euphausiacea (20)	0	19	1	0
Copepods**	+	+	+	+

 *CS — control station
  SD — sludge dump
  AD — acid—iron dump
  SW — slope water
**Various species — several hundred — not quantified.

(80/94=85%), Thysanopoda raschii (5/5=100%), other Euphausiacea
(4/20=20%), and were also found infesting Sergestidae (2/26=8%). The
number of ciliates varied per crustacean and no attempt was made to
quantify them. No histopathology was notable at attachment sites.

     Further gross examination of euphausiids showed evidence of
blackened gill areas (Fig. 3). In stained whole mounts and histological sections, these areas are shown to be of a cellular nature and
appeared melanized and seemingly necrotic (Fig. 4). Fifty—one of 94
(54%) E. krohnii and 17 of 20 (85%) other Euphausiacea had this gill
condition. Table 2 summarizes the percent occurrence of the black
gill condition and ciliate infestation in euphausiids examined.

     Abnormalities in other planktonic Crustacea were few. Three
specimens of Idotea metallica from control and dump stations, by
gross examination, showed patchy areas of the pleon lacking pigmentation. Histologic examination of the epidermis in some of these areas
showed apparent edema and hypertrophy which depressed the line of
pigmented cells away from the cuticle. The cause of this epidermal

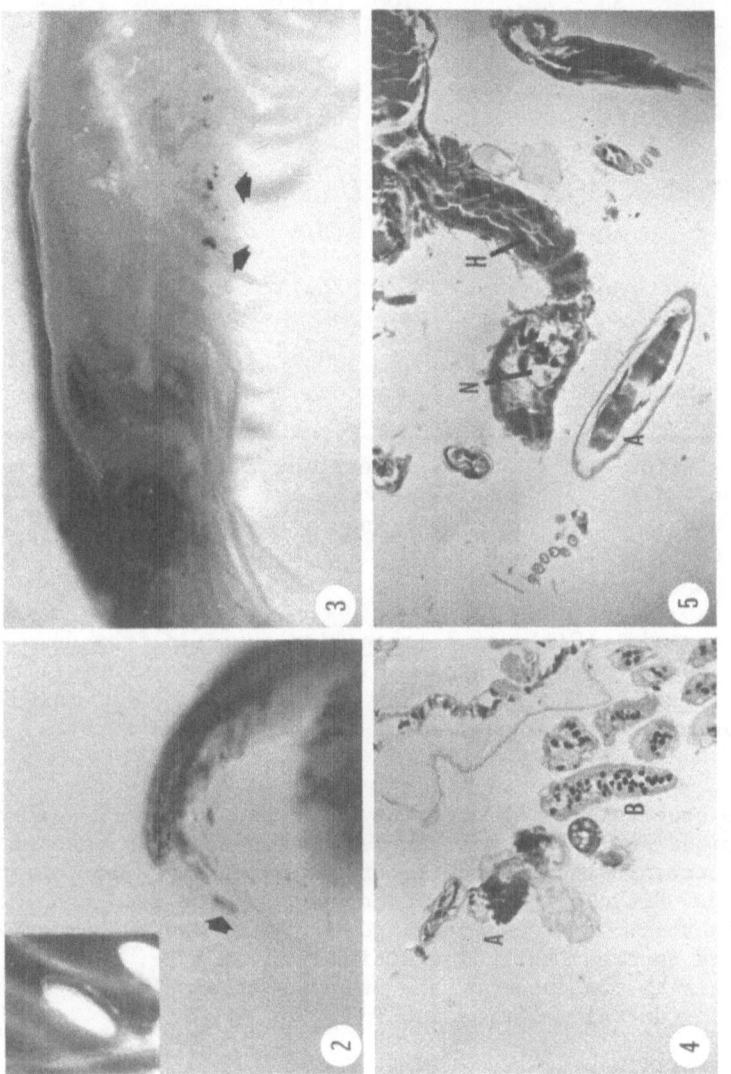

Fig. 2-5. Apostome ciliates on setae of euphausiid pereiopod. 2. Hematein-stained whole mount. 140X. Inset. Magnification of encysted apostome. Protargol. 350X. 3. Blackened gill foci of Euphausia krohnii. Unstained, preserved specimen. 15X. 4. Melanized and necrotic gill lamella (A) and normal lamella (B) of E. krohnii. Hematoxylin-eosin. 140X. 5. Abnormal swollen appendage of copepod with infiltration of hemocytes (H) and nodule formation (N). Note appearance of normal appendage (A). Hematoxylin-eosin. 140X.

Table 2. Percent ciliate infestation and gill melanization in
          euphausiids examined – July 1977.

Species	Station*	% ciliate infestation	% gill melanization
E. krohnii	SD – 2	100 (9/9)	56 (5/9)
	SD – 3	96 (50/52)	48 (25/52)
	AD – 3	64 (21/33)	64 (21/33)
Thysanopoda sp.	CS – 1	0 (0/3)	0 (0/3)
	AD – 3	83 (5/6)	0 (0/6)
Euphausiacea	SD – 1	0 (0/1)	0 (0/1)
	SD – 3	22 (4/18)	94 (17/18)
	AD – 3	0 (0/1)	0 (0/1)

*CS – control station   1 – neuston
 AD – acid dump   2 – bongo
 SD – sludge dump   3 – Isaacs-Kidd
 SW – slope water

swelling is uncertain. Gills of I. metallica examined from dump stations appeared swollen and histologically showed hemocytic infiltration and coagulation of hemolymph. One specimen of Idotea showed
cellular response to damaged cuticle at the base of the first
antenna.

The gross observations of a variety of copepods were unremark-
able. One copepod out of several hundred examined had an unusual
growth or structure on the basopodite of one appendage. As yet, this
is unidentified. Histologically, one copepod from an acid-iron station had an abnormal, enlarged appendage characterized by hemocytic
infiltration and apparent cuticular disintegration (Fig. 5). A
phagocytic nodule is indicated by the presence of swollen, encapsu-
lating cells and central melanization (N).

Mollusks, Ctenophores, and Chaetognaths. Histopathology of
these organisms is summarized in Table 3. The only significant pathologic finding in control samples were infections of two of nine heteropods by a helminth parasite. The larval worms were abundant in
the lumina of the digestive diverticula. Morphology of the parasite
suggests that it is a larval cestode (Figs. 6, 7). Heteropods and
ctenophores exhibited a cell abnormality characterized by enlarged

Table 3. Histopathology of organisms examined.

	Heteropods	Ctenophores	Chaetognaths	Squid
Control site				
Parasites	22% (2/9) (Helminths)	3% (1/30)	0 (0/4)	--
Cell abnormalities	11% (1/9)	33% (10/30)	0 (0/4)	--
Necrosis	11% (1/9)	23% (7/30)	0 (0/4)	--
Sewage dump				
Cell abnormalities	100% (1/1)	100% (1/1)	0 (0/12)	0 (0/12)
Necrosis	0 (0/1)	0 (0/1)	58% (7/12)	0 (0/12)
Parasites	0 (0/1)	0 (0/1)	0 (0/12)	25% (3/12)
Acid-iron dump				
Cell abnormalities	66% (4/6)	10% (2/20)	0 (0/3)	--
Necrosis	33% (4/6)	85% (17/20)	66% (2/3)	--

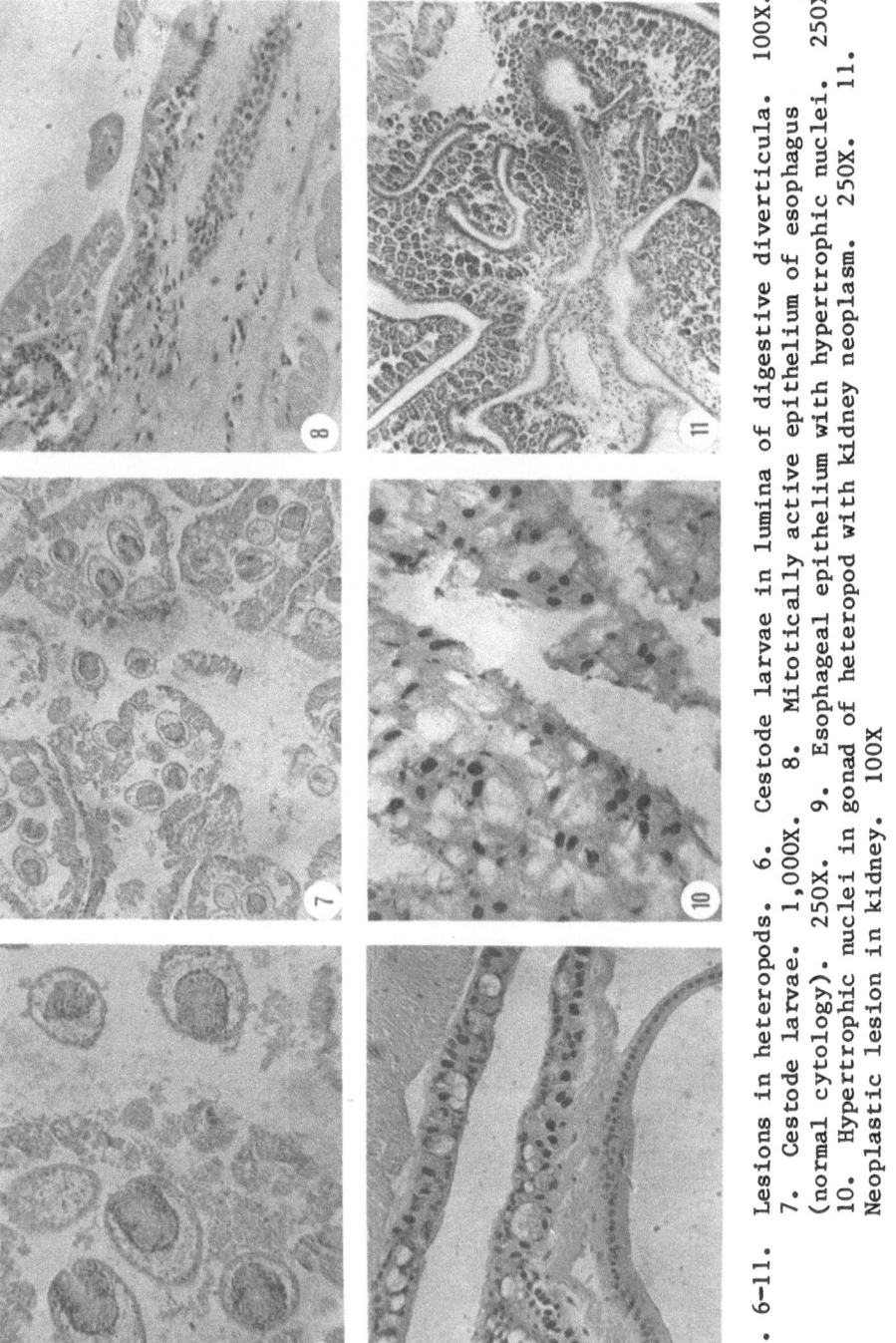

Fig. 6-11. Lesions in heteropods. 6. Cestode larvae in lumina of digestive diverticula. 100X. 7. Cestode larvae. 1,000X. 8. Mitotically active epithelium of esophagus (normal cytology). 250X. 9. Esophageal epithelium with hypertrophic nuclei. 250X. 10. Hypertrophic nuclei in gonad of heteropod with kidney neoplasm. 250X. 11. Neoplastic lesion in kidney. 100X

nuclei which contained finely granular feulgen-positive material
(Figs. 8, 9, 10). The cause of this condition is unknown at this
time. Necrosis (tissue death) was seen in samples from all sites but
appeared to be more prevalent in those from dump areas. Sampling
methods may have caused the appearance of this condition in control
sites but it is a cytologic indicator of dead tissue and may be a

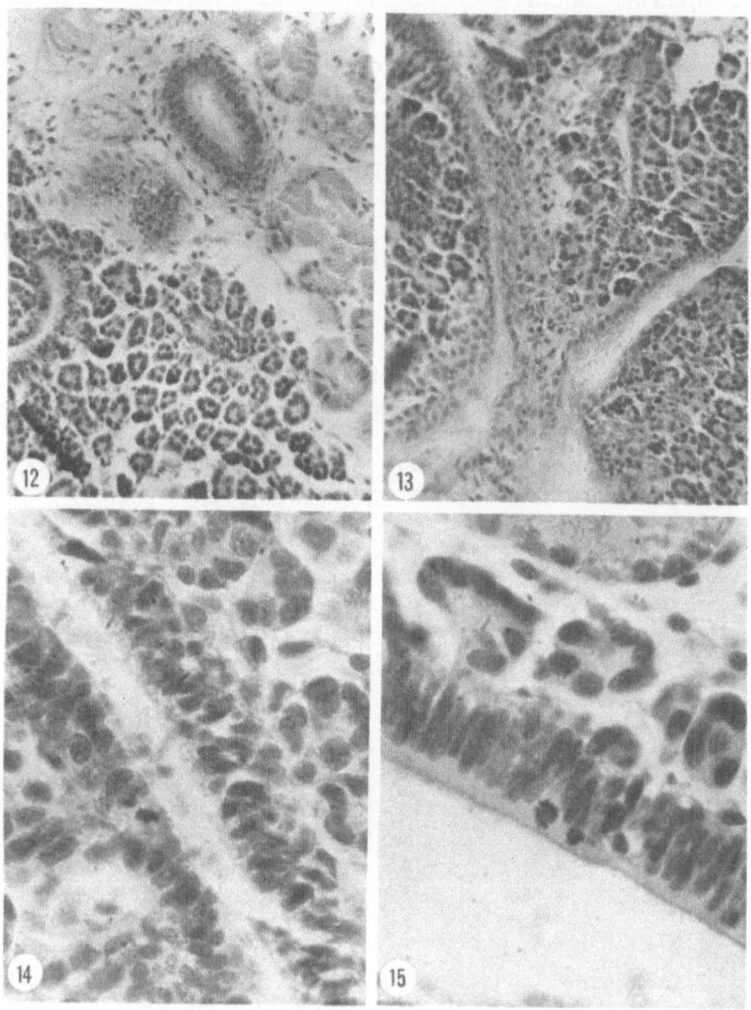

Fig. 12-15.   Neoplastic kidney lesion in heteropod. 12. Partially
              differentiated cells in kidney lesion apparently invad-
              ing digestive gland sinus. 250X. 13. A different re-
              gion of the neoplastic kidney lesion. 250X. 14. Mitot-
              ic figures in neoplastic epithelium. 1,000X. 15. Neo-
              plastic epithelium with polar view metaphase figure.
              1,000X.

significant acute feature of dump material effect on fauna.

Two heteropods from the acid-iron dump had lesions resembling
neoplasia. One had a partially differentiated growth of epithelial
tissue of kidney (Figs. 11, 12, 13) characterized by mitosis (Figs.
14, 15), variance in nuclear size, and loss of polarity. The lesion
was possibly invasive locally, but satellite lesions were not readily
apparent. Diffuse occurrence of individual abnormal cells was observed in the digestive gland, and hyperplasia and hypertrophy in
other tissues was also apparent (Fig. 10). The second animal had a
proliferative hyperplastic growth of gastrointestinal tract epithelium in the head region (Fig. 8).

Finfish. Very few larval fishes were obtained from the July
cruise. At least 80 species of juvenile or adult fishes were captured, however, with one exception, none were obtained in numbers
from both a control and a dump station. In addition, most of the
specimens captured in bongos or the Isaacs-Kidd trawl were badly

Table 4. Results of examination of leptocephali.

Station	Method of collection	No. of individuals	Lesions
AD-1	Neuston	2	Serous pericardial exudate, lateral displacement of cranial cartilage, protein-aceous exudate in ventricles of brain. Incomplete development of eye and brain.
AD-2	Bongo	4	1 dead larva, microvesicles in cornea (1 animal), protein-aceous exudate in ventricles of brain (3 animals), serous pericardial exudate (2 animals), unilateral derangement of olfactory pit epithelium (1 animal).
AD-3	Isaacs-Kidd	1	Proteinaceous exudate in ventricles of brain.
CS-3	Isaacs-Kidd	5	Serous exudate on epicardium, slight (1 animal). Proteinaceous exudate in ventricles of brain (5 animals).

damaged during capture and handling and were not useful for the study
of gross abnormalities. Many suffered extensive loss of scales (a
problem common to many species of myctophids), fin damage, loss of
epidermis, and collapse and enucleation of eyes.

In the absence of larval fishes, a myctophid (<u>Myctophum affine</u>)

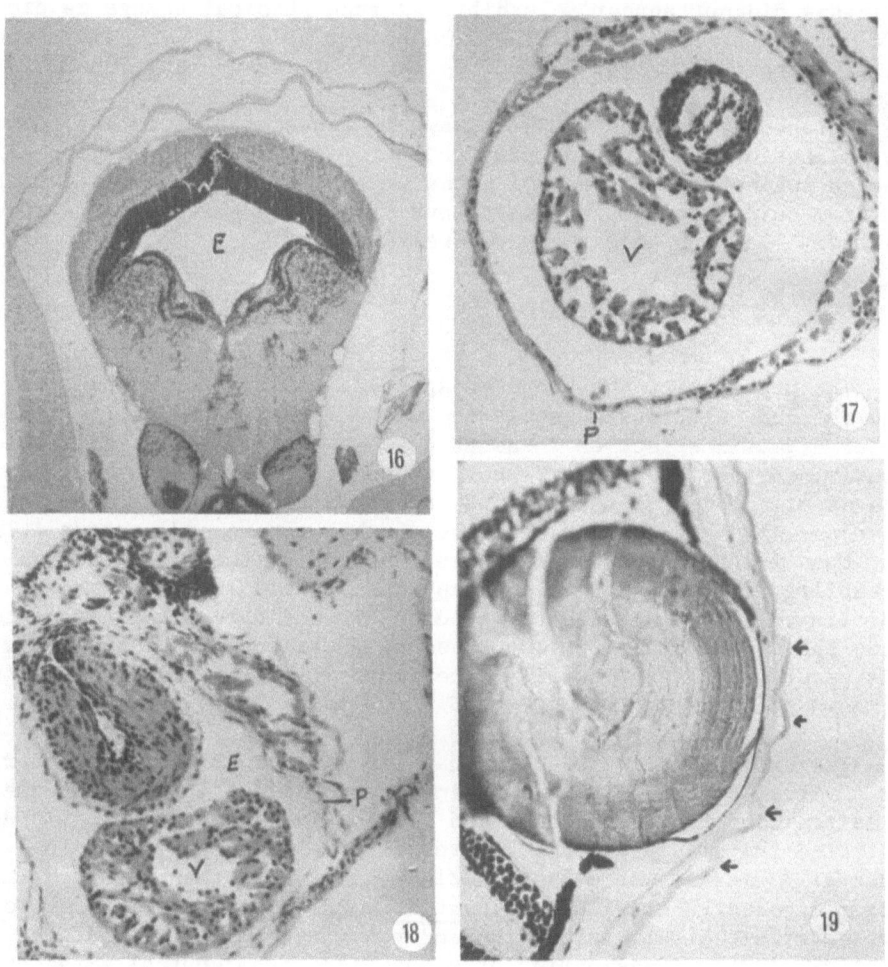

Fig. 16-19. Lesions of leptocephali. 16. Transverse section of
             brain. E = exudate. 17. Transverse section in region
             of heart. Normal animal. P = pericardium; V = ven-
             tricle. 18. Transverse section in region of heart.
             Pericardial exudate. P = pericardium; V = ventricle; E
             = exudate. 19. Microvesicles in epithelium covering
             lens.

which was present in fair numbers of neuston samples from both the
sludge dump and the control stations was processed for histological
examination. Six specimens from the sludge dump (6 h postdumping)
and seven individuals from the slope water control station were examined. No lesions were found.

For the acid-iron dump, leptocephali (Muraenidae) were chosen
for examination, even though they were available only in small numbers and mostly from bongo and Isaacs-Kidd samples. Grossly, these
organisms did not appear to exhibit as much physical damage as did
most other species taken by these methods (Table 4). These lesions
are illustrated in Figs. 16-19.

Mutagens. Results of the mutagenic tests of tissues and composite collections of plankton from this cruise indicate that generally
samples collected from control areas were negative while mixed plankton from acid-iron dump and sargassum from the sewage dump areas were
positive. Table 5 summarizes the results.

DISCUSSION

The presence of the encysted apostome ciliates on euphausiids
from the dump stations suggests the waste material had little immediate or dramatic effect on these protozoans at the time of collection.
However, further studies may prove these ciliates to be useful as indicator organisms. Lindley (1978) reported this apostome from various euphausiids in the boreal zones of the western and eastern Atlantic, thus illustrating the wide distribution of this organism and its
suitability for comparative geographic studies. In efforts to determine its value as an indicator species, it would be advantageous to
study the normal distribution, seasonal prevalence, and life cycle of
this apostome and subsequently experiment with its tolerance limits
to known pollutants.

The external position of euphausiid gills in direct contact with
the environment offers some exciting possibilities in terms of the
melanized condition. Lightner and Redman (1977) histochemically demonstrated that the brown-black material in penaeid shrimp having
black gill disease was melanin. Black gills have been reported in
shrimp exposed to cadmium (Nimmo et al., 1977) and a variety of foreign materials, i.e., turpentine and polyvinyl chloride (Fontaine and
Lightner, 1975). Examination of euphausiids from control sites and
determination of the normal prevalence and distribution of gill melanization, along with further studies in the dumpsite areas, are
needed to determine the potential of this condition as an indication
of environmental alteration. Laboratory experimentation with industrial wastes is necessary before a correlation between ocean dumping
and black gills in euphausiids can be made.

Table 5. Mutagenic testing of samples.

	Control sites	Sewage dump	Acid-iron dump
Mixed plankton (neuston)	Negative	Negative	--
Mixed plankton (bongo)	(50 m) Negative	(25 m) Negative	(100 m) Positive
Sargassum	Negative	Positive	--
Myctophids	Slight positive	--	--
Fish, Crustacea	--	Slight positive	--
Shrimp	--	--	(100 m) Negative

The prevalence of abnormal conditions in the biota sampled was not significantly related to ocean dumping activities. The cases of swelling, cellular responses, and melanization noted in the Crustacea indicate the potential of these species to elicit inflammatory responses. Further field and laboratory studies may show environmental factors to be involved in evoking such responses.

While lesions resembling neoplasia in heteropods are provocatively interesting in regard to conditions and prevalence, these numbers are inconclusive because of sample sizes. All of the larval fish lesions seen could have been a result of trawl-induced trauma, except for the one animal captured at AD1 which exhibits obvious developmental defects (Figs. 20-22). A recent paper (Dayle, P.G. and E.T. Garside (1980) Canadian Journal of Zoology, 58:27-43) describes transudates and pericardial effusions in the hearts of Atlantic salmon alevins exposed to acid water. Metaplastic and necrotic lesions of the brain were also among the changes noted by them.

Mutagenic (and possibly carcinogenic) compounds introduced into aquatic environments via man's activities is of increasing concern to fishery biologists and managers. Consequently, relatively rapid, simple, and inexpensive tests for the presence of xenobiotic mutagens and toxicants in the environment and tissues of biota would be helpful in attempts to explain disease etiology. In this current and very unrefined study it is unknown whether mutagens were present as incorporated or adherent matter on the organisms, or whether dump material itself was responsible for the positive test. Another unknown is whether these positive samples were reflecting effects of

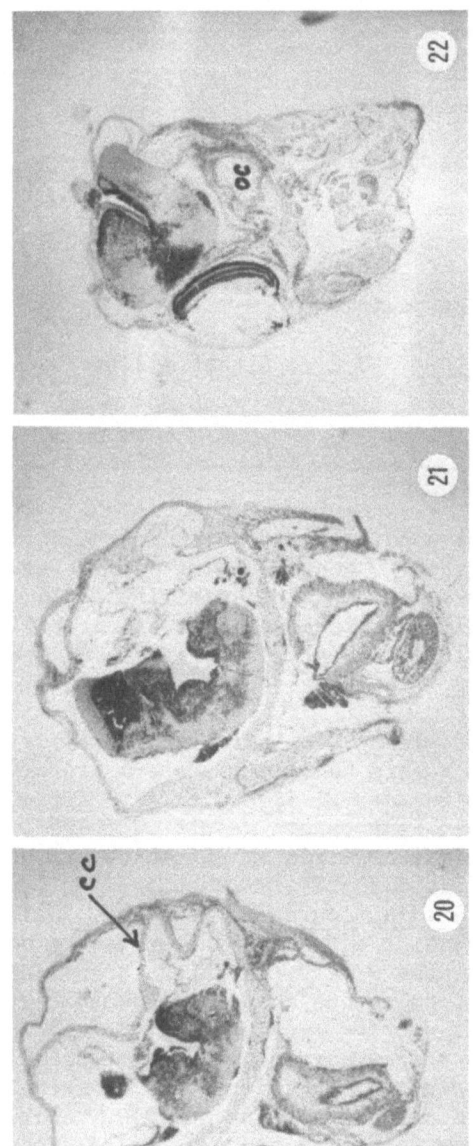

Fig. 20-22. Developmental defects. 20. Displacement of cranial cartilage. 21. Incomplete development of one side of brain. 22. Incomplete development of eye. Undeveloped optic cup (oc) on right; normally developed eye on left.

previous dumps in the areas. The positive sample from 100 m (acid-
iron waste) was theoretically taken from below the plume; however,
positive material could have been obtained as the net passed through
the plume during the setting and retrieval of the net.

Relatively little information is available on the pathologic
conditions of marine organisms other than cultured commercial species. Thus, another aspect of our study has been to establish the
beginning baseline on the histology and histopathology of planktonic
organisms. Our studies show that acquiring information on the normal
histology of these organisms is necessary in order to recognize the
subtle changes that may take place as a result of ocean dumping.

FUTURE CONSIDERATIONS

Emphasis these days is on the health of the oceans and determining adequate means of assessing this. One means is to identify key
organisms that can be used as field biological indicators. Although
at this point there is no apparent effect of dumping of the apostome
ciliates, the results are based on very limited sampling and examination of euphausiids. The phoronts of apostomes metamorphose and ex-
cyst into the feeding form (trophont) in response to the molting of
the crustacean. Lee and Nicols (1981) discussed molt inhibition of
planktonic crustacea exposed to wastes in the Puerto Rico dumpsite.
Although the wastes are very different chemically, euphausiids from
the DWD 106 may show molt inhibition and a resultant variation in
ciliate infestation. Apostomes also shed freeswimming forms that
find new hosts, settle, and develop into the encysted phoronts. Langlois (1975) found the settlement and thereby distribution of Vorti-
cella, a marine sessile ciliate, to be influenced by dissolved organics, phenols, and carbohydrates. Detailed studies on euphausiid exoskeletons exposed to known waste material and its effect on ciliate
settling rates may determine the value of these apostomes as field
monitors of environmental alteration. Similarly, the black gill foci
in euphausiids is worthy of closer examination and experimentation.

Our findings from this test dump situation are preliminary and
inconclusive and provide little hard evidence for or against ocean
dumping. It must be realized, though, that the test dumps were relatively insignificant in terms of the amount of material, the frequency of dumps, and the interaction of the various wastes and residues in the ocean that normally occur within a week at DWD 106. The
impedence of the thermocline on the downward movement of particulate
wastes suggests repeated exposure of diurnally migrating zooplankton
and chronic exposure of organisms comprising the thermocline community. The pathologic conditions we noted in planktonic Crustacea and
mollusks, then, may deserve closer examination in relation to long-
term effects of industrial waste disposal. In the future, perhaps
our studies should be directed toward the pycnocline community

and those organisms that may be exposed to higher concentrations of
waste over a longer period of time than neuston or vertically migrating zooplankton.

There is a critical need for chemical analyses of the planktonic
organisms in interpretation of the pathological and mutagenic findings. Ideally, a sample of each collection of organisms examined for
pathologic conditions or tested for mutagens should be analyzed for
heavy metals and specific waste organics in order to correlate biological findings with chemical analyses.

Our studies have focused on assessing the sublethal effects of
ocean dumping on organisms in the field, but there was no indication
that organisms exposed to dump materials were affected by the wastes.
Future laboratory and continued field studies are necessary to define
the pathological conditions observed in field-collected specimens and
the relationship of these conditions to waste materials and chronic
exposures.

ACKNOWLEDGEMENTS

The authors wish to thank Ms. L. C. Smith, Ms. D. Howard, and
Ms. P. Hambleton for processing and preparing tissues for light mi-
croscopy; and Ms. A. Charles and Ms. S. Cassanelli for sample collections and for providing other technical assistance.

REFERENCES

Farley, C. A. (1969) Probable neoplastic disease of the hematopoietic
        systems of oysters, Crassostrea virginica and Crassostrea
        gigas. National Cancer Institute Monograph, 31, 541-555.
Fontaine, C. T. and D. V. Lightner (1975) Cellular response to injury
        in penaeid shrimp. Marine Fisheries Review, 37, (5-6), 4-10.
Langlois, G.A. (1975) Effect of algal exudates on substratum
        selection by motile telotrochs of the marine peritrich ciliate
        Vorticella marina. Journal of Protozoology, 22, 115-123.
Lee, W. Y. and J. A. C. Nicol (1981) Toxicity of biosludge in pharma-
        ceutical wastes to marine invertebrates. In: Ocean Dumping of
        Industrial Wastes, B. H. Ketchum, D. R. Kester, and P. K. Park,
        editors, Plenum Press, New York. This volume, pp. 439-454.
Lightner, D. V. and R. Redman (1977) Histochemical demonstration of
        melanin in cellular inflammatory processes of penaeid shrimp.
        Journal of Invertebrate Pathology, 30, 298-302.
Lindley, J. A. (1978) Continuous plankton records: the occurrence of
        apostome ciliates (Protozoa) on Euphausiacea in the North
        Atlantic Ocean and North Sea. Marine Biology, 46, 131-136.
Nimmo, D. W. R., D. V. Lightner and L. H. Bahner (1977) Effects of
        cadmium on the shrimps Penaeus duorarum, Palaemonetes pugio, and

Palaemonetes vulgaris. In: Physiological Responses Of Marine Biota To Pollutants, F. J. Vernberg, A. Calabrese, F. P. Thurberg and W. B. Vernberg, editors, Academic Press, pp. 131-183.

Parry, J. M., D. J. Tweats and M. A. J. Al-Mossawi (1976) Monitoring the marine environment for mutagens. _Nature_, _264_, 538-540.

TOXICITY OF BIOSLUDGE AND PHARMACEUTICAL WASTES TO

MARINE INVERTEBRATES*

W. Y. Lee and J. A. C. Nicol

University of Texas
Marine Science Institute
Port Aransas Marine Laboratory
Port Aransas, Texas 78373

ABSTRACT

The toxicity of two industrial wastes to marine invertebrates
was assessed in the laboratory. The test animals were benthic and
planktonic, and included a sea anemone, a hydromedusa, a polychaete,
amphipods, an isopod, a crab and a shrimp. Owing to the difference
in their chemical composition, pharmaceutical waste was more toxic
than the biosludge. Based on the survival data, biosludge was acutely toxic at about 10% (by volume) and pharmaceutical waste at a level
$\leq$ 1%. Animals differed in their sensitivity to the toxic material,
juvenile being more sensitive than the adult. During chronic exposure, the reproduction of amphipods in diluted mixtures of biosludge
was affected at a lower concentration (< 10%) than that which induced
acute mortality. Besides survival and reproduction, a feasible
method for assaying toxicity at the cellular level was also suggested, involving cell aggregation in the sponge Microciona.

INTRODUCTION

Our previous research on the toxicity of industrial products to
marine invertebrates dealt with petroleum oils (Lee and Nicol, 1977;
Lee et al., 1977; Lee, 1978). We found deleterious action on survival, growth rate, fecundity, respiration, behavior, and feeding.

---

* The University of Texas Marine Science Institute, Contribution
  No. 327.

Based on our experiments with these techniques, we began a study of
the effects of two industrial wastes on several marine invertebrates,
including members of the benthos and zooplankton. Shell biosludge
and pharmaceutical wastes from Puerto Rico, destined for ocean dumping were used in this investigation. Biological parameters which
were examined included survival, reproduction and cell aggregation of
a red sponge.

MATERIALS AND METHODS

Biosludge. Four lots of biosludge were supplied in barrels by
the Shell Chemical Company. Due to difference in physical properties, they were designated as sample I, II, III and IV. Two kinds of
animals were included in the toxicity test: one pelagic and one benthic. Three types of toxic materials were used: whole biosludge, the
solid matter, and the supernatant.

Sample I contained much solid material and was filtered by the
Chemistry Department of the Port Aransas Laboratory. The solid material recovered by filtration was tested. It was centrifuged, and the
material in the pellet was suspended in seawater, 35 g in 965 ml of
seawater. The suspension was homogenized in Waring blender, and subsequently stirred to maintain the suspension. From this suspension,
as stock solution, dilutions were made (Fig. 1). Sample II contained
much finer solid material. It remained fairly homogeneous. It was
shaken vigorously before use and aliquots were taken as required for
testing. Sample III and IV were fairly homogeneous, resembling
sample II. They were left undisturbed for two weeks, whereupon solid
material settled, and the clean supernatant was siphoned off for use.

The animals used in toxicity tests of biosludge were Amphithoe
valida and Lucifer faxoni. A. valida is a benthic and intertidal
amphipod, with distribution ranging from New England coast to the
coast of the Gulf of Mexico. L. faxoni is a planktonic shrimp, mainly found in the subtropical coastal waters. Both species have been
successfully cultured through several generations in this Port Aransas Marine Laboratory, and methods for rearing the amphipod have been
described (Lee, 1977). Briefly, amphipods were kept in culture bowls
(19 cm in diameter) containing 1.5 l of seawater (30 $^o/_{oo}$ salinity).
Bowls were covered with PVC film, and gently aerated. They were fed
tropical fish food flake and dry sea lettuce (Ulva). Culturing L.
faxoni is much more difficult, requiring much time and patience.
Therefore, we limited ourselves to a survival study only. Shrimps
were collected at the pier near the laboratory and fed a mixture of
rotifers and newly hatched Artemia.

Pharmaceutical Wastes. The material which was received was the
combined wastes from five pharmaceutical companies and one petrochemical plant. Chemical compositions of the mixture were not fully

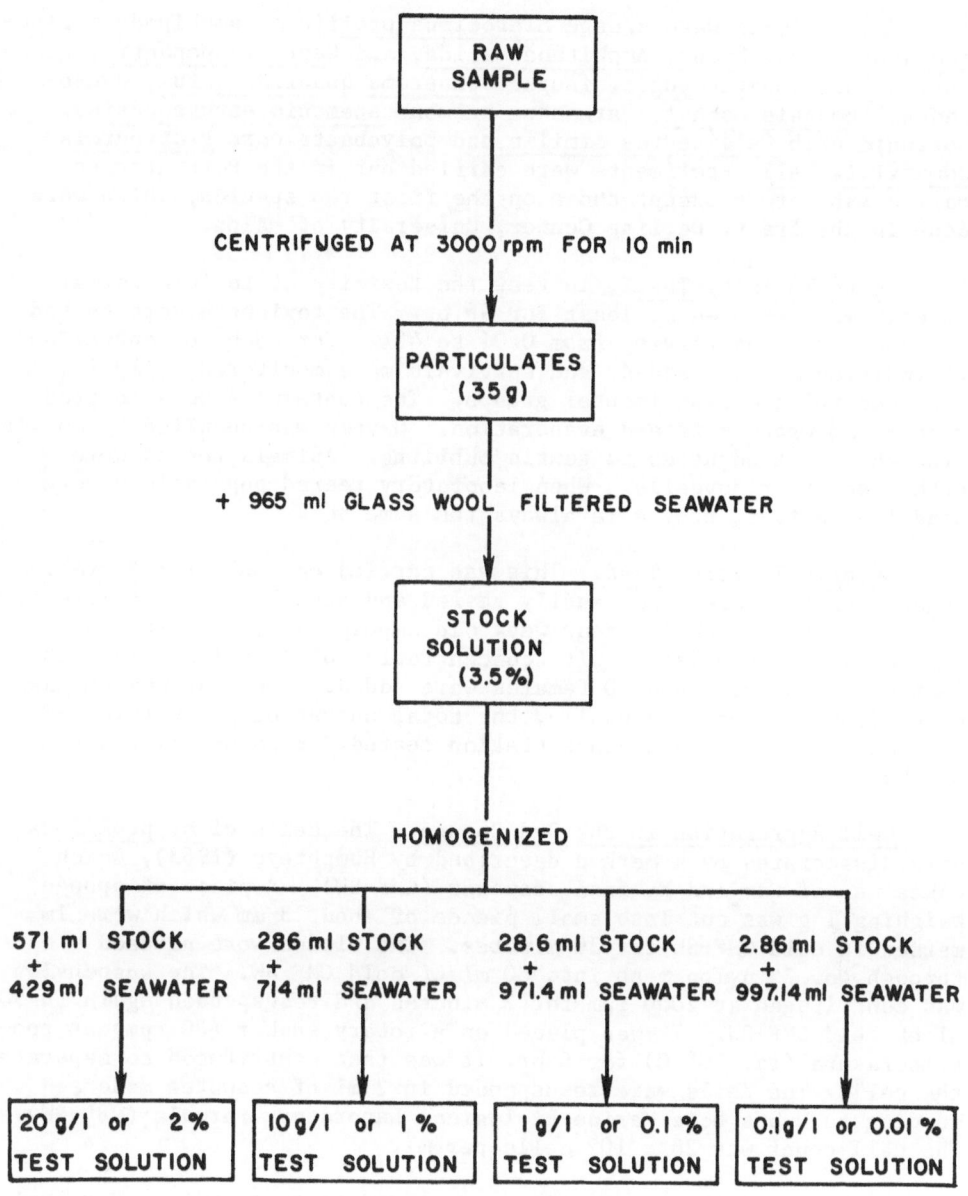

Fig. 1. Flow chart for the preparation of test medium. Biosludge was filtered, centrifuged, and then used as a toxicant.

known. The material was dark brown, contained fine black sediment, and had an offensive odor.

Test animals were sponge Microciona prolifera, amphipods Marino-gammarus finmarchicus, Amphithoe valida, and Caprella penantis, grass shrimp Palaemonetes pugio, isopod Sphaeroma quadridentatum, hydro-medusa Nemopsis bachei, sargassum anemone Anemonia sargassensis, portunid crab Callinectes similis and polychaete worm Platynereis dumerilli. All experiments were carried out in the Port Aransas Marine Laboratory except those on the first two species, which were done in the Ira C. Darling Center, University of Maine.

Acute Toxicity Test. To test the toxicity of toxic materials, animals were exposed at least for 96 hr. The toxicants were tested at concentrations varying from 0.01 to 40%. For each concentration, 20 individuals were added, and survivals were monitored daily for both control and experimental groups. The containers were covered with Saran wrap to retard evaporation. Oxygen was supplied by an air line which was adjusted to gentle bubbling. Animals tested were either adult or juvenile. When laboratory reared populations were used in the test, they were always the same age.

Chronic Toxicity Test. This was carried out only for A. valida, because this species was readily reared and also has a short life cy-cle (ca. 4 to 5 weeks). Four week old amphipods were tested in the supernatant of biosludge, with concentration of 1 to 10%. In each test bowl, 10 males and 10 females were added. The test medium was changed once a week, meanwhile, the total number of young released was determined for each concentration tested. Exposure lasted two months.

Cell Aggregation in the Red Sponge. The cells of M. prolifera were dissociated by a method described by Humphreys (1963), which makes use of Ca- and Mg-free seawater (CMF-SW). A piece of sponge weighing 1 g was cut into small pieces of about 3 mm which were im-mersed in cold CMF-SW for 30 minutes. The pieces were pressed through No. 25 nylon mesh into 80 ml of cold CMF-SW. The suspension was centrifuged at 2000 rpm for 2 minutes and resuspended again in 50 ml of cold CMF-SW. It was placed on a rotary shaker (80 rpm) at room temperature (ca. 20° C) for 6 hr. It was then centrifuged to separate the cells. The cells were resuspended in 2 ml of seawater made ac-cording to Woods Hole Marine Biological Laboratory formula (MBL-SW). The cell count was 28 x 10^6 cells per ml.

Aliquots of this suspension were added to test media. Material tested was a sample from pharmaceutical plants in Puerto Rico. The solids in the sample were removed by passing it sucessively through Whatman No. 2 and 4 filter papers. Three ml of media as follows were added to small Wheaton bottles: 0 (SW), 5, 1, 0.5, 0.1, 0.05, and 0.01 filtered Puerto Rican material in MBL-SW. An aliquot of 0.1 ml suspended sponge cells was added to each bottle, and the bottles were placed on the shaker for 8 hr.

As a variant of this procedure, cells were suspended in 50 ml of CMF-SW containing 5% filtered Puerto Rican material and placed on the shaker. A control in CMF-SW was also maintained.

Seawater. Offshore seawater was used in all experiments except that used in the sponge study. Sea water was first passed through glassfiber wool and then filtered through 3 pieces of Gelman filter (type A-E). Salinity of the seawater was then adjusted to 30°/$_{oo}$. Antibiotics were then added, namely Penicillin G and Streptomycin sulphate at 50 mg and 25 mg l^{-1}. The test animals were maintained at room temperature (ca. 24 $\pm$ 2°C).

Calculation of LC . The Litchfield and Wilcoxon (1949) method was used to calculate the median lethal concentration and its 95% confidence limits. The values of LC_{50} were useful in comparing the relative sensitivity of various animals to a toxicant.

RESULTS

Biosludge. Several concentrations (0.01 to 3%) of solid components were tested on the amphipod, A. valida. Times to kill half of the test population for 2.0% (20 g l^{-1}) and 1.0% (10 g l^{-1}) mixtures

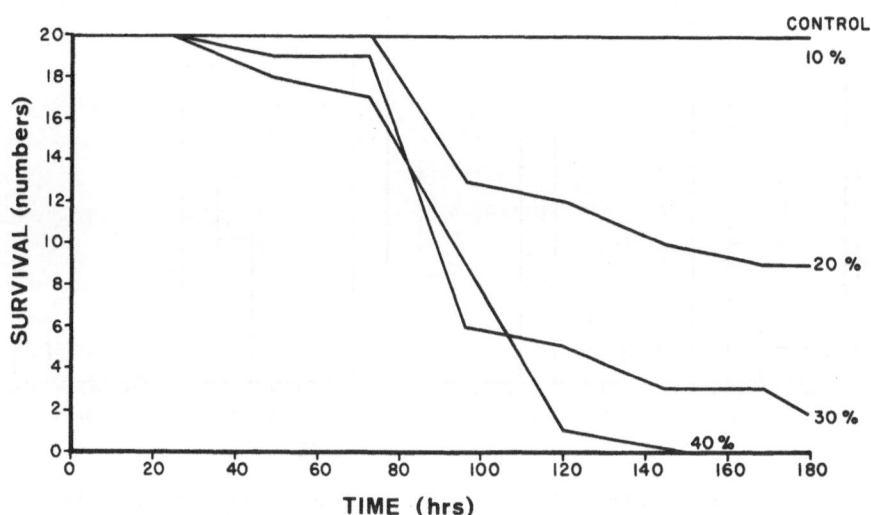

Fig. 2. Survival of Amphithoe valida (3 weeks old) in the super-natant of biosludge. The supernatant was autoclaved before the animals were introduced.

were 24 and 33 hr, respectively. Another test showed that at concentration of 2.0% the TL_{50} was delayed to 48 hr and there was no mortality at all for the population exposed to 1.0% during 4 days' exposure. The difference in sensitivity to the toxicant was probably caused by the ages used in the two series of tests. Amphipods of 10 weeks old were used in the first experiment and 6 weeks old in the second test.

Molts were recorded also. There was no difference between control and experimental groups in the test solution with concentrations < 0.01%. There were 11 molts in the control and one in the 0.1% mixture.

Three age groups of A. valida were used in the whole biosludge experiment, and they were 3, 5, and 9 weeks old. The results showed that amphipods in either young or old stage were more sensitive than the just matured group (5 weeks old). For example, in a 10% mixture, most of the test animals (> 90%) of age 5 weeks was still alive after 4 days exposure, while for the other two ages, under the same conditions, the mortality was $\geq$ 34%.

The toxicity of the supernatant from biosludge was tested in two ways, namely autoclaved and non-autoclaved, and with two kinds of animals, amphipods and planktonic shrimp.

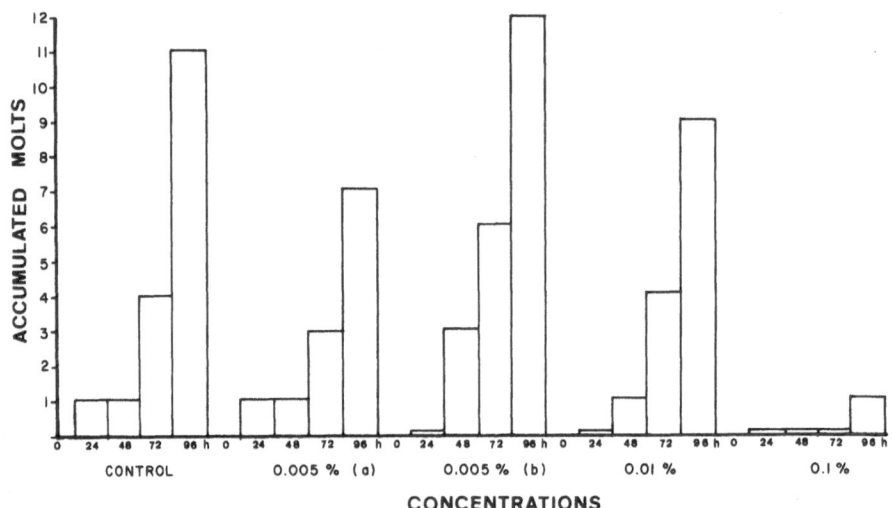

Fig. 3.   Effect of biosludge on the molting of Amphithoe valida (6 weeks old). Mixtures were made from the solid part of biosludge. Replicate samples were tested at 0.005% and they were designated as 0.005% (a) and 0.005% (b).

In the experiment with amphipods, the supernatant was autoclaved. Concentrations used were 10%, 20%, 30% and 40%, together
with a control. Exposure time extended to 8 days. No mortality was
observed in the 10% group. For the remaining groups (20, 30 and
40%), the TL_{50} was found to decrease with increasing concentration,
and in order 165, 105, and 81 hr, respectively (Fig. 2).

The planktonic shrimp, L. faxoni, was tested with autoclaved and
non-autoclaved supernatant. The non-autoclaved was toxic at 10% and
all test animals were dead in 15% at 48 hr. The latter result was
somewhat vitiated because of heavy bacterial growth. The autoclaved
supernatant was much less toxic and less than half of the test animals was killed in the 25% mixture during 180 hr exposure.

The molting data (Fig. 3) recorded during exposure to the mixture of solid component showed that adverse effect of this biosludge
on amphipods may be found at a much lower concentration, namely 0.1%,
than that of either 24 hr or 96 hr-LC_{50}. To verify this, an

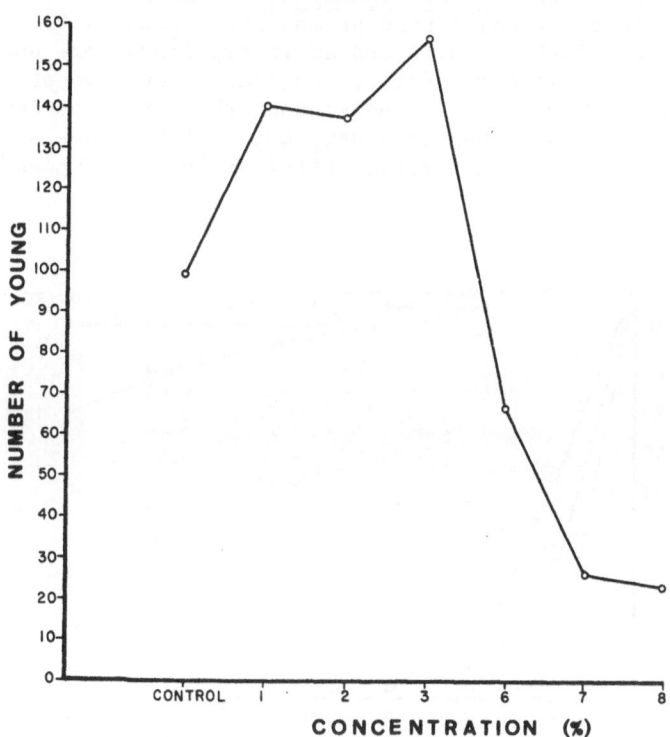

Fig. 4. Effect of biosludge on the fecundity of Amphithoe valida (4
         weeks old). Mixtures were made from the supernatant of
         biosludge.

experiment was carried out under conditions of low concentrations (<
10%)and long exposure. Amphipods four weeks old were introduced into
the following concentrations of biosludge supernatant, 1, 2, 3, 6, 7,
and 8%. A control was also provided as a check. The total number of
young released within 2 months for each concentration was shown in
Fig. 4. Two significant patterns could be recognized. The reproductive potential for these groups exposed to concentrations $\geq$ 6% was
significantly depressed as compared to that of the control. Yet the
amphipods which were exposed to concentrations $\leq$ 3%, released more
young than did the control. The largest number (157) of young produced was recorded at 3% and the least (23) at the 8%. The total
number of young produced in the control was 99.

Pharmaceutical Wastes. Dissociated sponge cells when placed in
MBL-SW containing Ca and Mg quickly reassembled. After 8 hr of rota-
tion, the suspension contained mostly spheres of aggregated sponge
cells. The dissociated cells which were placed in CMF-SW containing
Puerto Rican materials also aggregated into spheres while they were
being rotated; cells in the control did not form spheres.

In Maine, Marinogammarus finmarchicus was collected at mid-tide
level. The toxicant was tested at concentrations of 0.01, 0.05, 0.1,
0.5, 1 and 5%. Both juvenile and adult amphipods were used in the
test. Replicates were run for all concentrations except the 5% because the preliminary result showed that the latter concentration was
very poisonous. For juvenile amphipods, a significant difference
(χ^2 = 14.7, P < 0.01) in survival after seven days exposure was

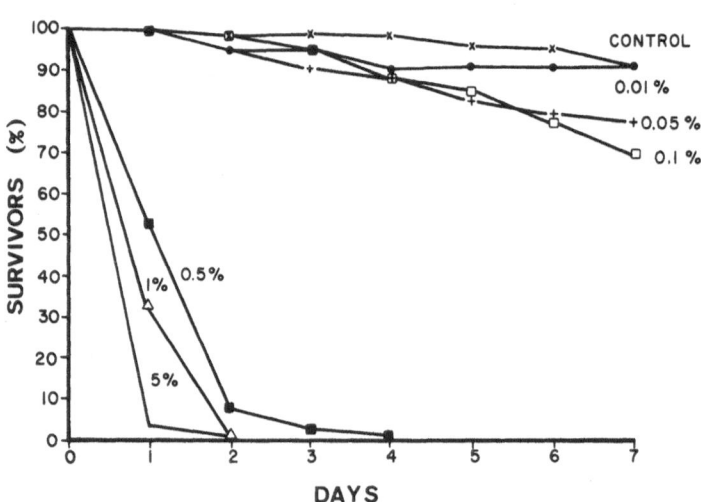

Fig. 5. Survival of juvenile amphipod, Marinogammarus
finmarchicus, in mixtures of pharmaceutical wastes.

observed between 0.05% and control. Mortalities were evident in 0.1%
and concentrations $\geq$ 0.5% were very toxic (Fig. 5). Mature animals
were more difficult to maintain than juveniles in the laboratory.
However, increasing mortality appeared in 0.05% solution and was
drastic in 0.1%. Amphipods in 0.5% and 1% solutions were all dead
within 48 hr (Fig. 6).

Along the Texas Gulf coast, eight species of invertebrates were
collected and used in the experiment. Dilutions of pharmaceutical
wastes were toxic to the adult shrimp Palaemonetes pugio at concentrations $\geq$ 4%; complete mortality was recorded within 48 hr for those
groups in 6% to 9%. Shrimps in dilutions $\leq$ 3% suffered no mortality
during 96 hr exposure. Dead shrimps were characterized by blanched
bodies and blackened gills.

At all concentrations, significant mortalities of P. pugio
(larva) were recorded after 48 hr (Fig. 7). Zoeas were all dead
after 72 hr in the 4% solution. At other concentrations including
the control, mortality was also high, being > 40% at 96 hr.

The isopods Sphaeroma quadridentatum were more sensitive than
adult shrimps. Total mortality was found in 2% to 9% solutions within 72 hr. At 1% and control, isopods suffered no mortality during

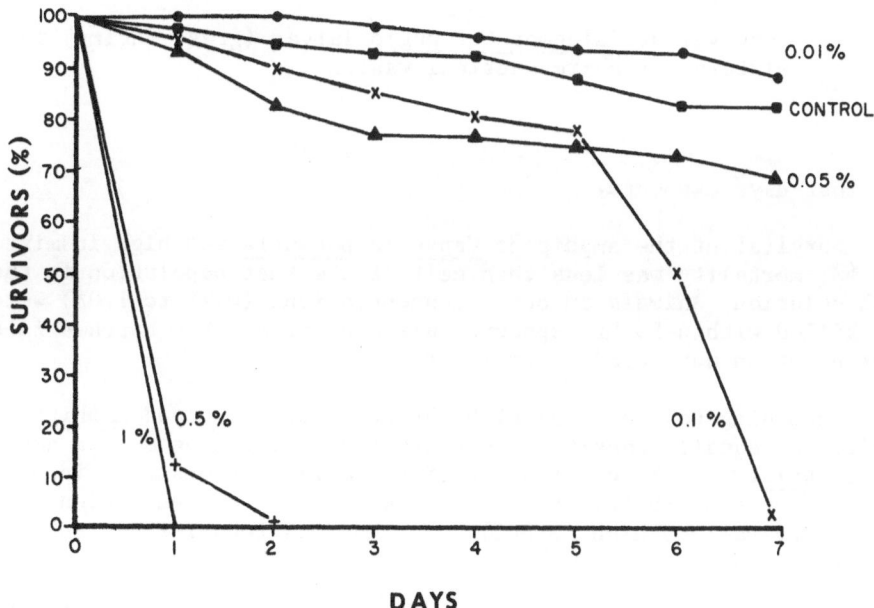

Fig. 6. Survival of adult amphipod, Marinogammarus finmarchicus,
in mixtures of pharmaceutical wastes.

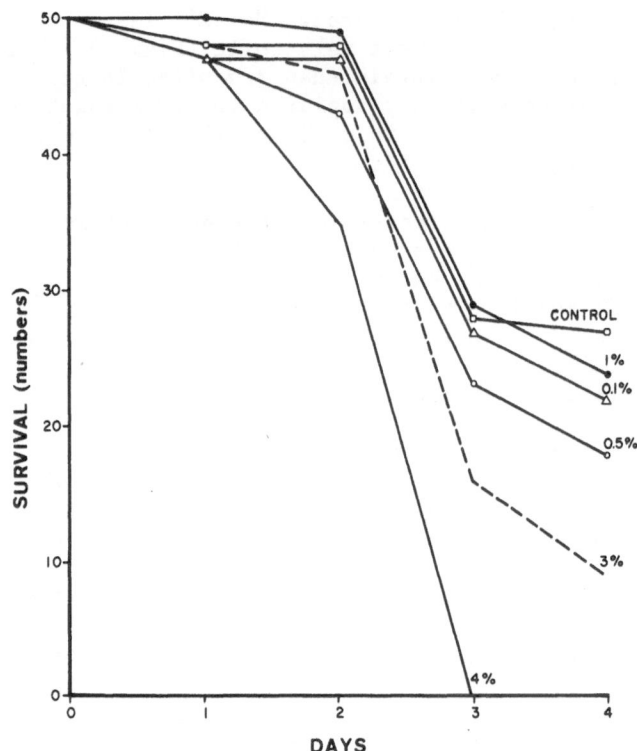

Fig. 7.   Survival of Palaemonetes pugio larvae (grass shrimp) in
mixtures of pharmaceutical wastes.

the four days exposure.

Survival of the amphipods Caprella penantis was high in mixtures
$\leq$ 0.5%; mortality was less than half of the test population in the
0.5% solution. Animals in other concentrations (0.7% to 1.0%) were
all killed within 96 hr exposure. Sixteen larvae that hatched in the
0.7% dilution were dead after 72 hr.

According to the value of 96 hr-LC_{50}, the amphipod Amphithoe
valida was equally sensitive to pharmaceutical wastes as the amphipod
C. penantis.   Complete mortality of A. valida was observed in 0.7% to
1.0% mixtures at 96 hr.   For those amphipods in concentrations $\leq$
0.5%, survival was high (> 80%).   No abnormal behavior was noted
during exposure.

The estimated 96 hr-LC_{50} for Anemonia sargassensis was 0.49%,
which is not significantly different from that obtained for the amphipods, C. penantis and A. valida.   Anemones in solutions of 0.6% to

1% experienced complete mortality at 96 hr and 50% loss for those in
0.5% dilution. However, all animals in concentrations above 3% manifested partial or complete tentacle withdrawal, and several individuals detached from the bottom. Many animals in solutions $\geq$ 0.6% had
thick mucus-like secretions covering the tentacles, body, and column
base. This condition was followed in less than 12 hr by complete
disintegration.

Test solutions ranged from 0.1% to 20%. Hydromedusae of Nemopsis bachei in concentrations $\geq$ 1% died within 96 hr. (Fig. 8). Even
in 0.1% solution, they were not active as the control; all animals in
the control were swimming actively. All experimental animals were
also darker than the control, and this became more obvious with increasing concentration. Individuals in mixtures $>$ 0.1% had withdrawn
tentacles and a shriveled appearance.

Survival of the crabs Callinectis similis was tested in

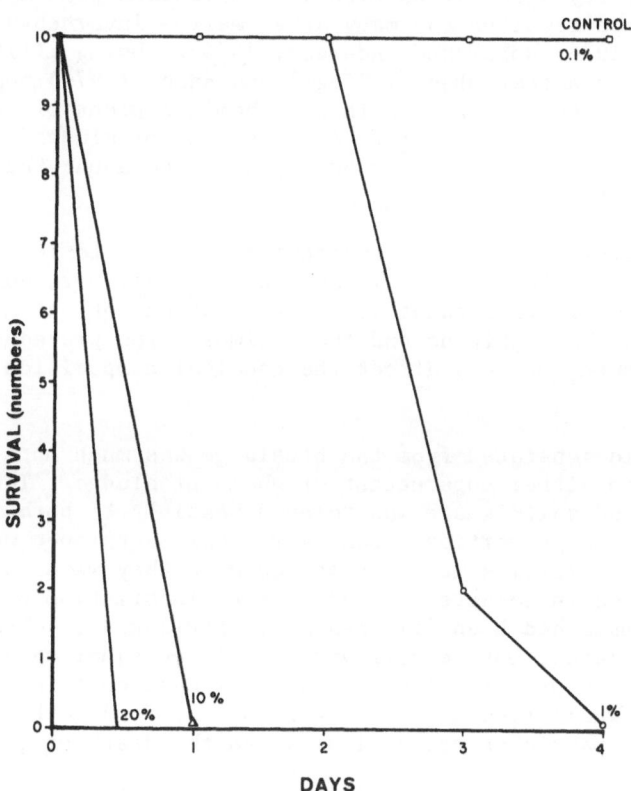

Fig. 8. Survival of the hydromedusa, Nemopsis bachei, in mixtures
of pharmaceutical wastes.

concentrations of 0.1%, 0.5%, 1.0% and 5%. Crabs in solutions $\leq$ 1%
suffered no mortality. In 5%, animals had only 25% survival at 48 hr
and the rest succumbed at 96 hr.

Platynereis dumerilii were tested in the same range of concentration as for crabs. All worms in 5% were sluggish compared to
groups of control, 0.1, 0.5 and 1%. Two individuals in the 5% never
formed tubes. When removed to clean seawater at the end of 4 day
exposure, all worms moved about actively except the two survivors
from 5%.

DISCUSSION

Biosludge. There were considerable differences between toxicities of the several samples of biosludge, perhaps because of 1) the
actual difference in the chemical composition of samples and 2) the
difference in method of preparation. For the amphipod, A. valida,
the whole biosludge was acutely toxic at about 10% by volume, and the
level of toxicity also varied with age; juveniles were more sensitive
than adults, as reported for many other marine invertebrates (Roesijadi et al., 1976; Rossi and Anderson, 1976). Using killfish (Fundulus similis) as a test object, Siegel and Rader (1974) reported that
the biosolid waste originating in the chemical plant biotreater was
toxic at about 9% (96 hr-LC_{50}). Although the biosludge used in the
two studies were different in sources, they had about the same toxicity to marine animals.

The supernatant from the suspension was also toxic at a level of
20%. Results for the planktonic shrimp L. faxoni were complicated by
bacterial growth. When autoclaved, the supernatant was much less
toxic to both the amphipods and the shrimps. The procedure controls
bacterial growth, but may affect the chemical composition of the supernatant.

The solid separated from the biosludge was much more toxic to
amphipods than either supernatant or whole biosludge. The higher
toxicity in the solid waste was related possibly to both its chemical and physcial properties. Some very toxic substance which adsorbed to the particles may be released when they were concentrated
and resuspended in seawater. There was no information on how much
volume of sample had been filtered and centrifuged to obtain 35 g of
stock particulates, but we believed that in an equal volume of test
media, a mixture prepared from solid stock must contain more solids
than did the whole biosludge. Clogging of the food-gathering appendages by silt particles could also hasten the death of test animals.

Chronic exposure of amphipods to low concentrations of biosludge
resulted in changes in both molting and fecundity. The change in
molting suggests that growth may be affected at a concentration as

low as 0.1%. However, data on reproduction showed that within the sublethal levels of supernatant, more young were released at a lower concentration (1% to 3%) than at higher concentration (6% to 8%). The total number of young produced in the control was between these two ranges. This pattern is not unusual, a similar pattern has been observed for the respiration of L. faxoni when exposed to the water soluble fractions of a No. 2 fuel oil (Lee et al., 1978).

Pharmaceutical Wastes. Eight invertebrates which were collected in the Gulf of Mexico, varied in sensitivity to the pollutant in the Puerto Rico material. If LC_{50} values were used to rank their sensitivities, they were (in order of increasing sensitivity): P. pugio (3.5%) > S. quadridentatum (1.42%) > C. penantis (0.55%) > A. valida (0.53%) > A. sargassensis (0.49%). Data for the other three species were insufficient for the calculation of 96 hr-LC_{50}. However, results did show that mortality increased with both the concentration and duration of exposure, and the P. dumerilii is probably the most resistant among the animals tested.

The eight test animals were all benthic except for the hydromedusa, N. bachei. Data on N. bachei were insufficient for further comparison of sensitivity between benthos and plankton. The present results showed that the amphipods, A. valida, and C. penantis, had about the same sensitivity to the Puerto Rican toxic materials, and that adult P. pugio was more tolerant than its larval stages. The latter phenomenon was also shown for the amphipods used in the toxicity test of biosludge.

Amphipods, M. finmarchicus, which were collected off the Maine coast, were very sensitive to the toxicant. Juveniles were less sensitive than the adults. This is possibly related to the difficulty encountered in maintaining the latter in the laboratory. The materials were toxic at 0.05 to 0.1%. Mortality significantly increased with time. All animals were killed by 1% within 48 hr.

When cells of some sponges are dissociated, they aggregate once more to form new sponges (Wilson, 1907). The process by which the several kinds of cells reassemble in ordered arrays involves selective cell adhesion and has been studied extensively (Galtsoff, 1925; Spiegel, 1954; Moscona, 1961; Humphreys, 1963; 1970). Processes are involved at the cellular level, and any interference with biochemical mechanisms or injury to cell membranes could delay or inhibit aggregation or regeneration. In the experiment on the red sponge, the dissociated cells reassembled normally in seawater containing Puerto Rico material. However, there was a significant interval during which the cells were suspended in seawater before they were placed in test solution, and the cells may aggregate during that interval. The result, therefore, was inconclusive. Cells suspended in CMF-SW containing Puerto Rico toxic material also aggregated. The chemical composition of the test toxicant was not available; it is conjectured

that it contained Ca and Mg of sufficient amounts to permit cell
aggregation.

    Ecological Consideration. To measure and monitor the ecological
effects of dumping industrial wastes into the open ocean is not an
easy task. The two selected Gulf of Mexico dumpsites were located
beyond the continental shelf and with an average depth > 1000 m.
Information on physical oceanography and biological structure of the
community in situ was insufficient to estimate either the residence
time of pollutant in a particular water mass or the responses of
benthos or plankton to the toxic materials. A toxic study, when carried out in the laboratory, only gives some answers to these questions such as: Are the dumped materials toxic to marine animals? If
they are toxic, what are the critical levels which would acutely affect the survival of animals or chronically affect the population
dynamics?

    When dumped into the ocean, the industrial wastes can be trans-
ported horizontally and vertically. Solid materials reaching the
bottom may smother or poison the benthos. During the course of sinking, nekton and zooplankton can be damaged through clogging of feeding apparatus or gills. The toxic materials carried horizontally by
currents may affect populations outside the dumpsites.

    Our study shows that the solid, supernatant or the whole bio-
sludge, is toxic to marine invertebrates when tested in the laboratory. The solids may be envisaged as the part which finally sinks to
the bottom, supernatant as the part which is transported beyond the
dumpsite, and the whole biosludge as that present only in the upper
water columm of the dumpsite. Pharmaceutical wastes may behave like
the biosludge, when dispersed, but they will be more toxic than the
biosludge. If this is the actual situation in the field, then the
benthos will be the most affected population, because solids are the
most toxic components in the biosludge, and the plankton outside the
dumpsite be the least affected, because the supernatant is the least
toxic component, and also is diluted during its horizontal transport.

    Our study also showed that if animals are exposed chronically to
a toxicant, the reproduction of amphipods will be hampered at concentrations lower than the levels which cause acute mortality. This information suggests that the intervals between disposal of industrial
wastes could be important and that several small discharges, if spaced at short intervals, may cause more damage to the community than
one discharge of a large amount. This phenomenon is not unusual, and
has been documented for the effect of oils on the saltmarsh (Baker,
1971a, b). Since the actual situations in the field are much more
complicated than that derived from a laboratory study, the laboratory
results when extrapolated to field conditions should be interpreted
with caution.

CONCLUSION

The industrial wastes came from different sources and hence varied greatly in chemical compositon and physical properties, and also differed considerably in their toxicity. Both the Shell biosludge and the pharmaceutical waste from Puerto Rico, which were destined to be dumped into the ocean, were acutely toxic to marine invertebrates and also differed in their toxicity by about an order of magnitude; biosludge was toxic at 10% and pharmaceutical waste at the level $\leq$ 1%. When animals were chronically exposed to biosludge, the reproduction of an amphipod, A. valida, was altered at a lower concentration than that which caused acute mortality. Therefore, many disposals of waste which are spaced at short time intervals may cause more damage to the ecosystem than a single discharge with a large amount of toxic material. This hypothesis is tentative and deserves more study either in the laboratory or in the field.

ACKNOWLEDGMENTS

This research was supported by the NOAA Ocean Dumping Program; part of the study (Puerto Rico pharmaceutical waste) was under the Grant No. 04-7-158-44053 and the other part (Shell's biosludge) was under the Grant No. PCM77-24358. The technical assistance of Ms. E. E. Payne and Mr. N. H. Hannebaum is gratefully acknowledged.

REFERENCES

Baker, J. M. (1971a) The effect of a single oil spillage. In: The Ecological Effects of Oil Pollution on Littoral Communities. E.B. Cowell, editor, Applied Science Publishers Ltd, pp.16-20.

Baker, J. B. (1971b) Successive spillages. In: The Ecological Effects of Oil Pollution on Littoral Communities. E.B. Cowell, editor, Applied Science Publishers Ltd, pp. 21-32.

Galtsoff, P. S. (1925) Regeneration after dissociation (an experimental study on sponges. I. Behavior of dissociated cells of Microciona prolifera under normal and altered conditions. Journal of Experimental Zoology, 42, 183-221.

Humphreys, T. (1963) Chemical dissolutions and in vitro reconstruction of sponge cell adhesions. I. Isolation and functional demonstration of the components involved. Developmental Biology, 8, 27-47.

Humphreys, T. (1970) Biochemical analysis of sponge cell aggregation. In: The Biology of the Porifera. W.G. Fry, editor, Academic Press, Symposia of the Zoological Society of London, 25, 325-334.

Lee, W. Y. (1977) Some laboratory cultured crustaceans for marine pollution studies. Marine Pollution Bulletin, 8, 258-259.

Lee, W. Y. (1978) Chronic sublethal effects of the water soluble
    fractions of No. 2 fuel oils on the marine isopod, Sphaeroma
    quadridentatum. Marine Environmental Research, 1, 5-17.
Lee, W. Y. and J. A. C. Nicol (1977) The effects of the water soluble
    fractions of No. 2 fuel oil on the survival and behavior of
    coastal and oceanic zooplankton. Environmental Pollution, 12,
    279-292.
Lee, W. Y., M. F. Welch and J. A. C. Nicol (1977) Survival of two
    species of amphipods in aqueous extracts of petroleum oils.
    Marine Pollution Bulletin, 8, 92-94.
Lee, W. Y., K. Winters and J. A. C. Nicol (1978) The biological
    effects of the water-soluble fractions of a No. 2 fuel oil on
    the planktonic shrimp, Lucifer faxoni. Environmental Pollu-
    tion, 15, 167-183.
Litchfield, J. T., Jr. and F. Wilcoxon (1949) A simplified method
    for evaluating dose effect experiments. Journal of Pharma-
    cology and Experimental Therapeutics, 96, 99-113.
Moscona, A. A. (1961) How cells associate. Scientific American,
    205, (3), 142-162.
Roesijadi G., S. R. Petrocelli, J. W. Anderson, C. S. Giam, and G. E.
    Neff (1976) Toxicity of polychlorinated biphenyls (Aroclor 1254)
    to adult, juvenile and larval stages of the shrimp Palaemonetes
    pugio. Bulletin of Environmental Contamination and Toxicolgy,
    15, 297-304.
Rossi S. S. and J. W. Anderson (1976) Toxicity of water soluble
    fractions of No. 2 fuel oil and south Louisiana crude oil to
    selected stages in the life history of the polychaete, Neanthes
    arenaceodentata. Bulletin of Environmental Contamination and
    Toxicology, 16, 18-24.
Siegel H. and W. E. Rader (1974) An analysis of biosolid waste from
    the Houston chemical biotreater. Technical Progress Report
    BRC-CORP 42-74-F, Project No. 41-8333, 32 pp.
Spiegel M. (1954) The role of specific surface antigens in cell
    adhesion. Part I. The reaggregation of sponge cells.
    Biological Bulletin, 107, 130-148.
Wilson H. V. (1907) On some phenomena of coalescence and regeneration
    in sponges. Journal of Experimental Zoology, 5, 245-258.

METABOLIC SENSITIVITY OF FISH TO OCEAN

DUMPING OF INDUSTRIAL WASTES

Donald E. Wohlschlag and Faust R. Parker, Jr.

University of Texas
Marine Science Institute
Port Aransas Marine Laboratory
Port Aransas, Texas 78373

Contribution No. 397

ABSTRACT

Respiratory metabolic stress studies of an ocean-dumped waste on Lutjanus campechanus, red snapper, indicated that sublethal concentrations of 0.2% v/v dilution of the ponded "biotreated" waste would have a depressing effect on swimming activity, on respiratory metabolism at maximum sustained swimming rates and on the metabolic scope for activity, which is the difference between the active and the standard (maintenance) metabolic rates. At 20°C the toxic effects were greater for the combined liquid and solid fractions than for the filtered liquid as a consequence of raising the standard, and reducing the active, metabolic rates. At 28°C the effects were lessened, presumably because the fish were already under some thermal stress and possibly because of the thermal lability and volatility effects on the toxicants.

An evaluation of the general technique of utilizing metabolic scope attenuation with stress reveals that weight and length effects on both respiratory metabolism and swimming propensities need careful future consideration, especially since larger fish may be the more stress sensitive. Several suggestions are made for increasing the sensitivity and precision of future applications of this technique.

The possibility of modifying the technique for a continuously monitoring system is advanced for cases in which the chemical nature, concentration, and/or uptake rates of the toxic materials by fish are

not known. The value of acquiring data from stress-metabolism
experiments that can also be energy-related to other population and
ecosystem characteristics is also emphasized.

INTRODUCTION

The purpose of this study is to determine whether aerobic respiratory metabolic responses of a marine fish are sufficiently sensitive at sublethal pollutant levels to provide a monitoring technique
in the absence of a priori information on the nature of a complex
pollutant.

The rationale of using respiratory scope -- the difference between oxygen consumption rates at maximum sustained aerobic activity
and at the maintenance or standard level -- for the assessment of environmental quality was suggested by Fry (1947) and subsequently
elaborated in general physiological terms (Fry 1957, 1971). Theoretical and empirical studies suggested that metabolic scope tends to be
reduced by stresses when standard rates may increase, activities decline, or both. Brett (1958, 1964, 1965, 1971), Brett and Glass
(1973), and Brett et al. (1969) have shown that at optimal temperatures, scope and swimming performances are also related to optima in
rations, assimilation, growth and related functions, often with marked departures at other temperatures.

Most fishes generally operate at a routine rate that lies between the standard and maximum, and is ecologically minimal at around
twice the standard level to account for about 1 unit body length per
second swimming (foraging) rate, specific dynamic action (assimilation) and other functions, excluding growth, spawning, extended mi-
grations, etc. (Fry 1971, Kerr 1971, Mann 1969, Winberg 1956,
Wohlschlag and Wakemen 1978). Stresses also can depress routine metabolic rates (Beamish 1964, Wohlschlag and Cameron 1967, Kloth and
Wohlschlag 1972, Cech and Wohlschlag 1975), although a depressed routine rate appears to be less definitive than scope for maximum sustained activity for species that may have a maximum swimming metabolic activity level 4-8 times standard levels (Randall, 1970).

For this study the red snapper (Lutjanus campechanus) was chosen
as a well known commercial and recreational species from both offshore and inshore waters. It is a relatively easy species to maintain in the laboratory.

The specific aims were to use the red snapper as a test organism:
     1. To identify metabolic effects at a very low (sublethal) tox-
icant level;
     2. To utilize the metabolic results at active and standard lev-
els for detection of scope diminution even though the chemical

composition of the toxicants could be considered unknown;

3. For suggesting a possible biological monitoring system that could operate with or without a detailed chemical knowledge of a toxicant, mixed toxicants, or interactions of toxicants; and

4. For acquisition of basic energetics data on a given species of general importance in fishery and ecological considerations.

METHODS

Throughout the red snapper, Lutjanus campechanus, was the fish of choice. Hook-and-line fishing at offshore "snapper banks" in 80-90 meters of water about 60 km offshore at Port Aransas provided specimens for 20°C experiments, while fish taken from shallower waters near the local Aransas Pass jetties and nearshore artificial reefs provided specimens for the 28°C experiments. At both locations natural salinities were continuously near 35°/oo.

Fish were held in live boxes with flowing seawater on board research vessels, and on shore in covered outdoor or indoor tanks with flowing seawater. Frequency of feeding was sufficient to promote growth. Before experiments, fish were held in temperature controlled, filtered water tanks at 35 °/oo and 20°C or 28°C for at least 48 hrs. Fish were fasted for at least 24 hrs before respiration measurements.

The pollutant biosludge examined in these experiments was obtained at the Shell Corporation Deer Park, Texas Biotreater Ponds on 9 November 1977. These wastes had been in the ponds considerably longer -- and possibly more biodegraded and less toxic -- than was usual for the more typical ocean dumped waste samples obtained in February and June 1977 for the exploratory and preliminary experiments. The biosludge contained both solid and liquid fractions with chemically recognizable components (Anderson, 1974; Siegel and Rader, 1974). Siegel and Rader (1974) noted that the 96-hr median lethal dosage for the killifish, Fundulus similis, was about 20% v/v dilution for the liquid fraction and about 9% for the centrifuged solid fraction. Because exploratory experiments had indicated that a 0.2% v/v dilution of the whole waste would be sufficient to produce readily detectible metabolic effects, and yet not have any direct mortality effects, all subsequent experiments were conducted at this dilution. (It should be noted that in exploratory experiments, the dilution of sludge to 0.1% v/v would also produce measurable effects.)

For the experiments at 20°C, either the whole sludge or the filtered liquid phase alone was used. In using the whole sludge, the liquid component was first filtered and the solid, flocculent portion then placed in a loosely woven gauze bag for suspension in the water circulating system. Otherwise the solid phase rapidly settled out or clung to the walls of the equipment and the gills of the fish to the

extent that the obvious metabolic impairment could have been caused
by the mechanical, as well as the toxic, properties of the biosludge.
At 28°C only the liquid phase was used. Controls with clean, settled
sea water were used at both temperatures and with both active and
resting metabolic determinations. For each of the experiments in
polluted waters, the fish were held 48 hrs under well oxygenated
conditions before metabolic measurements ensued.

Oxygen consumption rates were measured by withdrawal of small
samples for use in a Radiometer model E-5046 with a PHM 71 electrode
equipped with acid-base analyzer. Following completion of a set of
experimental oxygen consumption measurements, the fish were removed
and lengths and weights recorded.

Resting rates were determined by using 4 13.8 cm ID (15.2 cm OD)
diameter by 61 cm long acrylic tube flow-through chambers immersed in
a 450 l insulated, temperature controlled aquarium equipped with a
filtration system. Opaque plastic shields between the chambers and
black curtains around the entire mechanism prevented visually induced
excitement. Measurements of O_2 and flow rates at intakes and at
outlets were made over the course of 1 or 2 days to determine minimal
resting metabolism rates in well oxygenated waters.

Active metabolism rates were made in a 207 l Blazka chamber
(Blazka et al., 1960; Fry, 1971) as utilized by Wohlschlag and Wakeman (1978). The entire chamber was immersed in a 3,678 l tank which
was part of a temperature-salinity controlled system connected also
with the circular holding tank, filtration and cooling units. Fish
were maintained for one day swimming at low velocities (about 1 L
sec^{-1} where L is body length in cm) prior to active measurements.
After swimming in the chamber at an intermediate speed for 1 hr, the
velocity was increased gradually until the fish "broke" pace. At
this instant the velocity was lowered (usually quite slightly) to the
highest possible velocity at which normal swimming persisted without
breaking. With this "training" regimen, the maximum sustained swimming velocity (Webb, 1975) could be reproducible for each fish. The
maximum swimming velocity was determined at least twice to ascertain
consistency, after which the fish was tested for at least 1 hr for a
consistent maximum. Following the 1 hr or longer runs, the fish were
left in the chamber at intermediate or zero velocities with oxygen
rate measurements to detect any irregularities that could have resulted had the maximum swimming rates been associated with any unde-
sirable anaerobic metabolism. Because the swimming rate in general
was correlated the least with fish length when expressed as $L^{1/2}$
sec^{-1}, this expression will be used throughout.

Along with lengths, weights, oxygen consumption rates, and swimming rates expressed as $L^{1/2}$ sec^{-1}, salinities and temperatures
were recorded to 0.1 °/oo and 0.1°C for regressions at each control or
experimental condition in the form:

$$\hat{Y} = a + b_W X_W + b_V X_V$$

where: $\hat{Y}$ = expected O_2 consumption rate in log mgO_2hr^{-1},

a = constant,

X_W = weight in grams,

$X_V = L^{1/2} sec^{-1}$,

The various b values are the respective partial regression coefficients.

Similar procedures have been used by Wohlschlag and Juliano (1959) Wohlschlag and Cameron (1967), Wohlschlag and Cech (1970), and others.

RESULTS

Data in terms of average values and ranges of the variables for the flow-through chamber experiments are in Table 1 and for Blazka chamber experiments, in Table 2.

The regression equations for resting and active fish at 20°C or 28°C and at a salinity of 35°/₀₀ are in Table 3. Statistics and reference probability levels for these equations are in Table 4.

Because many of the fish in the flow-through chambers exhibited at least some nonlocomotory, spontaneous activity, the resting rate regressions (Equations 1-5) tend to be slightly higher than standard metabolic rates in spite of extensive precautions. Brett (1964) suggested utilizing a regression through the lowest respiratory rate (or rates) parallel and below the regression, which represents the resting rate for all the data and which is an approximation of a "true" standard rate as in Fig. 1. This procedure has been used in diverse studies; it compares well with several other time consuming physiological procedures of estimating standard rates that require elimination of spontaneous activity and that take into account diurnal respiratory fluctuations (Wohlschlag and Wakeman, 1978). Corresponding to resting rate Equations 1-5 in Table 3, Equations 1a-5a in Table 5 represent the corresponding standard metabolic rate equations estimated by Brett (1964) method.

The summary calculations of standard and active metabolic rates and of scopes at average weights and average maximum sustained swimming rates are in Table 6 for the various experiments.

The results of exploratory and preliminary experiments based on small numbers and with possible technical inadequacies will not be

Table 1. Average values and ranges of variables used in regression equations for resting metabolism in flow-through chambers. Petrochemical waste experiments on the red snapper.

Condition	N	Weight (grams)		Temperature (°C)		Salinity (°/oo)	
		Average	Range	Average	Range	Average	Range
Control	21	232.2	125.0 – 676.0	20.0	19.9 – 20.0	35.4	35.0 – 35.9
	12	208.0	132.0 – 370.0	28.0	28.0	35.2	34.9 – 35.7
Treated (Sludge)	8	233.3	151.0 – 290.0	20.0	20.0	35.4	35.1 – 35.6
(Liquid)	8	211.0	130.0 – 674.0	19.9	19.8 – 20.0	35.1	34.7 – 35.4
(Liquid)	8	209.6	125.0 – 349.0	28.0	28.0 – 28.1	35.3	35.1 – 35.4

Table 2. Average values and ranges of variables used in regression equations for active metabolism in the Blazka respirometer. Petrochemical experiments on the red snapper.

Condition	N	Weight (Grams)		Temperature (°C)		Salinity (°/∘∘)		Velocity[1]		
		Avg.	Range	Avg.	Range	Avg.	Range	Avg. $(L^{1/2}{}_s{}^{-1})$	Range $(L^{1/2}{}_s{}^{-1})$	Range (Ls^{-1})
Control	37	211.8	128.0-676.0	20.0	19.7-20.5	35.1	34.8-35.5	14.22	03.48-20.40	0.7-4.5
	29	254.2	128.0-690.0	28.0	27.1-28.5	35.0	34.5-35.5	22.37	00.00-28.88	0.0-5.8
Treated (Sludge)	10	239.2	151.9-290.0	20.0	20.0-20.1	35.1	35.0-35.5	17.44	15.94-18.92	3.2-4.2
(Liquid)	27	368.3	129.0-787.0	20.0	19.8-20.2	35.0	34.4-36.0	12.50	00.00-19.33	0.0-4.3
(Liquid)	23	229.6	123.0-661.0	27.8	27.0-28.4	35.1	35.0-36.0	20.69	00.00-29.44	0.0-5.9

[1] Swimming velocity data are given in both body length (L cm) measurements, $L^{1/2}sec$ and $Lsec^{-1}$. $L^{1/2}sec^{-1}$ is used throughout this study to reduce correlation of size and swimming rate. (See text.)

Table 3. Regression equations for resting and active red snapper
respiratory metabolism experiments. Control and petrochemical waste data for 20° and 28° C and salinity of
35 ppt. Waste concentration 0.2% v/v.

Resting Metabolism --- Flow-Through Chamber Experiments: Eq. No.

Control Water 20°C, N=21 $\hat{Y}= 0.1268 + 0.4802\ X_w$ (1)

Control Water 28°C, N=12 $\hat{Y}= -1.1114 + 1.1362\ X_w$ (2)

Sludge Treated 20°C, N=8 $\hat{Y}= -0.7016 + 0.8720\ X_w$ (3)

Liquid Treated 20°C, N=8 $\hat{Y}= -0.1607 + 0.6034\ X_w$ (4)

Liquid Treated 28°C, N=8 $\hat{Y}= -0.6560 + 0.9596\ X_w$ (5)

Active Metabolism --- Blazka Chamber Experiments:

Control Water 20°C, N=37 $\hat{Y}= -0.6864 + 0.8458\ X_w + 0.0382\ X_v$ (6)

Control Water 28°C, N=29 $\hat{Y}= 0.0163 + 0.7937\ X_w + 0.0104\ X_v$ (7)

Sludge Treated 20°C, N=10 $\hat{Y}= -1.3064 + 1.1221\ X_w + 0.0374\ X_v$ (8)

Liquid Treated 20°C, N=27 $\hat{Y}= -0.4734 + 0.8230\ X_w + 0.0271\ X_v$ (9)

Liquid Treated 28°C, N=23 $\hat{Y}= -0.2219 + 0.8940\ X_w + 0.0094\ X_v$ (10)

---

given, but will be mentioned in the following discussion for
comparative purposes.

DISCUSSION AND CONCLUSIONS

The results presented here should be quite conservative because
(1) toxicity of the wastes was probably diminished and (2) the scope
reduction measurements were more or less average and not maximum possible reductions in scope.

In the first instance, it should be noted that the 0.2% v/v experimental level -- at 1% of the 20% v/v 96-hr median lethal dosage
level determined by Siegel and Rader (1974) -- was very slightly toxic after two days at 28°C and at well aerated conditions with a scope
decline from 478 to 462 mg $O_2 kg^{-1} hr^{-1}$ compared to 365 to 206 mg O_2
$kg^{-1} hr^{-1}$ at 20°C (Table 6). It is also possible that the 20°C

Table 4. Regression statistics for resting and active red snapper metabolism equations. Probability (P) reference levels are: nonsignificant (n.s.) >0.05, 0.05, 0.025, 0.01, 0.005 and 0.001.

Eq. No.	N	Multiple Correlation Coefficient R	Of Estimate s_y	Standard Errors Weight Coefficient s_{b_w}	P	Activity Coefficient s_{b_v}	P
(1)	21	0.83	0.0578	0.0740	< 0.001	---	---
(2)	12	0.96	0.0447	0.1021	< 0.001	---	---
(3)	8	0.92	0.0414	0.1560	< 0.005	---	---
(4)	8	0.91	0.0673	0.1123	< 0.005	---	---
(5)	8	0.85	0.0854	0.2467	< 0.01	---	---
(6)	37	0.94	0.0798	0.0873	< 0.001	0.0029	< 0.001
(7)	29	0.91	0.0639	0.0713	< 0.001	0.0020	< 0.001
(8)	10	0.90	0.0576	0.2129	< 0.005	0.0169	n.s.
(9)	27	0.95	0.0882	0.0712	< 0.001	0.0025	< 0.001
(10)	23	0.98	0.0294	0.0351	< 0.001	0.0007	< 0.001

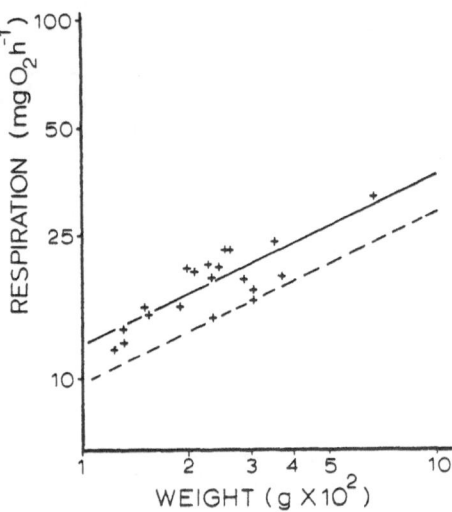

Fig. 1. Oxygen consumption rate and weight plot example at 20 C
         and 35 ppt. for Lutjanus campechanus in control water.
         Crosses represent observed data. Solid line drawn from
         Equation 1 in Table 3. Dashed line (Equation 1a) is for
         estimate of the standard level drawn parallel to the resting
         respiration line through the lowest measured values. (See
         Text.)

temperature is a more nearly optimal temperature offshore in deeper
waters for the red snapper against which a stress response may be
more marked than for the same species inshore in warmer waters.
Evidence for thermal stress to obscure other imposed stresses also
was suggested by increased morbidity, fin rot and other stress
manifestations when control fish were held for sustained periods at
28°C. Thus 0.2% v/v was a suitable level for testing at 20°C, but
doubtful at 28°C, most likely due to thermal lability and volatility
of some toxic components.

    From exploratory experiments there was also a good suggestion
that the initial Biotreater Pond retention times before ocean discharge was inversely related to toxicity because Lutjanus campechanus
at 20°C had definite decreases in scope at only 0.1% dilutions, although too few cases were investigated for statistical reliability.
In another preliminary trial, the whole sludge was dispersed throughout both the flow-through aquarium system and the Blazka system with
pronounced increases in standard metabolic rates and reductions in
active rates so that the scope at 20°C was reduced to 175 mg O_2 kg^{-1}
hr^{-1} compared to the 206 and 336 mg O_2 $kg^{-1}hr^{-1}$ scopes respectively
for the filtered liquid and the liquid plus the particulate fractions
retained in gauze. The particulate sludge fraction did have a

directly observable stress effect on the gills of L. campechanus for
the first few hours until a considerable flocculated fraction settled. The flocculated material on the gills would be at least part of
the cause for higher maintenance metabolic costs accompanying higher
"coughing" and gill pumping rates. Either or both impaired oxygen
uptake and toxicity associated with the particulate portion could
have been responsible for the greater decrease in active metabolism
than for the liquid fraction alone. One of the major reasons for the
greater toxicity of the wastes in the 21 June 1977 samples than in
the 9 November 1977 samples is that the Biotreater Ponds had been
frequently emptied before the earlier date. For this reason, the
later sample on which this study is based had been exposed to several
more months of biodegradation at summer temperatures. Such environmental and sampling variability must be taken into account along with
other variability inherent in the experiments themselves.

The problems of dealing with variability in studies of respiratory metabolism of fishes and other animals provide a large body of
knowledge, some of which is pertinent to this study. The general
variability is indicated by the tabulated ranges (Tables 1 and 2) and
statistical variability of the equations by the standard errors of
estimate, s_Y, and R values (Table 4). The statistical variability
of the weight and swimming velocity coefficients is indicated by the
respective standard errors in Table 4. The R values and standard errors are relatively small compared to similar studies, but on other
species. Stepwise regressions of multiple regression calculations of
expected log O_2 consumption rates dependent upon log weight and
several other variables will usually indicate that about 80-85% of
the total variability in O_2 consumption is related to weight variability. Hence the customary physiological procedure of expressing
the ratio of O_2 consumption per unit weight (in log form) for each
experiment yields ratios with unknown statistical properties, especially when there is a great weight range. For this reason, the use
of regressions to calculate O_2 consumption rates at an average

Table 5. Red snapper resting metabolic rate equations adjusted to
          standard (minimal) metabolic levels by Brett (1964)
          method.

(1a)	Control	20°C	$\hat{Y}_a = 0.0232 + 0.4802\ X_w$
(2a)	Control	28°C	$\hat{Y}_a = -1.1738 + 1.1362\ X_w$
(3a)	Sludge	20°C	$\hat{Y}_a = -0.7420 + 0.8720\ X_w$
(4a)	Liquid	20°C	$\hat{Y}_a = -0.2287 + 0.6034\ X_w$
(5a)	Liquid	28°C	$\hat{Y}_a = -0.7929 + 0.9596\ X_w$

weight (or other weight for comparative purposes) is desirable for
standard and active rate comparisons or scope calculations. When
utilizing scope measurements as a <u>difference</u> between minimum and maximum extremes of oxygen consumption rates, it is difficult to reconcile the usual standard error as a useful "two-sided" measure of variability about an extreme, unless a formal statistical expression for
selection of extremes were available. As a consequence, the scope
data of Table 6 are expressed without statistical measures of variability, but are based on calculations that utilize average weights
and average maximum sustained swimming velocity.

The maximum sustained swimming velocity also depends to a large
degree on size in both weight and length. Slight changes in the
"condition", or weight-length relationship, of a fish can occur quite
rapidly depending on nutrition and general health. Fish in poor condition may have greatly depressed active swimming and metabolic rates
(Wohlschlag and Wakemen, 1978). The swimming velocity and length

Table 6. Metabolic scope calculations. Red snappers under normal
and stressed conditions at 35 ppt. salinity. Number of
maximum sustained swimming speed observations in paren-
theses.

Equation		°C	Avg. wt. g.	Average $L^{1/2}s^{-1}$	$MgO_2 \ kg^{-1} \ h^{-1}$ Std.	Active	Scope
(1a)	Control	20	232	0	62	---	365
(6)	Control	20	212	17.70 (14)	---	427	
(3a)	Sludge	20	233	0	90	---	336
(8)	Sludge	20	239	17.28 (9)	---	426	
(4a)	Liquid	20	211	0	71	---	206
(9)	Liquid	20	368	17.05 (9)	---	342	
(2a)	Control	28	208	0	139	---	478
(7)	Control	28	254	26.06 (15)	---	617	
(5a)	Liquid	28	210	0	129	---	462
(10)	Liquid	28	230	26.10 (9)	---	591	

relationship is quite complicated and depends on many interrelated hydrodynamic and biological features. For example, Wakeman (unpublished University of Texas Ph.D. dissertation, 1978) noted that maximum velocity of the Cynoscion nebulosus (spotted seatrout) from 23.3 to 37.8 cm long, when expressed as cm sec^{-1}, increased with size; when expressed as lengths sec^{-1}, decreased with size; and when expressed as square root of length sec^{-1}, showed no significant change with size. Length raised to various powers also is an important, hydrodynamic consideration in determining drag force and work required for swimming (Webb, 1975, 1978; Beamish, 1978; Magnuson, 1978). From these hydromechanical considerations, and from biological observations of stress responses, it might be suggested that larger and older members of a fish species are the most stress sensitive. Exactly how the various morphological changes that accompany growth and length sequences affect drag and energetics of swimming performance, are not well understood for most species, however. While there may be a good theoretical basis for using $L^{1/2}$ sec^{-1}, the usage should be considered empirical until additional size-specific and species-specific swimming energetics are understood.

In Blazka equipment with appropriately acclimatized fish, excellent precision is possible in measuring the maximum sustained swimming rates in successive determinations on any given fish. However, among a group of healthy fish that have no apparent physical differences, there may usually be expected a wide range of swimming and metabolic rates as denoted by the standard errors for Equations 6-10 in Table 3. There is some possibility that the expected active oxygen consumption rates could be calculated for an overall average weight and for each observed swimming rate from these equations to make a plot of the "adjusted" active metabolism against swimming rate. From an equation of this plot, the average line could be adjusted upward through the single maximum swimming rate as a measure of a "record" high metabolic performance. If this record high metabolic value were combined with the analogously calculated standard value from the resting values to yield a new scope estimate, it would be observed that the scopes were about double the values in Table 6 and that the stress depression responses to the pollutant were about twice as sensitive. Also it should be emphasized that the average of the maximum swimming rates tend to decline with toxicity stress (Table 6) as might be expected, but the declines are not very abrupt with toxicity for this species and this pollutant. With even slight pollution stress it is therefore apparent that both swimming performance and the metabolic scope are reduced measurably.

The possibilities of adapting the techniques of stress evaluation by measuring metabolic scope diminution for a monitoring system also become apparent. The adaptation could readily be envisioned as a Blazka-type system modified to allow for a continuous flow-through system between oxygen consumption determinations. Suspected toxic materials could be added either in a single pulse or in a constant

low rate of influx sufficient to maintain some measurable level of
toxicity. A suitable sublethal level would be detectible by decreased swimming performance or metabolic scope or by obvious behavioral
manifestations that are not included in this study. For a monitoring
scheme, statistical comparisons from multiple regressions can be made
from about 20-30 runs when only 2 independent variables are involved
with the degree of variability illustrated by Table 4 standard error
values. Some additional study would be required to develop an insight into the statistical properties of maximum active and minimum
resting (standard) metabolic rates. In addition, some further study
on the hydromechanical and physiological relationships among metabolic rates, fish sizes, maximum sustained swimming speeds and energy
expenditures would be useful. In general, this study on fishes indicates that continuous monitoring of these metabolic relationships to
stresses is feasible and based on well established ecological and
physiological principles (Fry, 1971). Further, larger fish not only
have inherent advantages as sensitive experimental organisms, they
may be more sensitive than their smaller counterparts to sublethal
stresses (Wohlschlag and Cameron, 1967; Wohlschlag and Cech, 1970).

By using oxygen consumption units that are readily convertible
to other energy or work units, studies of this type are directly useful for energy cost of transport computations and for efficiency of
food conversion and growth studies. Energy appears as a common de-
nominator in theory and evaluation of general environmental optima
and stresses (Cody, 1974). Scope for activity evaluations have a direct application to growth studies (Kerr and Ryder, 1977). Metabolic energy expenditures can have direct relationships to growth-forag-
ing evaluations (Kerr, 1971). In general terms, respiratory or metabolic effects appear to be highly sensitive for heterotrophs in ecosystems exposed to common variables like temperature, e.g. (O'Neill,
1976). In this type of sensitivity, the longer living species like
the Lutjanus campechanus exposed to slight changes induced by sublethal, but chronic, stresses are now suspected of having capabilities
of further inducing dramatic effects on other ecosystem components
(Simenstad et al., 1978).

ACKNOWLEDGMENTS

This study was supported by National Oceanographic and Atmospheric Administration Grant No. 04-7-158-44053. We are especially
grateful for the aid of personnel from the Ocean Dumping Program of
the National Ocean Survey, especially Dr. Edward R. Meyer of that
program.

Special thanks are extended to Dr. John R. Burns and Julia A.
Kinney who conducted the exploratory technique and toxicity evalua-
tions prior to this study. The aid in laboratory and at sea by
Russell Vetter, Edgar Findley, Mark Dobbs and Michael Gunter is noted

with pleasure. The aid of Captains Don Gibson and crew of the R/V LONGHORN and of Captain Elgie Wingfield and crew of the R/V LORENE was indispensable.

REFERENCES

Anderson, J. W. (1974) Biological effects of spent caustic and biosolid wastes. Texas A&M University, processed.

Beamish, F. W. (1964) Respiration of fishes with special emphasis on standard oxygen consumption. Canadian Journal of Zoology, 42, 177-188.

Beamish, F. W. (1978) Swimming capacity. In: Fish Physiology. Locomotion. W.S. Hoar and D.J. Randall, editors, Academic Press, 7, New York, pp. 101-187.

Blazka, P., M. Volf, and M. Cepela (1960) A new type of respirometer for the determination of metabolism of fish in the active state. Physiologia Bohemoslovenica, 9, 553-558.

Brett, J. R. (1958) Implications and assessments of environmental stress. In: The Investigation of Fish-Power Problems. P.A. Larkin, editor, The H.R. MacMillan lectures in fisheries, Univ. British Columbia, Vancouver, pp. 69-83.

Brett, J. R. (1964) The respiratory metabolism and swimming performance of young sockeye salmon. Journal of the Fisheries Research Board of Canada, 21, 1183-1226.

Brett, J. R. (1965) The relation of size to rate of oxygen consumption and sustained swimming speed of sockeye salmon (Oncorhyncus nerka). Journal of the Fisheries Research Board of Canada, 22, 1491-1501.

Brett, J. R. (1971) Energetic responses of salmon to temperature. A study of some thermal relations in the physiology and freshwater ecology of sockeye salmon (Oncorphynchus nerka). American Zoologist, 11, 99-113.

Brett, J. R. and N. R. Glass (1973) Metabolic rates and critical swimming speeds of sockeye salmon (Oncorhyncus nerka) in relation to size and temperature. Journal of the Fisheries Research Board of Canada, 30, 379-387.

Brett, J. R., J. E. Shelbourn, and C. T. Shoop (1969) Growth rate and body composition of fingerling sockeye salmon, Oncorhynchus nerka, in relation to temperature and ration size. Journal of the Fisheries Research Board of Canada, 26, 2363-2394.

Cech, J. J. Jr. and D. E. Wohlschlag (1975) Summer growth depression in striped mullet, Mugil cephalus L. Contributions in Marine Science, 19, 91-100.

Cody, M. L. (1975) Optimization in ecology. Science, 183, 1156-1164.

Fry, F. E. J. (1947) Effects of the environment on animal activity. University of Toronto Studies of Biology, Ontario Fisheries Research Laboratory, 68, 1-62.

Fry, F. E. J. (1957) The aquatic respiration of fish. In: The

Physiology of Fishes. M.E. Brown, editor, Academic Press, New York, pp. 1-63.

Fry, F. E. J. (1971) The effect of environmental factors on the physiology of fish. In: Fish Physiology. Environmental Relations and Behavior. W.S. Hoar and D.J. Randall, editors, Academic Press, 6, New York and London, pp. 1-98.

Kerr, S. R. (1971) A simulation model of lake trout growth. _Journal of the Fisheries Research Board of Canada_, 28, 815-819.

Kerr, S. R. and R. A. Ryder (1977) Niche theory and percid community structure. _Journal of the Fisheries Research Board of Canada_, 34, 1952-1958.

Kloth, T. C. and D. E. Wohlschlag (1972) Size-related metabolic responses of the pinfish, _Lagodon rhomboides_ to salinity variations and sublethal pollution. _Contributions in Marine Science_, 16, 125-137.

Magnuson, J. J. (1978) Locomotion by scombroid fishes: hydromechanics, morphology, and behavior. In: Fish Physiology, Vol 7 Locomotion. W.S. Hoar and D.J. Randall, editors, Academic Press, New York, San Francisco and London, 7, pp. 239-313.

Mann, K. H. (1969) The dynamics of aquatic ecosystems. _Advances in Ecological Research_, 6, 1-81.

O'Neill, R. V. (1976) Ecosystem persistence and heterotrophic regulation. _Ecology_, 57, 1244-1253.

Randall, D. J. (1970) Gas exchange in fish. In: Fish Physiology. W.S. Hoar and D.J. Randall, editors, Academic Press, New York, pp. 253-292.

Siegel, H. and W. E. Rader (1974) An analysis of biosolid waste from the Houston chemical plant biotreater. Technical Progress Report BRC-CORP 42-74-F. Shell Development Company, Houston.

Simenstad, C. A., J. A. Estes, and K. W. Kenyon (1978) Aleuts, sea otters and alternate stable-state communities. _Science_, 200, 403-411.

Webb, P. W. (1975) Hydrodynamics and energetics of fish propulsion. _Bulletin of Fisheries Research Board of Canada_, 190, x + 158 pp.

Webb, P. W. (1978) Hydrodynamics: nonscombroid fish. In: Fish Physiology. Locomotion. W.S. Hoar and D.J. Randall, editors, Academic Press, 7, New York, San Francisco and London, pp. 189-237.

Winberg, G. G. (1956) Rate of metabolism and food requirements of fishes. _Fisheries Research Board of Canada Translation Series_, 194, 1-253.

Wohlschlag, D. E. and J. J. Cech (1970) Size of pinfish in relation to thermal stress response. _Contributions in Marine Science_, 5, 22-31.

Wohlschlag, D. E. and J. N. Cameron (1967) Assessment of a low level stress on the respiratory metabolism of the pinfish (_Lagodon rhomboides_). _Contributions in Marine Science_, 12, 160-171.

Wohlschlag, D. E. and R. O. Juliano (1959) Seasonal changes in bluegill metabolism. *Limnology and Oceanography*, 4, 195-209.

Wohlschlag, D. E. and J. M. Wakeman (1978) Salinity stresses, metabolic responses and distribution of the coastal spotted seatrout, *Cynoscion nebulosus*. *Contributions in Marine Science*, 22, 171-185.

PRELIMINARY STUDIES OF RESPONSES OF FISH

TO COMPLEX PHARMACEUTICAL AND PETROCHEMICAL WASTES

Robert G. Otto and S. Ian Hartwell

Chesapeake Bay Institute, The Johns Hopkins University
Baltimore, Md. 21218

ABSTRACT

Investigations were conducted to define the abilities of fish to survive or to detect and avoid exposure to complex organic waste materials from pharmaceutical and petrochemical sources. These wastes are currently barge-dumped in deep waters north of Puerto Rico. The 24-hr and 96-hr LC_{50}'s for juvenile spot Leiostomus xanthurus and killifish Fundulus heteroclitus and avoidance limits for juvenile spot and juvenile menhaden Brevoortia tyrannus were determined. Fish accepted exposures without mortality to waste concentrations substantively greater than are expected to occur in the field situation.

INTRODUCTION

This study deals with effects of ocean dumped pharmaceutical and petrochemical wastes on fish. These wastes are produced by a group of seven drug manufacturing facilities and a small refinery, all located in Puerto Rico. The aggregate wastes total approximately 3.2×10^6 m^3 of solutions and 6×10^6 metric tons of solids per year (National Academy of Sciences, 1975). They are collected at irregular intervals from the production sites, composited at a storage site at Arecibo and dispersed from a towed barge north of the island.

A variety of technical, economic, regulatory and political criteria for deciding which, if any, waste materials can safely be eliminated through ocean disposal have been enunciated in recent years (Goldberg, 1976; National Academy of Sciences, 1976; Zapatka and Hanna, 1976, 1977). Points of view range from support for virtually unrestricted dumping to advocacy of a complete moratorium. From a

purely technical standpoint, direct ocean disposal would seem to be
the method of choice for many waste products, assuming that three
general criteria can be met:
   1. That the wastes, once dumped, do not present a <u>direct health
hazard</u> to man;
   2. That the wastes do not contain materials which, through bioconcentration, metabolism or degradation, present an <u>indirect health
hazard</u> to man; and,
   3. That the materials will not result in <u>culturally unacceptable impacts</u> in terms of loss of resources such as food fishes, recreational space or other, more intangible measures of ecological
well-being in the receiving system.

   Most of the wastes considered in this study contain materials
which, in undiluted form, would be directly harmful to man. There is
an increasing awareness that complex organics and particularly some
of the cyclic or polycyclic hydrocarbons which commonly occur in petroleum based solvents may be carcinogenic or otherwise pathogenic.
However, except for individuals responsible for handling wastes between the source and dump sites and possibly researchers, the potential direct health hazard does not seem to be a suitable criteria for
rejecting ocean disposal of these wastes.

   Disposal takes place in the open ocean 64 km north of Puerto
Rico. The dump site has great depth and sufficient dilution is
available to reduce levels below that of a direct health hazard. The
nature of the dumping process and the availability of dilution waters
makes the occurrence of major cultural impacts or ecological effects
at the dump site unlikely. However, the potential for indirect
health effects is more interesting. Indirect health effects require
uptake and storage (bioconcentration) of harmful materials or metabolism of their precursors by aquatic organisms for future transmittal
through the food web to human consumers. Since the dumping of the
Puerto Rican wastes is irregular and since the dump site does not
currently support a major food fishery, it requires some imagination
to perceive a potential problem for the site. However, many of the
organic compounds known to be present in the wastes are common industrial byproducts including fermentation wastes and unrecovered solvents (National Academy of Sciences, 1975). Their fate in marine
systems and possible effects on man are of obvious generic interest.

   Uptake and bioaccumulation cannot take place without exposure to
the wastes. This exposure might be direct, in the sense that fish in
the dump site vicinity may physically encounter waste plumes resulting in uptake of waste materials across the body surface or, indirect, occurring when fish ingest organisms which have accumulated a
burden of waste material through previous direct exposure. We choose
to concentrate our preliminary studies on the problem of defining the
likelihood of direct contact of fish with the wastes. Specifically,
we set out to evaluate the role of fish behavior in setting exposure

limits to the wastes by (1) determining whether fish have the capability to detect the materials, (2) the concentrations, if any, which were rejected or avoided by fish, and (3) the relationship of detection and avoidance limits to toxicity. Results of the experiments on detection and avoidance, in conjunction with those of concurrent field studies on dispersal rates and laboratory studies on chemical composition and physiological response being conducted at other institutions are expected to provide information for a more comprehensive study of bioaccumulation and effects for specific target materials.

MATERIALS AND METHODS

Acquisition of waste samples representative of the dumped material has been a perplexing problem. Eight production facilities are involved and products and productions schedules vary for each. The individual wastes contain volatile materials which are expected to chemically or biologically degrade. This problem is compounded by inconsistency in waste storage time prior to disposal. One of the wastes (Squibb) is stored and dumped separately from the others. It would be difficult to say that any single sample of the composite waste from the Arecibo storage facility would be typical or representative over time.

Three sample collections have been made (Feb. 5, June 22, and Aug. 3, 1978). The first two were made by U. S. Coast Guard staff. Composite waste was collected in polyethylene carboys from the Arecibo storage tank and shipped by air to Baltimore. In each case, problems with the airline resulted in a five to six day transit period. It was obvious on inspection of the carboys that material was escaping through the container walls.

The latest sampling was made in person in hopes of circumventing some of the earlier transport difficulties. Composite wastes were drawn from the Arecibo storage tank and two individual wastes (Squibb and Pfizer) were sampled directly from tank trucks. Samples were collected in glass bottles and shipped by air to Baltimore for our work, to Texas A & M University for chemical analysis, and to the University of Texas for complementary biological studies.

In our laboratory, all three sets of samples were held at room temperature in a fume hood. Experiments were initiated immediately upon arrival of the wastes to minimize uncertainties about sample degradation.

Since our work to date has been preliminary in nature, we have not made an effort to use fish native to the dump site. The location and characteristics of the site make such collections logistically difficult as well as expensive. We have concentrated our efforts on

local species and have been guided in our choice not only by what was
readily available but by our experiences in other similar studies.
Test species which have broad distributions in U. S. coastal waters
and which are known to differ substantially in general susceptability
to environmental pollutants (EPA/Corps of Engineers, 1977) were
selected. Most tests have been conducted with the mummichog Fundulus
heteroclitus, spot Leiostomus xanthurus and menhaden Brevoortia
tyrannus.

Our work has emphasized two types of observations: tolerance as
measured by static bioassay and detection/avoidance measured in a
choice-type apparatus. Static bioassays were used to survey the
wastes for toxicity. Test fish were collected prior to obtaining
waste materials from the Arecibo Storage Facility. Fish were collected with beach seines or trawls from the Chesapeake Bay and held
in fiberglass tanks in which continuous currents maintained activity.
Salinity, temperature and photoperiod were controlled to approximate
collection conditions. All species adapted readily to the holding
conditions and accepted food within two to four days of collection.
Fish were held a minimum of 21 days prior to testing to ensure recovery from collection stress.

When we were satisfied that the fish had adjusted to the laboratory environment, waste collection was initiated and a preliminary
toxicity screen was conducted. Groups of three fish were transferred
directly to flasks containing pre-mixed solutions of wastes and observed for 24 hours. Results were used to set dose levels for 96-hr
bioassays.

Groups of five fish were placed in flasks containing water only.
Twenty-four hours were allowed for the fish to recover from transfer
stress after which waste was added and the fish observed for 96
hours. Tests were conducted on animals which had been starved for 24
hours prior to dosing and which were not fed during the test.

Detection and avoidance studies were conducted in an apparatus
designed to maintain a constant concentration difference between two
chambers while allowing fish to pass freely from one to the other
(Fig. 1). The test section was constructed of fiberglass and was 36
cm deep, 51 cm wide and 102 cm long. The volume of the test section
was 90 liters. The tank was bisected down the long axis by an opaque
acrylic wall with an opening near one end which served as a passage
between chambers for the fish. The size of this opening could be
varied. A reservoir tank was situated below the test chamber.

The apparatus was operated as a recirculating system. Water was
pumped from the reservoir through perforated standpipes in each test
chamber. Flows were regulated by needle valves and could be varied
between 0 and 22 liters per minute. Water flowed through each test
chamber and exited via perforated standpipes in the downstream ends.

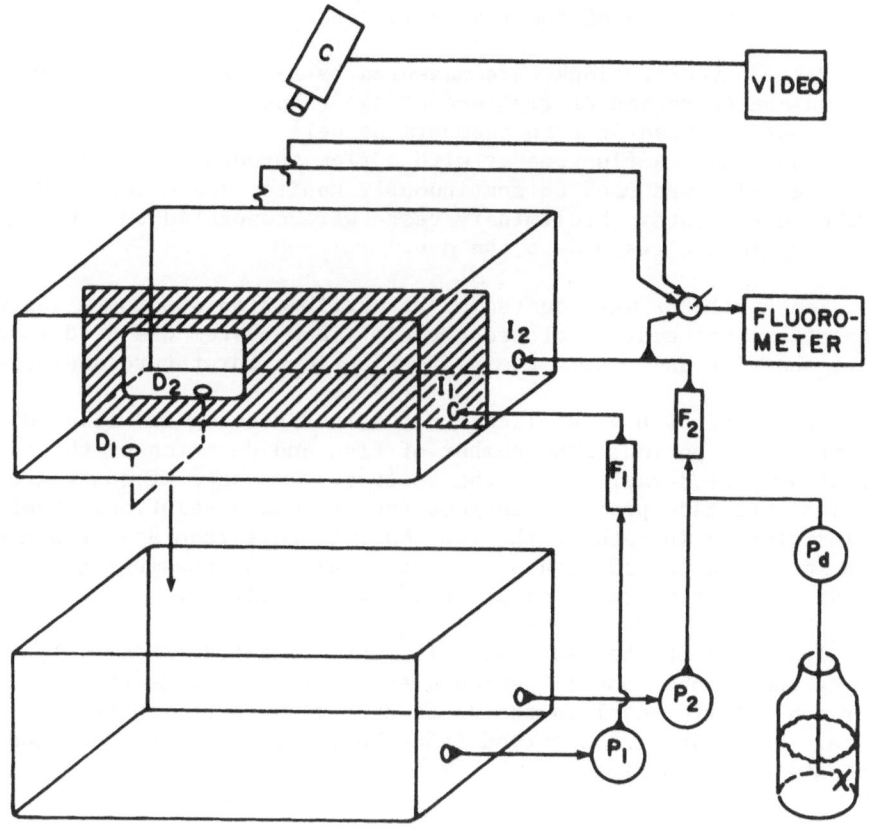

Fig. 1.   Choice-type apparatus for evaluating the avoidance response
of fish, P_1-P_2, pumps; P_d, waste metering Pump; F_1-F_2,
flowmeters; I_1-I_2, water inlets; D_1-D_2, water outlets; C,
video camera.

Screens prevented fish from congregating behind the drain pipes. The
opening in the acrylic divider was situated just upstream of the
screens. There was very little mixing across the divider when the
apparatus was properly adjusted. Water exiting the two test chambers
drained back into the reservoir by gravity flow where it was thoroughly mixed by a stirrer.

The concentration difference was achieved by continuously in-
jecting one of the chambers with a pre-mixed waste solution. Since
the system was designed to recirculate, waste concentrations gradually rose in both chambers with the dosed side always having the higher
level. The magnitude of the difference between chambers was variable
being determined by the concentration of the waste solution, the dose

rates and the volume of the reservoir tank.

Waste concentrations were measured by labeling the solution with a known concentration of fluorescent dye (Rhodamine Wt.). Test points were located in both chambers as well as in the reservoir and injection line. A fluorometer with a flow-through cell and strip chart recorder was used to continuously monitor waste concentrations at the test points. Preliminary tests were conducted to ensure that the fish did not respond to the dye.

The apparatus was mounted on shock/vibration absorbers and was completely enclosed. A closed circuit video system was used to observe the fish and record their distributions for future analysis.

Groups of fish were placed in the test section and allowed a familiarization period. The number of fish and duration of the familiarization period varied with the species. Two sets of control observations were made prior to introducing the waste solution. Numbers of individuals in each of the two chambers were recorded at one minute intervals for 30 minutes. Individual fish crossings between chambers were also recorded as an index of activity.

Injection of the waste solution was initiated immediately following the second control period with the rate of injection adjusted to reach a toxic level in the dosed chamber in approximately 2 1/2 hours. The fish were observed (and their distributions recorded) for the entire test period.

RESULTS AND DISCUSSION

Toxicity. Lethal concentrations (24-hr and 96-hr LC_{50}'s) of the wastes varied from 0.001% to 0.6%, depending on the test species

Table 1. The toxicity of pharmaceutical and petrochemical waste materials to fish. All wastes collected from the Arecibo, Puerto Rico dump facility on August 3, 1978.

Test Material	24-hr LC_{50} (%)		96-hr LC_{50} (%)	
	Spot	Killifish	Spot	Killifish
Squibb	0.06	0.35	0.05	0.35
Pfizer	0.08	0.35	0.075	0.35
Composite	0.05	0.5	0.05	0.6

and the nature of the material. The dose-response relationship was very sharp and either 0% or 100% mortality occurred in all but a few tests. Most (> 90%) of the mortalities occurred in the first 30-hr of exposure, regardless of test concentration. The acute nature of the response made interpolation of precise estimates of lethal concentrations impossible without greatly expanding the test design. This was impractical for our purposes and we have calculated less accurate graphic estimates of the lethal limits (Table 1).

Species differences in response were as expected with spot being substantially more sensitive to all materials than were killifish. Within species differences were less consistent. Spot were somewhat more susceptible to the Squibb waste than to the Pfizer or composite materials. This is of interest because the Squibb waste is dumped separately from the composite. Killifish were more tolerant of the composite waste. This may reflect the ability of this species to tolerate low dissolved oxygen concentrations as would be expected in the composite solutions which contained a high concentration of particulate fermentation waste.

Available information on waste composition (Schwab, Anderson) indicated that some waste constituents should be bioaccumulated. However, exposures to sublethal concentrations do not seem to have a delayed toxic effect. To test this, killifish were dosed at 25%, 50% and 100% of the estimated 96-hr LC_{50} for periods of 24 and 48 hours and observed for 20 days. Additional groups were held at the same

Table 2. Detection and avoidance responses of fishes exposed to pharmaceutical and petrochemical waste solutions.

Species	Waste Material	Detection Limit (%)	Avoidance Limit (%)
Menhaden	Squibb	0.040	N.R.[a]
	Pfizer	0.008	N.R.
	Composite[b]	0.010	0.035
Spot	Squibb	0.001	0.013
	Pfizer	N.R.	N.R.
	Composite[b]	N.R.	N.R.

a. N.R. = No response
b. Composite samples collected August 3, 1978

dose levels for 20 days. No concentration/dose/time combination resulted in mortalities which occurred more than 96-hr following the addition of the waste to the test tank nor did any of the combinations yield a greater estimate of the LC$_{50}$.

These results, although preliminary in nature, do provide guidance in assessing the potential impact of these wastes on local fishes. Rapid dilution (< 24-hr) of the wastes to field concentrations of less than 0.1% should eliminate concern for acute effects (mortality) due to direct exposure. This is equally true of fish which are found to avoid waste concentrations greater than 0.1% (see following section, however). The acute nature of the dose-response relationship and concentration of mortalities in the immediate post-dose period suggest that the waste constituents which are toxic at these dilutions are quite volatile.

Detection Avoidance. Behavioral responses to the various waste materials were inconsistent. Replicate tests with the earlier composite collections (shipped in plastic, permeable containers) demonstrated that fish response could not be predicted at dose rates ranging from 3 to 20 p.p.m. (increase) per minute or for concentrations of waste material as great as 0.1%. Positive responses were obtained

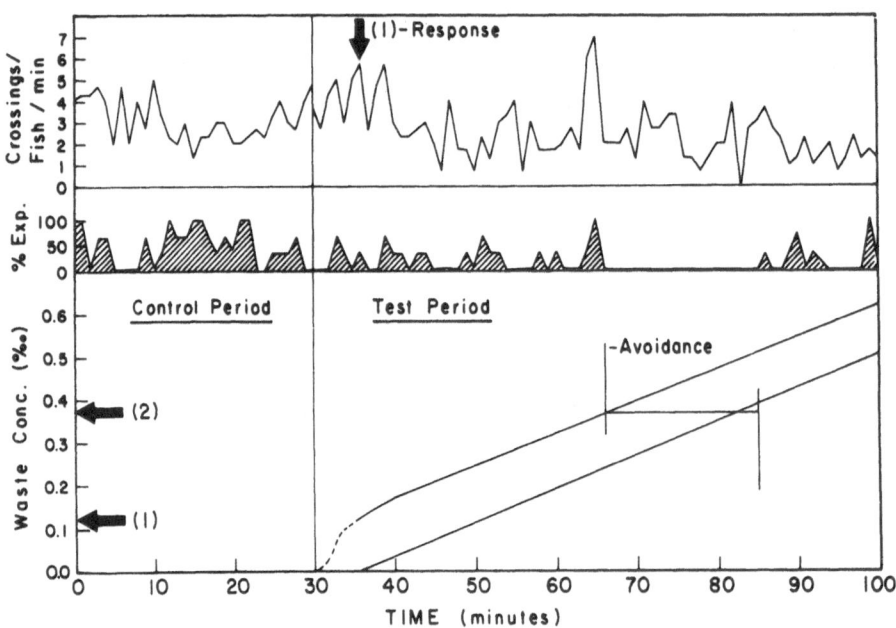

Fig. 2. Behavioral response of juvenile menhaden Brevoortia tyrannus
to a composite pharmaceutical and petrochemical waste.
Three fish. September 9, 1978.

in some of the later tests with waste materials which were collected and shipped in glass containers (Table 2). However, we place little confidence in the numerical results for those few tests in which a response was observed.

A partial summary of results for a test in which a response was observed is shown in Fig. 2. These data are for juvenile menhaden exposed to composite wastes from the third collection. Fish moved back and forth between the two chambers showing no preference for either side and crossed the center partition about 3 times per minute during the control period. Activity (measured as crossings of the center position) increased perceptibly in the first 10 minutes following the initiation of dosing. Fish crossed the center partition slightly more than 4 times per minute. Six minutes after the start of dosing (at a waste concentration of about 0.01%) behavioral patterns which indicated that the fish were aware of the presence of waste could be detected. This is a consistent but somewhat subjective sequence of responses in which fish abruptly cease to occupy the upstream end of the dosed chamber. Turning frequency increases and the fish begin to "nose" at the screen at the downstream end of the test chamber.

After the first 10 minutes of the dose period the number of fish crossings was slightly reduced while the proportion of time spent in the dosed chamber declined sharply. At a waste concentration of

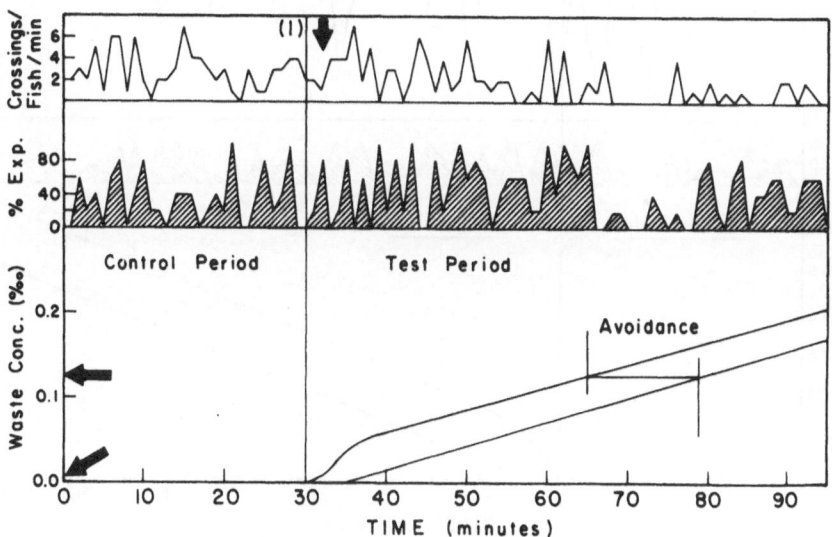

Fig. 3. Behavioral response of juvenile spot <u>Leiostomas</u> <u>xanthurus</u>
to a pharmaceutical waste (Squibb). <u>Five fish.</u> <u>August</u>
9, 1978.

about 0.035% there was a brief period of hyperactivity followed by a
period of complete rejection of the dosed chamber. Note that the
fish continued to sample the dosed chamber as indicated by continued
brief crossings of the partition. The fish continued to reject the
dosed chamber until the waste concentration in the second chamber
reached the avoidance level (0.037%).

A second example of an avoidance response is shown in Fig. 3.
In this case, five juvenile spot were exposed to the Squibb waste.
The fish moved freely between the two test chambers during the control period, spending about half of the time on each side. Response
to injection of waste was almost immediate. The fish continued to
move between the chambers without perceptible change in number of
crossings. However, behavior in the dosed chamber changed sharply as
the fish remained to the rear of the section, turning frequently and
nosing at the downstream screen.

During the first hour of the test period the proportion of time
spent in the dosed chamber was actually increased over the control
period. However, as waste concentration reached 0.012% the fish
moved out of the dosed chamber and continued to reject the area of
higher concentration until the avoidance level was exceeded in both

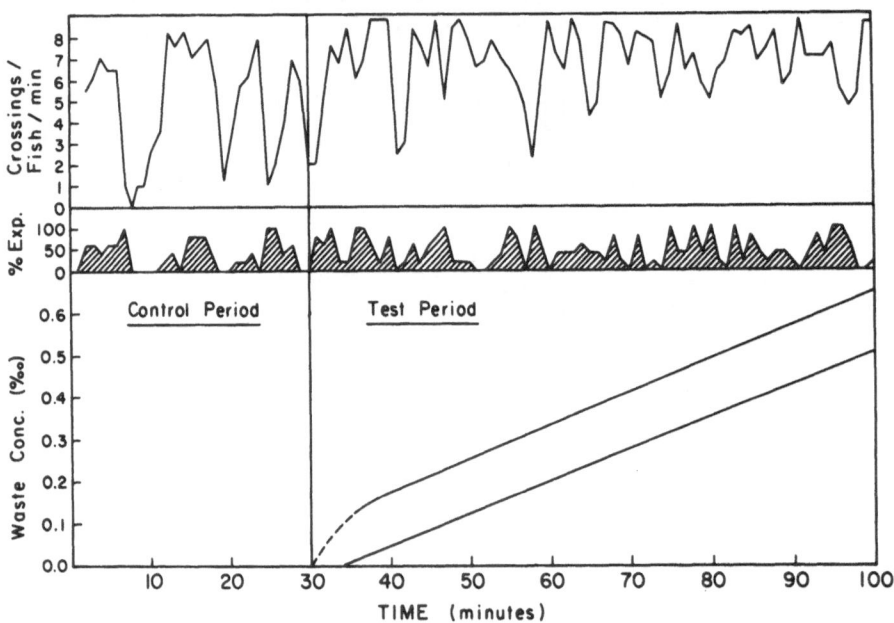

Fig. 4. Behavioral response of juvenile spot <u>Leiostomas</u> <u>xanthurus</u> to
a composite pharmaceutical and petrochemical waste. Five
fish. August 30, 1978.

chambers.

An example of a test for which no response was observed is given in Fig. 4. Juvenile spot were exposed to the composite wastes from the third collection. There was no indication that the fish could detect or avoid the material. Note that this same material was both detected and avoided by juvenile menhaden (Fig. 2).

The inconsistency of the avoidance response in conjunction with the known presence of organic solvents suggest that the wastes may be acting as a general anesthetic. The gradual but perceptible reduction in activity levels for dosed fish in the avoidance and bioassay chambers relative to undosed control fish is support for this speculation. Similar results were obtained by Ericksen Jones (1947) following exposure of fish to high concentrations of $CuSO_4$.

CONCLUSIONS AND RECOMMENDATIONS

While the results of the waste dispersion and analytical chemistry studies are not yet available to us in final form, it is unlikely that waste concentrations in the parts per thousand range will persist for even short periods outside the immediate vicinity of the barge outfall (see Csanady; Hatcher et al; and Kester et al., this volume). Also, a major proportion of the wastes are organic solvents or complex hydrocarbons which are lipophillic in nature and can be expected to bioaccumulate (Leo et al., 1971). Our preliminary results indicate that the wastes are neither toxic to or strongly avoided by fish at concentrations approaching the part per thousand level. In other words, fish will accept exposures to substantially greater concentrations of potentially bioaccumulated wastes than are expected to occur in the field situation without immediate or short term mortality.

On this basis, we recommend that further more detailed studies of toxicity or avoidance not be conducted beyond a possible confirmation of our general results for species resident to the dump site. A more useful effort in terms of applicability to the question of whether dumping at this site should be continued would be studies of the uptake and metabolism of target constituents as identified by the chemistry studies at exposure levels suggested by the dispersion studies.

REFERENCES

Environmental Protection Agency/Corps of Engineers (1977) Ecological evaluation of proposed discharge of dredge material into ocean waters. Rept. of Technical Comm. on Criteria for Dredged and Fill Material. Env. Effects Lab., U.S.A. Waterways Exp. Sta.,

Vicksburg, Miss. 19 pp. + Appendices.

Erickson Jones, J. R. (1947) The reactions of Pygosteus pungitius
    L. to toxic solutions. J. Exptl. Biol., 21, 110-122.

Goldberg, E. D. (1976) The Health of the Oceans. The UNESCO Press.
    Paris. 170 pp.

Leo, A., C. Hansch and D. Eckins (1971) Partition coefficients and
    their uses. Chem. Rev., 71, 525-615.

National Academy of Sciences (1975) Assessing potential ocean pol-
    lutants. Rept. to Ocean Affairs Bd., Comm. Nat. Resources, Nat.
    Res. Counc. Washington, D.C., 438 pp.

National Academy of Sciences (1976) Disposal in the marine environ-
    ment. Rept. to the U.S.E.P.A. by the Ocean Disposal Study
    Steering Committee. Washington, D.C., 76 pp.

Zapatka, T. F. and R. W. Hann, Jr. (1976) Evaluation of safety
    factors with respect to ocean disposal of waste materials.
    Texas A & M University Sea Grant Rept., 77-202., 93 pp.

Zapatka, T. F. and R. W. Hann, Jr. (1977) Technical and philosophical
    aspects of ocean disposal. Texas A & M University Sea Grant
    Rept., 78-203, 160 pp.

EFFECTS OF OCEAN DUMPING ON A TEMPERATE

MIDSHELF ENVIRONMENT

D. W. Lear, M. L. O'Malley, and S. K. Smith

U.S. Environmental Protection Agency, Annapolis
Field Office, Annapolis, Maryland 21401

ABSTRACT

The fate and effects of ocean dumped sewage sludge and acid iron waste were studied at two mid-continental shelf sites off the Delaware-Maryland coast. Materials accumulated in sediment, especially in winter, and tended to persist in topographic low areas. Ambient levels in sediments, of metals, organic carbon, and organohalogens were determined, and statistically significant areas of impaction delineated. Uptake of certain metals in scallops and clams was indicated. The mahogany clam, _Arctica islandica_, may have suffered mortalities due to dumping activities. Contamination of bottom sediments by sewage indicator bacteria was demonstrated. Studies of impact on bottom communities are continuing.

INTRODUCTION

In 1972 Congress passed the "Marine Protection, Research and Sanctuaries Act" (the "ocean dumping bill") which regulated ocean dumping.

In May 1973 Region III of the U.S. Environmental Protection Agency initiated a field program to determine the effects of ocean dumping under the aegis of the Region. Two dumpsites were studied, one an industrial acid waste site and the other a sewage sludge site, approximately 40 miles (72 km) off the Delaware-Maryland coast (Fig. 1).

The northerly site was designated for industrial wastes; metal-iferous acid wastes from a titanium ore extraction process were disposed at this site from 1968 until 1977. The other site,

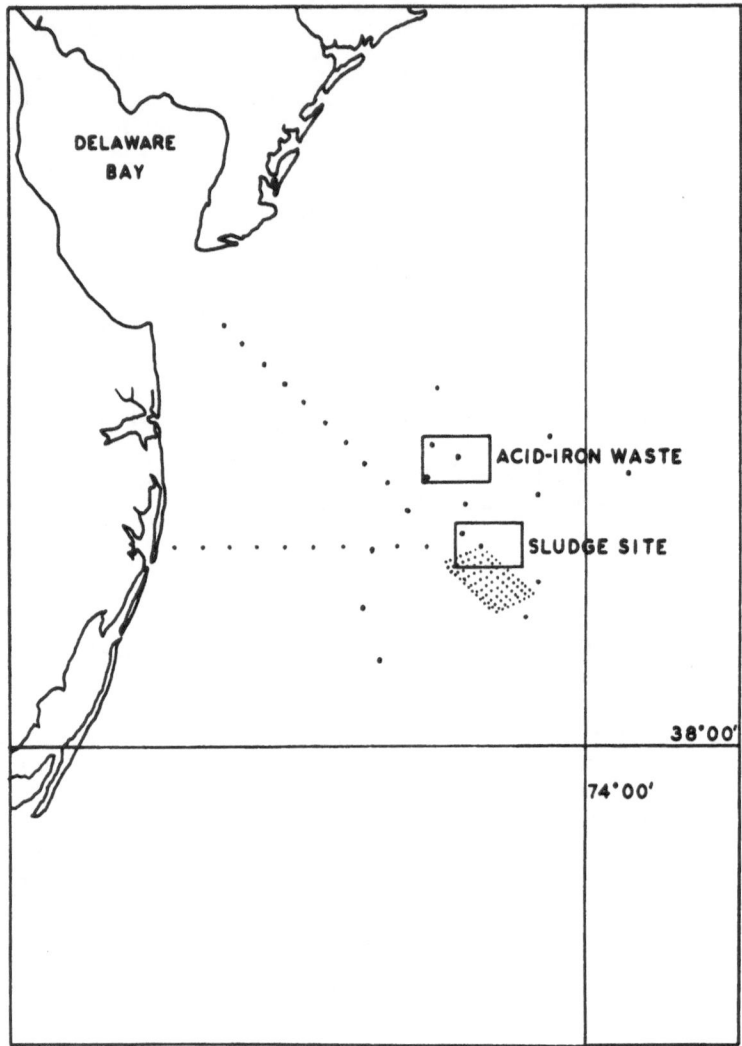

Fig. 1.   Dumpsite locations and monitoring station locations.   The
          nearfield "intensive grid" stations are located immediately
          south of the sewage sludge site, while some of the
          "farfield" stations are shown as isolated dots.

approximately eleven nautical miles (20 km) southeast of the acid
waste site, is used for disposal of municipal sewage sludge, also
with substantial metals concentrations.   Monitoring at the sludge
site began before dumping activities commenced in June 1973.

     The monitoring program evolved with two basic objectives: (1)

determine the fate of disposed materials in this environment, and (2) determine the effects of such disposal activities. The program was designed with emphasis on the more persistent effects on the benthic systems, rather than the more transient effects in the water column.

The rationale for the monitoring program was the establishment of a large grid, 2000 square nautical miles (6860 square kilometers), to (1) determine ambient levels of parameters, (2) search for anomalies and identify impacted areas, (3) to estimate the extent of translocation of materials, and (4) determine other possible inputs to the area.

In addition, a near-field sampling grid of one to three miles (1.8 to 5.6 km) intervals was established adjacent and south of the sewage sludge site to investigate apparent accumulations of sludge. Later monitoring strategies included transects to the shore and the Delaware estuary, and the addition of stations based on topography to more definitively sample contrasting midshelf ridge and swale habitats.

The study area was comprised of essentially a midshelf environment, between the 20-fathom and 35-fathom (36 to 63 m) isobaths, to minimize zonation effects on benthic indicator systems.

RESULTS

Hydrography. The water column was stratified in summer, with thermocline at 16 to 20 meters (Meyers, 1974; Lear et al., 1977). Temperatures and salinities below the thermocline were typical of the "cold pool" waters (Beardsley et al., 1973). Nearly isothermal conditions of the water column were found in winter, with an inshore-offshore gradient of cooler waters inshore (Bumpus et al., 1972,; Lear et al., 1977). A tongue of lower salinity water intruded from the west in summer, indicating a local influence from the Delaware estuary drainage.

There was indication that dumped materials may not immediately penetrate the pycnocline, and may be laterally transported from the release site. (E.I. du Pont de Nemours and Company, University of Delaware College of Marine Studies, Hydroscience, 1972).

Circulation. The fate of waste materials released into the aqueous environment was predicted by simulation modeling (Demenkow and Wiekramartane, 1976) and by Langrangian and eulerian current estimations. Earlier works (Beardsley et al., 1976; Bumpus, 1973; Ketchum, 1953; Norcross and Stanley, 1967) indicated a net southwesterly drift throughout this area.

Releases of Woodhead bottom drifters in May and November 1973

yielded recoveries, primarily on the Delmarva peninsula southwest of
the release areas (Fig. 2), generally confirming earlier observations.

Bottom current vectors from moored tilt vane current meters were
utilized to determine whether tidal currents could actively translo-
cate the indigenous sedimentary materials, and presumably concomitant
introduced contaminants (Palmer et al., 1976).

Telemetering drogues were employed by Klemas et al. (1977) to
follow water masses from the release area along with satellite
imagery. These results indicated that storm activity could markedly
affect the transport of dumped materials from the site, usually in

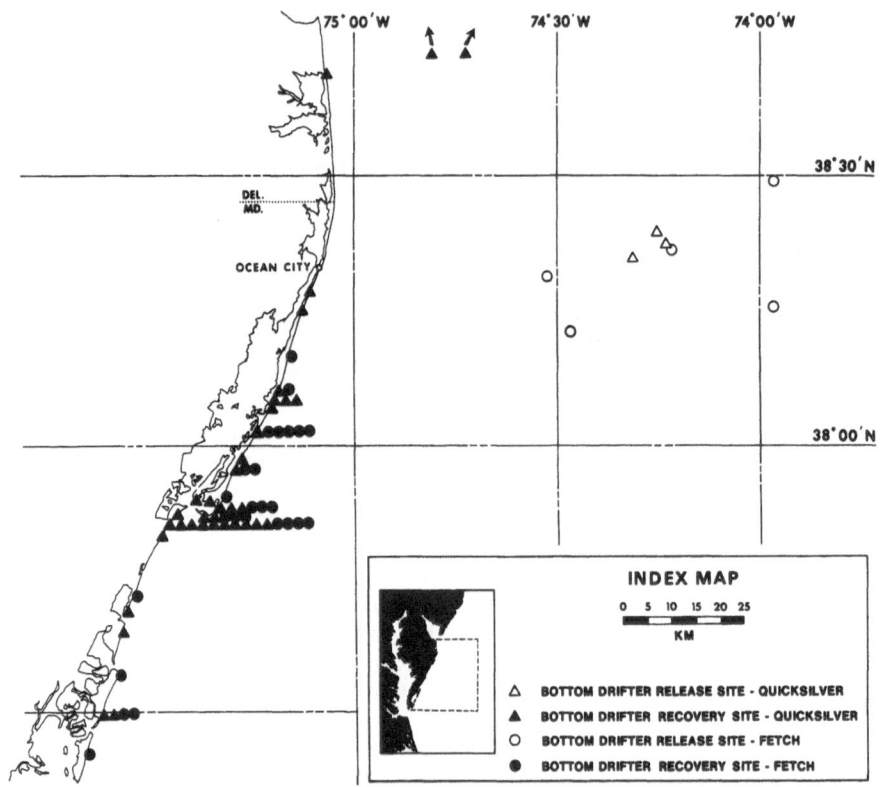

Fig. 2.   Releases and recoveries of bottom drifters from Operation
          Quicksilver (1-4 May 1973) and Operation Fetch (5-10
          November 1973) as of August 1974. Recoveries from Bradley
          Beach, N.J. and Marthas Vineyard, Mass. are indicated by
          arrows to the north.

the order of magnitude of 10 miles (18.5 km) from the release site, with the maximum distance measured during the study of 14.5 nautical miles (26.8 km) from the release site.

All above indications were that this was a hydraulically active environment, with several forces often superimposed, that can actively modify the distributions of dumped materials.

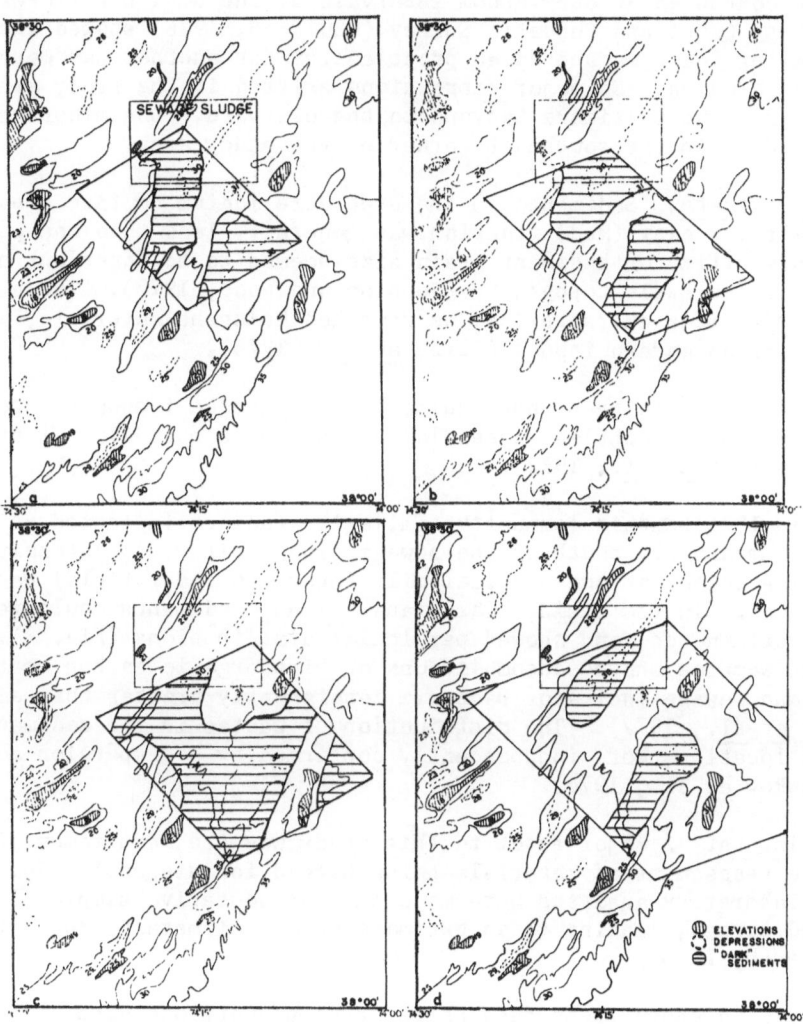

Fig. 3. Distribution of "dark" and "clean" sediments in intensive grid area. Fig. 3a, December 1975; Fig. 3b, August 1976; Fig. 3c, February 1977; Fig. 3d, August 1977. Note association of dark materials and topographic lows.

Plankton. Studies were made of the phytoplankton and zooplankton in the area. Generally, no major disruptions were noted, as
could be expected from the relatively vigorous hydrodynamics of the
area (Forns, 1977).

Topography. The bottom topography of the midshelf monitoring
area was dominated by subtle northeast-southwest trending ridges and
swales (Duane et al, 1972). Most of the bathymetric features were
not evident on standard navigation charts, but a special bathymetric
series contoured at one-fathom intervals by the U.S. Department of
Commerce, Coast and Geodetic Survey (0807N-56, 1976) showed the subtle relief features and added practicality for another parameter for
benthic studies. No major depressions existed in the study area
other than the Baltimore Canyons to the eastward. One minor scarp
was evident in the southeast corner of the study area.

Substrate. Sediments in the area were medium to fine sands.
The silt and clay (mud) fraction was generally less than one percent.
The variability of sediment grain size seemed to be dependent on the
subtle ridge-swale topography (Johnson and Wood, 1977). The sediments have been characterized as reworked Holocene materials, with
little or no modern inputs (Swift et al., 1976).

In the vicinity of both dumpsites the dredge frequently retriev-
ed wave-rounded cobbles, possible evidence of a post-glacial still-
stand (Swift et al., 1972).

In 1975 benthic community similarity indices indicated an aber-
rancy immediately south of the sewage sludge site. An intensive grid
sampling scheme at one nautical mile intervals (near field) was established. In the field, "dark" and "clean" sediments could readily
be discriminated, and showed particular distributions (Fig. 3). The
"dark" samples showed accumulations of high organic carbon, metals,
PCB, and sewage indicator bacteria consistently greater than ambients
(Lear et al., 1977). The distribution of wastes in the sediments at
these locations were independently confirmed by a dispersion model
(Demenkow et al., 1976).

Inputs. A major asset to this study was the requirement for accurate assessment of materials being barged for disposal. Comprehensive laboratory analyses were made on representative samples from
loaded barges, leading to an estimate of annual metals inputs (Lear
and Pesch, 1975).

Fig. 4 shows these data as relative proportions disposed by the
two major dumpers and total metric tons of metals disposed through
1974.

Metals in Sediments. Metals were of major concern in the materials dumped at both sites. The mean, standard deviation and range

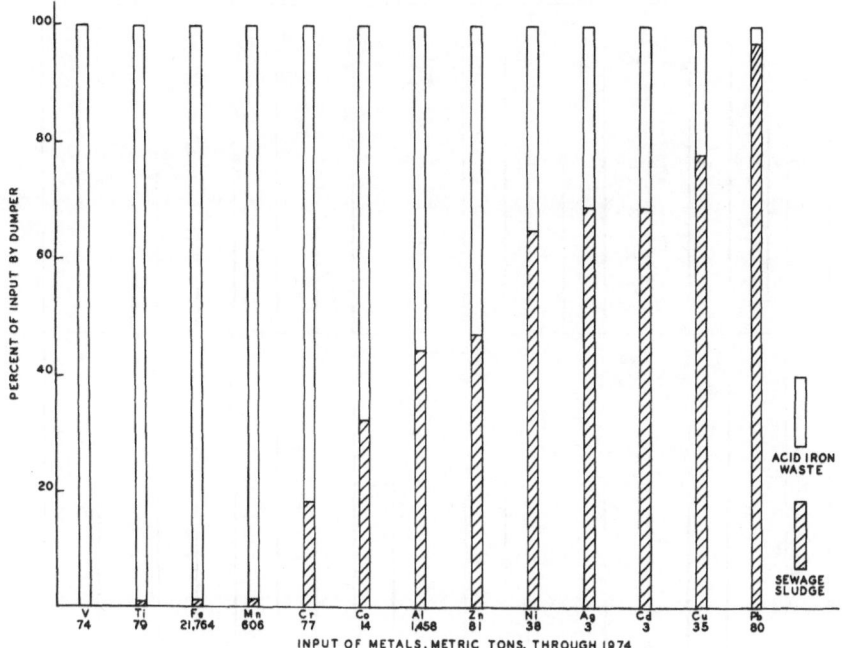

Fig. 4.   Total input of metals by ocean dumping, and the relative
          percentage contributed by the two major dumpers.

of selected metals of all stations are shown as plots in Fig. 5.   The
results from this wide-field sampling strategy indicate ambient sediment concentrations, the variability, as well as spatial and temporal
trends.

     Total Organic Carbon.   The distribution of total organic carbon
in sediments was determined in both the far-field and near-field sampling schemata.   Ambient concentrations of TOC, as estimated by mean,
standard deviation and range in the far-field grid, were in the order
of magnitude of 300 to 800 mg/kg dry weight (Fig. 5).   However, in
the intensive grid area associated with sewage sludge dumping, concentrations of 1000-4500 mg/kg dry weight were encountered at certain
locations.   One station in the acid waste site also showed consistent
and statistically higher concentrations (Lear et al., 1977).

     Organohalogens in Sediments.   Analyses of the sewage sludge re-
leased into this environment showed several organohalogens to be consistently present, consequently checks were made in bottom sediments
to determine ambient levels and examine for possible impaction areas.

     The data for polychlorinated biphenyls (PCB) as Arochlor 1254
are shown in Fig. 6.   As these data represent the far-field sampling

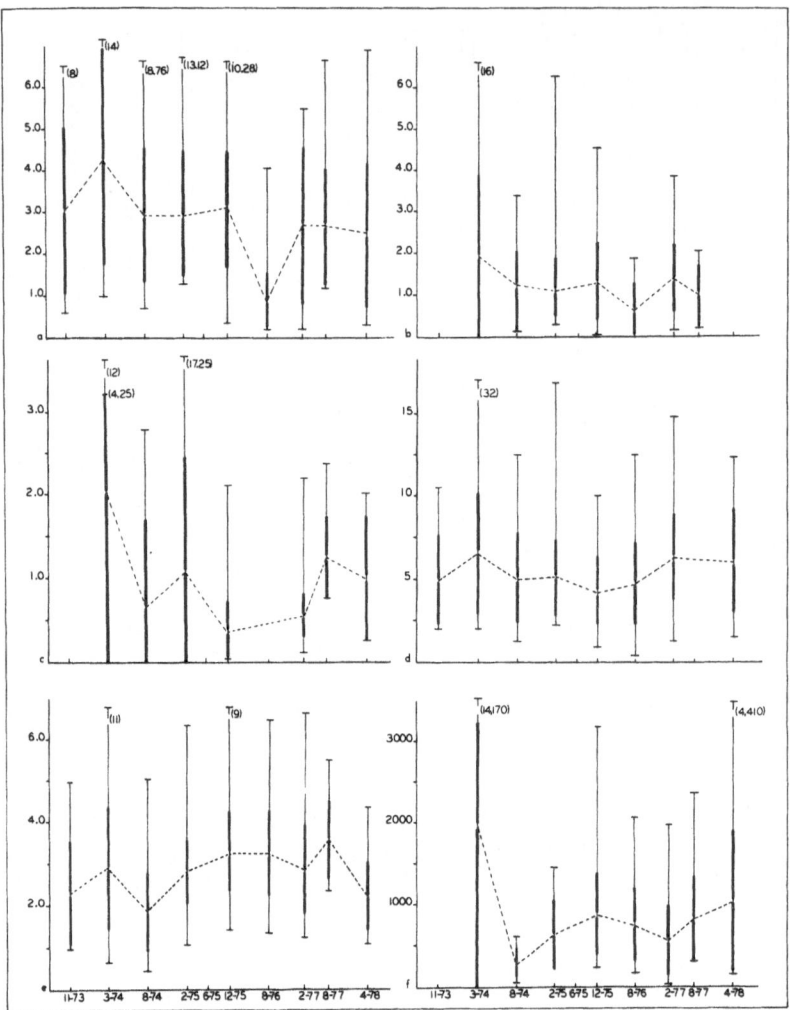

Fig. 5. Ambient midshelf concentrations of metals and TOC in
sediments. Means between sampling dates are connected;
standard deviation is shown by vertical heavy bars, and
vertical thin lines show range. Concentrations are mg/kg
dry wt. − a, Pb; b, Ni; c, Cu; d, Zn; e, Cr; and f, TOC.

grid, it is apparent that the concentrations in sediments fluctuate
in time over the whole study area. Local impacts could be noted as
atypically high concentrations (Lear and Pesch, 1975). Interpretation of the possible cyclic nature of these materials requires more
comparisons with source and transport data.

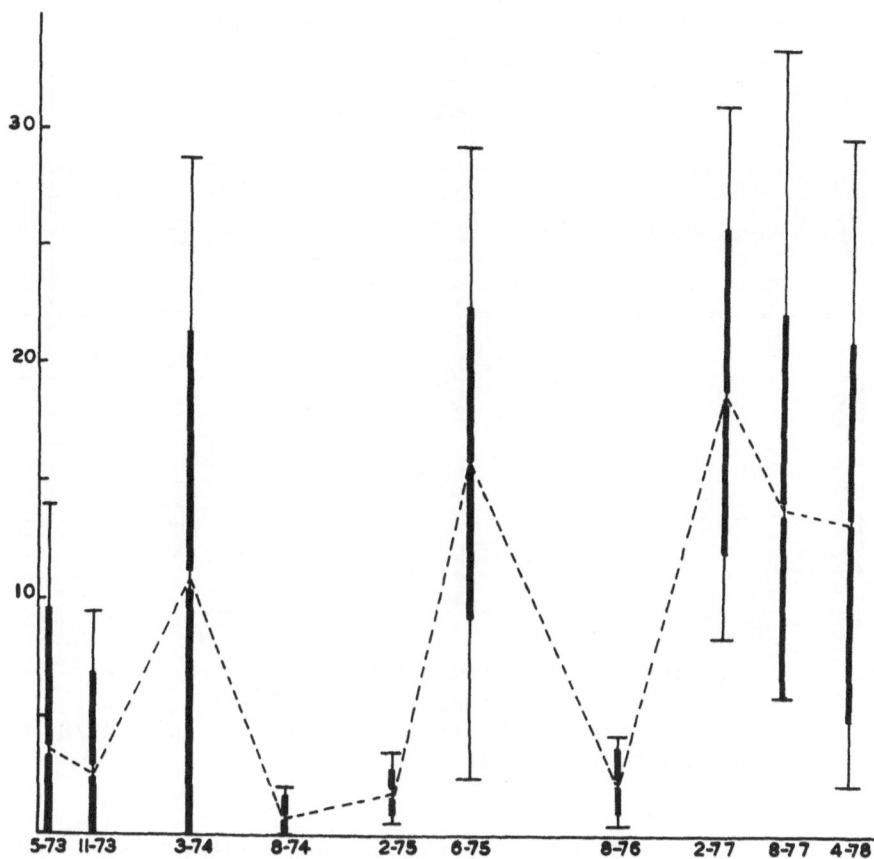

Fig. 6.   Ambient midshelf concentrations (μgm/gm dry wt.) of the
organohalogen PCB (Arochlor 1254) in sediments. Means
between sampling dates are connected; standard deviation is
shown by vertical heavy bars, and vertical thin lines show
range.

Bacteriology. Bacteriological analyses for coliform and fecal
coliform bacteria in water, sediments and shellfish were done in conjunction with the U.S. Food and Drug Administration Laboratory,
Davisville, Rhode Island (Lear and Pesch, 1975; Lear, 1974; Lear et
al., 1977).

In addition to the near-field and far-field station arrays described above, transects were run from the dumpsite to the mouth of
Delaware Bay and to the urban area at Ocean City, Maryland. Water
samples from approximately one meter from bottom consistently showed
negative results. There were insufficient samples of shellfish, Arctica islandica and Placopecten magellanicus, to indicate geographical

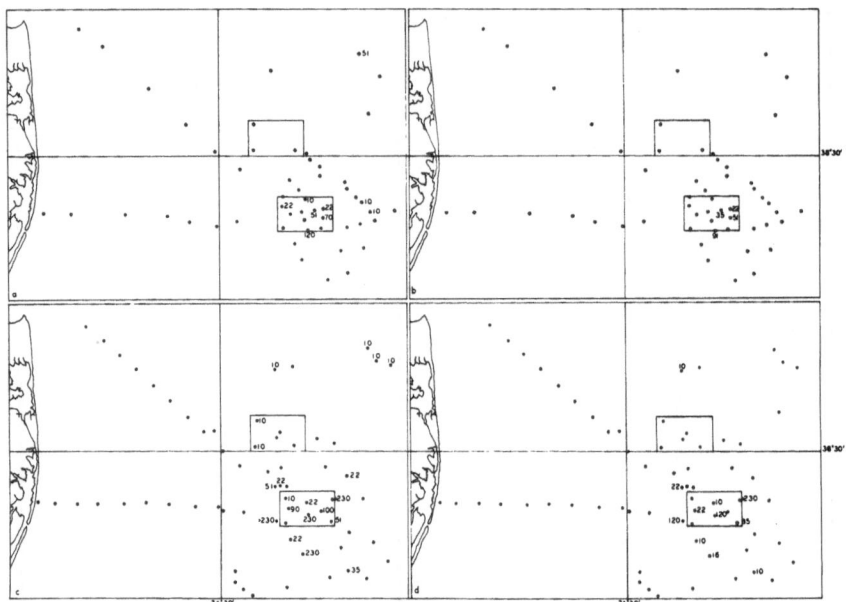

Fig. 7.   Distribution of coliform and fecal coliform bacteria in
          sediments. a. coliforms, September 1978; b. fecal
          coliforms, September 1978; c. coliforms, April 1978; d.
          fecal coliforms, April 1978.

contamination by indicator bacteria, but the results to date indicate
there is some degree of contamination, in need of further investigation.

        Results of the distribution of coliforms and fecal coliforms in
sediments frcm cruises in April and September 1978 are shown in Fig.
7. These data showed high concentrations of sewage bacteria associated with the sewage sludge dumpsite, with occasional sporadic low
background counts elsewhere. Negative results were generally found
on the transects towards possible landward sources.

        Benthic Biological Studies. Studies were conducted by the EPA
laboratory in Narragansett, Rhode Island, on metals uptake by the
mahogany clam, Arctica islandica, and the sea scallop, Placopecten
magellanicus. These data were subjected to rigorous statistical delineation using the Duncan's multiple range test. Significantly elevated concentrations of cadmium were found in the viscera and muscle
of scallops and the mahogany clam associated with the dumpsites.
Vanadium was found in scallops. Copper was found in muscle and viscera of scallops, and nickel in scallop muscle (Lear and Pesch, 1975;
Pesch et al., 1977).

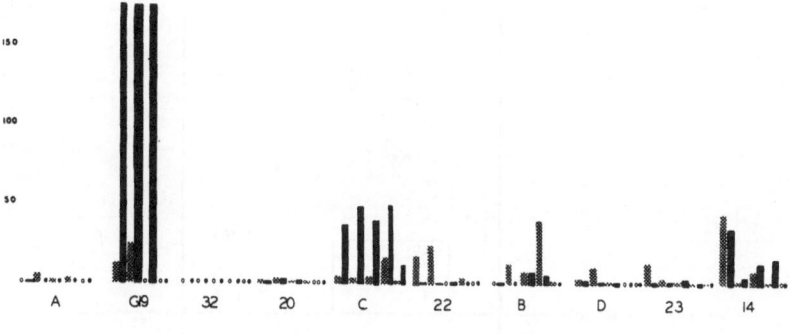

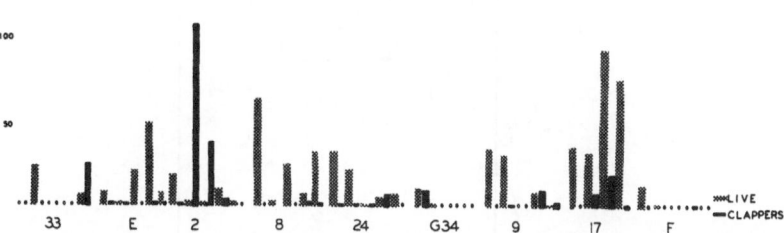

Fig. 8.   Apparent recent mortalities of mahogany clams, Arctica
          islandica, as indicated by live clams (solid bars) and
          hinged shells or "clappers" (hatched bars). Station
          locations are shown in Fig. 9.

A gross index of apparent mortalities of the mahogany clam, Arctica islandica, is the relative incidence of live, intact clams compared with empty valves or "clappers". "Clappers" were considered as
recently dead organisms. Unhinged, loose valves were not considered.
The data presented in Fig. 8 are the total numbers of live clams and
"clappers" found in duplicate dredge hauls, not percentages. Stations are shown in Fig. 9.

The low standing crops at Stations A, 32, 22, 23, and 9, all
near the 20-fathom (36 m) isobath, reflected the natural distribution
of this organism, which was generally found between the 20 and 30
fathom (36 and 54 m) isobaths. Stations F and G-34 were deeper than
30 fathoms and were generally sparse in Arctica.

The data shown indicate apparent recent mortalities on several
cruises and at several locations. Stations G-19, C, 14, and 2 showed
such indications.

Stations 2 and C were within dumpsites. Station G-19, approximately 20 nautical miles (37 km) northeast of the dumpsites,

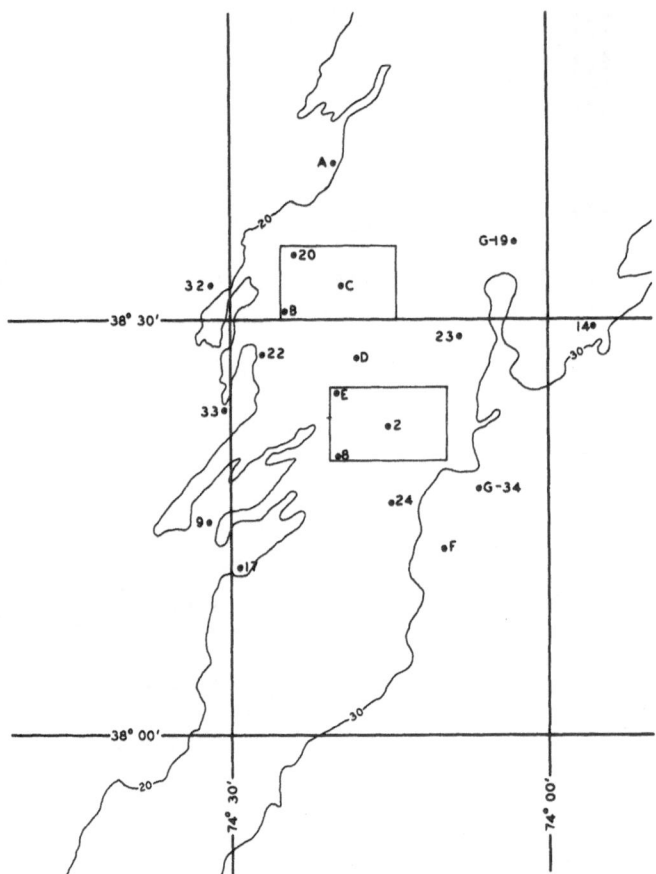

Fig. 9.   Station locations where <u>Arctica</u> <u>islandica</u> were sampled.

consistently showed significantly high concentrations of metals,
indicating deposition of materials. Station 14 was approximately 20
nautical miles (37 km) east of the dumpsites, showed indications of
mortalities, and has shown significantly high concentrations of
chromium and lead.

The data showed no indications of seasonal mortalities.

These data may also give some indication of the time required
for hinge ligaments to rot, whereby "clappers" become individual
valves. If the assumption is made that a single incident was responsible for a major mortality at Stations G-19 and 2, a plot of the incidence of "clappers" against time may give an order of magnitude estimate. Such a plot, shown in Fig. 10, indicates 12 to 14 months.
The apparent increase in numbers of live clams at Station 2 may

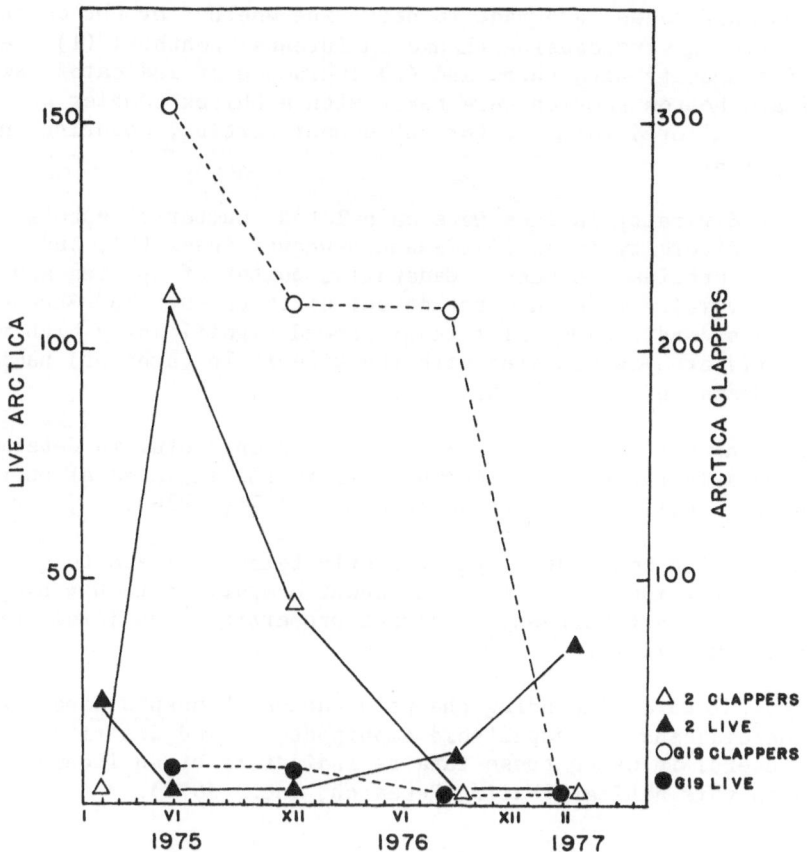

Fig. 10. Live Arctica islandica and hinged shells, "clappers", at
two apparently impacted locations, indicating the time for
hinge ligaments to decay in situ. Numbers are totals of
two replicate dredge licks, approximately 3383 m².

indicate a repopulation of an area once impacted.

The data from Station G-19 indicates a mortality previous to
June 1975, and no evidence of recovery.

Following observations of "fin rot" in the New York Bight area
(Murchelano and Ziskowski, 1976) and off southern California (Mearns
and Sherwood, 1977), examinations were made for external morphological evidence of disease or abnormalities. No incidences of "fin rot"
in fishes were found, but necrotic lesions in the exoskeletons of
Cancer crabs were noted in the vicinity of the sludge accumulation
areas.

Two approaches were made to determine whether or not ocean dumping activities were causing change in infaunal benthos: (1) examination of community structure, and (2) incidence of indicator taxa.
Triplicate bottom samples were taken with a Shipek sampler and preserved in buffered formalin for subsequent sorting, counting and interpretation.

Four diversity indices were calculated: number of species (S),
Simpson's diversity index (D), Shannon-Weaver index (W), and
Margalef's species richness. Generally, number of species and the
Simpson diversity index did not detect differences. The Shannon-
Weaver index and the Margalef index showed significantly high values
in the grid area as compared with the wide-field (ambient) samples
(Marine Research, Inc., 1978).

Percent similarity calculations have been useful in determining
differences in the infaunal communities in the impacted areas and the
surrounding shelf (Marine Research, Inc., 1975, 1976).

Clustering techniques are presently being utilized to determine
community distributions, with subsequent comparison to benthic parameters including topography, sediment properties, and levels of known
introduced materials.

Several taxa, including the polychaetes, Minuspio japonica, the
lumbrinereids and the ampeliscid amphipods, showed distributions that
may be useful in using these taxa as indicators of environmental
change in this habitat (Marine Research, Inc., 1978).

DISCUSSION

The dumped materials apparently have but transient effects on
the water column. The accumulation of disposed sludge in sediments
may, however, ultimately present oxygen demands on bottom waters and
possibly contribute to localized overenrichment of nutrients.

Sludge materials were observed to accumulate in areas related to
topography in and immediately to the south of the sewage sludge release site. A box canyon and a basin shaped depression were loci
where sludge materials were consistently found, while intervening
ridges apparently did not accumulate sludge. The implications here
are that the hydraulic regime is sufficiently energetic to keep
ridges swept clean. The persistence of accumulation in the depressions, however, is not completely understood, for high energy events
such as storm surges may be capable of redistributing or burying
these surficial materials. Evidence for burial were observations of
intermediate layering of dark bands with clean bands of sediment in
the bottom grab, found at stations up to 12 nautical miles (22 km)
west of the sludge release site (Lear, 1978). This indicated

cataclysmic activity rather than biological reworking. As this area
is described as geologically nondepositional (Swift et al., 1972), it
can be postulated that materials transported to the westward, a shallower, more energetic system, would be more subject to burial, while
materials transported eastward to deeper bottoms would tend to remain
as surficial deposits.

The identity of the deposited sludge was established by the significantly elevated concentrations of metals, the organohalogen PCB,
organic carbon, and sewage indicator bacteria.

The concentrations of metals in impacted area were significantly
higher than the ambient levels of metals in sediments in this midshelf environment. However, higher ambient metals in sediments have
been reported from estuaries (Johnson and Villa, 1976) as well as off
the continental shelf (Harris et al, 1977). These higher concentrations are apparently related to a larger proportion of silts and
clays in the sediments.

The identity of the deposits as disposed sludge was further confirmed by independently derived simulation modeling.

The persistence of the materials was shown for a three-year period but the depths of accumulation, rates of burial, and/or dispersal are not yet known.

The ubiquitous incidence of the organohalogen, PCB, in sediments
throughout the study area indicates a general contamination with localized areas of increased impaction due to ocean dumping practices.

Ocean dumping of wastes is affecting benthic habitats in certain
areas. These studies are continuing, but some effects have been
noted.

Sewage contamination of a large area of commmercial shellfishing
grounds has public health aspects as well as the economic effects of
closure of this area to clamming. Similarly, the demonstrated uptake
of metals in scallops and clams has public health and economic significance, and further the potential for ecological degradation. There
was indication of recent mortalities of Arctica islandica associated
with disposal activities. There is some indication of shell lesions
in Cancer crabs in the vicinity of the dumpsites.

Analyses of data of benthic infaunal communities are as yet incomplete, but changes in bottom community structure were indicated in
areas immediately south and adjacent to the sewage sludge site. Generally a change from normal faunal assemblages to more pollution tolerant taxa was found. These observations indicate a transition from
the ecologically healthy shelf environments towards the severely impacted benthic environment documented in the New York Bight. (Pearce

et al., 1976).

This study has had a temporal and spatial aspect. Rather than simple delineation of a "polluted" area, it has been an attempt to detect subtle changes over time and space on nearly pristine conditions from induced stresses on the ecosystem. The isolation of the study area from terrestrial or other sources of stress has contributed towards an understanding of induced ecological changes due to ocean dumping.

ACKNOWLEDGEMENTS

Many institutions, agencies and individuals actively contributed to this program. The United States Coast Guard, NOAA-NMFS at Oxford, Maryland, American University, U.S. Food and Drug Administration, States of Maryland and Virginia, University of Delaware, Virginia Institute of Marine Science are some of the institutions, but at least 29 institutions have contributed. The EPA Annapolis Field Office did most of the chemical analyses.

REFERENCES

Beardsley, R. C., W. C. Boicourt, and C. V. Hansen (1976) Physical oceanography of the Middle Atlantic Bight. In: Middle Atlantic Continental Shelf and New York Bight, M. Grant Gross, editor, Spec. Symposia, Amer. Soc. Limnol. and Oceanogr., 2, pp. 69-89.
Bumpus, D. F. (1973) A description of the circulation on the continental shelf of the east coast of the United States. In: Progress in Oceanography, 6, Mary Sears, editor, pp. 111-157.
Bumpus, D. F., R. E. Lynde, and D. M. Shaw (1972) Physical oceanography. In: Coastal and Offshore Environmental Inventory Cape Hatteras to Nantucket Shoals, Marine Publication Series No. 2, University of Rhode Island, pp. 1-72.
Demenkow, J. W. and P. Wiekramartane (1976) Far Field Sewage Release Simulations. Raytheon Corporation, Portsmouth, RI.
Duane, D. B., M. E. Field, E. P. Meisburger, D. J. P. Swift, and S. J. Williams (1972) Linear shoals on the Atlantic inner continental shelf, Florida to Long Island. In: Shelf Sediment Transport: Process and Pattern, D. J. P. Swift, D. B. Duane, and O. H. Pilkey, editors, Dowden, Hutchinson and Ross, Stroudsburg, PA, pp. 447-498.
E. I. du Pont de Nemours and Company (1972) Waste dispersion characteristics in an oceanic environment. University of Delaware College of Marine Studies, Hydroscience, Inc. Draft of a report to the Water Quality Program, U. S. Environmental Protection Agency, pp. 1-393.
Forns, J. M. (1977) Phytoplankton and zooplankton taxonomic investigations of two interim ocean dumpsites. EPA 68-01-3211, U. S.

Environmental Protection Agency, Region III, Philadelphia, PA, 40 pp.

Harris, R., R. Jolly, R. Huggett, and G. Grant (1977) Trace metals. In: Middle Atlantic Outer Continental Shelf Environmental Studies, VII-B, Chemical and Biological Benchmark Studies, U. S. Department of the Interior and Virginia Institute of Marine Science, Gloucester Point, VA, 8, pp. 1-60.

Johnson, P. G. and O. Villa (1976) Distribution of metals in Elizabeth River sediments. Tech. Rpt. No. 61, U. S. Environmental Protection Agency, Annapolis Field Office, Annapolis, MD, 31 pp.

Johnson, P. P. and S. A. Wood (1977) Seasonal variability of sediment texture in the Middle Atlantic region. In: Geologic Studies of Middle Atlantic Outer Continental Shelf, Vol. III Geologic Studies, D. W. Folger, editor, U. S. Geol. Survey, Woods Hole, MA, U. S. Dept. Interior, Bureau of Land Management, and Virginian Institute of Mar. Sci., Gloucester Point, VA.

Ketchum, B. H. (1953) Preliminary evaluation of the coastal water off Delaware Bay for disposal of industrial wastes. Woods Hole Ref. No. 53-51. (Unpublished manuscript).

Klemas, V., G. R. Davis, and R. D. Henry (1977) Satellite and current drogue studies of ocean-disposed waste drift. Journal Water Pollution Control Federation, 49, 757-763.

Lear, D. W. (1974) Environmental survey of two interim dumpsites, Middle Atlantic Bight. Supplemental Report, U. S. Environmental Protection Agency, Region III, EPA-903/9-74-010B.

Lear, D. W. (1978) Testimony at Public Hearing, City of Philadelphia permit, Georgetown, DE, May 1978.

Lear, D. W., M. L. O'Malley, and S. K. Smith (1977) Effects of ocean dumping activity Mid-Atlantic Bight 1976. Interim Report, Environmental Protection Agency, Region III, Philadelphia, PA, EPA-903/9-77-029, 168 pp.

Lear, D. W. and G. G. Pesch (1975) Effects of ocean disposal activities on mid-continental shelf environment off Delaware and Maryland. U. S. Environmental Protection Agency, Region III, EPA-903/9-75-015.

Lear, D. W., S. K. Smith, and M. L. O'Malley, editors (1974) Environmental survey of two interim dumpsites, Middle Atlantic Bight. U. S. Environmental Protection Agency, Region III, Philadelphia, PA, EPA-903/9-74-010A.

Marine Research, Inc. (1975a) Analysis of Operation "Deep Six" benthic invertebrates. Marine Research, Inc., Falmouth, MA.

Marine Research, Inc. (1975b) Analysis of Operation "Midwatch" benthic invertebrates. Marine Research, Inc., Falmouth, MA.

Marine Research, Inc. (1976a) Analysis of Operation "Dragnet" benthic invertebrates. Marine Research, Inc., Falmouth, MA.

Marine Research, Inc. (1976b) Analysis of Operation "Touchstone" benthic invertebrates. Marine Research, Inc., Falmouth, MA.

Marine Research, Inc. (1978) Analysis of Operation "Mogul" benthic invertebrates. Marine Research, Inc., Falmouth, MA.

Mearns, A. J. and M. J. Sherwood (1977) Changes in the prevalence of
    fin erosion off Los Angeles and Orange Counties. Annual Report,
    Southern California Coastal Water Research Project, El Segundo,
    CA, pp. 143-146.
Meyers, T. D. (1974) An observation of rapid thermocline formation in
    the Middle-Atlantic Bight. Estuarine and Coastal Marine
    Science, 2, 75-82.
Murchelano, R. A. and J. Ziskowski (1976) Fin rot disease studies in
    the New York Bight. In: Middle Atlantic Continental Shelf and
    New York Bight, M. Grant Gross, editor, Spec. Symposia Amer.
    Soc. Limnol. and Oceanogr., 2, 329-336.
Norcross, J. J. and E. M. Stanley (1967) Inferred surface and bottom
    drift, June 1963 through October 1964. In: Circulation of Shelf
    Waters off the Chesapeake Bight, W. Harrison, J. J. Norcross, N.
    A. Pore, and E. M. Stanley, editors, ESSA Prof. Papers, 3, (2),
    Washington, DC, pp. 11-42.
Palmer, H. D., J. R. Guala, and J. L. Nolder (1976) Current meter
    data reduction, with comments on bedload sediment transport:
    Middle Atlantic Bight. EPA Region III WD-6-99-0669B, Westing-
    house Electric Corporation, Oceanic Division, Annapolis,
    MD, 17 pp.
Pearce, J. B., J. V. Caracciolo, M. B. Halsey, and L. H. Rogers
    (1976) Temporal and spatial distributions of benthic macro-
    invertebrates in the New York Bight. In: Middle Atlantic Conti-
    nental Shelf and the New York Bight, M. Grant Gross, editor,
    Spec. Symposia Amer. Soc. Limnol, and Oceanogr., 2, 394-403.
Pesch, G., B. Reynolds, and P. Rogerson (1977) Trace metal from
    within and around two ocean disposal sites. Marine Pollution
    Bulletin, 8, (10), 224-228.
Stalling, D. L. and F. L. Mayer (1972) Toxicities of PCB's to fish
    and environmental residues. Environmental Health Perspectives,
    pp. 159-164.
Swift, D. J. P., F. L. Freeland, P. E. Gadd, G. Han, J. W. Lavelle,
    and W. L. Stubblefield (1976) Morphologic evolution and coastal
    sand transport, New York-New Jersey Shelf. In: Middle Atlantic
    Continental Shelf and the New York Bight, M. Grant Gross,
    editor, Spec. Symposia Amer. Soc. Limnol. and Oceanogr., 2,
    69-89.
Swift, D. J. P., J. W. Kofoid, F. P. Saulsbury, and P. Sears (1972)
    Holocene evolution of the shelf surface, central and southern
    Atlantic shelf of North America. In: Shelf Sediment Transport:
    Process and Pattern, D. J. P. Swift, D. B. Duane, and O. H.
    Pilkey, editors, Dowden, Hutchinson and Ross, Stroudsburg, PA,
    pp. 499-574.

V

FUTURE PROSPECTS OF OCEAN DUMPING

FUTURE PROSPECTS OF OCEAN DUMPING

Dana R. Kester, Bostwick H. Ketchum[*], and
P. Kilho Park[†]

Graduate School of Oceanography, University of Rhode
Island, Kingston, RI   02881

[*]P.O. Box 32, Woods Hole, MA   02543

[†]National Oceanic and Atmospheric Administration,
Rockville, MD   20852

ABSTRACT

The dumping of wastes in the ocean will be an important consideration in waste disposal management in the coming years. Maintaining the quality of the marine environment will require an improved
understanding of the fate and effect of wastes in the ocean. In addition to the industrial chemicals, dredged material, and sewage
wastes presently dumped in the ocean, consideration should be given
to the behavior in the ocean of radioactive wastes, ocean mining
wastes, and incineration residues. The concept of assimilative capacity provides a basis for identifying the physical, chemical, and
biological processes in the ocean which are important in waste disposal considerations, and one can recognize the distinctive behaviors
in the ocean of several types of chemicals such as biodegradable substances, toxic metals, and persistent synthetic organic chemicals.
There is a need to improve the assessment of the biological effects
of pollutants in the marine environment. Strategies for monitoring
waste disposal in the sea should provide an early warning of environmental degradation and should enhance the understanding of marine
processes in order to assure effective management of ocean dumping
practices.

INTRODUCTION

The preceding sections of this book have considered specific technical aspects of several industrial wastes that are dumped in the ocean. In this concluding chapter we will summarize a number of general considerations related to ocean dumping. Our approach will be a somewhat philosophical and idealistic one in which we bring together common thoughts on waste disposal in the marine environment, on the technical assessment of ocean dumping practices, on the approaches to monitor waste disposal, and on recommendations for future work.

WASTE DISPOSAL AS A LONG-TERM ENVIRONMENTAL PROBLEM

Management of waste disposal with contemporary technology is a major environmental issue. Several factors have converged within our society which require that attention be given to this problem now and for the coming decades. When viewed over the last century it is evident that localized increases in population density and extensive chemical industrialization have created new demands for waste disposal. Initially it appeared that the environment could assimilate these chemical wastes with little consequence. But in the past two decades it has become evident that poor waste disposal practices can lead to a deterioration in environmental quality which our society now considers unacceptable.

The capacity of the environment to assimilate wastes depends upon the rate of input of the waste and the environmental life-time of the waste before it is transformed into its constituents and processed by the natural bio-geochemical cycles. The eutrophication of natural waters produced by the disposal of phosphorus and nitrogenous nutrients is an example where the rate of input is a primary factor in exceeding the assimilative capacity. Synthetic organic chemicals such as DDT and its residues, polychlorinated biphenyl compounds (PCBs), and Kepone are examples of wastes for which persistence within the environment is a major factor in their impact. In seeking the best strategies for waste disposal it is necessary to consider the rates of input of specific wastes, their toxicity, the mechanisms in the environment for their assimilation, and their chemical and biological persistence.

One may identify several major alternatives for waste disposal which include containment on land, ocean dumping, and incineration. Recycling of materials and recovery of resources should make it possible to reduce in the future the volume of wastes presently generated by the once-through, disposable, approach which characterizes much of the present production and consumption practices. However, wastes will be generated even with the most extensive recycling, if not for economic reasons, then at least due to the requirements of the second law of thermodynamics -- that total entropy must increase

in a real cyclic process. We are now confronted with numerous examples in which disposal of wastes on land has been practiced without adequate regard for the long-term consequences. The permeation
of homes at Love Canal, New York by toxic chemicals, the enhanced
background radiation from Colorado uranium mine tailings, the interim
storage of radioactive wastes at several places in the United States,
and the unregulated dumping of hazardous chemicals in land-fills illustrate the need for better understanding of the long-term impact
and costs of waste disposal. Within the marine environment closing
waters for swimming and shell fishing represent examples of inadequate treatment of municipal wastes which has restricted the use of
marine resources.

     In the U.S. prior to 1973 increasing use was made of ocean dump-
ing to dispose of wastes during the time when steps were taken to reduce the inputs of toxic effluents to rivers, lakes, estuaries and
nearshore waters. Upon recognizing the increasing trend of ocean
dumping, it became the responsibility of the regulatory agencies such
as the U.S. Environmental Protection Agency to establish guidelines
and regulate waste disposal in the ocean. An objective was established to minimize ocean dumping around the United States by 1981,
and to ban the ocean dumping of wastes that could not meet specified
guidelines (U.S. Council on Environmental Quality, 1970). This objective was based more on ignorance about the possible consequences
of ocean dumping than on knowledge that such waste disposal was detrimental to society's overall use of the oceans.

     It is useful to distinguish the dumping of wastes in shallow
continental shelf waters from that in the deep ocean, because the
fate and impact of wastes in these two regions most likely are different, and because the resource value of the continental shelf is
typically greater than that of the deep ocean. There is generally
more experience and scientific information available concerning ocean
dumping on the continental shelf than in the deep ocean. Disadvantages of waste disposal on the continental shelf include the tendency
for substances to accumulate in the benthic organisms and sediments
(Pesch et al., 1977), and the possible degradation of continental
shelf resources such as fisheries, mineral deposits, and shoreline
use. Advantages of the continental shelf are the low to moderate
transportation costs and the localization of potential detrimental
effects. Disadvantages of deep ocean waste disposal include uncertainty about the ultimate fate and effect of wastes in this environment and the potentially large scale impact that could result. It is
likely that the planktonic and pelagic organisms will be more affected than the benthic organisms in deep ocean disposal. Advantages of
deep ocean dumping are the large dispersion and dilution of wastes
that can occur as well as apparently reducing the possible conflicts
in the utilization of other marine resources.

     The major types of wastes currently ocean-dumped are sewage

sludge, industrial chemical wastes, and dredged material. With sufficient expenditure of public funds it should be possible to eliminate sewage sludge dumping from the ocean, but safe alternative disposal methods are generally more expensive (National Research Council, 1978). Alternatives to ocean dumping probably can be found for
industrial chemicals, though the extra cost will most likely increase
the costs of the associated chemical products. There is need for the
continued ocean dumping of dredged material due to the large volumes
of material, the fact that it is mainly of marine origin, and much of
it is not contaminated with potentially harmful chemicals. The major
problems with dredged material disposal are to minimize the impact on
other resources and to provide for the handling of polluted sediment.
It is likely that some wastes not presently dumped in the ocean will
receive substantial consideration for ocean disposal in the future.
These could include radioactive wastes, ocean mining wastes, and incineration residues. To consider waste disposal in the ocean, and to
estimate what is acceptable and what must be avoided, are major challenges to the understanding of marine processes.

    Within the United States it may appear that ocean dumping is a
"dead" issue in view of the regulations and legislation to terminate
harmful ocean dumping by December 31, 1981. However, until the relative environmental merits and economic factors of the alternatives
to ocean dumping of present and future wastes have been demonstrated,
it is worthwhile to continue pursuing the scientific and societal assessment of ocean dumping activities. Irrespective of the course
taken by the U.S. regarding ocean dumping, it is evident that other
countries face different constraints in their consideration of ocean
waste disposal. Ocean dumping can be the basis for a conflict in international interests. If ocean waste disposal were to lead to a deterioration in ocean resources and quality, all mankind would share
the costs. Thus it is important that we achieve a good understanding
of the effect of wastes in the ocean, independent of specific national policies.

                A SCIENTIFIC BASIS FOR OCEAN DUMPING PRACTICES

    The concept of assimilative capacity provides a basis for asses-
sing the ability of the marine environment to absorb wastes without
unacceptable consequences (Goldberg, 1979). There are two aspects to
this concept. One may be termed the carrying capacity of the system
in which we include mainly the physical processes of removing a waste
from its source and dispersing it to unrecognizable levels of dilution. The second is the digestive capacity in which we consider the
internal chemical and biological processes which can alter wastes and
ultimately incorporate them into the natural bio-geochemical cycle.

    A schematic illustration of assimilative capacity can identify
the primary principles of this concept (Fig. 1). We may consider

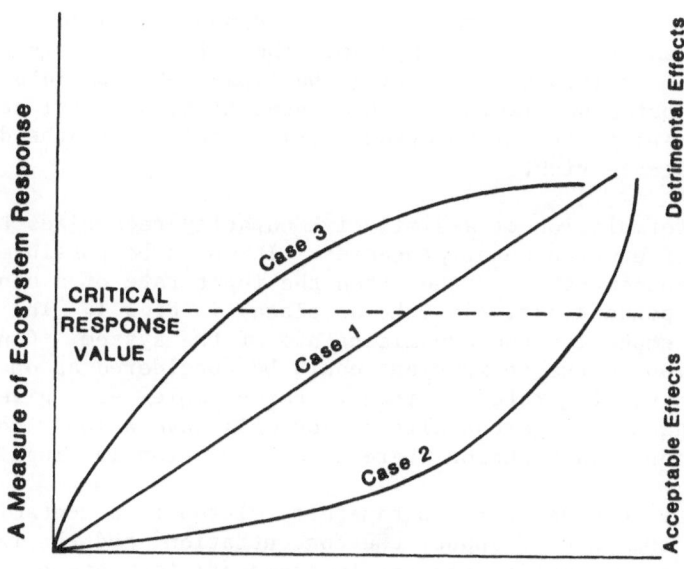

Rate of Contaminant Input

Fig. 1. Three examples are provided for the relationship between response and input rate. The carrying capacity if the input rate at which the response exceeds the critical value. (After Preston, 1977).

that a relationship exists between some measure of ecosystem response and the rate of contaminant input. For specific responses and contaminants the relationship could have a variety of shapes such as a proportional response (case 1), a delayed abrupt response (case 2), or a response leading to saturation (case 3). The delayed response (case 2) is the most insidious since inputs may occur for a long time with no obvious effect, but when some threshold is passed an incremental increase in the input produces a disproportionately large effect. We assume that some magnitude of response is acceptable whereas others are judged to be detrimental. The boundary between these represents the critical response value, and the input rate corresponding to the critical response is the assimilative capacity of the ecosystem for that contaminant. The relationship between response and input may be relatively simple or highly complicated by synergism among contaminants and environmental stresses. If we assume that such a response curve exists, then it is the task of environmental scientists to determine the appropriate ecosystem response to be measured and to define the response curve for important contaminants.

The definition of acceptable and detrimental responses and the

evaluation of a critical response value cannot be based solely on a
scientific assessment of the system. The scientist draws upon experience and observations in defining the impact of a pollutant on the
ecosystem, but other factors such as economics, societal and aesthetic values, and political interests will contribute to the definition
of an acceptable risk.

This formulation of assimilative capacity recognizes the dynamic
character of environmental processes. It would be possible to consider the concentration rather than the input rate of a contaminant
as the independent variable (the abscissa) in Fig. 1. The input rate
is used to emphasize the dynamic nature of the system. Contaminant
concentration in the environment could be considered as one ecosystem
response to varying rates of input. For example, at a given rate of
input a region of vigorous circulation will have a lower contaminant
concentration than a region where the circulation is sluggish.

One may consider three parameters related to a contaminant in an
ecosystem: the rate of input, the concentration, and the rate of removal. These parameters are not independent; the rate of removal is
generally a function of the concentration, and the concentration
varies when there is a difference between the input and the removal
rates. This dependence can lead to negative feedback in cases where
an increase in concentration increases the removal rate thereby tending to restore the concentration toward an initial value. When the
removal mechanisms are biological processes and when these procsses
are inhibited or reduced by an increase in contaminant concentration,
positive feedback may occur as the removal rate is diminished and the
concentration increases further.

In many cases we don't know what the appropriate ecosystem response is until we have exceeded the critical value and recognized a
detrimental effect. It is likely that the relationship between a
specific response and a particular contaminant input rate is not a
unique function. It probably will depend on the state of the ecosystem and the properties of the contaminant. A skeptic may argue
that such a poorly behaved function is not worth pursuing, but we
find it useful in considering the scientific factors which relate to
ocean dumping.

The physical dispersion of wastes in the ocean determines both
the region of impact and the concentrations to which organisms will
be exposed. Dispersion and mixing processes in the ocean are not
sufficiently understood to be predictable. Consideration must be
given to both lateral dispersion along surfaces of constant density
and to vertical mixing across isopycnal surfaces. The fluxes of material by lateral and vertical mixing can be comparable because even
though the lateral mixing coefficients are often several orders of
magnitude greater than the vertical mixing coefficients, the vertical
gradients are often several orders of magnitude greater than the

lateral gradients. The flux, or transport of a substance by mixing, is determined by the product of a mixing coefficient and a gradient.

Dispersion in the wake of a barge is a highly effective means of rapidly diluting a waste by a factor of 10^4 or more within a few hours. The persistence of a plume once it reaches a width of about 1 km and a dilution of 10^5 or 10^6 is poorly known. Vertical mixing in the surface ocean is restricted by the pycnocline and during stratified periods the slow rate of oceanic mixing may allow the plumes to maintain measurable concentrations for days or possibly weeks. Chemical measurements of plume persistence beyond three days has not been possible due to difficulties in tracking a plume over long periods of time.

A consideration of physical processes provides an initial basis for assessing the fate of wastes and the capacity of the environment to accommodate them. Chemical processes also may play a substantial role in determining the impact of wastes and their transfer through the marine ecosystem. In some cases the mixture of a liquid industrial waste with seawater results in the precipitation of solid phases which can adsorb and coprecipitate potentially toxic substances from the waste plume or from seawater. In other instances surface active components of a waste may concentrate at the sea surface leading to increased exposures for neuston. The physical and chemical form of a substance has a major effect on its toxicity and its tendency to accumulate in organisms. In many cases it is possible to draw upon a knowledge of marine chemical processes to predict the consequences related to waste dumping in the ocean. But these predictions should be verified by appropriate observations, and the present understanding of marine chemistry is not adequate to predict reliably the fate and effect of a waste in the ocean.

The biological consequences of ocean dumping are generally regarded as establishing the acceptable limits of waste disposal in the marine environment. Determining what biological parameters should be measured is a major scientific problem. This issue was considered in detail at an International Council for the Exploraton of the Seas (ICES) workshop on monitoring biological effects of pollution in the sea. Four classes of techniques were identified:
1. Bioassay measurements
2. Physiological techniques
3. Biochemical measurements
4. Ecological assessments

Bioassay measurements range from determining the concentration of a contaminant which is lethal to 50% of a group ·organisms after a fixed period (e.g., 96 hrs.) of exposure, which is known as the 96 hour LC_{50}, to more moderate effects such as the response of growth rate to various concentrations of a waste. These measurements are made in the laboratory under controlled conditions with one species

at a time and often with one pollutant. More extensve experiments
can examine the possible synergistic effects of multiple contaminants. Stebbing et al. (1980) suggested four bioassay techniques as
being most useful for examinng the effect of pollutants in the marine
environment. These included (1) the frequency of abnormal oyster
larvae after 48 hour exposure, (2) the developmental success of sea
urchin larvae after 15 hour exposure, (3) the growth of Skeletonema
costatum after 48 or 96 hour exposures, and (4) the growth of three
microalga species after 14 day exposures.

The report by Bayne et al. (1980) rated eight physiological
techniques for four types of organisms as being good, potentially
good, of limited use, and not useful. The techniques included measurements of growth, feeding rate, oxygen consumption, reproduction
rate or fecundity, osmotic regulation, haematology, nitrogen balance,
and heart beat or ventilation rate. The four types of organisms considered were fish, crustaceans, polychaets, and molluscs. Growth,
scope for growth, and feeding rate were judged to be best suited for
assessing the biological effects of contaminants for the four classes
of organisms.

The report by Uthe et al. (1980) on the use of biochemical measurements to assess sublethal effects on organisms emphasizes reproductive biochemistry, hormone metabolism, and blood chemical analyses. Considerable attention is given to problems associated with biochemical measurements related to pollutant effect assessment. These
problems include the appropriate selection of a species, the strain
within a species, and the effects of age, nutritional condition, and
sex on sensitivity to a toxicant.

Ecological assessment of a pollutant provides the most direct
and comprehensive consideration of a pollutant effect in the marine
environment. However, it is the most difficult type of measurement
to implemement, because natural variability can obscure pollutant effects and the effects may not be evident until a large perturbation
occurs (Gray et al. 1980). Studies of the overall ecosystem can
include an assessment of pollutant uptake, retention, and transfer
among organisms within the system. It may be possible to identify
indicator species for the effect of specific contaminants.

Various measurements of the biological effect of a substance
represent a progression from relatively simple short-term observations to complex long-range effects. In recent years there has been
a tendency to favor the chronic rather than the accute effects as a
measure of the biological impact of contaminants. It has generally
been necessary to strike a compromise between a measurement which is
practical on a routine basis and one which will provide the best indication of biological effects. Increased understanding of the
impact of wastes on marine ecosystems should lead to a better understanding for selecting the most appropriate biological measurements.

The impact of contaminants on marine biota generally provides the main criterion for limiting waste disposal in the ocean.   Considerations of impact criteria should not assume that marine biota are the only resource to be preserved. It is useful to recognize other possible criteria in considering the range of scientific problems which must be addressed. One approach is to set the limits of waste disposal in the ocean such that an effect on human health is unlikely. The present practice of attempting to avoid an impact on marine biota provides a more restrictive limitation on ocean waste disposal than if human health were the only criterion. A more general view is that ocean dumping should not conflict with other present and future uses of the marine environment which include aesthetic values related to marine recreation and natural resources which, in addition to fisheries, include those which presently may be regarded as of low economic value such as sand and gravel or potable water.

One issue which has not received much scientific assessment is a comparative evaluation of alternate ocean dumping strategies. It could be argued that the prevailing practice of dispersing ocean-dumped wastes above the thermocline in coastal and offshore waters maximizes the biological impact of these wastes by placing them in the photic zone of the ocean's most productive region. Alternatives could include disposal of wastes beneath the thermocline and in the deep sea. Information has not been brought together to identify the relative merits of different ocean dumping strategies.

MONITORING NEEDS AND STRATEGIES

The management of ocean dumping requires some approach to determining that impact criteria are not exceeded. The U.S. Environmental Protection Agency (EPA) has established a discharge permit procedure which can be characterized as a load assessment approach to assure that criteria will be met. The U.S. Bureau of Land Management has established a program to measure environmental baselines in coastal waters that could be affected by future offshore oil production. Presumably, future measurements could determine departures from these baselines permitting a trend assessment approach to identifying contaminant impacts. The U.S. EPA criteria tor natural waters specify the maximum permissible concentrations of some substances in these waters and in other cases it specifies a concentration relative to an LC_{50} for a "sensitive organism". These criteria could provide the basis for environmental monitoring of concentrations related to ocean-dumping activities. While the intent of a criterion or monitoring program related to a "sensitive organism" is reasonable, it is ambiguous in practice. The U.S. National Marine Fisheries Service programs known as Ocean Pulse and MARMAP represent a strategy to assess the "health" of fisheries resources based on periodic environmental measurements of selected parameters. Another approach is to identify and analyze an indicator organism which can serve as an

integrator of the accumulation of wastes by marine biota. The Mussel
Watch program (Goldberg et al., 1978) reported initial results for
four types of pollutants (heavy metals, transuranic elements, petroleum hydrocarbons, and halogenated hydrocarbons) which indicates the
ability of this approach to identify "hot spots" of marine pollution.

    In the absence of an effective monitoring strategy the manage-
ment of waste disposal in the marine environment becomes a matter of
crisis response. There have been enough examples of marine environmental crises so that we may hope to learn from them what types of
monitoring might be effective in the future. One of the clearest
cases in which waste disposal in the marine environment resulted in
human casualties is the Minimata Bay disease in which mercury poisoning occurred from methyl mercury discharge in the production of acet-
aldehyde (Goldberg, 1975). The effect of DDT on marine bird populations is another example in which a crisis led to recognition of the
problem. The accumulation of PCBs and Kepone in coastal sediments
and possibly organisms was realized only after considerable discharge
of contaminants to the marine environment. There is a common factor
in each of these environmental crises which if recognized and considered in future monitoring strategies could lead to earlier identification of impacts from ocean-waste disposal. In each case there was
a lack of understanding of the trajectory of the contaminant in the
marine environment and of the weak link or critical factor that would
result in an impact. With DDT the weak link was not that its concentration exceeded some unique value, but that it was transferred and
accumulated and magnified in higher trophic levels (birds) and resulted in a thinning of their egg shells thereby decreasing their
reproductive efficiency. The critical factor for PCBs and Kepone appears to be their tendency to accumulate in fatty tissues and to
adsorb onto solid substances, particulate matter, and accumulate in
sediments where they can be transferred to benthic organisms.

    A second common factor in these marine environmental crises is
that they either actually or potentially impacted on human health.
The mercury poisoning was a direct impact on human health. The synthetic organic contaminants were a potential impact on human health,
because of their known biochemical effects such as their ability to
cause cancer or disrupt reproductive processes. Plausible pathways
could also be identifed between the marine environment and man. Consequently there has been an effectve response to these crises in the
form of regulatng toxic metal discharges and prohibiting or minimizing the use of DDT and PCBs. In the future if the relationship between an environmental effect and a risk to human health is not evident, will the effect be regarded as a crisis requiring a response,
or will the effect be ignored?

    Monitoring should be designed to consider the trajectory of
specific contaminants through the marine ecosystem and to anticipate
or identify the weak link in the system for each contaminant. If a

program of repeated routine measurements is necessary, it should be supplemented by an experimental approach which will enhance the value of the data generated and increase the understanding of the marine ecosystem.

SUMMARY

The disposal of wastes in the ocean presents long-range problems requiring the attention of marine scientists. The behavior of wastes in the marine environment and their effect on the utilization of marine resources must be established through appropriate research. The information currently available on the impact of industrial wastes dumped at deep water sites off U.S. waters does not indicate that it is necessary to achieve zero input of all these wastes to the ocean. This conclusion, however, must be qualified by the realization that we do not know the long-term fate of these wastes nor do we know the capacity of these pelagic oceanic regions to assimilate wastes without detrimental effects.

In as much as marine biota represent one of the most valuable ocean resources which might be affected by ocean dumping, it is important to strive for improved methods of assessing the biological impact of wastes. The large variability in biological observations compared with physical or chemical measurements, limits the ability to reach reliable conclusions concerning the effects of contaminants on biological systems. The lack of replicability in biological experiments may be viewed as an inherent characteristic in which there is variability among individual organisms or populations of organisms. Alternatively, measurements made of systems with a large number of variables, some of which may be unknown, may lead to poor repeatability due to inadequate experimental control. Improved replicability of biological measurements may be possible through advances in experimental design and statistical assessment of the sources of inherent variability.

In chemical studies there is a need to increase the emphasis given to the behavior and effect of synthetic organic substances in marine systems. The behavior of particulate phase organic substances dispersed as micelles or aggregated into larger particles is poorly known. While considerable information is being obtained on the concentrations and fluxes of metals in the marine environment, there is a need to distinguish the chemical reactivity of metals in different forms.

The short-term physical processes which disperse wastes are fairly well described from a variety of tracking experiments and quantitative models. There is less certainty about the long term transport and mixing of wastes in the ocean. Do episodic events, such as storms provide the primary mechanism for long-term dispersion? What are the relative roles of vertical and horizontal transport and mixing processes in determining the fate of wastes? Answers to these questions are not unique to ocean dumping considerations; they are fundamental problems in physical oceanographic processes.

Research during the past decade has greatly enhanced the understanding of waste disposal in the ocean. Several directions may be identified for future efforts. In addition to gaining information on the impact in the ocean of presently dumped wastes (sewage sludge, dredged material, and industrial chemicals) consideration should be given to wastes of possible future importance (ocean mining operations, radioactive substances, and incineration residues). An ability to manage ocean disposal of wastes effectively is directly dependent on an overall understanding of fundamental biological, chemical, and physical processes in the marine environment.

ACKNOWLEDGEMENTS

We would like to thank Drs. F. X. Cameron, J. A. Knaus, and A. Mead for their comments on this manuscript. This work was supported by NOAA Grants NA-79-AA-D-00033 and 04-8-M01-192.

REFERENCES

Bayne, B. L. (chairman), J. Anderson, D. Engel, E. Gillfillan, D. Hoss, R. Lloyd, and F. P. Thunberg (1980) Physiological techniques for measuring the biological effects of pollution in the sea. In: International Council for the Exploration of the Sea, Rapports et Proces-Verbuax des Reunions, Vol. 179, Charlottenlund Slot, Denmark.

Goldberg, E. D. (1975) Marine pollution. In: Chemical Oceanography, 2nd edition, Vol. 3, J. P. Riley and G. Skirrow, editors, Academic Press, New York, pp. 39-89.

Goldberg, E. D. (1979) Assimilative Capacity of U.S. Coastal Waters for Pollutants. U.S. Department of Commerce, NOAA Environmental Research Laboratories, Boulder, CO 80303, 284 pp.

Goldberg, E. D., V. T. Bowen, J. W. Farrington, G. Harvey, J. H. Martin, P. L. Parker, R. W. Risebrough, W. Robertson, E. Schneider, and E. Gamble (1978) The mussel watch. Environmental Conservation, 5 (2): 101-125

Gray, J. S. (chairman), D. Boesch, C. Heip, A. Jones, J. Lassig, R. Vanderhorst, and D. Wolfe (1980) Ecological working group report to monitoring of biological effects of pollution in the sea. In: International Council for the Exploration of the Sea, Rapports et Proces-Verbaux des Reunions, Vol. 179, Charlottenlund Slot, Denmark.

National Research Council (1978) Multimedium Management of Municiple
    Sludge. National Academy of Sciences. Washington, D.C., 187 pp.
Pesch, G., B. Reynolds, and P. Rogerson (1977) Trace metals in
    scallops from within and around two ocean disposal sites.
    Marine Pollution Bulletin, 8: 224-228.
Preston, A. (1977) The study and control of environmental radio-
    activity and its relevance to the control of other environmental
    contaminants. Atomic Energy Review 15 (3): 374-405.
Stebbing, A. R. D. (chairman), B. Akesson, A. Calabrese, J. H.
    Gentile, A. Jensen, and R. Lloyd (1980) Bioassay report to
    monitoring the biological effects of pollution in the sea. In:
    International Council for the Exploration of the Sea, Rapports
    et Proces-Verbaux des Reunions, Vol. 179, Charlottenlund Slot,
    Denmark
Uthe, J. F., H. C. Freeman, S. Mounib, and W. L. Lockhart (1980)
    Selection of biochemical techniques for detection of environ-
    mentally induced sublethal effects in organisms. International
    Council for the Exploration of the Sea, Rapports et Proces-
    Verbaux des Reunions, Vol. 179, Charlottenlund Slot, Denmark
U.S. Council on Environmental Quality (1970) Ocean Dumping: A
    National Policy. U.S. Government Printing Office, Washington,
    D.C., 45 pp.